Lecture Notes in Mathematics

Edited by A. Dold and B. Eckmann

T0253637

1183

Algebra, Algebraic Topology and their Interactions

Proceedings of a Conference
held in Stockholm, Aug. 3–13, 1983,
and later developments

Edited by J.-E. Roos

Springer-Verlag
Berlin Heidelberg New York Tokyo

Editor

Jan-Erik Roos
Department of Mathematics, University of Stockholm
Box 6701, 113 85 Stockholm, Sweden

Mathematics Subject Classification (1980): 13-06, 13D03, 13E05, 13H99, 13J10, 14-06, 14F35, 16A24, 17B70, 18G15, 18G20, 20F05, 20F10, 55-06, 55P35, 55Q15, 55S30, 57-xx

ISBN 3-540-16453-7 Springer-Verlag Berlin Heidelberg New York Tokyo
ISBN 0-387-16453-7 Springer-Verlag New York Heidelberg Berlin Tokyo

This work is subject to copyright. All rights are reserved, whether the whole or part of the material is concerned, specifically those of translation, reprinting, re-use of illustrations, broadcasting, reproduction by photocopying machine or similar means, and storage in data banks. Under § 54 of the German Copyright Law where copies are made for other than private use, a fee is payable to "Verwertungsgesellschaft Wort", Munich.

© by Springer-Verlag Berlin Heidelberg 1986
Printed in Germany

Printing and binding: Beltz Offsetdruck, Hemsbach/Bergstr.
2146/3140-543210

A MATHEMATICAL INTRODUCTION

These notes contain the outcome and later developments arising from a Nordic Summer School and Research Symposium held in Stockholm, August 3-13th, 1983 on "ALGEBRA, ALGEBRAIC TOPOLOGY AND THEIR INTERACTIONS".

Let me first give a brief indication of the main ideas behind this symposium. During the last decade several striking analogies between algebraic topology (at least rational homotopy theory) and algebra (at least local algebra) had been observed. Let me just give two examples. (More examples and details can be found in the paper Through the looking glass: A dictionary between rational homotopy theory and local algebra by L. AVRAMOV and S. HALPERIN in these proceedings.)

First some preliminaries. Let X be a finite, simply-connected CW-complex, ΩX the space of loops on X and $H_*(\Omega X,\underline{\underline{Q}})$ the rational homology algebra of ΩX. (This algebra is even a Hopf algebra.) At the same time, let $(R,\underline{m})$ be a local commutative noetherian ring R with maximal ideal $\underline{m}$ and residue field $k = R/\underline{m}$, and let $\mathrm{Ext}_R^*(k,k)$ be the graded vector space $\bigoplus_{n\geq 0} \mathrm{Ext}_R^n(k,k)$ equipped with the algebra structure coming from the Yoneda composition $\mathrm{Ext}_R^i(k,k) \otimes \mathrm{Ext}_R^j(k,k) \longrightarrow \mathrm{Ext}_R^{i+j}(k,k)$. This Yoneda Ext-algebra $\mathrm{Ext}_R^*(k,k)$ is also a Hopf algebra and it is even the enveloping algebra of a certain graded Lie algebra $\pi^*(R)$ over k. On the other hand, it is also known that $H_*(\Omega X,\underline{\underline{Q}})$ is the enveloping algebra of the rational homotopy Lie algebra $\pi_*(\Omega X)\otimes_{\underline{\underline{Z}}}\underline{\underline{Q}}$. (Note that the Samelson product on this Lie algebra corresponds under the isomorphism $\pi_{n-1}(\Omega X)\cong\pi_n(X)$ to the Whitehead product on the $\pi_n(X)$.) We are now ready for the examples:

Example 1.- Let $F \longrightarrow E \longrightarrow B$ be a Serre fibration and

$$\ldots \longrightarrow \pi_{n+1}(B) \xrightarrow{\partial} \pi_n(F) \longrightarrow \pi_n(E) \longrightarrow \pi_n(B) \longrightarrow \ldots \qquad (1)$$

the corresponding homotopy exact sequence. In [7] Halperin proved (under some minor extra conditions) that, if $H^*(F,\underline{\underline{Q}})$ is finite dimensional, then (1) breaks up into exact sequences of 6 terms if we tensor it with $\underline{\underline{Q}}$. (More precisely, $\partial(\pi_{odd}(B))$ is torsion.) On the other hand, if $A \longrightarrow B$ is a homomorphism of local commutative noetherian rings such that B is A-flat and if $\overline{B} = B \otimes_A k$ is the "fibre" ring (assuming for simplicity that the local rings have the same residue field k) then, using earlier partial results of Gulliksen, Avramov proved in [3] that there is an exact sequence

$$\ldots \longrightarrow \pi^n(\overline{B}) \longrightarrow \pi^n(B) \longrightarrow \pi^n(A) \xrightarrow{\delta} \pi^{n+1}(\overline{B}) \longrightarrow \ldots \qquad (2)$$

where δ has properties similar to those of ∂. (This time we do not have to tensor (2) with anything.) It should be remarked that neither Avramov nor Halperin knew about the other's work at the time. By now there are much more complete results and a common explanation in terms of differential graded algebras.

Example 2.- It was asked by Serre whether the series

$$\sum_{n\geq 0} \dim_{\mathbb{Q}}(H_n(\Omega X,\mathbb{Q}))\cdot Z^n \,, \tag{3}$$

and by Kaplansky and Serre whether

$$\sum_{n\geq 0} \dim_k(\mathrm{Ext}^n_R(k,k))\cdot Z^n \tag{4}$$

behaved in a nice way, e.g. whether they were rational functions of Z. (X and R are as in the preliminaries above.) I proved in [13] that, for spaces X with dim X ≤ 4 and for local rings $(R,\underline{m})$ with $\underline{m}^3 = 0$, the two questions were equivalent (even more precise results were proved...). Thus, when Anick found a counterexample to the rationality of (3), it was immediately obvious how to produce a counterexample to the rationality of (4). In [13] the algebra structures of $H_*(\Omega X,\mathbb{Q})$ and $\mathrm{Ext}^*_R(k,k)$ were also related to each other. By now there are much more general results, at least about how the series (3) and (4) are related. It has turned out that even for X arbitrary (finite, simply-connected) and $(R,\underline{m})$ arbitrary (local noetherian), the series (3) and (4) are all "rationally related" [1] to corresponding series (4) of local rings $(S,\underline{n})$ with $\underline{n}^3 = 0$ and to corresponding series (3) of finite Y:s with dim Y ≤ 4 and thereby also "rationally related" to series $\sum_{n\geq 0} \dim_k(\Gamma^n)\cdot Z^n$, where Γ is a finitely presented graded (1,2)-Hopf algebra, i.e. the quotient of a free associative algebra $k< X_1,\dots,X_m >$ on generators $X_1,\dots,X_n$ of degree 1, by the two-sided ideal generated by some "quadratic Hopf relations" :

$$\sum_{i\leq j} c_{ij}[X_i,X_j] \quad , \text{ where } c_{ij} \in k \text{ and where } [X_i,X_j] = \begin{cases} X_iX_j + X_jX_i \;, & \text{if } i < j \\ X_i^2 \;, & \text{if } i = j. \end{cases}$$

For more details, cf. [2].[In what follows, if $\bigoplus_{n\geq 0} V_n$ is a graded k-vector space, with $\dim_k(V_n) < \infty$ for all $n \geq 0$, we will call $\sum_{n\geq 0} \dim_k(V_n)\cdot Z^n$ the Hilbert series of $V = \bigoplus_{n\geq 0} V_n$.]

With these and other examples in mind, it was clear that, if algebraists and (algebraical) algebraical topologists could meet for a longer period of time, then a fruitful interaction between their ideas might take place. Here are just a few examples of results obtained at or after the Stockholm conference that are published here for the first time:

In Bøgvad-Halperin´s paper an algebraist and an algebraical topologist cooperate to prove that, if $H_*(\Omega X,\mathbb{Q})$ (recall that X is a finite, simply-connected CW-complex) is noetherian (left or right noetherian does not matter, since we are dealing with a Hopf

---

[1] For the meaning of "rationally related", cf. "LOOKING AHEAD" below.

algebra), then there are only a finite number of non-zero rational homotopy groups of X. (The converse is evident.) On the other hand they also prove that, if $(R,\underline{m})$ is a local commutative noetherian ring with residue field k (no restrictions on k!), then $\mathrm{Ext}^*_R(k,k)$ is noetherian (if and) only if R is a local complete intersection. (In this case only the two lowest $\pi^i(R)$ can be different from 0.) The idea of the proof comes from algebraic topology. The Lusternik-Schnirelmann (L.-S.) category (an old topological concept from the 1930:s), which had been introduced quite recently in rational homotopy theory (and thus in the theory of differential graded algebras) is also used here for Avramov's minimal models in a nice way. [The paper by Lemaire in these proceedings contains an up-to-date survey of L.-S. category, that completes and goes beyond the beautiful earlier survey of I.M. James [9] from 1978.]

In order to present the next new result, I first have to recall an old result of Levin [10], which combined with later results of Avramov and Löfwall (cf. these proceedings) can be formulated as saying that, for any local commutative noetherian ring $(R,\underline{m})$, the Lie algebra $\pi^*(R)$ is closely related to the Lie algebra $\pi^*(R/\underline{m}^n)$ of the artinian ring $R/\underline{m}^n$, provided n is big enough (precisely how big n should be depends on the Artin-Rees lemma, which Levin uses in [10] in a very clever way). More precisely, if $n \geq$ some n(R), then the natural Lie algebra map

$$\pi^*(R/\underline{m}^n) \xrightarrow{\pi^*} \pi^*(R)$$

is onto, and the kernel of π^* is a free graded Lie algebra. (There are even more precise results.) [In technical terms one says that $R \longrightarrow R/\underline{m}^n$ is a Golod map. A very general theory of Golod maps is presented for the first time in the paper by Avramov in these proceedings.] Here is one algebraical topological version of all this (it is proved in the joint paper by Halperin and Levin in these proceedings): Let X be a simply-connected CW-complex (not necessarily finite) with a finite number of cells in each dimension and such that $H^*(X,\underline{Q})$ is a finitely generated algebra (i.e. a noetherian ring). Then there exists an $n_o(X)$ such that, for all $n \geq n_o(X)$, the rational homotopy fibre of the inclusion of the n-skeleton $X^n \longrightarrow X$ is a wedge of spheres. Note the analogy: the rational homology ring of the loop space of a wedge of spheres is a free associative algebra. Results of this type had previously been known only for X = BU(m) and, more generally, for X = certain products of Eilenberg-MacLane spaces. The earlier ideas of Levin are essential for the general proof.

There are many more examples of interaction between algebra and algebraic topology. The analogy is often not perfect, and this inevitably leads to more work if one wants to go from one side to another.

Let me say a few words about some other papers in these proceedings. Löfwall's paper is a corrected version of about one half of his 1976 thesis, and this half was never published, presumably because Löfwall first wanted to prove by his methods the rationality of the series (4) in general. Now, as we have said above, we know better as (4) is not always rational, but it was a genuine surprise when it turned out in

1984 [2] that the special cases studied by Löfwall, and in particular finitely presented graded (1,2)-Hopf algebras and their Hilbert series, were "rationally related" to the general series (4) for general local rings ($R,\underline{m}$). Thus with hindsight one might say that in a sense Löfwall's thesis did treat the most general case. Löfwall's thesis has been used by many workers in the field and, in particular, by Löfwall himself [11] in his construction of counterexamples to a conjecture by Kostrikin and Šafarevič. The papers by Anick-Löfwall and Fröberg-Gulliksen-Löfwall in these proceedings are recent studies of how finitely presented graded algebras and their Hilbert series can behave. In particular the last paper can be used to prove that there exists a finite simply-connected CW-complex X, whose $H_*(\Omega X,\mathbb{Z})$ has torsion of all orders (Anick and Avramov, to appear).

The reader may have noticed that, on the homotopy side, we often work over a field of characteristic 0, whereas, in local algebra, we can have residue fields of all characteristics. There are reasons for this (cf. however with "LOOKING AHEAD" below, where more optimistic comments are given). Indeed, in Torsten Ekedahl's paper in these proceedings, we find the first theorems showing that the beautiful Deligne-Griffiths-Morgan-Sullivan theory [6], that a Kählerian compact manifold is "formal" over $\mathbb{R}$, i.e. its real homotopy type is a formal consequence of its real cohomology ring, is false in characteristic p. For the remaining papers in these proceedings (some in algebra, some in algebraic topology and some being a mixture of both), we refer the reader to the table of contents.

LOOKING AHEAD

Here are some further directions of research that seem to be fruitful:

1) Two formal power series $P(Z) = \sum_{n\geq 0} p_n Z^n$ ($p_o = 1$, p_i integers) and $Q(Z) = \sum_{n\geq 0} q_n Z^n$ ($q_o = 1$, q_i integers) are said to be "rationally related" if there exists a 2 × 2 matrix $(A_{ik}(Z))$ whose entries are polynomials in Z with integral coefficients such that $\det(A_{ik}(Z)) \neq 0$ and such that

$$P(Z) = \frac{A_{11}(Z)Q(Z) + A_{12}(Z)}{A_{21}(Z)Q(Z) + A_{22}(Z)} \qquad \left(\text{thus } \frac{A_{11}(Z) + A_{12}(Z)}{A_{21}(Z) + A_{22}(Z)} = 1 \text{ if } Z = 0 \right).$$

These matrices modulo the diagonal ones $\begin{pmatrix} A(Z) & 0 \\ 0 & A(Z) \end{pmatrix}$ form a group under matrix multiplication and it would be interesting to try to classify the orbits of this group acting on, say, the set of power series that are rationally related Hilbert series of finitely presented graded (1,2)-Hopf algebras. The old question of Kaplansky-Serre mentioned above is equivalent to asking whether there is just one orbit. Now we know that there are many orbits. Could we find nice representatives for them? Is there an analog of the Serret theorem (cf. e.g. [12], p. 55)? Here we have only been talking about rational relationship between Hilbert series of graded algebras. Is there an underlying theory of "rational relationship" between the algebras themselves? If so, it might be easier to get more precise results about the $H_*(\Omega X,\mathbb{Q})$ than in the papers by Halperin et al. in these proceedings.

2) Torsten Ekedahl has recently developed the analog of rational homotopy theory for spaces "over $\mathbb{Z}$", using cosimplicial algebras. This theory seems very promising, but nothing has yet been published about it.

3) Growth series and growth algebras of groups.
Let G be a finitely generated group, with a fixed set of generators S, where we suppose that S is closed under the operation of taking inverses in G. Let k be a field and introduce a filtration on the group ring k[G] by means of $F^{-1}(k[G]) = 0$, $F^n(k[G])$ = the sub vector space of k[G], spanned by products of $\leq n$ $(n \geq 0)$ elements from S. Then

$$\bigoplus_{n\geq 0} F^n(k[G])/F^{n-1}(k[G]) \overset{\text{def}}{=} gr_S(k[G])$$

becomes a finitely generated graded algebra (the growth algebra of (G,S)) [4]. The Hilbert series of this graded algebra is the growth series of (G,S). Under some conditions (cf. e.g. [1] for the commutative noetherian case) there is a spectral sequence of algebras

$$E_1^* = \mathrm{Ext}^*_{gr_S(k[G])}(k,k) \Rightarrow gr\ \mathrm{Ext}^*_{k[G]}(k,k). \qquad (5)$$

Could (5) be useful in some cases to relate the growth series of G to the cohomology of G? Another problem: It is known that, if G is finitely presented, then $gr_S(k[G])$ is not necessarily so. Indeed, if $gr_S(k[G])$ is finitely presented, then its Hilbert series is primitive recursive [8] and then [5] G must have a solvable word problem. But there are finitely presented groups whose word problem is unsolvable. Thus we are led to the following

PROBLEM: Is it true that the Hilbert series of finitely presented graded algebras are always rationally related to growth series of finitely presented groups with a solvable word problem (and conversely) ?

Stockholm, autumn 1985

JAN-ERIK ROOS

REFERENCES:

[1] R. ACHILLES - L. AVRAMOV, Relations between properties of a ring and its associated graded ring, Seminar Eisenbud, Singh, Vogel, vol. 2, Teubner-Texte der Mathematik, vol. 48, 1982, 5-29, Teubner, Leipzig.

[2] D. ANICK - T. GULLIKSEN, Rational dependence among Hilbert and Poincaré series, Journ. of Pure and Appl. Algebra, 38, 1985, 135-157.

[3] L. AVRAMOV, Homology of local flat extensions and complete intersection defects, Math. Ann., 228, 1977, 27-37.

[4] N. BILLINGTON, Growth of groups and graded algebras, Commun. in Algebra, 12, 1984, 2579-2588.(Correction later in the same journal.)

[5] J.W. CANNON, The growth of the closed surface groups and the compact hyperbolic Coxeter groups (preprint, cf. Theorem 9.1).

[6] P. DELIGNE - Ph. GRIFFITHS - J. MORGAN - D. SULLIVAN, Real homotopy theory of Kähler manifolds, Invent. Math., 29, 1975, 245-274.

[7] S. HALPERIN, Rational fibrations, minimal models and fibrings of homogeneous spaces, Trans. Amer. Math. Soc., 244, 1978, 199-224.

[8] C. JACOBSSON - V. STOLTENBERG-HANSEN, Poincaré-Betti series are primitive recursive, Journ. London Math. Soc., ser. 2, 31, 1985, 1-9.

[9] I.M. JAMES, On category, in the sense of Lusternik-Schnirelmann, Topology, 17, 1978, 331-348.

[10] G. LEVIN, Local rings and Golod homomorphisms, Journ. of Algebra, 37, 1975, 266-289.

[11] C. LÖFWALL, Une algèbre nilpotente dont la série de Poincaré-Betti est non rationnelle, Comptes rendus Acad. Sc. Paris, 288, série A, 1979, 327-330.

[12] O. PERRON, Die Lehre von den Kettenbrüchen, Band I, Dritte Aufl., 1954, Teubner, Stuttgart.

[13] J.-E. ROOS, Relations between the Poincaré-Betti series of loop spaces and of local rings, Lecture Notes in Mathematics, 740, 1979, 285-322, Springer-Verlag, Berlin, Heidelberg, New York.

Jan-Erik Roos
Department of Mathematics
University of Stockholm
Box 6701
S-113 85 STOCKHOLM (SWEDEN)

ACKNOWLEDGEMENTS AND GENERAL INFORMATION

The Nordic Summer School and Research Symposium on "ALGEBRA, ALGEBRAIC TOPOLOGY AND THEIR INTERACTIONS" received support from two sources:

1) The Swedish Natural Science Research Council (NFR)

and

2) The Nordic Governments, through "Nordiska Forskarkurser", which supports The Nordic Summer School of Mathematics, an organization with one director from each of the Nordic Countries and which works "with a minimum of bureaucracy" (those are the words of the founder of the school) and selects subjects and sites for Summer Schools. The founder and main animator is Professor Lars Gårding (Department of Mathematics, University of Lund, LUND, Sweden). Since the start in 1966, 13 summer schools, covering the following other subjects have been arranged: Harmonic Analysis (twice), Several Complex Variables, Algebraic Topology, Pseudodifferential operators and applications to index problems, Algebraic Geometry (twice), Discrete Groups and Quasiconformal maps, Operator Algebras and their Applications to Quantum Mechanics and Group representations, Singularities, Value Distribution of Holomorphic maps into Complex Projective Space (the Cartan-Ahlfors-Weyl theory) and Differential Geometry.

I wish to thank both NFR and "Nordiska Forskarkurser" for their generous support. I also wish to thank Lars Gårding for his original (1966) initiative, which has turned out to be so extremely useful and valuable.

The Summer School and Research Symposium took place at the University of Stockholm in Frescati, August 3 - August 13th, 1983. The morning sessions consisted mainly of survey lectures, intended to bring the audience to the level of the research symposium (in the afternoons), which successively grew more and more advanced. The following survey lecture series were given:

David ANICK, Basic algebraic topology.
Luchezar AVRAMOV, Local algebra and algebraic topology.
David EISENBUD, Commutative algebra through examples in algebraic geometry.
Tor H. GULLIKSEN, Local algebra and differential graded algebra.
Stephen HALPERIN, Rational homotopy theory.
Melvin HOCHSTER, The homological conjectures for local rings.
Christer LECH, Relations between a local ring and its completion.
Jean-Michel LEMAIRE, Lusternik-Schnirelmann category and related topics.
Rodney Y. SHARP, Basic commutative algebra.
Richard STANLEY, Commutative algebra and combinatorics.

In the afternoons there were both problem sessions (exercises) for some of the morning lectures as well as lectures in the research symposium. The following research symposium lectures did not lead to a publication in these proceedings:

R. FRÖBERG, Koszul algebras and Veronese embeddings.

M. FIORENTINI, Algèbres graduées associées aux suites régulières.

T. OGOMA, A note on unmixed domains, using Poincaré series.

D. EISENBUD, Linear series on reducible curves and applications.

A.R. KUSTIN, Deformation and linkage of Gorenstein algebras.

A. HOLME, Chern numbers of smooth codimension 2 subvarieties of P^N ($N \geq 6$)

M. HOCHSTER, Modules of finite homological dimension with negative intersection multiplicities.

R.Y. SHARP, Generalized fractions and the monomial conjecture.

A.R. PRINCE, Local rings and finite projective planes.

M. BRODMANN, Remarks on the connectedness of algebraic varieties.

A. SLETSJØE, Toroidal embeddings and Poincaré series.

K. BEHNKE, Infinitesimal deformations of cusp singularities.

R. STANLEY, Symmetric functions and representations of $SL(n,\mathbb{C})$.

A. BJÖRNER, On the Stanley-Reisner ring of a Tits building.

N. SUZUKI, d-sequences.

I wish to thank all the participants (in total about 100 people) and in particular all the lecturers for their interest in this meeting. I also wish to give special thanks to the following people who helped with practical details before, during and after the conference: Maje ARONSSON, Jörgen BACKELIN, Rickard BØGVAD, Torsten EKEDAHL, Iréne FLODÉN, Ralf FRÖBERG, Inez HJELM, Clas LÖFWALL, June YAMAZAKI and Calle JACOBSSON (main organizer of an excursion by boat in the Stockholm archipelago). I also thank Hubert SHÜTRICK for linguistic help.

Finally I wish to thank Springer-Verlag for their cooperation. I hope their patience will be rewarded.

Stockholm autumn 1985

JAN-ERIK ROOS

TABLE OF CONTENTS

THROUGH THE LOOKING GLASS: A DICTIONARY BETWEEN RATIONAL HOMOTOPY THEORY AND LOCAL ALGEBRA

by

Luchezar Avramov and Stephen Halperin

§ 0. INTRODUCTION

"Now, if you'll only attend, Kitty,... I'll tell you all my ideas about Looking-glass House. First, there's the room you can see through the glass - that's just the same as our drawing room, only the things go the other way."

Alice [C]

Homological methods, originally invented as tools for algebraic topologists, have almost from their inception played an important role in the study of rings. This has led to any number of analogies between the two subjects and to a certain overlap of terminology.

More recently it has developed that if one restricts attention to <u>rational homotopy theory</u> (within topology) and to <u>commutative rings</u> (within algebra) one gets a particularly coherent analogy of unusual scope and power. This has made it possible to use intuition and techniques from topology to prove theorems in algebra, and conversely.

In $[Av_1]$ for example homotopy-theoretic ideas are applied to the study of local rings in an essential way. That article also contains a brief sketch of the analogy.

Bøgvad's article in these proceedings is another instance; there the idea of Lusternik-Schnirelmann category in topology is used in algebra to prove that a local ring with noetherian Ext-algebra is a complete intersection. (That article restricts to characteristic zero and defers the positive characteristic case to later.) [1)]

An example in the other direction is the result of Levin and Avramov in 1978 ([Le-Av]) on factoring the socle in a local Gorenstein ring. Translated (first for formal manifolds by Avramov $[Av_1]$, then for general manifolds by Halperin and Lemaire [Ha-Le]) it becomes the following theorem in topology: Suppose $M = X \cup_\alpha e^n$ is a closed 1-connected manifold with X an $n-1$ complex. Assume $H^*(M;\mathbb{Q})$ is not singly generated. If the Lie algebra $\pi_*(\Omega X) \otimes \mathbb{Q}$ is filtered by the "powers" of the ideal generated by α then the associated graded Lie algebra is the free product of $\pi_*(\Omega M) \otimes \mathbb{Q}$ and the free Lie algebra $\mathbb{L}(\alpha)$ on α.

Our aim here is to develop the main ingredients of this analogy with some care, so that the reader who wishes may be in a position to continue translating on his own.

---

1) Note by the editor: The general case (arbitrary characteristic) is now solved, and Bøgvad's paper has been replaced by a joint paper by Bøgvad and Halperin in these proceedings.

From time to time we summarize our conclusions in the form of a dictionary.

In particular we have tried to avoid overlap with [Av_1], for which this article might well be considered as preparation. We should also make clear that much of the mathematical substance presented here is not new and is deliberately in elementary form. This is because of our mutual experience that an algebraist's (resp. topologist's) standard triviality confuses a topologist (resp. algebraist) and because we are hoping for comprehension from both.

We refer the reader to [Su], [B-G] and [Ha] for the missing details in rational homotopy, and to [Av_2] for its algebraic analogue. The basic homotopy theory and commutative algebra we use can be found in [W] and in [A-M] and [Ma].

To describe the central ideas behind the analogy we need first to recall the following essential observation of Quillen ([Q_1]) which we state as

Quillen's principle: Differential graded algebras ought not to be regarded merely as a tool for the calculation of (co) homology. In fact a reasonable DGA category will also carry a "homotopy theory" and with it a number of other invariants.

Here this principle is implemented twice, once for topology and once for algebra. In each case we shall arrive at a category of DGA's (commutative in the graded sense) with a homotopy theory. The strong similarities between the two categories will then form the basis for our analogy.

(This homotopy theory (on the algebra side) does not seem to have been described before, although the ideas are clearly present in [Q_1]. Homotopy theory has been used by André [A_2] and Quillen [Q_2] in a simplicial context for the study of commutative rings. This yields a theory which in positive characteristic is distinct from the one presented here.)

On the topology side we use the Thom-Sullivan Functor A_{PL} to pass from topological spaces and continuous maps to DGA's and DGA morphisms (defined over $\mathbb{Q}$, commutative in the graded sense, graded non-negatively upstairs). This faithfully converts rational homotopy theory to Quillen's algebraic homotopy theory.

On the algebra side we may simply regard commutative rings as DGA's concentrated in degree zero. When we come to homomorphisms $\phi : R \to S$ (say of local rings with residue field $\Bbbk$), however, we are faced with the graded algebra $\mathrm{Tor}^R(S,\Bbbk)$. It turns out that this is the common homology algebra of a certain canonical class of DGA's F (again commutative in the graded sense, but now non-negatively graded downstairs).

The DGA's F then turn out to contain other information about ϕ, which is lost when we pass to homology. In particular we can form $\mathrm{Tor}^F(\Bbbk,\Bbbk)$ which turns out to "measure the difference" between $\mathrm{Tor}^R(\Bbbk,\Bbbk)$ and $\mathrm{Tor}^S(\Bbbk,\Bbbk)$, a measure which disappears if we look at $\mathrm{Tor}^{H(F)}(\Bbbk,\Bbbk)$ instead. (Recall $H(F) = \mathrm{Tor}^R(S,\Bbbk)$.)

Thus on the algebra side we implement Quillen's principle by expanding the category of commutative rings to a certain category of DGA's, large enough to permit the construction of such F's. As in the topological case we get a homotopy theory (although it is not clear that all the Quillen axioms hold).

Before going further we emphasize three points:

(i) In order to get a proper analogy we need to consider topological spaces on the one hand and a certain category of DGA's (containing commutative rings) on the other.

(ii) Our analogy is contravariant (for instance it converts pullbacks to pushouts and reverses arrows!) because A_{PL} is a contravariant functor.

(iii) The two sides (topology and algebra) are not simple mirror images. A theorem on one side may translate to a false statement or a triviality on the other and even when it does translate to a good theorem its proof may not translate to a proof. In addition, the normal imprecision on any lexicon is present here. A good translation may not always be possible or may depend on the context, and the taste of the translator.

With this in mind we look at the analogy at work. Homomorphisms correspond to continuous maps and the class of DGA's determined by $\phi:R \to S$ corresponds to the "homotopy fibre" of a continous map. Passing to homology we get $\mathrm{Tor}^R(S,k)$ corresponding to the cohomology of the fibre.

At a more complicated level the Ext algebra of a local ring (or suitable DGA) is the universal enveloping algebra of a graded Lie algebra corresponding to the homotopy Lie algebra of a space.

Finally, two examples we did not include in the body of the paper: the universal cover of a topological space corresponds to the Koszul complex of a local DGA and a topological space with rational Poincaré duality corresponds to a Gorenstein ring.

In fact the caveats above notwithstanding, the basic idea works pretty well. It goes back at least to Quillen's homotopical algebra in the late sixties. Since then any number of mathematicians have worked at its elaboration (including Anick, Felix, Lemaire, Löfwall and Thomas to name a few).

But the real credit for the exploitation of this idea is surely due to Roos, beginning with his fundamental article [R] and continuing up through his organizing of this conference. Aside from this the two of us have, both individually and collectively, and personally as well as professionally reason to be grateful to Jan-Erik. We take this opportunity to say thank-you.

§ 1. DIFFERENTIAL GRADED ALGEBRAS (DGA's)

In the literature a DGA is a $\mathbb{Z}$-graded ring, R, together with a differential, d, ($d^2=0$) which is a degree-1 ($d: R_n \to R_{n-1}$) derivation ($d(ab)=(da)b+(-1)^{\deg a}a(db)$). Its <u>homology algebra</u>, H(R), is the graded ring Kerd/Imd. We adopt the convention $R^n = R_{-n}$; when degrees are written upstairs then d has degree +1.

In any given context, however, one is usually interested in some particular subcategory, and it is normal under those circumstances to reserve DGA for objects in that subcategory. Here we shall be simultaneously interested in one of several categories, depending on whether we are in the algebra, or topology context.

For reasons of space and simplicity we adopt the device of permitting a single expression to have two meanings: one for each of our two contexts. Thus we establish the

1.1 <u>Convention.</u> Graded objects are graded over the non-negative integers. But

(a) <u>In the algebra context</u>: (i) There is no restriction as to coefficient, and we write $\mathbb{A} = \mathbb{Z}$.

(ii) With rare (and clearly indicated) exceptions the grading is downstairs: $R = \bigoplus_{n\geq 0} R_n$.

(b) <u>In the topology context</u>: (i) The coefficients are always $\mathbb{Q}$ and we write $\mathbb{A} = \mathbb{Q}$.

(ii) With rare (and clearly indicated) exceptions the grading is upstairs: $R = \bigoplus_{n\geq 0} R^n$.

In either case we write $\otimes$ and Hom for $\otimes_{\mathbb{A}}$ and $\mathrm{Hom}_{\mathbb{A}}$.

1.2 <u>Definition</u> (1). A DGA is a graded $\mathbb{A}$-algebra R, together with a differential d such that

(i) $ab = (-1)^{\deg a \deg b} ba \quad (a,b \in R)$,

(ii) $a^2 = 0 \quad (a \in R, \deg a \text{ odd})$, and

(iii) d is a derivation of degree -1 (resp. $+1$) in the algebra (resp. topology) context.

(2) The DGA's in the algebra (resp. topology) context are called DGA_*'s (resp. DGA^*'s).

(3) In any category with differentials a morphism ϕ for which $H(\phi)$ is an isomorphism ($\cong$) will be called a <u>homology isomorphism</u> and denoted by $\xrightarrow[\phi]{\simeq}$. A DGA-morphism which is a homology isomorphism will be called a <u>quism</u>.

(4) Sometimes we shall use R (instead of (R,d)) to denote a DGA; in this case the underlying graded $\mathbb{A}$-algebra will be denoted $R^{\#}$.

Now it turns out that the DGA^*'s we can handle most successfully are those satisfying $H^0 = \mathbb{Q}$. On the other hand we are primarily interested in those DGA_*'s which satisfy two additional conditions - which are automatic for $\mathbb{Q}$-algebras. One of these (existence of divided powers) is standard; the other is less well known but essential for us to have a homotopy theory. Explicitly, we make the

1.3 <u>Definition</u>: (i) A $DG\Gamma_*$ is a DGA_*, (R,d), together with an assigned system $\{\gamma^p\}$ of divided powers in R for which $d(\gamma^p a) = (da).\ \gamma^{p-1}(a)$.

(ii) By a quism between $DG\Gamma_*$'s we shall mean a DGA_* quism preserving divided powers.

(iii) An h-$DG\Gamma_*$ is a $DG\Gamma_*$, (R,d), such that each $a \in R_+$ is in the image of some $DG\Gamma_*$ morphism $R' \to R$ for which $H_+(R') = 0$.

(iv) An h-DGA^* is a DGA^* for which $H^o = \mathbb{Q}$.

(v) An h-DGA is either an h-DGA^* or an h-$DG\Gamma_*$.

For the convenience of the reader we briefly review divided powers. A divided powers (Γ-) algebra is a graded algebra for which

(i) $ab = (-1)^{\deg a\cdot\deg b}ba$ and $a^2 = 0$ if $\deg a$ is odd $(a,b \in R)$.

(ii) For $p \geq 0$, $n \geq 1$ there are assigned set maps $\gamma^p\colon R_{2n} \to R_{2np}$, (or $R^{2n} \to R^{2np}$) such that $\gamma^o(a) = 1$, $\gamma^1(a) = a$ and, among other properties, $p!\gamma^p(a)=a^p$, $p \geq 2$. (For a complete definition see [G-L] for instance.) When R is a $\mathbb{Q}$-vector space $\gamma^p(a) = \frac{1}{p!}\cdot a^p$ is the unique divided powers structure.

The <u>free Γ-algebra</u> $\mathbb{A}\langle x\rangle$ on a symbol x of degree n is given by

(i) If $\deg x = 0$, $\mathbb{A}\langle x\rangle$ is the polynomial algebra $\mathbb{A}[x]$.

(ii) If $\deg x$ is odd then $\mathbb{A}\langle x\rangle = \mathbb{A} \oplus \mathbb{A}\cdot x$ is the exterior algebra on x .

(iii) If $\deg x$ is even and >0 then $A\langle x\rangle = \bigoplus_{p\geq 0} A\cdot\gamma^p(x)$; $\gamma^p(x)\cdot\gamma^q(x)=\binom{p+q}{q}\gamma^{p+q}(x)$.

In any case $\mathbb{A}\langle x\rangle$ is a free graded $\mathbb{A}$-module.

More generally, if $\{x_\alpha\}$ is a basis for a free graded A-module X then the free Γ-algebra $\mathbb{A}\langle X\rangle$ on X is the tensor product: $\bigotimes_\alpha \mathbb{A}\langle x_\alpha\rangle$; we also write $\mathbb{A}\langle\{x_\alpha\}\rangle$. If R is any Γ-algebra, then $R\langle X\rangle = R \otimes \mathbb{A}\langle X\rangle$ inherits a Γ-structure; more generally [Av_2] if $R \to S,T$ are Γ-algebra morphisms then a Γ-structure is determined in $S \otimes_R T$. Finally, in the topology context ($\mathbb{A}=\mathbb{Q}$) it is common to write $\mathbb{Q}\langle X\rangle = \Lambda X$ and

$$R\langle X\rangle = R \otimes \Lambda X = R \otimes (\text{exterior algebra } X^{odd}) \otimes (\text{symmetric algebra } X^{even}) .$$

The essential role of condition (ii) in definition 1.3 will become clear in § 3. For the moment we simply observe that it implies that if R is an h-$DG\Gamma_*$ and $x \in R_{2n+1}$, $n \geq 1$ then $\gamma^p(dx)$ is a boundary $(p \geq 1)$. Thus $H(R)$ inherits a Γ-structure from R .

1.4 <u>Homotopy type</u>. Two topological spaces M and N have the same <u>homotopy type</u> if there are maps $\phi\colon M \to N$, $\psi\colon N \to M$ such that $\phi\psi$ and $\psi\phi$ are homotopic to the identity; ϕ and ψ are called <u>inverse homotopy equivalences</u>.

A weaker notion is that of <u>weak homotopy equivalence</u>; these are the continuous maps which yield isomorphisms $\pi_i(M) \xrightarrow{\cong} \pi_i(N)$ ($i \geq 0$, each base point). By a theorem of Whitehead, these two notions coincide for CW complexes. But for general spaces weak equivalences may not have "inverses" and so two spaces M,N are said to have the same <u>weak homotopy type</u> if they are connected by a chain

$$M = M_o \leftarrow \ldots\ldots \rightarrow M_n = N$$

of weak equivalences.

A second theorem of Whitehead asserts that a map between simply connected spaces is a weak homotopy equivalence if and only if it induces an isomorphism of integral homology. By analogy, it is called a <u>rational homotopy equivalence</u> if it induces an isomorphism of rational homology. Two simply connected spaces M,N have the same rational homotopy type if they are connected by a chain of rational equivalences.

If one thinks of DGA's (resp. morphisms) as spaces (resp. continuous maps) then quisms will correspond to rational equivalences. Thus we say that two DGA's R,S have the same <u>homotopy type</u> if they are connected by a chain of quisms

$$R = R(0) \rightarrow \ldots\ldots \leftarrow R(n) = S\ ;$$

this definition applies equally to: DGA^*'s, DGA_*'s, $DG\Gamma_*$'s and h-$DG\Gamma_*$'s, but the $R(i)$ are then required to belong to the appropriate subcategory.

In particular, we can begin the dictionary with

<u>Topology</u>		<u>Algebra</u>
Topological spaces, continuous maps	$\longleftrightarrow$	h-$DG\Gamma_*$'s, $DG\Gamma_*$-morphisms
Rational homotopy equivalence	$\longleftrightarrow$	Quism
Rational homotopy type	$\longleftrightarrow$	Homotopy type

1.5 <u>DGA's in topology</u>. In topology DGA^*'s were first envisaged as a means to the calculation of the cohomology ring of a space (in our case, with rational coefficients). The standard "differential algebra" used is the algebra of singular cochains, which fails to be commutative in our sense. This failure is in some sense necessary in positive characteristic, because of the Steenrod operations, and it is for this reason we restrict to rational coefficients in topology.

Once this restriction is made, however, we get the very good (contravariant) functor A_{PL} from topological spaces to DGA^*'s. Originally described by Thom and rediscovered over a decade later by Sullivan [Su], A_{PL} has the property of all cochain constructions: The algebra $H(A_{PL}(\ ))$ is naturally isomorphic with rational (singular) cohomology.

Briefly, if M is a space then an element $\Phi \in A_{PL}^q(M)$ is a function on the singular simplices of M. If σ is an n-simplex, then $\Phi(\sigma)$ is a differential q-form on the standard n-simplex, Δ_n. The collection $\{\Phi(\sigma)\}$ is required to satisfy two conditions:

(i) $\Phi(\sigma) \in$ subalgebra generated over Q by the barycentric functions, x_i, and their exterior derivatives, dx_i .

(ii) $\{\Phi(\sigma)\}$ is compatible with the face and degeneracy operators.

Addition, multiplication and differentiation are defined simplexwise:

$$(\Phi+\Psi)(\sigma) = \Phi(\sigma) + \Psi(\sigma); \ (\Phi\wedge\Psi)(\sigma) = \Phi(\sigma) \wedge \Psi(\sigma); \ (d\Phi)(\sigma) = d(\Phi(\sigma)) .$$

Now the fact which makes A_{PL} valuable is that it converts rational homotopy theory to algebraic homotopy theory. In fact, since A_{PL} sends rational homotopy equivalences to quisms (clearly) it induces a map

$$(1.6) \qquad \left\{\begin{array}{l}\text{rational homotopy types of}\\ \text{simply connected spaces}\\ \text{with each } \dim H^p(\ ;\mathbb{Q}) \text{ finite}\end{array}\right\} \rightarrow \left\{\begin{array}{l}\text{homotopy types of } DGA^*\text{'s}\\ \text{with } H^o = \mathbb{Q},\ H^1 = 0 \text{ and}\\ \text{each } \dim H^p \text{ finite .}\end{array}\right\}$$

1.7 <u>Theorem</u> ([Su], [B-G]). The map 1.6 is a bijection.

□

At a more computational level, all the rational invariants of a space M (eg. $\pi_*(M) \otimes \mathbb{Q}$, rational Whitehead products, ...) can be extracted from any DGA^* in the homotopy type of $A_{PL}(M)$.

1.8 <u>Base points, path connected spaces and local rings.</u> It turns out that $A_{PL}(pt)=\mathbb{Q}$ and so A_{PL} converts a base point inclusion, $* \hookrightarrow M$, to an augmentation $A_{PL}(M) \to \mathbb{Q}$. Moreover M is path connected if and only if $H^o = \mathbb{Q}$, which in turn is equivalent to: H^o has a unique maximal ideal (H^o is <u>quasi-local</u>). By analogy a DGA^* R is called <u>connected</u> if $R^o = \mathbb{Q}$; then M is path connected if and only if $A_{PL}(M)$ has the homotopy of a connected DGA^* .

On the algebra side we consider augmented h-$DG\Gamma_*$'s $\varepsilon: R \to \mathbb{k}$ such that $\mathrm{Im}\,\varepsilon$ generates the field $\mathbb{k}$. Given one such we set $\underline{m} = (\ker \varepsilon)_o$ and localize R at $\underline{m}$ by inverting the elements of $R_o - \underline{m}$. There results a new h-$DG\Gamma_*$, $R_{\underline{m}}$, and a morphism $R \to R_{\underline{m}}$ which is a quism if and only if $H_o(R)$ is quasi local. (In this case $\mathrm{Im}\,\varepsilon$ is surjective.)

We now extend our dictionary:

Pointed spaces	$\longleftrightarrow$	Augmented h-$DG\Gamma_*$'s
Path connected spaces	$\longleftrightarrow$	h-$DG\Gamma_*$'s with H_o quasi-local
Connected DGA^*'s	$\longleftrightarrow$	h-$DG\Gamma_*$'s R with R_o quasi-local.

Finally, a space M is <u>simply connected</u> if it is path connected and $\pi_1(M) = 0$. For such spaces $A_{PL}(M)$ has the homotopy type of a DGA^* R satisfying $R^o = \mathbb{Q}$ and $R^1 = 0$; such DGA^*'s are called <u>simply connected</u>.

Since path connected spaces correspond to h-$DG\Gamma_*$'s R with $H_o(R)$ quasi-local, it seems appropriate to make simply connected spaces correspond to h-DGΓ_*'s with $H_o(R)$ a field. Such h-DGΓ_*'s have the same homotopy type as an h-DGΓ_* R with R_o a field, cf. (2.18). Thus our dictionary continues as follows:

Simply connected spaces ↔ h-DGΓ_*'s R with $H_o(R)$ a field
Simply connected DGA*'s ↔ h-DGΓ_*'s R with R_o a field.

1.9 Finiteness conditions: A graded ring, R, is called piecewise noetherian if R_o is noetherian and each R_i is a noetherian R_o-module. A graded ring is called local if R_o is quasi-local and R is piecewise noetherian. A graded vector space has finite type if each homogeneous component has finite dimension. A topological space has finite $\mathbb{Q}$-type if its rational cohomology has finite type. Our dictionary continues with:

Path connected spaces M of finite Q-type ↔ h-DGΓ_*'s R with H(R) local
DGA*'s R of finite type with $R^o = \mathbb{Q}$ ↔ h-DGΓ_*'s R which are local

A space (resp. a DGA*, a DGA$_*$) will be said to have formal dimension n if its rational cohomology (resp. homology) is non-zero in degree n and vanishing in all higher degrees. Thus

Path connected spaces of finite formal dimension ↔ h-DGΓ_*'s R of finite formal dimension.

1.10 Differential Tor. We end with a brief review of the Eilenberg-Moore extension of Tor to modules over a DGA ([M], [G-M]). Suppose that R is either a DGA$_*$ or a DGA* such that $H^o(R) = \mathbb{Q}$.

1.11 Definition (1) An R-module, M, is a graded $R^{\#}$-module together with a differential, d, of degree (-1 in algebra, +1 in topology) such that $d(r\cdot m) = dr\cdot m + (-1)^{\deg r}r\cdot dm$.

(2) The tensor product $M\otimes_R N$ is the $R^{\#}$-module, $M\otimes_{R^{\#}}N$, together with $d(m\otimes n) = dm\otimes n + (-1)^{\deg m}m\otimes dn$.

(3) An R-module is called free if it is free as an $R^{\#}$-module on a basis $\{e_\alpha\}$ of cycles ($de_\alpha = 0$).

(4) An R-module, C, is called semi-free if it admits a filtration $0 = F_{-1}\subset F_o\subset \dots$ by R-submodules such that $C = \bigcup_i F_i$ and each F_i/F_{i-1} is free. (In this case the filtration is called semi-free as well.) Note that a semi-free R-module will always be free as an $R^{\#}$-module but may not be a free R-module.

(5) A semi-free filtration F_i of C is called an Eilenberg-Moore filtration if

$$H(F_o) \twoheadrightarrow H(C) \quad \text{and} \quad \ker\{H(F_i) \to H(C)\} = \ker\{H(F_i) \to H(F_{i+1})\}\ ,\ i \geq 0\ .$$

(6) A <u>semi-free resolution</u> of an R-module M is a homology isomorphism of R-modules: $\rho\colon C \to M$ with C semi-free. If C admits an Eilenberg-Moore filtration we call ρ an <u>Eilenberg-Moore resolution.</u>

Exactly as in the classical case we have the easy exercises: <u>Lifting property:</u> In a diagram of R-modules

$$\begin{array}{ccc} B & \xrightarrow{\ \psi\ } & M \\ {\scriptstyle i}\big\downarrow & & \big\downarrow{\scriptstyle \simeq\ \pi} \\ C & \longrightarrow & N \end{array}$$

with i an inclusion, C/B semi-free and π a surjective homology isomorphism, there is a morphism $C \to M$ extending ψ and lifting ϕ .

<u>Existence property</u>: Every morphism $M \to N$ of R-modules factors as $M \hookrightarrow P \xrightarrow{\simeq} N$ with P/M admitting an Eilenberg-Moore resolution and $P \to N$ a surjective homology isomorphism. (When R is a DGA^* this requires $H^o(R) = \mathbb{Q}$!).

The same argument as in the classical case now establishes

1.12 <u>Lemma</u>. Suppose $\phi\colon C \xrightarrow{\simeq} C'$ is a homology isomorphism of semi-free R-modules. There is then a homology isomorphism $C' \xrightarrow{\simeq} C$ which is "homotopy inverse" to ϕ . In particular, for any R-module N , $\phi \otimes \mathrm{id}\colon C \otimes_R N \to C' \otimes_R N$ is also a homology isomorphism.

1.13 <u>Definition</u> (cf [G-M], [M]). If M,N are R-modules then the differential torsion functor is defined by

$$\mathrm{Tor}^R(M,N) = H(C \otimes_R N)$$

where $C \to M$ is a semi-free resolution.

1.14 <u>Remark.</u> It follows as in the classical case from (1.12) that this is independent of the choice of resolution, functorial in all three variables and symmetric in M and N .

Finally, suppose $\{F_i\}$ is a semi-free filtration of C. The E_1-term of the resulting spectral sequence, together with $H(F_o) \to H(C)$ has the form

$$\cdots \xrightarrow{d^1} H(F_{i+1}/F_i) \xrightarrow{d^1} H(F_i/F_{i-1}) \xrightarrow{d^1} \cdots \xrightarrow{d^1} H(F_o) \to H(C) \to 0\ .$$

Since $H(F_i/F_{i-1})$ is a free $H(R)$-module, the condition that $\{F_i\}$ be an E-M resolution is precisely that this sequence be a free $H(R)$ resolution of $H(C)$ (in the classical sense).

Now if U and V are graded modules over a graded ring K then U admits a resolution $\cdots \to X_i \xrightarrow{d_i} X_{i-1} \to \cdots$ by free graded K modules such that $\deg d_i = 0$. Thus $\mathrm{Tor}^K(U,V) = H(X_* \otimes_K V)$ acquires a second gradation, called the <u>internal degree</u>. We write $\mathrm{Tor}^K_{p,q}(U,V)$; here p is the homological degree and q is the internal degree. The sum, p+q is called the <u>total degree</u>. In the resolution above $H_j(F_i/F_{i-1})$ has homological degree i, internal degree j-i, and total degree j.

In particular, suppose $C \to M$ is a resolution with EM filtration $\{F_i\}$ of C. The filtration $\{F_i \otimes_R N\}$ of $C \otimes_R N$ (N a second R-module) then yields a spectral sequence independent of the choice of resolution or EM filtration, and symmetric in M and N. It is called the <u>Eilenberg-Moore spectral sequence</u> (E.M.s.s.). Moreover,

<u>for DGA_*'s:</u> The E.M.s.s. is a first quadrant homology spectral sequence with $E^2_{p,q} = [\mathrm{Tor}^{H(R)}_{p,q}(H(M), H(N))]$, naively convergent to $\mathrm{Tor}^R(M,N)$.

<u>for DGA^*'s:</u> Write $F^{-i} = F_i$ to identify the E.M.s.s. as a second quadrant (in fact third octant) cohomology spectral sequence with $E_2^{-p,q} = [\mathrm{Tor}^{H(R)}_{p,-q}(H(M), H(N))]$. Naive convergence may fail, but $d_i = 0$ in $E_i^{-p,q}$ for $i > p$. Set $E_\infty^{-p,q} = \varinjlim_{i>p} E^{-p,q}$, then E_∞ is the bigraded $H(R)$ module associated with the induced filtration of $\mathrm{Tor}^R(M,N)$. In this sense the E.M.s.s. is convergent.

Finally, either directly or via the E.M.s.s. we get

1.15 <u>Theorem</u>. Let $R \to R'$ be a quism, either of DGA_*'s or of DGA^*'s with $H^0 = \mathbb{Q}$. Suppose $M \to M'$, $N \to N'$ are homology isomorphisms of R'-modules. Then

$$\mathrm{Tor}^R(M,N) \to \mathrm{Tor}^{R'}(M',N')$$

is an isomorphism.

□

§ 2. FREE EXTENSIONS AND HOMOTOPY

Recall the definition of $R\langle X\rangle$ for a graded Γ-algebra, R. Now suppose R is a DGA^* or a $DG\Gamma_*$ with underlying ring $R^\#$.

2.1 <u>Definition</u>: A <u>free extension</u> of R is a morphism $R \to S$ in which

(i) $S^\# = R^\#\langle X\rangle$

(ii) X admits a well ordered homogeneous basis, x_α, such that $dx_\alpha \in R^\#\langle\{x_\beta\}_{\beta<\alpha}\rangle$.

We shall abuse notation and write $R\langle X\rangle$ for S and for $S^\#$.

Condition (ii) may be unfamiliar to algebraists, because it is automatic for $DG\Gamma_*$'s: simply order the x_α so $\alpha < \beta$ if $\deg x_\alpha < \deg x_\beta$. On the other hand it is essential for DGA^*'s: $\mathbb{Q}\langle x_1,x_2,x_3\rangle$ with $\deg x_i = 1$, $dx_1 = x_2x_3$, $dx_2 = x_3x_1$, $dx_3 = x_1x_2$

is not a free extension of $\mathbb{Q}$.

Free extensions were introduced by Tate [T] for the homological study of local rings. Their systematic use in rational homotopy theory is due to Sullivan.

Free extensions have lifting and existence properties quite analogous to those of resolutions, and almost as easily established.

<u>Lifting property</u>: Suppose in a commutative square of DGA^*'s or $DG\Gamma_*$'s,

$$\begin{array}{ccc} R & \xrightarrow{\psi} & S \\ {\scriptstyle i}\downarrow & & \downarrow{\scriptstyle \pi}\ \simeq \\ R\langle X\rangle & \xrightarrow[\phi]{} & T \end{array}$$

that i is a free extension and π is a surjective quism. Then there is a morphism $R\langle X\rangle \to S$ extending ψ and lifting ϕ through π.

<u>Existence property</u>: Suppose $\phi: R \to S$ is a morphism either between $DG\Gamma_*$'s or between DGA^*'s satisfying $H^0 = \mathbb{Q}$. Then ϕ factors as the composite $R \xrightarrow{i} R\langle X\rangle \xrightarrow[\simeq]{m} S$ of a free extension and a quism; moreover this factorization can be chosen so m is surjective. The morphism $m: R\langle X\rangle \xrightarrow{\simeq} S$ is called a <u>free model</u> of ϕ.

Next suppose $R \to R\langle X\rangle$ is a free extension of DGA^*'s or of $DG\Gamma_*$'s. For any morphism $R \to T$ (in the respective category) we get a pushout square

$$\begin{array}{ccc} R & \longrightarrow & T \\ \downarrow & & \downarrow \\ R\langle X\rangle & \longrightarrow & T\langle X\rangle, \end{array} \quad T\langle X\rangle = T \otimes_R R\langle X\rangle ,$$

in which both vertical arrows are free extensions.

2.2 <u>Lemma</u>: (i) $R\langle X\rangle$ is a semi-free R-module.

(ii) Assume
$$\begin{array}{ccccc} R\langle X\rangle & \longleftarrow & R & \longrightarrow & T \\ \simeq\downarrow & & \simeq\downarrow & & \simeq\downarrow \\ R'\langle X'\rangle & \longleftarrow & R' & \longrightarrow & T' \end{array}$$

is a commutative diagram of DGA^* or $DG\Gamma_*$ morphisms, in which the vertical arrows are quisms and the left arrows are free extensions. Then the induced morphism

$$T\langle X\rangle \to T'\langle X'\rangle$$

is a quism.

<u>Proof</u>: (i) follows by induction on the well-ordered basis $\{x_\alpha\}$ of X, cf. (2.7). This implies that the morphism $H(T\langle X\rangle) \to H(T'\langle X'\rangle)$ can be identified with the map $\mathrm{Tor}^R(R\langle X\rangle,T) \to \mathrm{Tor}^{R'}(R'\langle X'\rangle,T')$. But this is an isomorphism by Theorem 1.15.

□

2.3 Elementary properties of h-$D\Gamma_*$'s. For the moment we work only on the algebra side; i.e., with $D\Gamma_*$'s. Thus $\mathbb{A} = \mathbb{Z}$. Define $\mathbb{A}\langle x,dx\rangle \xrightarrow{\phi} \mathbb{A}$ by $x,\ dx \to 0$; this is a quism if $\deg x = 1$ or $\deg x = 2n$, $n \geq 1$. In any case we may choose a free model $\mathbb{A}\langle C(x)\rangle = \mathbb{A}\langle x,dx\rangle\langle X\rangle \xrightarrow{\simeq} \mathbb{A}$ for ϕ. This is an example of a free extension $\mathbb{A}\langle C\rangle$ admitting a quism $\mathbb{A}\langle C\rangle \to \mathbb{A}$; such free extensions are called acyclic.

An element x in a $D\Gamma_*$, R, is called admissible if it is in the image of a morphism $R' \to R$ with $H_+(R') = 0$. If $\deg x > 0$ then x is admissible if and only if the morphism $\mathbb{A}\langle x,dx\rangle \to R$ extends to $\mathbb{A}\langle C(x)\rangle \to R$. Since acyclic free extensions are closed under (infinite) tensor products we deduce

2.4 Lemma. (i) The admissible elements in a $D\Gamma_*$, R, form a sub $D\Gamma_*$, $\hat{R}$, which is in fact the maximal h-$D\Gamma_*$ contained in R.

(ii) $\hat{R}$ contains R_{2n} $(n \geq 0)$, R_1, and the cycles in R.

(iii) There is an acyclic free extension $\mathbb{A}\langle C\rangle$ and a morphism $\mathbb{A}\langle C\rangle \to R$ whose image contains $\hat{R}_+ \oplus dR_1$.

The lemma shows that a $D\Gamma_*$, R, is an h-$D\Gamma_*$ if and only if there is an acyclic free extension $\mathbb{A}\langle C\rangle$ and a morphism $\mathbb{A}\langle C\rangle \to R$ whose image contains $R_+ \oplus d(R_1)$. Such a morphism will be called an acyclic cover for R.

2.5 Proposition: Suppose $\phi\colon R \to S$ is a $D\Gamma_*$ morphism.

(i) If R is an h-$D\Gamma_*$ and ϕ_+ is surjective then S is an h-$D\Gamma_*$.

(ii) If S is an h-$D\Gamma_*$ and ϕ is a quism then R is an h-$D\Gamma_*$.

(iii) In a commutative $D\Gamma_*$ diagram

$$\begin{array}{ccc} T & \xrightarrow{\alpha} & R \\ {\scriptstyle i}\downarrow & & \downarrow{\scriptstyle \phi}\,\simeq \\ T\langle X\rangle & \xrightarrow[\beta]{} & S \end{array}$$

in which ϕ is a quism, i is a free extension and S is an h-$D\Gamma_*$ we have: α extends to $\alpha'\colon T\langle X\rangle \to R$.

Proof: (i) is trivial. For (ii) choose an acyclic cover $\mathbb{A}\langle C\rangle \to S$. Tensored with ϕ this gives a surjection $\pi\colon R\langle C\rangle \xrightarrow{\simeq} S$. Now $\mathbb{A}\langle C\rangle$ has an $\mathbb{A}$-basis of the form $\{1,\omega_\alpha,d\omega_\alpha\}$ and it follows that $R \xrightarrow{j} R\langle C\rangle$ is a quism. The quism $\mathbb{A}\langle C\rangle \xrightarrow{\simeq} \mathbb{A}$ then defines an inverse quism $\rho\colon R\langle C\rangle \to R$.

Now for $x \in R_n$ $(n \geq 1)$ we get $\sigma\colon \mathbb{A}\langle x,dx\rangle \to R$; because S is an h-$D\Gamma_*$, $\phi\alpha$ extends to $\beta\colon \mathbb{A}\langle C(x)\rangle \to S$. Since $\phi = \pi j$ is a quism, so is π. Thus β lifts to a $\gamma\colon \mathbb{A}\langle C(x)\rangle \to R\langle C\rangle$ extending $j \circ \sigma$. Hence $\rho\sigma\colon \mathbb{A}\langle C(x)\rangle \to R$ extends σ and shows x is admissible.

To prove (iii) construct a surjective quism $R\langle C\rangle \xrightarrow{\simeq} S$ extending ϕ as in (ii).

Lift β to $\gamma: T\langle X\rangle \to R\langle C\rangle$ and set $\alpha' = \rho\gamma$.

□

Note that proposition 2.5 (iii) holds for all DGA^*'s because, since $\mathbb{Q}\langle x,dx\rangle$ is always acyclic, every DGA^* has an acyclic cover.

2.6 Corollary. (i) If $R \to S,T$ are $DG\Gamma_*$ morphisms and S,T are h-$DG\Gamma_*$'s then $S \otimes_R T$ is also an h-$DG\Gamma_*$.
(ii) If $R\langle X\rangle \xrightarrow{\simeq} S$ is a free model of a $DG\Gamma_*$ morphism $R \to S$ and if S is an h-$DG\Gamma_*$ then so is $R\langle X\rangle$.

2.7 Example. Let R be a local ring (concentrated in degree zero) with maximal ideal $\underline{m}$ and residue field $\mathbb{k}$. The Koszul complex K^R is the DGA_* $R\langle x_1,\ldots,x_n\rangle$, where $\deg x_i = 1$ and $\{dx_i \in \underline{m}\}$ represent a basis for $\underline{m}/\underline{m}^2$. It satisfies $H_o(K^R) = \mathbb{k}$.

Of course R is itself an h-$DG\Gamma_*$; it turns out that so is its Koszul complex. Indeed if $y_1,\ldots,y_n$ are symbols of degree zero then the $DG\Gamma_*$ $\mathbb{A}\langle\{y_i,x_j\}\rangle$ $(dx_i = y_i)$ is acyclic. Thus $K^R = R \otimes_{\mathbb{A}\langle\{y_i\}\rangle} \mathbb{A}\langle\{y_i,x_j\}\rangle$ is an h-$DG\Gamma_*$.

More generally, if R is any h-$DG\Gamma_*$ with $H_o(R)$ a local ring we can construct an h-$DG\Gamma_*$

$$K^R = R\langle x_1,\ldots,x_n\rangle, \quad \deg x_i = 1$$

such that $H_o(R)\langle x_1,\ldots,x_n\rangle = K^{H_o(R)}$. This will be called a Koszul complex for the h-$DG\Gamma_*$, R. It satisfies $H_o(K^R) = \mathbb{k}$.

2.8 Homotopy. Fix an h-DGA morphism $\alpha: W \to R$ and let $m: W\langle Y\rangle \xrightarrow{\simeq} R$ be a free model for α. Consider

$$m\cdot m: W\langle Y\rangle \otimes_W W\langle Y\rangle \to R$$

and extend it to a free model ($W\langle Y\rangle \otimes_W W\langle Y\rangle = Q\langle Y,Y\rangle$)

$$m_I: I_W R = W\langle Y,Y\rangle\langle Z\rangle \xrightarrow{\simeq} R .$$

Let $\lambda_o,\lambda_1: W\langle Y\rangle \to I_W R$ denote the left and right inclusions.

We follow Quillen and call $(I_W R,\lambda_o,\lambda_1)$ a relative cylinder object for α. When $W = \mathbb{A}$ we write IR; it corresponds (as we shall see in (3.3) to the space of paths in a topological space rather than to the cylinder $I \times M$; this is again due to the contravariance of A_{PL}.

2.9 Definition. Two h-DGA morphisms $\phi_o,\phi_1: R \to S$ are homotopic (rel W) - we write $\phi_o \sim \phi_1$ (rel W) - if for some free model of α and some relative cylinder object $I_W R$ there is a morphism $\Phi: I_W R \to S$ such that $\Phi \circ \lambda_i = \phi_i \circ m$. (Note that this implies that $\phi_1 \circ \alpha = \phi_o \circ \alpha$!) When $W = \mathbb{A}$ we write $\phi_o \sim \phi_1$ and call ϕ_o and ϕ_1 homotopic.

If $W\langle Y\rangle \to W\langle Y'\rangle$ is a quism restricting to id_W then Lemma 2.2 (ii) implies that $W\langle Y\rangle \otimes_W W\langle Y\rangle \to W\langle Y'\rangle \otimes_W W\langle Y'\rangle$ is also a quism. Thus the next proposition is a straightforward consequence of the existence and lifting properties for free extensions, and of (2.5 (iii)). The arguments are very similar to those of Quillen [Q_1].

2.10 <u>Proposition</u>. (i) Homotopy (rel W) is an equivalence relation, independent of the choices of free model and cylinder object.

(ii) If $\psi: S \to T$ is an h-DGA morphism then $\phi_o \sim \phi_1$ (rel W) implies $\psi\phi_o \sim \psi\phi_1$ (rel W). The reverse implication holds if ψ is a quism.

(iii) If α factors as $W \xrightarrow{\beta} T \xrightarrow{\psi} R$ then $\phi_o \sim \phi_1$ (rel W) implies $\phi_o\psi \sim \phi_1\psi$ (rel W). The reverse implication holds if ψ is a quism.

(iv) If $\phi_o \sim \phi_1$ (rel W) then $\phi_o \sim \phi_1$.

□

As an immediate corollary we deduce (recall an h-DGA is either an h-DGA* or an h-DGΓ_*.)

2.11 <u>Theorem</u>. Let

$$\begin{array}{ccc} W & \xrightarrow{\alpha} & R \\ {\scriptstyle i}\downarrow & & \downarrow{\scriptstyle \simeq\,\phi} \\ W\langle Y\rangle & \xrightarrow[\beta]{} & S \end{array}$$

be a commutative diagram of h-DGA's in which i is a free extension and ϕ is a quism. Then α extends to $\alpha': W\langle Y\rangle \to R$ such that $\phi\alpha' \sim \beta$ (rel W). If α'' is a second such extension then $\alpha' \sim \alpha''$ (rel W).

□

2.12 <u>Homotopy pushouts</u>. Suppose $R \to S$, $R \to T$ are h-DGA morphisms and choose a free model $m: R\langle X\rangle \xrightarrow{\simeq} S$. If $m': R\langle X'\rangle \xrightarrow{\simeq} S$ is a second free model then by Theorem 2.11 we obtain a unique homotopy class (rel R) of quisms $R\langle X\rangle \xrightarrow{\omega} R\langle X'\rangle$ such that $m'\omega \sim m$ (rel R).

These homotopy classes are closed under composition and so provide a unique (up to homotopy (rel R)) identification of the homotopy types of the $R\langle X\rangle$. Performing $T \otimes_R$ yields unique homotopy classes (rel T) of quisms

$$T\langle X\rangle \xrightarrow{\simeq} T\langle X'\rangle \ . \tag{2.13}$$

2.14 <u>Definition</u>. The common homotopy type of the $T\langle X\rangle$ is called the <u>homotopy push-out</u> of $S \to R \to T$.

Of course we would equally well have considered free models $R\langle Y\rangle \xrightarrow{\simeq} T$ and deduced unique homotopy classes (rel S) of quisms $S\langle Y\rangle \xrightarrow{\simeq} S\langle Y'\rangle$. By Lemma 2.2 we have quisms

$$(2.15)\qquad T\langle X\rangle \xleftarrow{\simeq} R\langle X,Y\rangle \xrightarrow{\simeq} S\langle Y\rangle$$

which are compatible with the identifications (up to homotopy) on either side. In particular we can canonically use any $S\langle Y\rangle$ or $R\langle X,Y\rangle$ to represent the homotopy pushout.

Finally, we remark that (by Lemma 2.2 (i)) the homology of the homotopy pushout is given by

$$(2.16)\qquad H(T\langle X\rangle) = \mathrm{Tor}^R(S,T)$$

and that the identification $H(T\langle X\rangle) = \mathrm{Tor}^R(S,T) = H(T\langle X'\rangle)$ is precisely the isomorphism induced from the homotopy class of quisms (2.13). Similarly the identification $\mathrm{Tor}^R(S,T) = \mathrm{Tor}^R(T,S)$ is induced by the quisms (2.15).

It follows, in particular, that $\mathrm{Tor}^R(S,T)$ has naturally the structure of a Γ-algebra, and that the symmetry identification is an isomorphism of Γ-algebras.

2.17 <u>Coproducts</u>: If R and S are h-$DG\Gamma_*$'s their tensor product $R \otimes S$ is just the coproduct, or pushout, with respect to the initial object $\mathbb{A}$.

On the other hand, we have just defined the <u>homotopy pushout</u> of $\mathbb{A} \to R,S$; it is the homotopy type canonically represented by $\mathbb{A}\langle X\rangle \otimes \mathbb{A}\langle Y\rangle$ where these are respectively free models of R and S. It is important to note that the canonical morphism

$$\mathbb{A}\langle X\rangle \otimes \mathbb{A}\langle Y\rangle \to R \otimes S$$

need <u>not</u> be a quism. Indeed when $R = S = \mathbb{Z}_p$ concentrated in degree zero then also $R \otimes S = \mathbb{Z}_p$. On the other hand R (and S) has a free model $\mathbb{Z}\langle x\rangle$ with $\deg x = 1$ and $dx = p$; it follows that

$$H(\mathbb{Z}\langle x\rangle \otimes \mathbb{Z}\langle x\rangle) \simeq \mathbb{Z}_p\langle x\rangle .$$

2.18 <u>Fields</u>. We are frequently faced with the problem that the homology $H_o(R)$ of some h-$DG\Gamma_*$ R contains a field $\mathbb{k}$, although R itself does not. (For example, R might be a Koszul complex, cf. (2.7).) Free extensions permit us to replace R by another h-$DG\Gamma_*$ of the same homotopy type, which does contain $\mathbb{k}$. We need the

2.19 <u>Lemma</u>. Every field $\mathbb{k}$ has a free model of the form $\mathbb{Z}[X_o]\langle X_1\rangle$.

<u>Proof</u>: The lemma follows via a direct limit argument from the following three observations: (i) $\mathbb{Z}\langle x\rangle$, $dx = p$ is a free model of $\mathbb{Z}_p$; (ii) $\mathbb{k}\langle a,x\rangle$, $dx = f(a)$ is a free model for an algebraic extension $\mathbb{k}(a)$; (iii) if $R = R_o$ is an integral domain and $b \in R$ then $R\langle a,x\rangle$, $dx = ab - 1$ is a free model for $R(1/b)$.

□

Now suppose R is an h-$DG\Gamma_*$ and $H_o(R) \supset$ a field $\mathbb{k}$. Choose a free model $\rho\colon \mathbb{Z}\langle X_o,X_1\rangle \xrightarrow{\simeq} k$ and lift ρ to $\phi\colon \mathbb{Z}[X_o] \to R_o$ so that $[\phi x] = \rho x$. Then for $y \in X$

$$[\phi dy] = \rho dy = 0$$

and so $\phi dy \in dR_1$. Because X_1 is $\mathbb{Z}$-free we can extend ϕ to a morphism

$$\phi\colon \mathbb{Z}\langle X_0, X_1\rangle \to R\ .$$

Finally, extend ϕ to a free model of R . Since ρ is a quism we can apply Lemma 2.2 (ii) to get

$$k\langle Y\rangle \xleftarrow{\simeq} \mathbb{Z}\langle X_0, X_1\rangle\langle Y\rangle \xrightarrow{\simeq} R$$

in which the left hand morphism is also a quism. Thus R has the same homotopy type as $k\langle Y\rangle$.

§ 3. FIBRATIONS AND FREE EXTENSIONS

Fibrations play the same role in the topological category that free extensions do for DGA's. We recall that by definition a fibration is a continuous map $\pi\colon E \to B$ with the

Lifting property: In a commutative square of continuous maps

$$\begin{array}{ccc} A & \xrightarrow{\psi} & E \\ {\scriptstyle i}\downarrow & & \downarrow{\scriptstyle \pi} \\ X & \xrightarrow[\phi]{} & B \end{array}$$

in which A is a closed strong deformation retract of X, ψ extends to a continuous map $X \to E$ which lifts ϕ .

As with free extensions we have the

Existence property: Suppose $\phi\colon B_1 \to B$ is a continuous map. Then ϕ factors as the composite $B_1 \xrightarrow{j} E \xrightarrow{\pi} B$ of a homotopy equivalence j, and a fibration π . (The construction runs as follows: let B^I be the space of continuous paths $I \to B$ with the compact open topology. Then $E \subset B^I \times B_1 = \{(f,y)\,|\,f(0) = \phi y\}$; $\pi(f,y) = f(1)$; $j(y) = (f_{\phi y}, y)$ where $f_{\phi y}$ denotes the constant path at ϕy.)

If $E \xrightarrow{\pi} B$ is a fibration and $Y \xrightarrow{\phi} B$ is a continuous map, then in the pullback square

$$\begin{array}{ccc} Y\times_B E & \longrightarrow & E \\ {\scriptstyle \pi_Y}\downarrow & & \downarrow{\scriptstyle \pi} \\ Y & \longrightarrow & B\ , \end{array} \qquad Y\times_B E = \{(y,e)\,|\,\phi y = \pi e\}\ ,$$

π_Y is a fibration. If (cf. Lemma 2.2 (ii))

$$\begin{array}{ccccc} E & \longrightarrow & B & \longleftarrow & Y \\ \uparrow & & \uparrow & & \uparrow \\ E' & \longrightarrow & B' & \longleftarrow & Y' \end{array}$$

is a commutative square of continuous maps in which the vertical arrows are weak homotopy equivalences and the left arrows are fibrations then

$$Y \times_B E \longleftarrow Y' \times_{B'} E'$$

is a weak homotopy equivalence.

Suppose now (as in (2.12)) that $X \to E \xrightarrow{\pi} B$ and $X \to E' \xrightarrow{\pi'} B$ are two factorizations of the same continuous map $X \to B$ as the composite of a homotopy equivalence and a fibration. There is then a uniquely determined homotopy class of weak homotopy equivalences $\alpha: E \xrightarrow{\simeq} E'$ (such that $\pi'\alpha = \pi$) and hence (for any continuous $Y \to B$) a unique homotopy class of weak equivalences

(3.1) $$Y \times_B E \xrightarrow{\simeq} Y \times_B E' .$$

The common weak homotopy type of these spaces is called the <u>homotopy pullback</u> of $X \to B \leftarrow Y$. Had we factored $Y \to B$ as $Y \xrightarrow{\simeq} E_1 \xrightarrow{\pi_1} B$ (π_1 a fibration) then we would have obtained canonical weak equivalences

(3.2) $$Y \times_B E \xrightarrow{\simeq} E_1 \times_B E \xleftarrow{\simeq} E_1 \times_B X$$

so that any of these spaces naturally represents the homotopy pullback. Of course (3.1) and (3.2) are the precise analogues of (2.13) and (2.15), with homotopy pullback corresponding to homotopy pushout for h-DGΓ_*'s.

Finally, we note that the projections $X \to pt \leftarrow Y$ are fibrations and that their homotopy pullback is just the ordinary product $X \times Y$. Thus the product of topological spaces corresponds <u>not</u> to the <u>coproduct</u> of h-DGΓ_*'s but to their <u>homotopy pushout</u>.

3.3 <u>Example:</u> <u>Free path space and homotopy</u>. If M is a topological space recall that the space M^I of all continuous maps $I \to M$ is called the <u>free path space</u> on M. The projection

$$\pi: M^I \to M \times M;\ f \mapsto (f(0), f(1))$$

is a fibration and can be identified with the fibration obtained from the diagonal map $M \to M \times M$.

By definition, two continuous maps $\phi_0, \phi_1: N \to M$ are homotopic if and only if there is a map $\Phi: N \to M^I$ such that

$$\pi\Phi = (\phi_0, \phi_1) .$$

Now consider the case of $h\text{-}DG\Gamma_*$'s. If R is one such with free model $m: \mathbb{A}\langle X\rangle \xrightarrow{\simeq} R$ then we translate the product $M \times M$ to the homotopy pushout $\mathbb{A}\langle X\rangle \otimes \mathbb{A}\langle X\rangle$ (see just above) while the diagonal $M \to M \times M$ translates to

$$R \leftarrow \mathbb{A}\langle X\rangle \otimes \mathbb{A}\langle X\rangle : m\cdot m \ .$$

Converting $M \to M \times M$ to a fibration means that we should construct a free model of $m\cdot m$,

$$R \xleftarrow{\simeq} (A\langle X\rangle \otimes A\langle X\rangle)\langle Z\rangle = IR \ .$$

The projections $M \times M \to M$ correspond to the inclusions $\lambda_i: \mathbb{A}\langle X\rangle \to \mathbb{A}\langle X\rangle \otimes \mathbb{A}\langle X\rangle$, which we may think of as the inclusions $\lambda_i: \mathbb{A}\langle X\rangle \to IR$.

If now we were to translate the topological definition of homotopy we would say two morphisms $\phi_i: R \to S$ were homotopic precisely when there was a morphism $\Phi: IR \to S$ such that $\Phi \circ \lambda_i = \phi_i \circ m$. This is of course exactly the definition we gave in (2.8). Thus in our dictionary

Two continuous maps are homotopic $\longleftrightarrow$ Two $h\text{-}DG\Gamma_*$ morphisms are homotopic.

3.4 <u>The model for a fibration</u>. We have seen that fibrations (in topology) and free extensions (in algebra) have analogous properties. Since our dictionary is based on the passage A_{PL}: topology $\rightsquigarrow$ DGA^*'s together with the comparison DGA^*'s vs. $h\text{-}DG\Gamma_*$'s, one might hope that A_{PL} mapped fibrations to free extensions.

This, however, fails. Instead we consider fibrations $E \xrightarrow{\pi} B$ of path connected spaces and form the free model

$$m: \mathbb{A}_{PL}(B)\langle X\rangle \xrightarrow{\simeq} \mathbb{A}_{PL}(E)$$

of $A_{PL}(\pi)$. Then if $Y \to B$ is any continuous map the morphism $A_{PL}(B) \to A_{PL}(Y)$, together with m, determines a morphism

$$\phi: A_{PL}(Y)\langle X\rangle \to A_{PL}(Y \times_B E) \ .$$

In [Ha; § 20] is proved the

3.5 <u>Theorem.</u> Suppose that

(i) The fibre, F, of π is path connected.

(ii) Either F or both Y and B have finite $\mathbb{Q}$-type.

(iii) $\pi_1(B)$ acts nilpotently on each $H^p(F;\mathbb{Q})$.

Then the morphism ϕ is a quism.

3.6 <u>Corollary</u>. $H^*(Y \times_B E;\mathbb{Q}) \cong \mathrm{Tor}^{A_{PL}(B)}(A_{PL}(Y), A_{PL}(E))$.

One certainly has the right to expect that theorems on $h\text{-}DG\Gamma_*$'s and their homotopy pushouts will correspond to theorems on $h\text{-}DGA^*$'s and their homotopy pushouts. On the other hand, the effect of Theorem 3.5 is to assure that the homotopy pushout of $h\text{-}DGA^*$'s is a model for the homotopy pullback of topological spaces. It is for this reason that it is possible to translate theorems between commutative rings and topological spaces.

One other historical remark: Corollary 3.6 is a variant of the original theorem of Eilenberg and Moore which "began" the subject of differential homological algebra.

3.7 Fibre and homotopy fibre. Fix a path connected space, B. The fibre of a fibration $E \to B$ at $b \in B$ is the pull-back $\{b\} \times_B E = \pi^{-1}(b)$. Its homotopy type does not depend on b .

More generally, if $\phi\colon Y \to B$ is any continuous map then the homotopy fibre of ϕ is the homotopy pull-back of ϕ and (any) inclusion $\{b\} \to B$. It depends only on the homotopy class of ϕ and is an extremely important invariant. A representative space is constructed in one of two ways:

(i) Convert ϕ to a fibration $E \xrightarrow{\simeq} B$ and take a fibre, $\pi^{-1}(b)$.

(ii) Convert $\{b\} \to B$ into a fibration $P \to B$ and take the pull-back $P \times_B Y$.

The analogue for $h\text{-}DG\Gamma_*$'s should be clear. The inclusion $\{b\} \to B$ corresponds to an augmentation $R \xrightarrow{\varepsilon} \mathbb{k}$ of an $h\text{-}DG\Gamma_*$ (cf. 1.8) such that $\operatorname{Im} \varepsilon$ generates $\mathbb{k}$. Thus the fibre of a free extension $R<Y>$ at ε is the pushout $\mathbb{k}<Y>$. If $\phi\colon R \to S$ is any $h\text{-}DG\Gamma_*$ morphism then the homotopy pushout (2.12) of ϕ and ε is called the homotopy fibre of ϕ . As in topology it is an extremely important invariant. It has received less attention, presumably because even when R,S are classical commutative rings the homotopy fibre is usually a genuine $h\text{-}DG\Gamma_*$.

A representative for this homotopy fibre can be constructed in two ways:

(i) Choose a free model $R<Y> \xrightarrow{\simeq} S$ of ϕ; then $\mathbb{k}<Y>$ represents the homotopy fibre

(ii) Choose a free model $R<X> \xrightarrow{\simeq} \mathbb{k}$ of ε; then $S<X>$ represents the homotopy fibre.

The fact that $\mathbb{k}<Y>$ and $S<X>$ have the same homotopy type plays an important role in proving some theorems.

3.8 The Serre spectral sequence. Let $E \xrightarrow{\pi} B$ be a fibration with E path connected and B simply connected. Form a free model $A_{PL}(B)<X> \xrightarrow{\simeq} A_{PL}(E)$. Filtering the left hand side by $F^p = A_{PL}^{\geq p}(B)<X> \overset{\mathrm{def}}{=} A_{PL}^{\geq p}(B) \otimes_{A_{PL}(B)} A_{PL}(B)<X>$ we get a first quadrant cohomology spectral sequence converging to $H^*(E;\mathbb{Q})$. Its E_2-term is just

$$E_2^{p,q} = H^p(B) \otimes H^q(\mathbb{Q}<X>) .$$

If the hypotheses of Theorem 3.4 $(Y = \mathrm{pt.})$ are satisfied then $H(\mathbb{Q}<X>) = H^*(F;\mathbb{Q})$ and this is the Serre spectral sequence.

Analogously suppose $R \xrightarrow{\phi} S$ is an $h\text{-}DG\Gamma_*$ morphism with R augmented to $\mathbb{k}$

and $H_o(R) = \mathbb{k}$, cf. (1.8). Let $R\langle X\rangle \xrightarrow{\simeq} S$ be a free model for ϕ and filter the H-DGΓ_*, $R\langle X\rangle$, by $F_p = R \otimes [\mathbb{A}\langle X\rangle]_{<p}$. (Notice that $R_{<p}\langle X\rangle$ is not stable under d!)

This produces a first quadrant homology spectral sequence in which $E^1_{p,q} = H_q(R) \otimes [\mathbb{A}\langle X\rangle]_p = H_q(R) \otimes_{\mathbb{k}} \mathbb{k}\langle X\rangle_p$, and

$$E^2_{p,q} = H_q(R) \otimes_{\mathbb{k}} H_p(\mathbb{k}\langle X\rangle) .$$

This spectral sequence converges to H(S); by analogy we also call it a Serre spectral sequence.

Observe that if the requirement that B is simply connected is removed then the E_2-term of the topological spectral sequence becomes "cohomology with twisted coefficients". The identical phenomenon occurs if the requirement that $H_o(R) = \mathbb{k}$ is replaced by $H_o(R) \supset \mathbb{k}$. This is another reason for the parallel: simple connectivity $\leftrightarrow$ $H_o(R) = \mathbb{k}$ in (1.8).

The analogy between the two sides is reinforced by the well known

3.9 Proposition. Let $\pi: E \to B$ be a fibration with B simply connected, E path connected, and either the fibre F, or B of finite $\mathbb{Q}$-type. Let $R \to S$ be an h-DGΓ_* morphism with R augmented to $\mathbb{k}$ and $H_o(R) = \mathbb{k}$. The following assertions are then equivalent:

(i) The Serre spectral sequence collapses at E_2 .

(ii) The morphism $H^*(E;\mathbb{Q}) \to H^*(F;\mathbb{Q})$ (resp. $H(S) \to \mathrm{Tor}^R(S;\mathbb{k})$) is surjective.

(iii) $H^*(E;\mathbb{Q})$ is a free $H^*(B;\mathbb{Q})$ module (resp. H(S) is a free H(R)-module).

Proof: In the topological case denote $A_{PL}(B) \to A_{PL}(E)$ also by $R \to S$. Then $H^*(E) \to H^*(F)$ is identified with $H(S) \to \mathrm{Tor}^R(S;\mathbb{k})$. In either case, this is one edge homomorphism for the spectral sequence; the other is $H(R) \to H(S)$. Now clearly (i) $\Rightarrow$ (ii) and (iii).

To show (ii) $\Rightarrow$ (i) we need to reduce to the case $R^o = \mathbb{Q}$ (or $R_o \supset \mathbb{k}$). On the topology side we may simply replace R by an appropriate free model; on the algebra side we apply 2.18. Let $\mathbb{k}$ stand for $\mathbb{Q}$ on the topology side. Then on either side the spectral sequence arises from a model $R \otimes_{\mathbb{k}} \mathbb{k}\langle X\rangle$.

Now choose cycles $\Omega_i \in R \otimes_{\mathbb{k}} \mathbb{k}\langle X\rangle$ which project to a $\mathbb{k}$-basis of $H(k\langle X\rangle)$; let Ω be their span and observe that $R \otimes_{\mathbb{k}} \Omega \to R \otimes_{\mathbb{k}} \mathbb{k}\langle X\rangle$ gives an isomorphism of spectral sequences from E^2 (E_2) on.

To see that (iii) $\Rightarrow$ (ii) we note that if (iii) holds the bigraded algebra $\mathrm{Tor}^{H(R)}_{**}(H(S),\mathbb{k})$ is concentrated in bidegrees $(0,*)$. Thus the Eilenberg-Moore spectral sequence (cf. (1.10)) collapses and $H(S) \to \mathrm{Tor}^R(S;\mathbb{k})$ is identified with $H(S) \to \mathrm{Tor}^{H(R)}(H(S);\mathbb{k})$, which in this case is surjective.

□

A fibration satisfying the conclusions of (3.9) is said to have TNCZ fibre (TNCZ = totally non cohomologous to zero). A morphism $R \to S$ satisfies the conclusions precisely when $H(S)$ is $H(R)$-flat. Thus these conditions are analogous.

3.10 <u>The dictionary continued</u>. We summarize § 3 by the following table in our dictionary:

Fibrations	$\longleftrightarrow$	Free extensions
Homotopy pullback	$\longleftrightarrow$	Homotopy pushout
Products $M \times N$ of M and N	$\longleftrightarrow$	Homotopy pushouts $\mathbb{A}\langle X\rangle \otimes \mathbb{A}\langle Y\rangle$ of $\mathbb{A} \to R,S$
Cohomology algebra of the homotopy pullback	$\longleftrightarrow$	The Γ-algebra $\mathrm{Tor}^R(S,T)$
Homotopy fibre	$\longleftrightarrow$	Homotopy fibre
Cohomology of the homotopy fibre	$\longleftrightarrow$	$\mathrm{Tor}^R(S;\mathbf{k})$
Homotopy fibre has finite formal dimension	$\longleftrightarrow$	S has finite flat dimension over R
Serre spectral sequence	$\longleftrightarrow$	Serre spectral sequence
Homotopy fibre is TNCZ	$\longleftrightarrow$	$H(S)$ is $H(R)$-flat.

§ 4. LOOP SPACES

4.1 <u>Topology</u>. Let $(M,*)$ be a pointed space. The inclusion of $*$ in M is homotopy equivalent to the path space fibration $\pi\colon PM \to M$:

$$PM = \{f\colon I \to M \mid f(0) = *\} \ ; \quad \pi f = f(1).$$

Its fibre, ΩM, is the space of pointed maps $(S^1,*) \to (M,*)$ and is called the <u>loop space</u> on M. By definition, ΩM is the homotopy pullback of $* \to M \to *$.

Evidently any continuous map $\phi\colon (M,*) \to (N,*)$ determines $\Omega\phi\colon \Omega M \to \Omega N$ in the obvious way.

In addition to being a topological space (pointed by the constant loop e_* at $*$) ΩM admits a continuous multiplication $\Omega M \times \Omega M \to \Omega M$:

$$(f\cdot g)(t) = \begin{cases} f(2t) & 0 \le t \le 1/2 \\ g(2t-1) & 1/2 \le t \le 1 \end{cases}, \quad f,g \in \Omega M .$$

It is homotopy associative, and e_* acts as a homotopy identity.

If we pass to rational homology we get in $H_*(\Omega M;\mathbb{Q})$ the structure of a graded cocommutative Hopf algebra. On the other hand, if ΩM has finite $\mathbb{Q}$-type, then $H^*(\Omega M \times \Omega M;\, \mathbb{Q}) = H^*(\Omega M;\mathbb{Q}) \otimes H^*(\Omega M;\mathbb{Q})$; in this case we get a (dual) commutative Hopf algebra with diagonal $H^*(\Omega M;\mathbb{Q}) \to H^*(\Omega M;\mathbb{Q}) \otimes H^*(\Omega M;\mathbb{Q})$ arising from the multiplication

More generally, let $E \to B$ be any fibration with fibre F over $* \in B$ and consider the pullback diagram

$$\begin{array}{ccccc} F & \longrightarrow & E\times_B PB & \longrightarrow & E \\ \downarrow & & \downarrow & & \downarrow \\ e_* & \longrightarrow & PB & \longrightarrow & B \end{array} \;;$$

it gives a homotopy equivalence $F \xrightarrow{\simeq} E\times_B PB$. On the other hand, the composite $E\times_B PB \to B$ is a fibration with fibre $F \times \Omega B$; thus

(4.2) $$F \xrightarrow{\simeq} E\times_B PB \leftarrow F \times \Omega B$$

defines (up to homotopy) a continuous map

$$F \times \Omega B \to F \;; \qquad (x,f) \mapsto x\cdot f \;.$$

Up to homotopy this is an action of ΩB on F ($x\cdot fg \sim (x\cdot f)\cdot g$ and $x\cdot e_* \sim x$) and when E is itself the path space fibration the resulting map $\Omega B \times \Omega B \to \Omega B$ is homotopy equivalent to the multiplication defined above.

Finally, if we pass to rational homology in (4.2) we get an action

(4.3) $$H_*(F;\mathbb{Q}) \otimes H_*(\Omega B;\mathbb{Q}) \to H_*(F;\mathbb{Q})$$

of the Hopf algebra $H_*(\Omega B;\mathbb{Q})$ on $H_*(F;\mathbb{Q})$.

This action is the central object of study in the article [F-T] of Felix-Thomas in these proceedings.

4.4 <u>Algebra</u>. In analogy with (4.1) we consider an augmented h-DGΓ_* $\varepsilon: R \to \Bbbk$ with Im ε generating $\Bbbk$. The path space fibration corresponds to a free model

$$R\langle X\rangle \xrightarrow{\simeq} \Bbbk$$

for ε and its fibre is just the h-DGA$_*$ $\Bbbk\langle X\rangle = \Bbbk \otimes_R R\langle X\rangle$ whose homology is $\mathrm{Tor}(\Bbbk,\Bbbk)$. As observed in (3.12), a second free model leads to a unique homotopy class (rel $\Bbbk$) of quisms $\Bbbk\langle X\rangle \to \Bbbk\langle X'\rangle$; we can therefore make the

4.5 <u>Definition</u>. The h-DGΓ_*, $\Bbbk\langle X\rangle$, will be written ΩR and called the <u>loop DGA</u> for R .

To understand the analogue of loop space multiplication and of loop space action on a fibre, we note that the analogue of the pullback diagram in (4.1) is, cf. (2.15),

$$\begin{array}{ccccc} \Bbbk\langle Y\rangle & \xleftarrow{\simeq} & R\langle Y\rangle\langle X\rangle & \longleftarrow & R\langle Y\rangle \\ \uparrow & & \uparrow & & \uparrow \\ \Bbbk & \xleftarrow{\simeq} & R\langle X\rangle & \longleftarrow & R \end{array}$$

where $R\langle Y\rangle$ is a free model of $\phi: R \to S$ and $R\langle Y\rangle\langle X\rangle = R\langle Y\rangle \otimes_R R\langle Y\rangle$.

On the other hand the inclusion $F \times \Omega B \to E\times_B PB$ corresponds to the projection $R\langle Y\rangle\langle X\rangle \to \Bbbk\langle Y\rangle\langle X\rangle$ and we note

$$\Bbbk\langle Y\rangle\langle X\rangle = \Bbbk\langle Y\rangle \otimes_{\Bbbk} \Bbbk\langle X\rangle \qquad \text{(as } DG\Gamma_*\text{'s)}.$$

Thus altogether we get

(4.6) $$\Bbbk\langle Y\rangle \xleftarrow{\simeq} R\langle Y\rangle\langle X\rangle \longrightarrow \Bbbk\langle Y\rangle \otimes \Bbbk\langle X\rangle$$

in analogy with (4.2).

Here (cf. (2.14)) $\Bbbk\langle Y\rangle$ is the homotopy fibre of ϕ. Observe as well that a second choice of free models leads to a second version of (4.6); the two versions are then connected by unique homotopy classes of quisms (rel R or $\Bbbk$) making the resulting diagram homotopy commute.

Passing to homology in (4.5) yields the standard

(4.7) $$\mathrm{Tor}^R(S,\Bbbk) \to \mathrm{Tor}^R(S,\Bbbk) \otimes \mathrm{Tor}^R(\Bbbk,\Bbbk);$$

when $S = \Bbbk$ this is an associative comultiplication in $\mathrm{Tor}^R(\Bbbk,\Bbbk)$ which makes $\mathrm{Tor}^R(\Bbbk,\Bbbk)$ into a Γ-Hopf algebra. For general S (4.7) defines a co-action of this Hopf algebra on the Γ-algebra $\mathrm{Tor}^R(S,\Bbbk)$.

Dualizing with respect to $\Bbbk$ converts $\mathrm{Tor}^R(S,\Bbbk)$ into $\mathrm{Ext}_R(S,\Bbbk)$ - this works for DGA_*'s as well as for rings. In particular, the diagonal above dualizes to a graded algebra structure in $\mathrm{Ext}_R(\Bbbk,\Bbbk)$ and this is eaxctly the <u>Yoneda algebra</u>. More generally, the map (4.7) dualizes to make $\mathrm{Ext}_R(S;\Bbbk)$ into a module over $\mathrm{Ext}_R(\Bbbk,\Bbbk)$.

(The Yoneda algebra is defined as follows: $\mathrm{Hom}_R(R\langle X\rangle, R\langle X\rangle)$ has a product given by composition - passing to homology defines a product in $H(\mathrm{Hom}_R(R\langle X\rangle, R\langle X\rangle)) = \mathrm{Ext}_R(\Bbbk,\Bbbk)$.)

When H(R) is piecewise noetherian then $\mathrm{Tor}^R(\Bbbk,\Bbbk)$ will be finite dimensional in each degree and in this case the multiplication in $\mathrm{Tor}^R(\Bbbk,\Bbbk)$ dualizes to a comultiplication in $\mathrm{Ext}_R(\Bbbk,\Bbbk)$ - which is then a Hopf algebra.

4.8 <u>Remark</u>. If R is a supplemented $\Bbbk$-algebra then the quism $R\langle Y\rangle\langle X\rangle \xrightarrow{\simeq} \Bbbk\langle Y\rangle$ has a unique homotopy (rel $\Bbbk$) inverse $\Bbbk\langle Y\rangle \to R\langle Y\rangle\langle X\rangle$. In this case we get an h-$DG\Gamma_*$ morphism

(4.9) $$\Bbbk\langle Y\rangle \to \Bbbk\langle Y\rangle \otimes \Bbbk\langle X\rangle$$

whose (rel $\Bbbk$) homotopy class is unique. When $S = \Bbbk$ the resulting diagonal is homotopy associative (rel $\Bbbk$); in general it is a homotopy co-action (rel $\Bbbk$).

In this case, however, the bar construction can be used to provide a strictly associative version of (4.9).

4.10 <u>The dictionary, again</u>. We have

Loop space B	⟷	Loop DGA $\Omega R = \Bbbk\langle X\rangle$, where $R\langle X\rangle \xrightarrow{\simeq} \Bbbk$
Homology algebra, $H_*(\Omega B;\mathbb{Q})$	⟷	Yoneda algebra, $\mathrm{Ext}_R(\Bbbk,\Bbbk)$
Action of $H_*(\Omega B;\mathbb{Q})$ on $H_*(F;\mathbb{Q})$; F the homotopy fibre of $E \to B$	⟷	Action of $\mathrm{Ext}_R(\Bbbk,\Bbbk)$ on $\mathrm{Ext}_R(S,\Bbbk)$.

4.11 Warning. (i) The graded spaces $H_*(M;\mathbb{Q})$ (rational homology of a topological space) and $\mathrm{Ext}_R(S;\Bbbk)$ are two of the examples where we grade downstairs in topology and upstairs in algebra.

(ii) Although $\mathrm{Ext}_R(\Bbbk,\Bbbk)$ corresponds to $H_*(\Omega B;\mathbb{Q})$ where $\mathrm{Tor}^R(\Bbbk,\Bbbk)$ corresponds to $H^*(\Omega B;\mathbb{Q})$ it is important to observe that Ext is the dual of Tor and H^* is the dual of H_*. This distinction is necessary when these graded spaces do not have finite type because then products do not dualize to coproducts. In this case

$H^*(\Omega B;Q)$ is a Hopf algebra; $\mathrm{Ext}_R(\Bbbk,\Bbbk)$ is only an algebra

and

$H^*(\Omega B;Q)$ is only an algebra; $\mathrm{Tor}^R(\Bbbk,\Bbbk)$ is a Γ-Hopf algebra.

(iii) If M is a simply connected topological space and $k_M \geq 1$ is the first integer such that $H^{k_M}(M;Q) \neq 0$ then $k_{\Omega M} = k_M - 1$. If R is an h-DGΓ_* with $H_0(R)=\Bbbk$ and if $k_R \geq 1$ is the first integer such that $H_{k_R}(R) \neq 0$ then $k_{\Omega R} = k_R + 1$.

The observations 4.11 (ii) and 4.11 (iii) are examples in which simple minded translation between the two sides fails. It is because of this kind of phenomenon that the translator needs to be very careful.

§ 5 THE HOMOTOPY LIE ALGEBRA

In this chapter M will always denote a simply connected topological space of finite $\mathbb{Q}$-type while R will always denote an h-DGΓ_* such that H(R) is local (cf. (1.9)) with residue field $\Bbbk$.

Now the rational homotopy groups $\pi_i(M) \otimes \mathbb{Q}$, equipped with the Whitehead product almost define a graded Lie algebra. In fact, when this product is transferred to $\pi_*(\Omega M) \otimes \mathbb{Q}$ via the canonical (degree-1) isomorphism it does give a graded Lie structure (the Samelson product); the result is the rational homotopy Lie algebra of M.

To translate this to rings and h-DGΓ_*'s we need two further identifications. Recall that $H_*(\Omega M;\mathbb{Q})$ is a graded Hopf algebra. The Cartan-Serre theorem asserts that the Hurewicz homomorphism $\pi_*(\Omega M) \otimes \mathbb{Q} \to H_*(\Omega M;\mathbb{Q})$ is an isomorphism onto the primitive subspace (of elements $\alpha \to \alpha \otimes 1 + 1 \otimes \alpha$ under the diagonal). Almost by definition this is Lie isomorphism when the primitive subspace P_* is given the Lie bracket $[\alpha,\beta] = \alpha\beta - (-1)^{\deg\alpha\deg\beta}\beta\alpha$.

Under the duality between $H_*(\Omega M;\mathbb{Q})$ and $H^*(\Omega M;\mathbb{Q})$ P_* is dual to the space $Q^* = H^+/H^+ \cdot H^+$ of indecomposables. The diagonal in $H^*(\Omega M;\mathbb{Q})$ yields a Lie comultiplication in Q^* dual to the bracket in P_*.

To translate this to R we recall (§4) that $H^*(\Omega M;\mathbb{Q})$ corresponds to $\mathrm{Tor}^R(\mathbb{k},\mathbb{k})$, which has the natural structure of a connected Γ-Hopf algebra. Now for any connected Γ-algebra A we will let $I(\gamma^p)$ denote the linear span of the elements in $\mathrm{Im}\,\gamma^p$ and call

$$Q_\Gamma A = A_+/I \ ; \quad I = A_+\cdot A_+ + \sum_{p\geq 2} I(\gamma^p)$$

the space of <u>Γ-indecomposables</u> for A. In particular we set

$$\pi_*(R) = Q_\Gamma(\mathrm{Tor}^R(\mathbb{k},\mathbb{k})) \ ;$$

to corresponds to the space of indecomposables $Q^*(H^*(\Omega M;\mathbb{Q}))$.

Dualizing we get a graded subspace

$$\pi^*(R) \subset \mathrm{Ext}_R(\mathbb{k},\mathbb{k})$$

corresponding to $\pi_*(\Omega M) \otimes \mathbb{Q}$. Now $\pi^*(R)$ is always a subspace of the primitive subspace P_* of $\mathrm{Ext}_R(\mathbb{k},\mathbb{k})$ and is in fact a sub Lie algebra, (with bracket dual to the diagonally induced Lie comultiplication in π^*). However, when char $\mathbb{k} > 0$ we usually have $\pi_* \neq P_*$.

One justification for this choice resides in the following theorem (cf. Milnor-Moore [M-M], André [A_1], Sjödin [Sj]):

5.1 <u>Theorem</u> (i) The Hopf algebra $H_*(\Omega M;\mathbb{Q})$ is the universal enveloping algebra of $\pi_*(\Omega M) \otimes \mathbb{Q}$.

(ii) The Hopf algebra $\mathrm{Ext}_R(\mathbb{k},\mathbb{k})$ is the universal enveloping algebra of $\pi^*(R)$.

For this reason we call $\pi^*(R)$ the <u>homotopy Lie algebra</u> of R. When $R = R_o$ then the numbers $\varepsilon_i = \dim \pi_i(R)$ are called the <u>deviations</u> of R.

5.2 <u>Fibration sequences</u>. Let $F \to E$ be the homotopy fibre of a continuous map $E \to B$ (B path connected); then the sequence $\pi_i(B) \leftarrow \pi_i(E) \leftarrow \pi_i(F)$ is exact. Transferring to loop spaces yields an exact sequence of Lie algebra maps,

(5.3) $$\pi_*(\Omega B) \leftarrow \pi_*(\Omega E) \leftarrow \pi_*(\Omega F).$$

Now suppose $R \to S$ is an h-DGΓ_* morphism in which $H(R)$, $H(S)$ are local with residue fields $\mathbb{k} \hookrightarrow \ell$, and let $S \to F$ be a homotopy fibre. Then ([Av_2])

(5.4) $$\pi^*(R) \otimes_{\mathbb{k}} \ell \to \pi^*(S) \to \pi^*(F)$$

is also an exact sequence of graded Lie algebras.

For topological spaces one gets the long exact homotopy sequence from (5.3) as follows. Recall the action (4.2) of ΩB on F. This restricts to a map $\Omega B \to F$ which turns out to be the homotopy fibre of $F \to E$. Thus we get an infinite sequence

$$B \xleftarrow{\pi} E \xleftarrow{j} F \leftarrow \Omega B \xleftarrow{\Omega\pi} \Omega E \xleftarrow{\Omega j} \Omega F \leftarrow \dots \tag{5.5}$$

in which each map is the homotopy fibre of the map to its left. Passing to homotopy and using $\pi_i(B) = \pi_{i-1}(\Omega B)$ yields the long exact sequence

$$\pi_i(B) \leftarrow \pi_i(E) \leftarrow \pi_i(F) \leftarrow \pi_{i+1}(B) \leftarrow \dots \,. \tag{5.6}$$

In algebra one can mimic everything except the last step. Let $R \xrightarrow{\phi} S$ be as above and replace it by a free model $R \to R\langle Y\rangle$. If $R\langle X\rangle \xrightarrow{\simeq} \mathbb{k}$ is a free model of the augmentation then the homotopy fibre of ϕ is represented by $j\colon R\langle Y\rangle \to R\langle Y\rangle\langle X\rangle$ and the co-action (4.6) yields in an obvious way

$$R\langle Y\rangle\langle X\rangle \to \ell\langle X\rangle = \Omega R \otimes_{\mathbb{k}} \ell$$

as the homotopy fibre of j. This leads to the analogue

$$R \to R\langle Y\rangle \to R\langle Y,X\rangle \to \Omega R \otimes_{\mathbb{k}} \ell \xrightarrow{\Omega\phi} \Omega S \to \Omega F \quad \dots \tag{5.7}$$

of (5.5) in which each morphism is the homotopy fibre of the one before.

The last step of the analogy fails, however. There is a natural surjection $\pi^*(\Omega R) \to \pi^{*-1}(R)$ which is an isomorphism if char $\mathbb{k} = 0$; when char $\mathbb{k} > 0$ the kernel can easily be infinite dimensional. Thus "usually" there is no analogue for (5.6), although (cf. [Av_1]) there are important and useful occasions when such an analogue can be established.

REFERENCES

[A_1] M. André, Hopf algebras with divided powers, J. Algebra 18 (1971), 19-50.

[A_2] M. André, Homologie des Algèbres Commutatives, Springer, Berlin 1974.

[A-M] M.F. Atiyah and I.G. Macdonald, Introduction to Commutative Algebra, Addison-Wesley, 1969.

[Av_1] L.L. Avramov, Local algebra and rational homotopy, in Homotopie Algébrique et Algèbre Locale, Astérisque 113/114 (1984), 15-43.

[Av_2] L.L. Avramov, Homotopy Lie algebras for commutative rings and DG algebras, to appear.

[B-G] A.K. Bousfield and V.K.A.M. Gugenheim, On the PL De Rham Theory and rational homotopy type, Memoirs Amer. Math. Soc. 179 (1976).

[C] L. Carroll, Through the Looking Glass and What Alice Found There, Macmillan 1871.

[F-T] Y. Felix and J.-C. Thomas, Sur l'opération de l'holonomie rationnelle, these proceedings.

[G-L] T.H. Gulliksen and G. Levin, Homology of Local Rings, Queen's papers in Pure and Applied Mathematics - No. 20 Queen's University, Kingston, Ontario, 1969.

[G-M] V.K.A.M. Gugenheim and J.P. May, On the Theory and Application of Differential Torsion products, Memoirs of the Amer. Math. Soc. 142 (1974).

[Ha] S. Halperin, Lectures on Minimal Models, Mém. de la Soc. Math. de France 9/10, 1983.

[Ha-Le] S. Halperin and J.-M. Lemaire, Suites inertes dans les algèbres de Lie graduées, preprint. (To appear in Math. Scand.)

[L-Av] G. Levin and L.L. Avramov, Factoring out the socle of a local Gorenstein ring, J. Algebra 55 (1978), 74-83.

[M] J.C. Moore, Algèbre homologique et homologie des espaces classifiants, Séminaire H. Cartan, École Normale Supérieure, 1959-1960, Exposé 7, Secrétariat Math., Paris, 1961.

[Ma] H. Matsumura, Commutative Algebra 2nd ed., Benjamin/Cummings, Reading, Mass. 1980.

[M-M] J.W. Milnor and J.C. Moore, On the structure of Hopf algebras, Annals Math. 81 (1965), 211-264.

[Q_1] D. Quillen, Homotopical Algebra, Lecture Notes in Math. 43 (1967), Springer Verlag.

[Q_2] D. Quillen, On the (co)-homology of commutative rings, Proc. Symp. Pure Math. (17), Amer. Math. Soc. 1970, 65-87.

[R] J.-E. Roos, Relations between the Poincaré-Betti series of loop spaces and of local rings, Lecture Notes in Math. 740, 285-322, Springer Verlag, Berlin 1979.

[Sj] G. Sjödin, Hopf algebras and derivations, J. Algebra 64 (1980), 218-229.

[Su] D. Sullivan, Infinitesimal computations in topology, Publ. Math. IHES 47 (1978), 269-331.

[T] J. Tate, Homology of noetherian rings and local rings, Illinois J. Math. 1 (1957), 14-27.

[W] G. Whitehead, Elements of Homotopy Theory, Springer Verlag, 1981.

L. Avramov
Institute of Mathematics
University of Sofia
ul. "Akad. G. Bončev" Bl. 8
1113 Sofia, Bulgaria

S. Halperin
Department of Mathematics
University of Toronto
Toronto
Canada M5S 1A1

A RATIONAL HOMOTOPY ANALOG OF WHITEHEAD'S PROBLEM

by
David J. Anick

This note will state and prove a theorem in rational homotopy which is an analog of the famous unsolved problem due to J.H.C. Whitehead as to whether or not subcomplexes of aspherical two-dimensional CW complexes are aspherical [9].

We first rephrase Whitehead's question so that it has a natural generalization to higher homotopy. Let Y be an aspherical two-dimensional CW complex and let X be a subcomplex. It is well-known that we need only consider the case where X and Y share the same 1-skeleton W and base point w_0, so we may assume that Y is obtained by attaching 2-cells to X, which in turn is gotten by attaching 2-cells to W. W has the homotopy type of a wedge of circles, so the asphericity of X (resp. Y) is equivalent to the surjectivity of the homomorphism $(i_{WX})_\#$ (resp. $(i_{WY})_\#$) induced on $\pi_*(\ )$ by the inclusion $i_{WX}: W \to X$ (resp. i_{WY}).

Whitehead's question becomes the following.

<u>Question 1</u>. Let W be a wedge of S^1's and let X be obtained by attaching 2-cells to W and Y by attaching 2-cells to X. If $(i_{WY})_\#: \pi_*(W,w_0) \to \pi_*(Y,w_0)$ is surjective, is $(i_{WX})_\#: \pi_*(W,w_0) \to \pi_*(X,w_0)$ necessarily surjective ?

In rational homotopy, we generally consider simply connected spaces only and tensor all homotopy groups with Q. In place of a wedge of circles we get a wedge of spheres $\bigvee_{\alpha\in I} S^{d_\alpha}$, $d_\alpha \geq 2$, and each attached cell may have any dimension three or greater. The natural analog to Question 1 is

<u>Qustion 2</u>. Let $W = \bigvee_{\alpha\in I} S^{d_\alpha}$, where I is any indexing set and $d_\alpha \geq 2$. Suppose Y is obtained by attaching cells to W, $Y = W \cup_f (\bigcup_{\beta\in J} e^{(b_\beta)})$ for some indexing set J and dimensions $b_\beta \geq 3$, and suppose X is a subcomplex of Y containing W. If $(i_{WY})_\#: \pi_*(W,w_0) \otimes Q \to \pi_*(Y,w_0) \otimes Q$ is surjective, is $(i_{WX})_\#: \pi_*(W,w_0) \otimes Q \to$ $\to \pi_*(X,w_0) \otimes Q$ necessarily surjective ?

We may answer Question 2 affirmatively for locally finite Y by the following argument. Let (A_W,d_W) denote the Adams-Hilton model [2] over Q for W so that $H_*(A_W,d_W) \approx H_*(\Omega W;Q)$, and likewise for (A_X,d_X) and (A_Y,d_Y). Using the equivalence between the rational homology of the loop space and the universal enveloping algebra of the rational homotopy Lie algebra of a space [see e.g. 5], we see that $(i_{WY})_\#$ (resp. $(i_{WX})_\#$) surjects if and only if $(i_{WY})_*: H_*(A_W,d_W) \to H_*(A_Y,d_Y)$

(resp. $(i_{WX})_*$) surjects.

The latter condition is a familiar one to rational homotopy theorists. It is discussed in [7] and by [7] and [3, Theorem 2.9] we know that $(i_{WY})_*$ is onto if and only if the images $(\Omega f_\beta)_*(z_\beta)$ of the generators $z_\beta \in H_{b_\beta-2}(\Omega S^{b_\beta-1};Q)$ of spherical loop space homology under the attaching maps $f_\beta\colon \dot{e}^{(b_\beta)} \to W$ constitute a "strongly free set" in $H_*(\Omega W;Q)$. Because by [3, Lemma 2.7] a subset of a strongly free set is strongly free, the collection $\{(\Omega f_\beta)_*(z_\beta)\,|\,\beta \in K\}$, where J indexes the cells of $Y - W$ and the subset $K \subseteq J$ indexes the cells of $X - W$, also has this property. This is equivalent to the surjectivity of $(i_{WX})_*$.

When Y is not locally finite, the argument still works, but the proofs in [3], which relied on Hilbert series, must be replaced by more general ones. Happily this is easily accomplished and we will only outline how. For B a set of homogeneous elements in the (possibly locally infinite) connected graded k-algebra H, B should be classified as "strongly free" if and only if $(1 \amalg \rho)\colon k < B > \amalg (H/HBH) \to H$ is one-to-one for some (equivalently, every) choice of graded vector space homomorphism ρ: $(H/HBH) \to H$ which is a right inverse to the projection $p\colon H \to H/HBH$. [3, Lemma 2.7] is easily reproved and the proof of [3, Theorem 2.9] remains valid. We have shown

Theorem 1. The answer to Question 2 is "yes".

In the rational homotopy case we can take this further by relaxing the requirement that W be a wedge of spheres. Qustion 1 is trivially seen to be replaced by an equivalent question if W is permitted to be any twodimensional CW complex, and the answer to the original Whitehead question becomes "no" if W is allowed to have dimension three [1]. For Question 2, however, the effect of these substitutions is less clear. We therefore formulate

Question 3. Same as Question 2, except that W may be any simply connected CW complex.

To answer this, we will use the following lemma.

Lemma. Let (L,δ) be any associative connected differential graded algebra over a field k, and let $B \subseteq \ker(\delta)$ be any subset of homogeneous elements. Let $\tilde{L} = (L \amalg k < C >, \tilde{\delta})$, where $\tau\colon C \to B$ is a one-to-one correspondence with $\deg(\tau(x)) = \deg(x) - 1$ and $\tilde{\delta}$ extends δ via $\tilde{\delta}(x) = \tau(x)$. Writing $G = H_*(L,\delta)$ and $\tilde{G} = H_*(\tilde{L},\tilde{\delta})$, there is a natural map $i_*\colon G \to \tilde{G}$ induced by the inclusion $i\colon L \to \tilde{L}$ of chain algebras. Then i_* is onto if and only if the set of cycles $\overline{B} = \{\overline{y}\,|\,y \in B\} \subseteq G$ is strongly free in G.

Note. The "if" direction of the lemma is essentially proved in [8]. In his talk at the 1983 Nordic Summer School, J.-M. Lemaire proved this lemma under the non-essential restriction that B be countable. His proof appears in [6, Section 2].

We offer here a simple, completely general proof based on [3].

Proof of Lemma. Let $\tilde{L}_{(0)} = L$ and $\tilde{L}_{(n)} = (LCL)^n$, where LCL denotes Span $\{uxv | u \in L, x \in C, v \in L\} \subseteq \tilde{L}$ and likewise for $(LCL)^n$. Writing $\tilde{L}_{p,q} = \tilde{L}_{(p)} \cap L_{q+p}$, we have

$$\tilde{L} = \bigoplus_{p,q\geq 0} \tilde{L}_{p,q} \quad \text{and} \quad \tilde{\delta}(\tilde{L}_{p,q}) \subseteq L_{p-1,q} \oplus \tilde{L}_{p,q-1} .$$

We obtain from this bigraded complex a spectral sequence $E^*_{*,*}$ which converges to $\tilde{G} = H_*(\tilde{L},\tilde{\delta})$. Its first terms and differentials are $E^0_{p,q} = \tilde{L}_{p,q}$ with $\tilde{\delta}_0(C) = 0$, $\tilde{\delta}_0|_L = \delta$, and $E^1 = G \amalg k\langle C\rangle$ with $\tilde{\delta}_1(G) = 0$, $\tilde{\delta}_1(x) = \overline{\tau(x)} \in G$ for $x \in C$. $\tilde{\delta}_r$ has bidegree $(-r, r-1)$.

If $\overline{B}$ is strongly free in G, then by [3, Theorem 2.9] $E^2_{p,*} = 0$ for $p > 0$ and the spectral sequence degenerates, yielding

$$\tilde{G} = E^\infty = E^2 = E^2_{0,*} = E^\infty_{0,*} = \mathrm{im}(i_*) ,$$

as desired. If instead $\overline{B}$ is not strongly free in G, by [3, 2.9] $E^2_{1,*} \neq 0$. The same reasoning which shows that $E^2_{1,*} \neq 0$ also shows, when s is the minimal degree having $E^2_{1,s} \neq 0$, that $E^2_{p,q} = 0$ for $p \geq 2$ and $q < s$. Thus $E^2_{1,s}$ persists to $E^\infty_{1,s}$, and $\mathrm{im}(i_*) = E^\infty_{0,*}$ is not the whole of $\tilde{G}$.

Theorem 2. The answer to Question 3 is "yes".

Proof of Theorem 2. As before, it suffices to show that the map $(\Omega i_{WX})_*\colon H_*(\Omega W;Q) \to$ $\to H_*(\Omega X;Q)$ surjects when we know that $(\Omega i_{WY})_*$ surjects. Using [2], $(\Omega i_{WY})_*$ is precisely the map i_* of the Lemma if we set $(L,\delta) = (A_W, d_W)$, $(\tilde{L},\tilde{\delta}) = (A_Y, d_Y)$, and $\overline{B} = \{(\Omega f_\beta)_*(z_\beta)\}_{\beta\in J}$. Since $(\Omega i_{WY})_*$ surjects, $\overline{B}$ is strongly free in $G =$ $= H_*(\Omega W;Q)$. Again we deduce that $\{(\Omega f_\beta)_*(z_\beta)\}_{\beta\in K}$ is strongly free in G, so $(\Omega i_{WX})_*$ is onto as well.

As a final remark we notice that the answer to Question 2 becomes "no" if we allow the cells of $Y - X$ to attach to X instead of to W. An example is $Y = S^2 \times S^2 \times S^2$, $X = Y^4$, $W = Y^2$.

REFERENCES

[1] J.F. Adams, A new proof of a theorem of W.H. Cockcroft. J. London Math. Soc. 30 (1955), 482-488.

[2] J.F. Adams and P.J. Hilton, On the chain algebra of a loop space. Comm. Math. Helv. 30 (1955), 305-330.

[3] D.J. Anick, Non-commutative graded algebras and their Hilbert series. J. Algebra 78 (1982), 120-140.

[4] H. Cartan and S. Eilenberg, Homological Algebra. Princeton Univ. Press, Princeton N.J., 1956.

[5] S. Halperin, Lectures on Minimal Models, Publications de l'U.E.R. Mathématiques Pures et Appliquées, Université des Sciences et Techniques de Lille, Vol. 3 (1981). (Also published as Mem. de la Soc. Math. de France 9/10 (1983).

[6] S. Halperin and J.-M. Lemaire, Suites inertes dans les algèbres de Lie graduées. Publications de l'Univ. de Nice, 1984, No. 22; also scheduled to appear in Math. Scand.

[7] J.-M. Lemaire, Algèbres connexes et homologie des espaces de lacets, Lecture Notes in Math. No. 422, Springer-Verlag, Berlin-Heidelberg-New York, 1974.

[8] J.-M. Lemaire, Autopsie d'un meurtre dans l'homologie d'une algèbre de chaines. Ann. Sci. École Norm. Sup (4), 11 (1978), 93-100.

[9] J.H.C. Whitehead, On adding relations to homotopy groups. Ann. Math. 42 (1941), 409-428.

D.J. Anick
Department of Mathematics
Massachusetts Institute of Technology
Cambridge, MA 02139, U.S.A.

HILBERT SERIES OF FINITELY PRESENTED ALGEBRAS

by

David ANICK and Clas LÖFWALL

Summary

Let $\mathscr{S}$ denote the collection of all Hilbert series of finitely presented connected graded algebras over a field k. What can we say about the set $\mathscr{S}$? This paper addresses itself to that question. In 1974 Govorov [Go-2] conjectured that only rational power series belonged to $\mathscr{S}$. This conjecture was first disproved by Shearer [Sh], using methods which we will generalize and extend in this paper. We will also show that the set $\mathscr{S}$ is countable and derive some of its properties.

Definitions

Let k denote any field. A graded vector space (over k) is any countable direct sum $A = \bigoplus_{n=0}^{\infty} A_n$ of finite-dimensional k-spaces. A graded vector space A is a (connected) graded algebra (over k) if there is an isomorphism $\varepsilon: A_0 \xrightarrow{\cong} k$ and there are associative pairings $\mu_{mn}: A_m \otimes A_n \longrightarrow A_{m+n}$ for each m and n and if $\mu_{0n} \circ (\varepsilon^{-1} \otimes 1): k \otimes A_n \longrightarrow A_n$ and $\mu_{n0} \circ (1 \otimes \varepsilon^{-1}): A_n \otimes k \longrightarrow A_n$ agree with scalar multiplication. Viewing the A_n's as subspaces of A, we say that a non-zero element $x \in A$ is homogeneous of degree n, written $|x| = n$, if $x \in A_n$. The elements of positive degree are written A^+. The Hilbert series of a graded space $A = \bigoplus_{n=0}^{\infty} A_n$ is denoted $A(z)$ and is defined to be the formal power series $A(z) = \sum_{n=0}^{\infty} \operatorname{rank}(A_n) z^n$. If A and B are graded vector spaces, then $A \otimes B$ is a graded vector space where $(A \otimes B)_n = \bigoplus_{i+j=n} A_i \otimes B_j$. The Hilbert series $A \otimes B(z)$ is equal to $A(z)B(z)$. If A and B are graded algebras, $A \otimes B$ is a graded algebra by the rule $(a_1 \otimes b_1)(a_2 \otimes b_2) = a_1a_2 \otimes b_1b_2$ (in this paper we have no need for the usually introduced sign in the definition of the product). If α is a finite subset of A consisting of linearly independent homogeneous elements, $\alpha(z)$ denotes $\operatorname{span}(\alpha)(z) = \sum_{w \in \alpha} z^{|w|}$, which is a polynomial in z. If A is also a graded algebra, $\langle \alpha \rangle$ denotes the two-sided ideal of A generated by α. In this case, $\langle \alpha \rangle$ is a graded space and $A/\langle \alpha \rangle$ is a graded algebra, both of them inheriting their gradations directly from A.

For a finite set S , $k\langle S \rangle$ denotes the free associative k-algebra on S . If each element of S is assigned a positive integral degree, then $k\langle S \rangle$ becomes a graded algebra in a natural way. An algebra A which is a quotient of such a $k\langle S \rangle$ by a two-sided ideal I is said to be finitely generated and if $I = \langle \alpha \rangle$ and α is finite, we say A is finitely presented (henceforth abbreviated "f.p."). Furthermore, if S can be chosen to consist only of degree-one generators, we say A is degree-one generated and if in addition each relation $w \in \alpha$ has degree two, we call A a one-two algebra. Let $\mathcal{A}^{(k)}$ denote the collection of all f.p. algebras over k . Except in theorem 1, k will be taken as fixed, and $\mathcal{A}^{(k)}$ will be shortened simply to $\mathcal{A}$. Likewise $\mathcal{A}_1 = \mathcal{A}_1^{(k)}$ is the collection of all degree-one generated algebras (over k) and $\mathcal{A}_{12} = \mathcal{A}_{12}^{(k)}$ consists of all one-two algebras (over k). Lastly, $\mathcal{S} = \{A(z) \mid A \in \mathcal{A}\}$ and $\mathcal{S}_1 = \{A(z) \mid A \in \mathcal{A}_1\}$ and $\mathcal{S}_{12} = \{A(z) \mid A \in \mathcal{A}_{12}\}$ are the corresponding collections of Hilbert series.

Local rings

Let us briefly mention why it is interesting to study $\mathcal{A}_{12}$. In the theory of local rings, the Yoneda Ext-algebra is defined. This is a graded algebra, but in general not finitely generated [Ro] . But if the local ring is graded, the subalgebra generated by the one-dimensional elements is an object in $\mathcal{A}_{12}$. And if moreover the cube of the maximal ideal of the ring is zero, then this subalgebra determines the whole Ext-algebra [Lö], [Ro]. Now, this subalgebra is more than an object in $\mathcal{A}_{12}$, it is also a Hopf algebra. Let $\mathcal{H}_{12}$ denote the Hopf-algebras in $\mathcal{A}_{12}$, and let $\mathcal{L}_3$ denote local algebras with the cube of the maximal ideal equal to zero. The construction above gives a bijective correspondence between $\mathcal{L}_3$ and $\mathcal{H}_{12}$. Now Anick [An -1] has defined a map $\mathcal{A}_{12} \longrightarrow \mathcal{H}_{12}$ which transforms the Hilbert series in a certain exponential manner (see theorem 5(a) in this paper) and in theorem 6 in this paper we define a map $\mathcal{A} \longrightarrow \mathcal{A}_{12}$ which in a sense does not change the Hilbert series. The composite map $\mathcal{A} \longrightarrow \mathcal{L}_3$ was used by Jacobsson [Ja] to disprove a conjecture by Lemaire.

Properties of $\mathcal{S}$

As mentioned, we are concerned in this paper with describing the set $\mathcal{S}$. It seems unlikely that there is any easy analytic way to characterize the elements of $\mathcal{S}$, for we shall see in theorem 5 that $\mathcal{S}$ is closed under some rather complicated operations. Of special interest, however, are the rates of growth of the sequences $\{\mathrm{rank}(A_n)\}_{n\geq 0}$, and these rates are reflected in the analytic properties of the series $A(z)$. Such sequences generally grow exponentially, and the radii of convergence of their Hilbert series give us our crudest measure of their rates of growth. A more subtle measure is in the nature of the singularity at the radius of convergence r . A first step toward examining this singularity is taken in [An-2, thm 4] , where it is proved that $\{\mathrm{rank}(A_n)r^n\}$ cannot approach zero and that $A(z)$ goes to infinity, as z approaches r , at least as fast as a first order pole.

We first prove that $\mathcal{S}$ is countable. In our next four results, we show that $\mathcal{S}$ is closed under certain simple operations, as well as certain complicated ones involving infinite products.

Theorem 1 The set $\bigcup \mathcal{S}^{(k)}$, the union taken over all fields k , is countable.

Proof We construct a countable set which maps surjectively onto $\bigcup \mathcal{S}^{(k)}$. Firstly, the prime fields, Z_p , p prime, and the rationals, are countably many countable fields. Let F be any countable field, then a simple extension of F is either of the form $F(x)$, x transcendent over F , or of the form $F[x]/(p)$ where p is an irreducible polynomial. Since $F[x]$ is countable, there are only countably many simple extensions of F of this standard form and all these fields are countable. Repeating this we find that the set $\mathcal{F}$ = {finite extensions in standard form of the prime fields} is countable. Now, again, let F be any countable field. A finite presentation over F is given by positive integers $n, d_1, \ldots, d_n$ and a finite set of homogeneous elements of the graded algebra $F\langle t_1, \ldots, t_n\rangle$, $|t_i| = d_i$. Since $F\langle t_1, \ldots, t_n\rangle$ is countable there are countably many finite presentations over F . Hence the following set is countable:

$$\mathcal{M} = \{\text{finite presentations over } F \ ; \ F \in \mathcal{F}\}.$$

A map $\mathcal{M} \longrightarrow \bigcup \mathcal{S}^{(k)}$ is defined by taking the Hilbert series of the algebra derived from the finite presentation. In order to prove that this map is surjective, let $A(z) \in \bigcup \mathcal{S}^{(k)}$ where A is f.p. over a field k . Since in the presentation of A , only a finite number of elements in k are involved, there is a subfield $k_0 \subset k$ such that $A = B \otimes_{k_0} k$ where B is a f.p. algebra over k_0 and k_0 is isomorphic to a field $F \in \mathcal{F}$. Since $\mathrm{rank}_k(A_n) = \mathrm{rank}_{k_0}(B_n)$ the Hilbert series for A is equal to the Hilbert series for a f.p. algebra over F and hence the map $\mathcal{M} \longrightarrow \bigcup \mathcal{S}^{(k)}$ is surjective.

<u>Theorem 2</u> Let $A(z) \in \mathcal{S}$, $B(z) \in \mathcal{S}$. Then $A(z) + B(z) - 1 \in \mathcal{S}$, $A(z)B(z) \in \mathcal{S}$, and $(A(z)^{-1} + B(z)^{-1} - 1)^{-1} \in \mathcal{S}$. The same results hold if $\mathcal{S}$ is replaced throughout by $\mathcal{S}_1$ or by $\mathcal{S}_{12}$.

<u>Proof</u> Write $A = k\langle S \rangle/\langle \alpha \rangle$, $B = k\langle T \rangle/\langle \beta \rangle$, where S, α, T, β are finite sets and S and T are disjoint. Let $\gamma = \{st, ts \mid s \in S,\ t \in T\} \subseteq$ $\subseteq k\langle S \cup T\rangle$ and $\delta = \{st - ts \mid s \in S,\ t \in T\} \subseteq k\langle S \cup T\rangle$. If $C = k\langle S \cup T\rangle/\langle \alpha \cup \beta \cup \gamma\rangle$, then $C_0 \cong k$ and $C_j \cong A_j \oplus B_j$ for $j>0$, hence $C(z) = A(z) + B(z) - 1$. Letting $D = A \otimes B$, then $D(z) = A(z)B(z)$ and D has the presentation $k\langle S \cup T\rangle/\langle \alpha \cup \beta \cup \delta\rangle$ (this follows e.g. by an application of the general method given below on page 6). If $E = k\langle S \cup T\rangle/\langle \alpha \cup \beta\rangle$ then $E \cong A \amalg B$ is the free product of A and B and its Hilbert series is given by the formula $E(z)^{-1} = A(z)^{-1} + B(z)^{-1} - 1$ (see [Le]). Finally, as to $\mathcal{S}_1$ and $\mathcal{S}_{12}$, note that if S and T consist of degree-one generators, then C, D, and E are degree-one generated and γ and δ are in degree two. If α and β are in degree two as well, then $C, D, E \in \mathcal{A}_{12}$.

The <u>Segre product</u> of two graded spaces $A = \bigoplus_{n \geq 0} A_n$ and $B = \bigoplus_{n \geq 0} B_n$ is a graded space which is denoted by $A \bullet B$ and defined by $(A \bullet B)_n = A_n \otimes B_n$.

For homogeneous subsets $S \subset A$, $T \subset B$ we also define $S \bullet T$ as $\{s \otimes t \in A \bullet B \ ;\ s \in S,\ t \in T,\ |s| = |t|\}$. If both A and B are graded algebras then $A \bullet B$ is a graded algebra in which multiplication is defined as for $A \otimes B$. Note that if we write $A(z) = \sum_{n=0}^{\infty} a_n z^n$ and $B(z) = \sum_{n=0}^{\infty} b_n z^n$, then $(A \bullet B)(z) = \sum_{n=0}^{\infty} a_n b_n z^n$. I.e., the Hilbert series of a Segre product is the <u>Hadamard product</u> of the Hilbert series.

<u>Theorem 3</u> Let $A, B \in \mathcal{A}_1$. Then $A \bullet B \in \mathcal{A}_1$. The same result holds if $\mathcal{A}_1$ is replaced throughout by $\mathcal{A}_{12}$. Furthermore, if $A(z) = \sum_{n=0}^{\infty} a_n z^n$ and $B(z) = \sum_{n=0}^{\infty} b_n z^n$ belong to $\mathcal{S}_1$ (or $\mathcal{S}_{12}$), then $\sum_{n=0}^{\infty} a_n b_n z^n$ belongs to $\mathcal{S}_1$ (or $\mathcal{S}_{12}$). If the radii of convergence of $A(z)$ and $B(z)$ are r_A and r_B, then the radius of convergence of $(A \bullet B)(z)$ is $r_A r_B$.

<u>Proof</u> Write $A = k\langle S\rangle/\langle \alpha \rangle$, $B = k\langle T\rangle/\langle \beta \rangle$, where S, T are finite sets of degree-one generators.Assign every element of $S \times T$ to degree one. There is an obvious map of graded algebras $k\langle S \times T\rangle \longrightarrow k\langle S\rangle \bullet k\langle T\rangle$.

This is an isomorphism its inverse ϕ is defined inductively on elements of the form $a \otimes b$, where a and b are words in S and T and $|a|=|b|$, by $\phi(a \otimes b) = (s,t)\phi(a' \otimes b')$ where $s \in S$, $t \in T$ and $a=sa'$, $b=tb'$.
The surjections $\pi_A: k\langle S\rangle \longrightarrow A$ and $\pi_B: k\langle T\rangle \longrightarrow B$ induces a surjection $k\langle S\rangle \cdot k\langle T\rangle \longrightarrow A \cdot B$, this is an algebra map and its kernel is $\langle \alpha \rangle \cdot k\langle T\rangle + k\langle S\rangle \cdot \langle \beta \rangle$. This ideal is generated by the finite set $\alpha \cdot W_T \cup W_S \cdot \beta$ where W_T, W_S are homogeneous k-basis for $k\langle T\rangle$, $k\langle S\rangle$. Since $k\langle S\rangle \cdot k\langle T\rangle$ is freely one-generated, it followa that $A \cdot B \in \mathcal{A}_1$. Also, if α and β are concentrated in degree two, $\alpha \cdot W_T$ and $W_S \cdot \beta$ are in degree two and hence $A \cdot B \in \mathcal{A}_{12}$.

The remaining claims of theorem 2 are dispensed with easily. The fact that $\mathcal{S}_1$ (or $\mathcal{S}_{12}$) is closed under Hadamard products is an immediate consequence of the closure of $\mathcal{A}_1$ (or $\mathcal{A}_{12}$) under Segre products. Lastly, it follows from [An-2, see lemma 1 and proof of thm 2] that whenever $C(z) = \sum_{n=0}^{\infty} c_n z^n \in \mathcal{S}_1$, $\lim_{n\to\infty}\{c_n^{1/n}\}$ exists and equals r_C^{-1} if r_C is the radius of convergence of $C(z)$. Application of this fact to $A(z)$, $B(z)$ and $(A \cdot B)(z)$ allows us to deduce $r_{A \cdot B}^{-1} = \lim_{n\to\infty}(a_n b_n)^{1/n} = (\lim_{n\to\infty} a_n^{1/n})(\lim_{n\to\infty} b_n^{1/n}) = r_A^{-1} r_B^{-1}$.

<u>Remark</u> It is essential to assume that $A, B \in \mathcal{A}_1$. Indeed, if $A = B = k\langle s,t\rangle$, $|s| = 2$, $|t| = 1$, then $A \cdot B$ is not finitely generated, since $\{ts^n \otimes s^n t \; ; \; n \geq 0\}$ is a set of indecomposable elements.

To illustrate the idea that a wide range of series operations are encompassed by Hadamard products, we have the following corollary.

<u>Theorem 4</u> Let $A(z) \in \mathcal{S}_1$ (or $\mathcal{S}_{12}$). Then $d/dz(zA(z)) \in \mathcal{S}_1$ (or $\mathcal{S}_{12}$).

<u>Proof</u> Writing $A(z) = \sum_{n=0}^{\infty} a_n z^n$, we have

$$d/dz(zA(z)) = d/dz \sum_{n=0}^{\infty} a_n z^{n+1} = \sum_{n=0}^{\infty} (n+1) a_n z^n .$$

In view of theorem 3, we need only show that $\sum_{n=0}^{\infty} (n+1) z^n$ belongs to $\mathcal{S}_1$ (or $\mathcal{S}_{12}$). But $\sum_{n=0}^{\infty} (n+1) z^n = B(z)$, where $B = k[x,y] = k\langle x,y\rangle / \langle xy-yx\rangle$ is a free commutative polynomial ring on two variables of degree one and $B \in \mathcal{A}_{12}$.

We now want to discuss a general method of computing Hilbert series of f.p. algebras, which will be used in the proof of theorem 5. The method applies also to the problem of deciding whether an algebra has a certain presentation.

Let $G \in \mathfrak{A}$. We want to find $G(z)$. There are two ways to attack this problem. The first is to find a spanning set $S \subsetneq G$ and then prove that S is linearly independent and hence $G(z) = S(z)$. That S spans G is usually proved by an easy induction. The proof of independence is however normally harder to get directly. The other way to find $G(z)$ consists of defining a graded vector space S (with a known series $S(z)$) and a multiplication on S such that S becomes a graded associative algebra. Then one proves that S has the same presentation as G . This, again, consists of two parts, one easy and one hard. The easy one is to find generators for S and a set of obvious relations, which, hopefully, constitute all the relations in G . The hard part is to prove that there are no more relations in S . We now are very lucky, because we can combine the two methods and just do the easy parts of them and yet be done. In fact the easy part of the first method gives the coefficient-wise inequality $S(z) \geq G(z)$ while the easy part of the second method gives $S(z) \leq G(z)$. The following propositions are helpful when the second method above is carried out. The construction in proposition 1 may be seen as a generalization of the notion of semi-tensor product for Hopf algebras.

Proposition 1 Let A, B be (graded) algebras and $\phi\colon B^+ \otimes A^+ \longrightarrow A \otimes B$ be a map of (graded) vector spaces. The map ϕ may be extended to a map $B \otimes A \longrightarrow A \otimes B$ (also called ϕ) by means of the maps

$$k \otimes A \xrightarrow{T} A \otimes k \xrightarrow{1 \otimes \eta} A \otimes B$$

and

$$B \otimes k \xrightarrow{T} k \otimes B \xrightarrow{\eta \otimes 1} A \otimes B$$

where T is the natural twisting isomorphism and η is the natural embedding. Suppose the following diagrams are commutative (μ stands for multiplication):

$$\begin{array}{ccc}
B^+ \otimes A^+ \otimes A^+ & \xrightarrow{1 \otimes \mu_A} & B^+ \otimes A^+ \\
\downarrow{\scriptstyle \phi \otimes 1} & & \\
A \otimes B \otimes A^+ & & \downarrow{\scriptstyle \phi} \\
\downarrow{\scriptstyle 1 \otimes \phi} & & \\
A \otimes A \otimes B & \xrightarrow{\mu_A \otimes 1} & A \otimes B
\end{array}
\qquad
\begin{array}{ccc}
B^+ \otimes B^+ \otimes A^+ & \xrightarrow{\mu_B \otimes 1} & B^+ \otimes A^+ \\
\downarrow{\scriptstyle 1 \otimes \phi} & & \\
B^+ \otimes A \otimes B & & \downarrow{\scriptstyle \phi} \\
\downarrow{\scriptstyle \phi \otimes 1} & & \\
A \otimes B \otimes B & \xrightarrow{1 \otimes \mu_B} & A \otimes B
\end{array}$$

Then an associative multiplication on $A \otimes B$ may be defined as

$$(A \otimes B) \otimes (A \otimes B) \xrightarrow{1 \otimes \phi \otimes 1} A \otimes A \otimes B \otimes B \xrightarrow{\mu_A \otimes \mu_B} A \otimes B$$

making $A \otimes B$ to a (graded) algebra.

Moreover A and B may be considered as subalgebras of A ⊗ B in the natural way and the resulting left A-module and right B-module structures of A ⊗ B are the natural ones.

Proof The last statement follows directly from the definition of the product. The first statement follows if we can prove the associativity of the product. Firstly, the diagrams will commute also after we have removed the plus signs. Associativity means that the composition of the maps in the top rows of the following two commutative diagrams are equal.

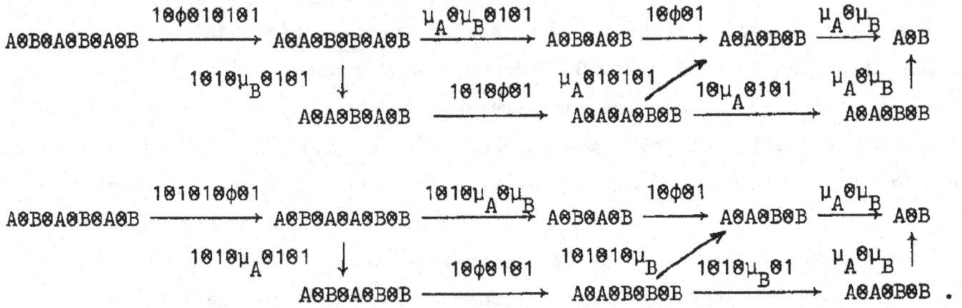

Going by the lower rows of the diagrams instead, we see that the first A and the last B is unchanged (except for the last map). Hence it is sufficient to prove that the composition of the maps of the top rows of the two diagrams below are equal. From the assumptions it follows that the diagrams are commutative and since the lower rows are equal the proof is finished.

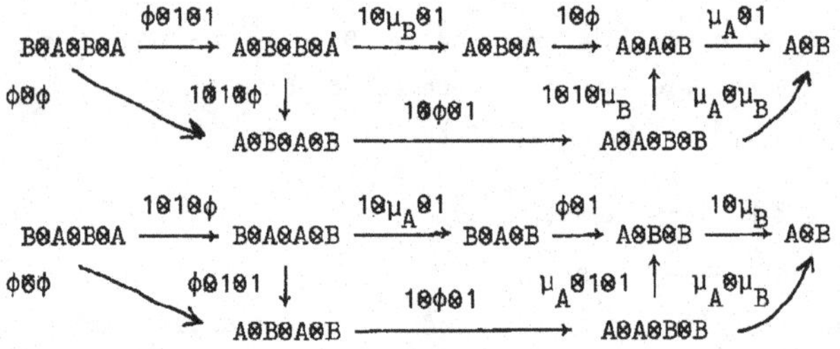

It may be hard in general to check that the conditions of Proposition 1 are satisfied. The next proposition considers a case when this could be done by a machine (if the algebras involved are finitely presented). Before stating the theorem we introduce some conventions.

If S and T are disjoint sets, the k-linear map $\mu: k\langle S\rangle \otimes k\langle T\rangle \longrightarrow k\langle S \cup T\rangle$ given by $\mu(x \otimes y) = xy$ is injective and we will identify $k\langle S\rangle \otimes k\langle T\rangle$ with its image under μ as a subspace of $k\langle S \cup T\rangle$. In the same way $k\langle T\rangle \otimes k\langle S\rangle$ is identified with a subspace of $k\langle S \cup T\rangle$. Also $k\langle S\rangle$ and $k\langle T\rangle$ are identified with subspaces of $k\langle S\rangle \otimes k\langle T\rangle$.

<u>Proposition 2</u> Suppose S and T are disjoint sets and $\varphi: T \times S \rightarrow k\langle S\rangle \otimes k\langle T\rangle$ is a map such that

$$\mathrm{im}(\varphi) \subset 1 \otimes k\langle T\rangle \oplus \mathrm{span}(S) \otimes k\langle T\rangle$$

<u>or</u> $$\mathrm{im}(\varphi) \subset k\langle S\rangle \otimes 1 \oplus k\langle S\rangle \otimes \mathrm{span}(T) \qquad (1)$$

(span(S) means the k-space spanned by S). Then φ has a unique extension $\phi: k\langle T\rangle \otimes k\langle S\rangle \rightarrow k\langle S\rangle \otimes k\langle T\rangle$ making $k\langle S\rangle \otimes k\langle T\rangle$ to an associative algebra (as in Proposition 1). Suppose also $\alpha \subset k\langle S\rangle^+$ and $\beta \subset k\langle T\rangle^+$ are sets of homogeneous elements (the elements of S and T having degree one) and let $A = k\langle S\rangle/\langle\alpha\rangle$ and $B = k\langle T\rangle/\langle\beta\rangle$. The imbedding $\mu: k\langle S\rangle \otimes k\langle T\rangle \rightarrow k\langle S \cup T\rangle$ induces a k-linear map

$$\overline{\mu}: A \otimes B \rightarrow G = k\langle S \cup T\rangle / \langle\, \alpha \cup \beta \cup \{ts - \varphi(t,s);\ s \in S,\ t \in T\}\, \rangle .$$

The map $\overline{\mu}$ is <u>surjective</u>. Moreover $\overline{\mu}$ is injective if and only if

$$\begin{aligned} &\phi(b \otimes s) \text{ and } \phi(t \otimes a) \text{ are contained in} \\ &\langle\alpha\rangle \otimes k\langle T\rangle + k\langle S\rangle \otimes \langle\beta\rangle \quad \text{for all } a \in \alpha,\ b \in \beta,\ s \in S,\ t \in T . \end{aligned} \qquad (2)$$

If this is satisfied, then ϕ induces a map $B \otimes A \rightarrow A \otimes B$ making $A \otimes B$ to an algebra (as in Proposition 1), and the map $\overline{\mu}$ above to an algebra isomorphism.

The proof makes use of the following two lemmas.

Lemma 1 Suppose S and T are disjoint sets and $\varphi: T \times S \longrightarrow k<S> \otimes k<T>$ satisfies condition (1) of Proposition 2. Then there is a unique k-linear map $\phi: k<S \cup T> \longrightarrow k<S \cup T>$ which satisfies the following properties

(a) $\phi(x) = x$ if $x \in k<S> \otimes k<T>$

(b) $\phi(ts) = \varphi(t,s)$ for $s \in S$, $t \in T$

(c) $\phi(xyz) = \phi(x\phi(y)z)$ for all x,y,z in $k<S \cup T>$.

Moreover, the image of ϕ is $k<S> \otimes k<T>$, the restriction of ϕ to $k<T>^+ \otimes k<S>^+ \longrightarrow k<S> \otimes k<T>$ satisfies the conditions of Proposition 1, and hence ϕ defines an associative algebra structure on $k<S> \otimes k<T>$.

Proof This is an application of Bergman's Diamond lemma [Be]. In the case $im(\varphi) \subseteq k<S> \oplus (k<S> \otimes Span(T))$, we may define a semigroup partial order on $<S \cup T>$ (= the set of monomials on $S \cup T$) as follows. Any monomial on $S \cup T$ may be factored uniquely as $w = \sigma u_1 u_2 \ldots u_q$, where $\sigma \in <S>$ and $u_i \in T<S>$. Assign to w a degree as the ordinal number (ω means the first infinite ordinal number)

$$length(\sigma) + length(u_1)\cdot\omega + length(u_2)\cdot\omega^2 + \ldots + length(u_q)\cdot\omega^q .$$

The induced partial order on $<S \cup T>$ is easily seen to be compatible with the semigroup structure and also compatible with the reduction rules defined by φ, since the degree of ts is $2\cdot\omega$ while $\varphi(t,s)$ is a linear combination of monomials each of which has a degree of at most $a + \omega$ where a is finite. There are no inclusion or overlap ambiguities among $\{ts;\ (t,s) \in T \times S\}$, so by [Be] φ extends uniquely to a reduction rule on the entire free algebra $k<S \cup T>$. The reduced words are obviously precisely $k<S> \otimes k<T>$, so properties (a) - (c) follow from [Be]. At the same time property (c) shows that the diagrams of Proposition 1 commute, proving that ϕ defines an associative algebra structure on $k<S> \otimes k<T>$, which also follows from [Be]. The case $im(\varphi) \subseteq k<T> \oplus (Span(S) \otimes k<T>)$ may be handled similarly.

<u>Lemma 2</u> Suppose S, T, φ and ϕ are as in Lemma 1 and suppose $\alpha \subset k\langle S\rangle^{+}$ and $\beta \subset k\langle T\rangle^{+}$ are sets of homogeneous elements (the elements of S and T having degree one) and let $A = k\langle S\rangle/\langle\alpha\rangle$ and $B = k\langle T\rangle/\langle\beta\rangle$. Consider ϕ as a map $k\langle T\rangle \otimes k\langle S\rangle \rightarrow k\langle S\rangle \otimes k\langle T\rangle$. Then ϕ induces a map $B \otimes A \rightarrow A \otimes B$ if and only if condition (2) of Proposition 2 is satisfied.

<u>Proof</u> The "only if" part is obvious. Suppose (2) is satisfied. We consider only the case where φ satisfies the first part of condition (1) (the other case is similar). Let p be the natural projection $k\langle S\rangle \otimes k\langle T\rangle \rightarrow A \otimes B$ and consider the set $I = \{x \in k\langle S \cup T\rangle;\ p \circ \phi(x) = 0\}$. The lemma follows if we can prove that I is a two-sided ideal, since $\alpha \subset I$ and $\beta \subset I$ then also gives that $k\langle T\rangle \otimes \langle\alpha\rangle + \langle\beta\rangle \otimes k\langle S\rangle \subset I$, which is a reformulation of the claim. First observe that $x \in I \Rightarrow sx \in I$ and $xt \in I$ for $s \in S$ and $t \in T$, also by (2) we have $\beta S \subset I$, $T\alpha \subset I$. Next we claim that it is enough to prove that $T\langle\alpha\rangle \subset I$ and $\langle\beta\rangle S \subset I$. Because, if this is true, then it is proved by induction on the length of $x \in \langle T\rangle$ that $x\langle\alpha\rangle \subset I$. Indeed, suppose $x\langle\alpha\rangle \subset I$, then

$$\phi(tx\langle\alpha\rangle) = \phi(t\phi(x\langle\alpha\rangle)) \subset \phi(t(\langle\alpha\rangle k\langle T\rangle + k\langle S\rangle\langle\beta\rangle)) \subset \phi(t\langle\alpha\rangle)k\langle T\rangle + \phi(tk\langle S\rangle)\langle\beta\rangle \subset$$
$$\subset \langle\alpha\rangle k\langle T\rangle + k\langle S\rangle\langle\beta\rangle .$$

By a similar argument it is proved that $\langle\beta\rangle k\langle S\rangle \subset I$ provided $\langle\beta\rangle S \subset I$.

Now consider the filtration of $k\langle S \cup T\rangle$ induced by

$$F^{r}\langle S \cup T\rangle = \{w \in \langle S \cup T\rangle;\ w \text{ contains at most } r \text{ elements from } S\} .$$

The map ϕ preserves this filtration.

<u>Claim:</u> $I \cap F^{1}k\langle S \cup T\rangle$ is a left $k\langle T\rangle$ - module.

<u>Proof of claim:</u> Suppose $x \in I \cap F^{1}k\langle S \cup T\rangle$ and suppose $t \in T$. Then

$\phi(x) \in F^1k\langle S \cup T\rangle \cap (\langle\alpha\rangle \otimes k\langle T\rangle + k\langle S\rangle \otimes \langle\beta\rangle) = F^1\langle\alpha\rangle \otimes k\langle T\rangle + F^1k\langle S\rangle \otimes \langle\beta\rangle$,

since α consists of homogeneous elements. But $F^1\langle\alpha\rangle \subset \alpha$ hence,

$$\phi(x) = \Sigma a_i \otimes b_i + \Sigma a_i' \otimes b_i'$$

where $a_i \in \alpha$, $b_i \in k\langle T\rangle$, $a_i' \in k\langle S\rangle$ and $b_i' \in \langle\beta\rangle$.

Now, $\phi(tx) = \Sigma\phi(ta_i)b_i + \Sigma\phi(ta_i')b_i'$ and hence $p \circ \phi(tx) = 0$ since $p \circ \phi(ta_i) = 0$ by (2) and $p(\phi(ta_i')b_i') = p \circ \phi(ta_i')\cdot p(b_i') = 0$.

Now we are able to prove that $\langle\beta\rangle S \subset I$. By the claim above it is enough to prove that if $b \in \langle\beta\rangle$, $t \in T$ and $bS \subset I$ then also $btS \subset I$. But, with $s \in S$, $\phi(bts) = \phi(b\phi(ts))$ and $\phi(ts) \in F^1k\langle S \cup T\rangle$, hence $p \circ \phi(b\phi(ts)) = 0$ since $p \circ \phi(\langle\beta\rangle) = 0$ and $p \circ \phi(bS) = 0$ by assumption.

The next step is to prove that for any r,

$$T\langle\alpha\rangle \cap F^r \subset I \Rightarrow k\langle T\rangle\langle\alpha\rangle \cap F^r \subset I$$

(we use F^r as short for $F^rk\langle S \cup T\rangle$).

Suppose $b \in k\langle T\rangle$ and $ba \in I$ for all $a \in \langle\alpha\rangle \cap F^r$, then

$$\phi(ba) \in F^r \cap (\langle\alpha\rangle \otimes k\langle T\rangle + k\langle S\rangle \otimes \langle\beta\rangle) = F^r\langle\alpha\rangle \otimes k\langle T\rangle + F^rk\langle S\rangle \otimes \langle\beta\rangle .$$

Hence $\phi(ba) = \Sigma a_ib_i + \Sigma a_i'b_i'$ where $a_i \in \langle\alpha\rangle \cap F^r$, $b_i \in k\langle T\rangle$, $a_i' \in k\langle S\rangle$ and $b_i' \in \langle\beta\rangle$. Then for $t \in T$,

$$\phi(tba) = \phi(t\phi(ba)) = \Sigma\phi(ta_i)b_i + \Sigma\phi(ta_i')b_i'$$

and this is mapped to zero by p since, by assumption, $ta_i \in I$ and $b_i' \in \langle\beta\rangle$.

The last step is to prove that for all r

$$T\langle\alpha\rangle \cap F^{r-1} \subset I \Rightarrow T\langle\alpha\rangle \cap F^r \subset I .$$

So suppose $T\langle\alpha\rangle \cap F^{r-1} \subset I$ and $a \in \langle\alpha\rangle \cap F^r$. Then $a = \Sigma s_ia_i + a_i's_i' + a''$ where $a_i, a_i' \in \langle\alpha\rangle \cap F^{r-1}$ and $s_i, s_i' \in S$ and $a'' \in \text{Span}(\alpha)$. Then with $t \in T$

$$\phi(ta) = \Sigma\phi(\phi(ts_i)a_i) + \Sigma\phi(\phi(ta_i')s_i') + \phi(ta'') .$$

By (2)i, $p \circ \phi(ta'') = 0$ and if $\phi(ts_i) = \Sigma x_jb_j$, $x_j \in k\langle S\rangle$, $b_j \in k\langle T\rangle$, then $\phi(\phi(ts_i)a_i) = \Sigma x_j \phi(b_j a_i)$ and by assumption and from the previous step $p \circ \phi(b_j a_i) = 0$. Also $\phi(ta_i')s_i' \in I$, since $ta_i' \in I$ by assumption and hence $\phi(ta_i') \in \langle\alpha\rangle \otimes k\langle T\rangle + k\langle S\rangle \otimes \langle\beta\rangle$ from which it follows that $\phi(ta_i')s_i' \in I$ since we have proved that $\langle\beta\rangle S \subset I$.

The proof is completed by observing that $T\langle\alpha\rangle \cap F^0 = 0$ and $T\langle\alpha\rangle = \bigcup_{r=1}^{\infty} T\langle\alpha\rangle \cap F^r$.

<u>Proof of Proposition 2.</u>

Lemma 1 proves the first part of the proposition. Since $ts = \phi(ts)$ in G it follows that $x = \phi(x)$ in G for all x and hence $\bar{\mu}$ is surjective. Consider the k-linear map $p \circ \phi\colon k\langle S \cup T\rangle \longrightarrow A \otimes B$, where p is the projection $k\langle S\rangle \otimes k\langle T\rangle \longrightarrow A \otimes B$. Suppose condition (2) is satisfied. In the proof of Lemma 2 it is proved that $\ker(p \circ \phi)$ is a two-sided ideal in $k\langle S \cup T\rangle$. Hence $p \circ \phi$ induces a k-linear map $G \longrightarrow A \otimes B$ and since $\phi(x) = x$ in G it is obvious that this map and $\bar{\mu}$ are mutually inverses of each other. Suppose on the other hand that (2) is false. Then there is b and s such that $p \circ \phi(bs) \neq 0$ (or there is a and t such that $p \circ \phi(ta) \neq 0$) but $\bar{\mu}(p \circ \phi(bs)) = bs = 0$ (resp. $\bar{\mu}(p \circ \phi(ta)) = ta = 0$) so $\bar{\mu}$ is not injective.

The last assertion follows from Lemma 1 and 2 .

<u>Remark</u> The proof is valid under the weakened assumption that α (resp. β) consists of homogeneous elements if the first (resp. second) row of condition (1) is satisfied.

The following corollary to the proposition will be useful in the proof of theorem 6.

Corollary Suppose L, R and A are algebras such that L and R have trivial multiplication (i.e. $(L^+)^2 = (R^+)^2 = 0$). Suppose also given a map $\psi: R^+ \otimes L^+ \longrightarrow A$ of graded vector spaces. Then $L \otimes A$ is an algebra by means of the zero map $A^+ \otimes L^+ \longrightarrow L \otimes A$ and $L \otimes A \otimes R$ is an algebra by means of a map $R^+ \otimes (L \otimes A)^+ \longrightarrow L \otimes A \otimes R$ defined as zero on $R^+ \otimes (k \otimes A^+)$ and on $R^+ \otimes (L^+ \otimes A)$ as the composition

$$R^+ \otimes L^+ \otimes A \xrightarrow{\psi \otimes 1} A \otimes A \xrightarrow{\mu_A} A \hookrightarrow L \otimes A \otimes R .$$

Moreover, if G_L, G_R and G_A are minimal generating sets for L, R and A respectively and A has the presentation $k\langle G_A\rangle/\langle\alpha\rangle$, then $L \otimes A \otimes R$ has the presentation

$$k\langle G_L \cup G_A \cup G_R\rangle/\langle \alpha \cup \{al,\ ll',\ rr',\ ra,\ rl - \phi(r\otimes l);\ l,l' \in G_L, a \in G_A, r,r' \in G_R\}\rangle.$$

Proof A trivial application of Proposition 2 shows that $L \otimes A$ is an algebra with presentation

$$k\langle G_L \cup G_A\rangle/\langle\alpha \cup \{ll',\ al\ ;\ l,l' \in G_L\ ,\ a \in G_A\}\rangle.$$

Moreover R has the presentation $k\langle G_R\rangle/\langle rr'\ ;\ r,r' \in G_R\rangle$. The map ψ may be restricted and lifted to a map $\tilde{\psi} : G_R \times G_L \rightarrow k\langle G_A\rangle$ which defines a map

$$\varphi : G_R \times (G_L \cup G_A) \rightarrow k\langle G_L \cup G_A \cup G_R\rangle$$

by sending $G_R \times G_A$ to zero and using $\tilde{\psi}$ on $G_R \times G_L$. We apply Proposition 2 to the map φ. The claim of the corollary follows from the fact that the extension ϕ of φ satisfies condition (2) of Proposition 2. This again follows by an explicit computation in a few cases.

All the operations on $\mathcal{S}$ we have discussed so far have the property that if we start with rational Hilbert series,we end up with a rational series. In theorem 5 we discuss some operations for which this is not the case. In each of the constructions of theorem 5 , we obtain a Hilbert series which equals a rational function times a transcendental infinite product, possibly times a power of the original series.

Theorem 5 Let $G \in \mathfrak{A}$ and write $G = k\langle T\rangle/\langle\alpha\rangle$, where $T = \{t_1,\ldots,t_N\}$ and $\alpha = \{u_1,\ldots,u_M\}$. Let $g_n = \mathrm{rank}(G_n)$, so that $G(z) = \sum_{n=0}^{\infty} g_n z^n \in \mathcal{S}$. There exist A, B, $C \in \mathfrak{A}$ and if $G \in \mathfrak{A}_1$ there exist D, $E \in \mathfrak{A}_1$ such that

$$\text{(a)}\quad A(z) = \begin{cases} f(z)\prod_{n=1}^{\infty}(1 - z^n)^{-g_n} & \text{if char}(k) = 2 \\ f(z)\prod_{n=1}^{\infty}(1+z^{2n-1})^{g_{2n-1}}/(1-z^{2n})^{g_{2n}} & \text{if char}(k) \neq 2 \end{cases}$$

where $f(z) = (1 - T(z) - z^{-1}T(z)^2)^{-1}(1 - z - T(z))^{-1}$ is rational.

$$\text{(b)}\quad B(z) = (1 - z^2)^{-1}(1 - z)^{-1}G(z)^2\prod_{n=1}^{\infty}(1 + z^nG(z))$$

$$\text{(c)}\quad C(z) = (1 - z^2)^{-1}(1 - z)^{-1}G(z)^2\prod_{n=1}^{\infty}(1 - z^nG(z))^{-1}$$

$$\text{(d)}\quad D(z) = (G\cdot G)(z^2)G(z)\prod_{n=0}^{\infty}(1 + g_nz^{n+1})$$

$$\text{(e)}\quad E(z) = (G\cdot G)(z^2)G(z)\prod_{n=0}^{\infty}(1 - g_nz^{n+1})^{-1}\ .$$

Proof (a). See [An-1, prop. 8.4] . It is also possible to use the general method above. In this case A is the enveloping algebra of a graded Lie algebra $\mathfrak{g}$ and $\mathfrak{g}$ may be seen as an abelian extension of basic Lie algebras.

(b). This is a generalization of Shearer's example [Sh, see "note added in proof"] . Let $G' = k\langle T'\rangle/\langle\alpha'\rangle$ be a copy of G . This means that $T' = \{t'_1,\ldots,t'_N\}$ is a set of generators disjoint from but identical to T , i.e., $|t'_i| = |t_i|$ for $i = 1,\ldots,N$. Likewise, $\alpha' = \{u'_1,\ldots,u'_M\}$ is the same set of relations as α , but among the $\{t'_i\}$ instead of the $\{t_i\}$. The algebras G' and G are obviously isomorphic.

The desired algebra B has a presentation as

$$B = k\langle T\cup T'\cup\{a,b,c\}\rangle/\langle\alpha\cup\alpha'\cup\beta\rangle$$

where $|a| = |b| = 1$ and $|c| = 2$ and $\beta = \{ac-ca,\ bc-aba,\ b^2\}\ \cup$ $\cup\ \{at_i-t_ia,\ at'_i-t'_ia,\ ct_i-t_ic,\ ct'_i-t'_ic,\ t_it'_j-t'_jt_i,\ bt'_i-t_ib\ ;\ 1 \le i \le N\}$. We compute the series $B(z)$ by means of the "general method" above. Let W be a k-basis for G with $1 \in W$ and let W' be the corresponding k-basis for G' . A spanning set for B is found by induction to be all words of the form

$$a^{q_0}c^p w'w_{i_0}ba^{q_1}w_{i_1}ba^{q_2}w_{i_2}\ldots ba^{q_r}w_{i_r}$$

where $q_1 > q_2 > \ldots > q_r \ge 0$ and p, q_0 are arbitrary, $w' \in W'$, $w_{i_0},\ldots,w_{i_r} \in W$. It is easy to see that this set has the series we look for. We now apply Proposition 2 to define an algebra $L \otimes R$ with the right series and which is a quotient of B . Put

$$R = k\langle\{w(i);\ w \in W,\ i \geq 1\}\rangle / \langle\{w_1(i)w_2(j);\ i \leq j\ ,\ w_1, w_2 \in W\}\rangle$$

where $|w(i)| = |w| + i$, and put

$$L = k\langle a,c\rangle / \langle ac-ca\rangle \otimes G \otimes G'$$

where $|a| = 1$, $|c| = 2$. The algebra structure on L is defined as the tensor product of algebras (defined in "Definitions"). For $g \in G$ let $g(i)$ denote the element $\Sigma\lambda_j w_j(i)$ in R where $g = \Sigma\lambda_j w_j$ with $\lambda_j \in k$, $w_j \in W$.

A map

$$\{w(i);\ w \in W,\ i \geq 1\} \times (\{a,c\} \cup T \cup T') \longrightarrow k\langle\{a,c\} \cup T \cup T'\rangle \otimes k\langle\{w(i);\ w \in W,\ i \geq 1\}\rangle$$

is defined by

$$(w(i),\ a) \longrightarrow w(i+1)$$
$$(w(i),\ c) \longrightarrow a \otimes w(i+1)$$
$$(w(i),\ t) \longrightarrow (wt)(i)$$
$$(w(i),\ t') \longrightarrow t \otimes w(i) \quad (t' \in T' \text{ corresponds to } t \in T).$$

This map satisfies condition (1) of Proposition 2 (in fact both cases). The extended map satisfies condition (2). The verification of this involves a control of 12 equations, since R has one group of generators and one group of relations and L has 4 groups of generators and 8 groups of relations. We give 3 of them below.

If $i \leq j$, $w(i)w(j) \otimes c$ is mapped to $1 \otimes w(i+1)w(j+1)$ which is zero in $L \otimes R$. If $u \in \alpha$, $w(i) \otimes u$ is mapped to $1 \otimes (wu)(i)$ which is zero in $L \otimes R$. If $t \in T$ and $t' \in T'$ correspond to each other, $w(i) \otimes (ct' - t'c)$ is mapped to $(at - ta) \otimes w(i+1)$ which is zero in $L \otimes R$.

We may now apply Proposition 2 and get an algebra structure on $L \otimes R$. It is now easy to verify that this algebra is a quotient of B (a and c are mapped to a and c in L, b is mapped to $1(1) \in R$, T is mapped to $G \subset L$ and T' to $G' \subset L$). Since also the series $L \otimes R(z)$ is the right one, (b) is proved.

(c). The algebra C is nearly identical to B , except that the relation b^2 is replaced by b^2a . We obtain the same generating set for C as for B

except that now words are allowed with $q_1 \geq q_2 \geq \dots \geq q_r$. The algebra R is changed to

$$R = k\langle\{w(i);\ w \in W,\ i \geq 1\}\rangle / \langle\{w_1(i)w_2(j);\ i<j\ ,\ w_1,w_2 \in W\}\rangle$$

but otherwise the definition of $L \otimes R$ is unchanged. By similar reasoning we obtain the stated product formula for $C(z)$.

(d). This is a somewhat different generalization of the same example of Shearer. Let G^{op} be the algebra which is equal to G as a graded vector space, but with reversed multiplication. A presentation of G^{op} is $k\langle T\rangle/\langle \alpha^{op}\rangle$ with an obvious definition of the set α^{op} (e.g. $(t_1t_2t_3)^{op} = t_3t_2t_1$). By theorem 3 $G \cdot G^{op} \in \mathcal{O}_1$ since, by assumption, $G \in \mathcal{O}_1$, and then also $G^{op} \in \mathcal{O}_1$. From the proof of theorem 3 we have (W is a k-basis for G with $1 \in W$)

$$G \cdot G^{op} \cong k\langle T\rangle \cdot k\langle T\rangle / \langle \alpha \cdot W \cup W \cdot \alpha^{op}\rangle \cong k\langle T \times T\rangle / \langle \beta \rangle$$

where β is a finite set and the elements of $T \times T$ are of degree one. Let $(T \times T)^{(2)}$ (rep. $\beta^{(2)}$) denote the set $T \times T$ (resp. β) with the elements of degree two. Let also b be a variable of degree one. Put

$$D = k\langle T \cup (T \times T)^{(2)} \cup \{b\}\rangle / \langle \alpha \cup \beta^{(2)} \cup \{(t_1,t_2)^{(2)}t - t(t_1,t_2)^{(2)},\ b^2,\ b(t_1,t_2)^{(2)} - t_1bt_2;\ t,t_1,t_2 \in T\}\rangle .$$

Since $G \cdot G^{op} \cong k\langle T \times T\rangle / \langle \beta \rangle$, there is a map (φ): $G \cdot G^{op} \longrightarrow D$ (φ is a map of algebras which doubles the degree.) A spanning set for D as a k-space consists of all elements of the form

$$(w_{j_1} \cdot w_{j_2})^{(2)} w_{i_0} bw_{i_1} bw_{i_2} \dots bw_{i_r}$$

where $w_{j_1}, w_{j_2}, w_{i_0}, \dots, w_{i_r} \in W$, $|w_1| = |w_2|$, $|w_{i_1}| > \dots > |w_{i_r}| \geq 0$ and $r \geq 0$.

This is easily seen by induction. Also, this set of elements has a series which is less than or equal to the indicated series. For the second step of the "general method" define

$$R = k\langle \tilde{W}\rangle / \langle\{\tilde{w}_1 \tilde{w}_2;\ |w_1| \leq |w_2|\}\rangle$$

where $\tilde{W} = \{\tilde{w};\ w \in W$ and $|\tilde{w}| = |w| + 1\}$ and let

$$L = (G \cdot G^{op})^{(2)} \otimes G$$

where $(G \cdot G^{op})^{(2)}$ is $G \cdot G^{op}$ but the degree is doubled.

$$\tilde{W} \times ((T \times T)^{(2)} \cup T) \rightarrow k\langle (T \times T)^{(2)} \cup T\rangle \otimes k\langle \tilde{W}\rangle$$

is defined by

$$(\tilde{w}\ ,\ (t_1,t_2)^{(2)}) \rightarrow t_1 \otimes \widetilde{t_2w}$$

$$(\tilde{w}\ ,\ t_1) \rightarrow 1 \otimes \widetilde{wt_1}$$

(If in G, $g = \Sigma\lambda_i w_i$, $\lambda_i \in k$, $w_i \in W$, then $\tilde{g}$ means $\Sigma\lambda_i\tilde{w}_i$).

Proposition 2 applies and an algebra structure on $L \otimes R$ is induced, since the extended map satisfies the following.

If $|w_1| \leq |w_2|$ then $\tilde{w}_1\tilde{w}_2(t_1,t_2)^{(2)} \rightarrow 1 \otimes (\widetilde{w_1t_1})(\widetilde{t_2w_2}) = 0$ in $L \otimes R$

and $\tilde{w}_1\tilde{w}_2t \rightarrow 1 \otimes \tilde{w}_1(\widetilde{w_2t}) = 0$ in $L \otimes R$.

From the proof of Theorem 3 we get the presentation of $(G\cdot G^{op})^{(2)}$ as $k\langle(T\times T)^{(2)}\rangle/\langle\{(a\cdot x)^{(2)}, (x\cdot a^{op})^{(2)} ; a \in \alpha, x \in k\langle T\rangle, |a| = |x|\}\rangle$.

We have for $a \in \alpha$ and $x \in k\langle T\rangle$ and $|a| = |x|$,

$$\tilde{w}(a\cdot x)^{(2)} \rightarrow a \otimes \widetilde{xw} = 0 \text{ in } L \otimes R$$

$$\tilde{w}(x\cdot a^{op})^{(2)} \rightarrow x \otimes \widetilde{aw} = 0 \text{ in } L \otimes R$$

$$\tilde{w}a \rightarrow 1 \otimes \widetilde{wa} = 0 \text{ in } L \otimes R .$$

Finally, if $t_1,t_2,t_3 \in T$, $\tilde{w}(t_1,t_2)^{(2)}t_3$ and $\tilde{w}t_3(t_1,t_2)^{(2)}$ are both mapped to $t_1 \otimes \widetilde{t_2wt_3}$.

Hence $L \otimes R$ is an algebra and the series $L \otimes R(z)$ is the right one and $L \otimes R$ is a quotient of D ($t \rightarrow t \in L$, $(t_1,t_2)^{(2)} \rightarrow (t_1\cdot t_2)^{(2)} \in L$ and $b \rightarrow \tilde{1} \in R$) and this ends the second step and proves (d).

(e). The algebra E is like D, with b^2 replaced by $\{b^2t;\ t \in T\}$. A k-generating set for E consists of words of the same form as for D, except that we allow

$$|w_{i_1}| \geq |w_{i_2}| \geq \ldots \geq |w_{i_r}| \geq 0 .$$

The product formula for this set follows. Also $L \otimes R$ is unchanged except that

$$R = k\langle \tilde{W}\rangle/\langle\{ \tilde{w}_1 \tilde{w}_2;\ |w_1| < |w_2| \}\rangle .$$

The product formula for $L \otimes R(z)$ follows and $L \otimes R$ is a quotient of E.

Remark Let r be the radius of convergence (r.c.) of $G(z)$. The r.c. of $A(z)$ is less than r, since $f(z)$ has a pole for $z<r$. Also the r.c. of $C(z)$ and $E(z)$ are (usually) smaller than r. However r is r.c. of $B(z)$ and $D(z)$, hence the growth of the coefficients in the series for $\prod_1^\infty (1+z^nG(z))$ and $\prod_1^\infty (1+g_nz^{n+1})$ reflects on the growth of the coefficients of $B(z)$ and $D(z)$ respectively.

Four of the five constructions in theorem 5 explicitly involve a generator or generators of degree two. Thus they leave open the question of whether or not similar constructions exist in $\mathcal{S}_1$ or in $\mathcal{S}_{12}$. Likewise we may wonder whether theorem 3, which only holds in $\mathcal{S}_1$, can be generalized to all of $\mathcal{S}$. That such constructions and generalizations do exist is a consequence of our next theorem. Theorem 6 will show, at least as far as rates of growth are concerned that the properties of any series occuring in $\mathcal{S}$ are approximated by a series belonging to $\mathcal{S}_1$ and, less closely, by a member of $\mathcal{S}_{12}$.

Theorem 6 (a). Let $A \in \mathfrak{A}$. Then there exists $B \in \mathfrak{A}_1$ and polynomials $P_1(z)$, $P_2(z)$ with non-negative integer coefficients such that

$$B(z) = P_1(z)A(z) + P_2(z) .$$

(b). For any $B \in \mathfrak{A}_1$, there exists $C \in \mathfrak{A}_{12}$ and a polynomial $Q(z)$ with non-negative integer coefficients such that

$$B(z) \leq C(z) \leq Q(z)B(z)$$

inequality being coefficient-wise for these power series.

Proof (a). This is shown in [An-2, proof of thm 4]. A brief outline of the proof is as follows. Write $A = k\langle S\rangle/\langle \alpha \rangle$ where $S = \{s_1,\dots,s_N\}$ and let T consist of the degree-one generators $\{t_{ij};\ i=1,\dots,N\ ,\ j=1,\dots,|s_i|\}$. There is a natural injection $\phi\colon k\langle S\rangle \longrightarrow k\langle T\rangle$ given by $\phi(s_i) = t_{i1}\cdot\ldots\cdot t_{i|s_i|}$. The induced map $\overline{\phi}\colon A \longrightarrow G$, where $G = k\langle T\rangle/\langle\phi(\alpha)\rangle$, is also one-to-one, hence A may be viewed as a subalgebra of G. The algebra B with the desired Hilbert series is obtained as being a quotient algebra of G by monomials in T, which is minimal with respect to the property that A embeds in it as a subalgebra.

(b). Proof is by induction on the "complexity" of B. Define the complexity of a finite set of relations β in a free degree-one generated algebra $k\langle T\rangle$ to be $\mathrm{co}(\beta) = \sum_{w\in\beta} (|w| - 2)$. Let the complexity of a f.p. degreee-one generated algebra B be $\mathrm{co}(B) = \mathrm{co}(\beta)$ where $B = k\langle T\rangle/\langle \beta \rangle$ is a minimal presentation. This is well-defined, indeed we have

$$\mathrm{co}(B) = \sum_{i=3}^{\infty} (i - 2)\,\mathrm{rank}(\mathrm{Tor}^B_{2,i}(k,k)) .$$

We have $\mathrm{co}(B) \geq 0$ and $\mathrm{co}(B) = 0$ if and only if $B \in \mathfrak{A}_{12}$. Thus complexity is a measure of how much an algebra deviates from being a one-two algebra and theorem 6(b) is trivially true if $\mathrm{co}(B) = 0$.

Suppose now $\mathrm{co}(B) = N > 0$ and that theorem 6(b) is true for algebras in $\mathfrak{A}_1$ whose complexity is smaller than N. We will show that there is an algebra $D \in \mathfrak{A}_1$ with $co(D) < N$ and a polynomial $Q_1(z)$ with non-negative integer coefficients such that

$$B(z) \leq D(z) \leq Q_1(z)B(z)$$

coefficient-wise. Since our inductive assumption applied to D gives us an algebra $C \in \mathfrak{A}_{12}$ with

$$D(z) \leq C(z) \leq Q(z)D(z)$$

for a suitable polynomial $Q(z)$, we obtain

$$B(z) \leq C(z) \leq (Q(z)Q_1(z))B(z)$$

as desired.

Let $B = k\langle T\rangle/\langle\beta\rangle$ be a minimal presentation with $T = \{t_1,\ldots,t_p\}$ and $|t_i| = 1$ and with $co(B) = co(\beta) = N > 0$. Let $y \in \beta$ be of degree ≥ 3, and let $\hat{B} = k\langle T\rangle/\langle\beta - \{y\}\rangle$ and let $\hat{y}$ denote the image of y in $\hat{B}$, so $\hat{B}/\langle\hat{y}\rangle = B$. To obtain the promised algebra D, we shall adjoin new variables to $\hat{B}$, along with relations of degree two and a single relation y' of degree $|y| - 1$. This is done such that the quadratic terms t_it_j all have a common left factor and the relation y' is obtained from y by deleting this factor.

Define algebras L, R as $L = k \oplus L_1$ and $R = k \oplus R_1$ where a basis for L_1 is $\{u_{ij};\ 1\leq i,j\leq p\}$ and a basis for R_1 is $\{v\}$ (the algebra structure is defined by $L_1^2 = R_1^2 = 0$). We apply the Corollary to the map $\hat{f}\colon R^+ \otimes L^+ \longrightarrow \hat{B}$ where $\hat{f}(v,u_{ij}) = t_it_j$ to get an algebra $\hat{G} = L \otimes \hat{B} \otimes R$. In the same way $G = L \otimes B \otimes R$ is an algebra and $\hat{G}/\langle\hat{y}\rangle = G$. Suppose $y = \Sigma c_\tau t_\tau$ where $c_\tau \in k$ and t_τ denotes $t_{\tau_1}t_{\tau_2}\ldots t_{\tau_l}$ when $\tau=(\tau_1\ldots,\tau_l)$, $1\leq\tau_i\leq p$. For $\tau = (\tau_1,\ldots,\tau_l)$, let τ' denote $(\tau_3,\ldots,\tau_l)$ and define $y' \in \hat{G}$ as

$$y' = \Sigma c_\tau u_{\tau_1\tau_2} \otimes t_{\tau'}\ .$$

Now, put $D = \hat{G}/\langle y'\rangle$. Since, in $\hat{G}$, $v\cdot y' = \hat{y}$ and $\hat{G}/\langle\hat{y}\rangle = G$, we have

$$D = \hat{G}/\langle y'\rangle = \hat{G}/\langle\hat{y},y'\rangle = G/\langle\overline{y}'\rangle$$

where $\overline{y}'$ is the image of y' under the map $\hat{G} \longrightarrow G$. Hence

$$D(z) \leq G(z) = (1 + p^2z)(1 + z)B(z)\ .$$

<u>Claim</u>: $co(D) \leq N-1$.

From Corollary we get a presentation of $\hat{G}$ as

$$k\langle\{u_{ij}\}\ \ \{v\} \cup T\rangle / \langle(\beta - \{y\}) \cup \{v^2,\ vt_i,\ u_{ij}u_{mn},\ t_i u_{mn},\ vu_{ij} - t_i t_j\}\rangle$$

and hence $co(\hat{G}) = co(\hat{B}) = N - (|y| - 2)$

$$co(D) \leq N - (|y| - 2) + (|y'| - 2) = N - 1$$

and the claim is proved.

It remains to show that B embeds in D , which will imply the remaining inequality $B(z) \leq D(z)$. We have

$$\ker(B \longrightarrow D = G/\langle\bar{y}'\rangle) = \langle\bar{y}'\rangle \cap B$$

where B is identified with the subalgebra $k \otimes B \otimes k$ of G . But $\bar{y}' \in L^+ \otimes B \otimes R$ and this is a right ideal in G , and $G^+ \bar{y}' = 0$ (since $vy' = \hat{y}$ in $\hat{G}$ and $G = \hat{G}/\langle\hat{y}\rangle$), so in G it holds that $\langle\bar{y}'\rangle = \bar{y}'G$. Hence

$$\langle\bar{y}'\rangle \cap B \subset L^+ \otimes B \otimes R \cap k \otimes B \otimes k = 0$$

and this completes the proof.

A few words are in order on the interpretation of theorem 6(b). The result must unfortunately be left in terms of an inequality rather than an explicit formula like 6(a), because the explicit formula involves the Hilbert series of a certain right ideal. In special cases it may be possible to obtain an exact formula, and in general it is possible to complete the construction without introducing as many new variables as we used here. For example, see Löfwall's reduction of Shearer's algebra to a one-two algebra in the appendix of [Ja] .

In spite of the inequality we may still use 6(b) to make some strong claims about the relative rates of growth of $B(z)$ and $C(z)$. For example, if $B(z) = \sum_{n=0}^{\infty} b_n z^n$ and $C(z) = \sum_{n=0}^{\infty} c_n z^n$, and we allow the weak additional assumption that $\{b_n\}$ is monotonic or even that $\inf\{b_{n+1}/b_n\}$ is positive, then we deduce at once that there is some fixed constant q such that $b_n \leq c_n \leq qb_n$ for all n .

The set of radii of convergence

We conclude with a few observations about the possible exponential rates of growth for the coefficients of a Hilbert series. As noted, the radius of convergence (r.c.) of $A(z) = \sum_{n=0}^{\infty} a_n z^n$ gives the least subtle measure of the growth of the sequence $\{a_n\}$. Let $\mathcal{R} = \{r \in (0,1];\ r$ is the r.c. of some $A(z) \in \mathcal{A}\}$ and likewise for $\mathcal{R}_1$ and $\mathcal{R}_{12}$.

Theorem 7 The sets $\mathcal{R}$, $\mathcal{R}_1$ and $\mathcal{R}_{12}$ are equal.

Proof Clearly $\mathcal{R}_{12} \subset \mathcal{R}_1 \subset \mathcal{R}$ so we need only show that $\mathcal{R} \subset \mathcal{R}_{12}$. Let $r \in \mathcal{R}$ and choose any $A(z) \in \mathcal{A}$ with r.c. equal to r. By theorem 6(a) we may choose $B(z) \in \mathcal{A}_1$ with $A(z) \leq B(z) = P_1(z)A(z) + P_2(z) \leq P_3(z)A(z)$, where $P_3(z) = P_1(z) + P_2(z)$ has non-negative integer coefficients. By theorem 6(b) we obtain $C(z) \in \mathcal{A}_{12}$ with coefficient-wise inequality

$$A(z) \leq B(z) \leq C(z) \leq Q(z)B(z) \leq P(z)A(z)$$

where $P(z) = P_3(z)Q(z)$ has non-negative integer coefficients. Since the inequalities hold coefficient-wise and all coefficients are non-negative and $A(z)$ converges for $|z| < r$, $C(z)$ must converge for $z \in [0,r)$, and consequently for all $|z| < r$. Hence, r is less than or equal to the r.c. of $C(z)$. But, since $A(z) \leq C(z)$, we also get the reversed inequality. The r.c. of $C(z)$ is therefore exactly r, and $r \in \mathcal{R}_{12}$, as desired.

Theorem 8 $\mathcal{R}$ is a divisible submonoid of $(0,1]$ under multiplication and a countable dense subset of $(0,1]$ in the usual topology.

Proof We must show that $\mathcal{R}$ is closed under multiplication and that $r^{1/d} \in \mathcal{R}$ if $r \in \mathcal{R}$ and d is a positive integer. That $\mathcal{R} = \mathcal{R}_1$ is closed under multiplication follows from theorem 3. Now let $d \geq 1$ and let r

be the r.c. of $A(z) = \sum_{n=0}^{\infty} a_n z^n \in \mathcal{S}$. Consider an algebra B which is identical to A except that all generators (and hence all relations) have their degrees increased uniformly by a factor of d. The $n^{\underline{th}}$ graded component of B will be zero unless d divides n, and when it does, $B_n \cong A_{n/d}$. It follows that

$$B(z) = \sum_{d|n} a_{n/d} z^n = \sum_{n=0}^{\infty} a_n z^{dn} = A(z^d)$$

which has radius of convergence $r^{1/d}$.

That $\mathcal{R}$ is countable follows immediately from theorem 1. For density, note that $\mathcal{R}$ contains every α, where α is the smallest positive root of an equation $1-e_1z-e_2z^2-\ldots-e_dz^d = 0$ with $e_i \geq 0$ integers. To see this, choose $A = k\langle T\rangle$ with $T(z) = \sum_{i=1}^{d} e_i z^i$, then $A(z) = 1/(1-T(z))$. For $T(z) = 2^p z^q$ we get $2^{-p/q} \in \mathcal{R}$ for any positive rational p/q and these points are themselves dense in $(0,1]$.

As to how Hilbert series behave near their smallest singularity, we have the following,

<u>Theorem 9</u> Let $r \in \mathcal{R}$. Then there is an algebra $C \in \mathcal{O}_{12}$ such that r is the r.c. of C(z) and r is an essential singularity of C(z).

<u>Proof</u> Since $r \in \mathcal{R} = \mathcal{R}_1$, we may choose some algebra $A \in \mathcal{O}_1$ such that A(z) has r.c. equal to r. In [An-2, see lemma 1 and proof of theorem 2] it is shown that, if $A(z) = \sum_{n=0}^{\infty} a_n z^n$, then $a_n \geq r^{-n}$ for each n. Let H be Shearer's algebra, as described in [Sh, see "note added"]. This H has the property that its r.c. is unity but $\lim_{z\to 1^-} (1-z)^d H(z) = \infty$ for any d, hence the singularity at 1 is essential. Using theorem 6(a) we construct a degree-one generated algebra G with this same property. Let $B = A \cdot G$. By theorem 3, $B \in \mathcal{O}_1$ and the r.c. of B(z) is also r. Writing $G(z) = \sum_{n=0}^{\infty} g_n z^n$, we have $B(z) = \sum_{n=0}^{\infty} a_n g_n z^n$ and inequalities $a_n g_n \geq g_n r^{-n} \geq 0$ for each n. For $z \in [0,r)$ it follows that $B(z) \geq \sum_{n=0}^{\infty} g_n (zr^{-1})^n$, so

substituting $\lambda = zr^{-1}$, we obtain

$$\lim_{z\to r^-} (r - z)^d B(z) \geq \lim_{\lambda\to 1^-} r^d(1 - \lambda)^d \sum_{n=0}^{\infty} g_n\lambda^n = r^d \lim_{\lambda\to 1^-} (1 - \lambda)^d G(\lambda) = \infty$$

for any fixed d . Using theorem 6(b) we get an algebra $C \in \mathfrak{A}_{12}$, with the r.c. of $C(z)$ also equal to r and $\lim_{z\to r^-} (r - z)^d C(z) = \infty$ for any fixed d . Thus $C(z)$ has an essential singularity at r .

The reverse question to theorem 9, whether or not for every $r \in \mathcal{R}$ there is an $A(z) \in \mathcal{S}$ converging for $|z| < r$ and with a simple pole at r , remains open. If true, it would follow from theorem 4 that there are series with poles of any desired order at r .

We close by mentioning one more open question about the set $\mathcal{R}$. Does the set $\mathcal{R}$ contain any algebraic number which is not the reciprocal of an algebraic integer?[1)] In particular, does 2/3 belong to $\mathcal{R}$? This question is motivated by the observation that when $A(z)$ is rational, it always has the form $P(z)/Q(z)$, where $Q(z)$ has constant coefficient unity and $P, Q \in Z[z]$. Consequently the r.c. of $A(z)$, which coincides with the smallest root of $Q(z)$, is the reciprocal of an algebraic integer. A r.c. of 2/3 would mean an algebra in which the sequence $\{\mathrm{rank}(A_n)\}$ grows like $\{(3/2)^n\}$. It would be of interest to see how closely the coefficients could approximate such a sequence.

REFERENCES

[An-1] ANICK, D., A counterexample to a conjecture of Serre, Ann. Math. 115, 1982 1-33. Correction: Ann. Math., 116, 1983, 661.

[An-2] ANICK, D., The smallest singularity of a Hilbert series, Math. Scand., 51, 1982, 35-44.

[Be] BERGMAN, G.M., The diamond lemma for ring theory, Advances in Math., 29, 1978, 178-218.

[Go-1] GOVOROV, V.E., Graded algebras, Math. Notes of the Acad. Sc. of the USSR, 12, 1972, 552-556.

[Go-2] GOVOROV, V.E., On the dimension of graded algebras, Math. Notes of the Acad. Sc. of the USSR, 14, 1973, 678-682.

1) The answer is now known to be yes (added in proof).

[Ja] JACOBSSON, C., On the double Poincaré series of the enveloping algebras of certain graded Lie algebras, Math. Scand. 51, 1982, 45-58.

[Le] LEMAIRE, J.-M., Algèbres connexes et homologie des espaces de lacets, Lecture Notes in Mathematics, 422, 1974, Springer-Verlag, Berlin, Heidelberg, New York.

[Lö] LÖFWALL, C., On the subalgebra generated by the one-dimensional elements in the Yoneda Ext-algebra, these proceedings.

[Ro] ROOS, J.-E., Relations between the Poincaré-Betti series of loop spaces and of local rings, Lecture Notes in Mathematics, 740, 1979, 285-322, Springer-Verlag, Berlin, Heidelberg, New York.

[Sh] SHEARER, J.B., A graded algebra with non-rational Hilbert series, Journ. of Algebra, 62, 1980, 228-231.

David ANICK
Department of Mathematics
Mass. Institute of Technology
Cambridge, Mass. 02139
(USA)

Clas LÖFWALL
Department of Mathematics
University of Stockholm
Box 6701
S-113 85 STOCKHOLM
(SWEDEN)

ON ENDOMORPHISM RINGS OF CANONICAL MODULES

(joint work with Shiro Goto)

Yoichi Aoyama
Department of Mathematics
Faculty of Science
Ehime University
Matsuyama, 790 Japan

The purpose of this note is to show the main result of the paper [3]. A ring will always mean a commutative noetherian ring with unit. Let R be a ring, $\underline{a}$ an ideal of R and T an R-module. $E_R(T)$ denotes an injective envelope of T and $H^i_{\underline{a}}(T)$ is the i-th local cohomology module of T with respect to $\underline{a}$. We denote by ^ the Jacobson radical adic completion over a semi-local ring. Q(R) denotes the total quotient ring of R and we define $\dim_R 0$ to be $-\infty$. First we recall the definition of the canonical module.

Definition 1([5, Definition 5.6]). Let R be an n-dimensional local ring with maximal ideal $\underline{n}$. An R-module C is called the canonical module of R if $C\otimes_R \hat{R} \cong \mathrm{Hom}_R(H^n_{\underline{n}}(R), E_R(R/\underline{n}))$.

When R is complete, the canonical module C of R exists and is the module which represents the functor $\mathrm{Hom}_R(H^n_{\underline{n}}(\ ), E_R(R/\underline{n}))$, that is, $\mathrm{Hom}_R(H^n_{\underline{n}}(M), E_R(R/\underline{n})) \cong \mathrm{Hom}_R(M,C)$ (functorial) for any R-module M ([5, Satz 5. 2]). For elementary properties of the canonical module, we refer the reader to[4, §6], [5, 5 und 6 Vorträge] and [2, §1]. If R is a homomorphic image of a Gorenstein ring, R has the canonical module C and it is well known that $C_{\underline{p}}$ is the canonical module of $R_{\underline{p}}$ for every $\underline{p}$ in $\mathrm{Supp}_R(C)$ ([5, Korollar 5.25]). On the other hand, as was shown by Ogoma [7, §6], there exists a local ring with canonical module and non-Gorenstein formal fibre, hence not a homomorphic image of a Gorenstein ring. But, in general, the following fact holds.

Theorem 2([2, Corollary 4.3]). Let R be a local ring with canonical module C and let $\underline{p}$ be in $\mathrm{Supp}_R(C)$. Then $C_{\underline{p}}$ is the canonical module of $R_{\underline{p}}$.

Let R be a ring, M a finitely generated R-module and t an integer. We say that M is (S_t) if depth $M_{\underline{p}} \geq \min\{t, \dim M_{\underline{p}}\}$ for every $\underline{p}$ in $\mathrm{Supp}_R(M)$. Throughout this note A denotes a d-dimensional local ring with maximal ideal $\underline{m}$ and canonical module K, $H = \mathrm{End}_A(K)$ and let h be the natural map from A to H. We put $U_A = \bigcap \underline{q}$ where $\underline{q}$ runs through all the primary components of the zero ideal in A such that $\dim A/\underline{q} = d$. We have $\mathrm{ann}_A(K) = U_A$ (cf. [2, (1.8)].

Lemma 3([7, Lemma 4.1] and [3]). If A is (S_2), then $\dim A/\underline{p} = d$ for every $\underline{p}$ in Ass(A).

(Proof) We proceed by induction on d. If $d \leq 2$, then the assertion is obvious. Let $d \geq 3$ and let $(0) = \underline{q}_1 \cap \cdots \cap \underline{q}_t$ be a primary decomposition of the zero ideal in A such that $\dim A/\underline{q}_i = d$ if and only if $i \leq s$ $(1 \leq s \leq t)$. We put $\underline{a} = \underline{q}_1 \cap \cdots \cap \underline{q}_s$ and $\underline{b} = \underline{q}_{s+1} \cap \cdots \cap \underline{q}_t$. Let $\underline{p}$ be a non-maximal prime ideal in $\mathrm{Supp}_A(K)$. Then $U_{A_{\underline{p}}} = 0$ by the induction hypothesis because $K_{\underline{p}}$ is the canonical module of $A_{\underline{p}}$. Since $U_{A_{\underline{p}}} = (U_A)_{\underline{p}}$ by [2, (1.9)], we have $\underline{p} \not\supseteq \underline{b}$. Suppose that $s < t$. Then $\underline{a} + \underline{b}$ is an $\underline{m}$-primary ideal and we have a contradiction from the exact sequence $0 \to A \to A/\underline{a} \oplus A/\underline{b} \to A/\underline{a}+\underline{b} \to 0$. (q.e.d.)

Proposition 4([1], [7, Proposition 4.2] and [3]). The following are equivalent:

(a) The map h is an isomorphism.

(b) $\hat{A}$ is (S_2).

(c) A is (S_2).

(Proof) (a)$\Rightarrow$(b) follows from [2, (1.10)] and (b)$\Rightarrow$(c) is well known. We show (c)$\Rightarrow$(a) by induction on d. If $d \leq 2$, then A is Cohen-Macaulay and the assertion is known ([5, 6 Vortrag]). Let $d > 2$. By the induction hypothesis and Theorem 2, we have $\mathrm{Coker}(h_{\underline{p}}) = 0$ for every prime ideal $\underline{p} \neq \underline{m}$. By Lemma 3, we have $\mathrm{Ker}(h) = \mathrm{ann}_A(K) = U_A = 0$. Hence we have Coker(h) = 0 from the exact sequence $0 \to A \to H \to \mathrm{Coker}(h) \to 0$. (q.e.d.)

Corollary 5([3]). Assume that $\dim A/\underline{p} = d$ for every minimal prime ideal $\underline{p}$. Then the (S_2)-locus $\{\underline{p} \in \mathrm{Spec}(A) \mid A_{\underline{p}} \text{ is } (S_2)\}$ is open in Spec(A).

Theorem 6([3]). Let R be an A-algebra with structure homomorphism f. Then the following are equivalent:

(a) $R \cong H$ as A-algebras.

(b) R satisfies the following conditions

  (i) R is (S_2) and a finitely generated A-module,

  (ii) For every maximal ideal $\underline{n}$ of R, $\dim R_{\underline{n}} = d$,

  (iii) $\dim_A \mathrm{Coker}(f) \leq d-2$ and $\dim_A \mathrm{Ker}(f) \leq d-1$.

(Proof) (a)$\Rightarrow$(b):([2, Theorem 3.2]) Since $H = \mathrm{End}_A(K) = \mathrm{End}_{A/U_A}(K)$ and K is the canonical module of A/U_A by [2, (1.8)], we may assume that A is unmixed. (It is obvious that $\dim_A U_A < d$.) Let $\mathrm{Ass}(A) = \{\underline{p}_1, \ldots, \underline{p}_t\}$ and $S = A \setminus \bigcup_{i=1}^{t} \underline{p}_i$. Since K is torsion free ([2, (1.7)]), H is also torsion free and the natural map from H to $S^{-1}H$ is injective. Since $S^{-1}K \cong \bigoplus_{i=1}^{t} E_A(A/\underline{p}_i)$ by [2, Lemma 3.1], we have $S^{-1}H \cong \mathrm{Hom}_A(S^{-1}K, S^{-1}K) \cong \bigoplus_{i=1}^{t} A_{\underline{p}_i} \cong S^{-1}A = Q(A)$. Hence H is commutative. Since K is a finitely generated (S_2) A-

module, the condition (i) is satisfied. Since A is unimixed and H is an integral extension of A contained in Q(A), the condition (ii) can be proven by virtue of [6, (34.6)]. To show $\dim_A \mathrm{Coker}(h) \leq d-2$, we may assume that A is complete. Let $\underline{p}$ be a prime ideal of heght one. Then $K_{\underline{p}}$ is the canonical module of $A_{\underline{p}}$ ([5, Satz 5.22]) and $A_{\underline{p}}$ is Cohen-Macaulay. Hence we have $\mathrm{Coker}(h_{\underline{p}}) = 0$, that is, $\dim_A \mathrm{Coker}(h) \leq d-2$.
(b)⇒(a):([3, Theorem]) By the conditions, we can prove $\mathrm{Ker}(f) = U_A$. We may assume that $U_A = 0$ and f is injective because K is the canonical module of A/U_A and $H = \mathrm{End}_{A/U_A}(K)$. We put $L = \mathrm{Hom}_A(R,K)$. Then $L_{\underline{n}}$ is the canonical module of $R_{\underline{n}}$ for every maximal ideal $\underline{n}$ of R by [5, Satz 5.12]. Since $\dim_A R/A \leq d-2$, $\mathrm{Hom}_A(R/A,K) = 0$ and $\mathrm{Ext}_A^1(R/A,K) = 0$ by [2, (1.10)]. Hence we have an isomorphism $L = \mathrm{Hom}_A(R,K) \xrightarrow{\sim} \mathrm{Hom}_A(A,K) \cong K$ from the exact sequence $0 \to A \to R \to R/A \to 0$. From this isomorphism, we obtain an A-algebra isomorphism from H to $\mathrm{End}_A(L)$. It is obvious that $\mathrm{End}_A(L) = \mathrm{End}_R(L)$. Since R is (S_2), $R \cong \mathrm{End}_R(L)$ by Proposition 4. Hence we have $R \cong H$ as A-algebras. (q.e.d.)

For a relation between H and ideal transforms, and the Cohen-Macaulayness of H, we refer the reader to the paper [3].

Finally we note the following facts. They can be proven by using our Theorem 2.

Theorem 7. Assume that $H_{\underline{m}}^i(A)$ is of finite length for $i \neq d$. Then A is a homomorphic image of a Gorenstein ring.

Corollary 8. If a Buchsnaum local ring has the canonical module, then it is a homomorphic image of a Gorenstein ring.

Proposition 9. If $d = 2$ and $\dim A/\underline{p} = 2$ for every minimal prime ideal $\underline{p}$, then A is a homomorphic image of a Gorenstein ring.

Acknowledgement. The author was partially supported by Grant-in-Aid for Co-operative Research.

References

[1] Y. Aoyama, On the depth and the projective dimension of the canonical module, Japan. J. Math., 6(1980), 61 - 66.

[2] Y. Aoyama, Some basic results on canonical modules, J. Math. Kyoto Univ., 23(1983), 85 - 94.

[3] Y. Aoyama and S. Goto, On endomorphism rings of canonical modules, Preprint.

[4] A. Grothendieck, Local cohomology, Lect. Notes in Math. 41, Springer Verlag, 1967.

[5] J. Herzog, E. Kunz et al., Der kanonische Modul eines Cohen-Macaulay -Rings, Lect. Notes in Math. 238, Springer Verlag, 1971.

[6] M. Nagata, Local rings, Interscience, 1962.

[7] T. Ogoma, Existence of dualizing complexes, J. Math. Kyoto Univ., 24(1984), 27 - 48.

GOLOD HOMOMORPHISMS

by

Luchezar L. Avramov(*)

There is a growing understanding that theorems and constructions from the rational homotopy category reflect and are reflected by the homological and homotopical properties of local noetherian rings. Accordingly, maps which fibre as wedges of spheres have been perceived as the topological ghosts of local Golod homomorphisms. However, a closer inspection reveals a rather unsatisfactory materialization on the algebraic side of the looking-glass.

Indeed, in topology such a fibre F is up to rational equivalence uniquely characterized by either its homotopy, or its cohomology, or the Pontrjagin ring structure of its loop space. According to anybody's dictionary, one should look for similar properties displayed by the homotopy, homology, or Yoneda algebra of some ring, (or, more generally, DG algebra with divided powers), which arises from a local homomorphism. The trouble comes from the fact that besides imposing such conditions on the fibre, algebraists have found it necessary to make some additional assumption on the map itself, and often this is harder to verify than the properties of F. In this paper we make the extra work unnecessary, by establishing criteria involving only the fibre. In fact, as a particular case of theorems (2.3), (3.4) and (4.6), one has:

Theorem. Let $f: R \to S$ be a local homomorphism, and let X be a DG algebra with divided powers, which also is a free resolution of the residue field k of R. Suppose that the homology $H(F) = \mathrm{Tor}^R(S,k)$ of the fibre $F = S \otimes_R X$ has $\mathrm{length}_S \neq 2$. Then the following are equivalent:

(1) $\mathrm{Tor}^R(S,k)$ has trivial Massey products of all orders ≥ 2;

(2) the homotopy Lie algebra $\pi^*(F)$ is free;

(3) the Poincaré series of R and S are connected by:

$$P_S(t) = \frac{P_R(t)}{1+t-\sum_{i\geq 0} \mathrm{length}_S \mathrm{Tor}_i^R(S,k)\, t^{i+1}} .$$

As an immediate corollary one obtains:

For a local ring R, the universal enveloping algebra U of $\pi^{\geq 2}(R)$ has global dimension ≤ 1 if and only if R is a Golod ring.

(*) During the preparation of this paper the author was a G.A. Miller Visiting Scholar at the University of Illinois (Urbana), partially supported by the National Science Foundation of the United States; and a Visiting Professor at the University of Toronto, supported by the National Science and Engineering Research Council of Canada.

This represents a first step in answering a question of Roos [Ro, §10, Problem III]: he proved that if the completion of R can be reached from a regular ring through a sequence of n surjective Golod maps, then gl dim U $\leq$ n [ibid, Theorem 5], and has asked whether the converse holds.

The first two sections contain mostly definitions and the yoga of Massey products, needed for Golod's construction; for the reader's and the author's convenience, the messy part has not been skipped. Section 3 puts the Golod conditions in the perspective of the homotopy Lie Algebra theory of [Av_4]. The fourth section deals with Golod homomorphisms proper, and this is followed by a last section containing some miscellaneous remarks. The reader should be warned that our use of the "Golod lexicon" is somewhat different from that found in previous publications. At the referee's request, I have included as Remarks (2.5), (4.7). and (5.4) detailed comparisons of the different notions.

I should like to thank Steve Halperin, Gerson Levin, and Jan-Erik Roos for their interest in this paper. The financial help and excellent working conditions, provided by the Universities of Illinois (Urbana) and of Toronto, are gratefully acknowledged.

1. SOME DEFINITIONS

(1.1) We shall need to manipulate "series" in an indeterminate t, whose coefficients are either non-negative integers, or the symbol ∞, restricted by the condition that only a finite number of negative degrees occur with non-zero coefficients. In this case the standard rules $n+\infty = \infty + \infty = m \cdot \infty = \infty \cdot \infty = \infty$ for $n \geq 0$ and $m \geq 1$, along with $0 \cdot \infty = 0$, allow addition and multiplication of "series" in the usual fashion. Coefficientwise comparison, denoted $\preccurlyeq$, can also be attempted, with the understanding that $\infty \leq \infty$ and $n < \infty$ for any $n \in \mathbb{N}$.

(1.1.1) If V' and V'' are graded vector spaces over some field ℓ, such that $V'_i = 0 = V''_i$ for i sufficiently small, then the Hilbert "series" $\mathrm{Hilb}_{V'}(t) = \Sigma \dim_\ell V'_i t^i$ is defined, and $\mathrm{Hilb}_{V'\oplus V''}(t) = \mathrm{Hilb}_{V'}(t) + \mathrm{Hilb}_{V''}(t)$, $\mathrm{Hilb}_{V'\otimes V''}(t) = \mathrm{Hilb}_{V'}(t) \cdot \mathrm{Hilb}_{V''}(t)$. If V' is a subfactor of V'', then $\mathrm{Hilb}_{V'}(t) \preccurlyeq \mathrm{Hilb}_{V''}(t)$.

(1.1.2) More generally, Hilbert "series" can be defined by means of length functions, for graded modules over an arbitrary ring. The formulas above still hold, except for the one involving tensor products.

(1.2) Differential graded (= DG) algebras will be, unless specified, considered non-negatively graded, skew-commutative, with differentials of degree -1. A graded algebra F is said to be piecewise noetherian, if F_0 is a noetherian ring, and for each i F_i is a finitely-generated F_0-module. A DG algebra F is augmented, if a surjective homomorphism ε to a field ℓ is fixed, such that $\varepsilon d_1 = 0$; we set $IF = \mathrm{Ker}\, \varepsilon$. DG modules are non-negatively graded, and torsion products are taken in the sense of Eilenberg and Moore [Mo]. In particular, if X is a F-module, such that

$X^{\#}$ (= X with trivial differential) is $F^{\#}$-free, and $X \to M$ is a quasi-isomorphism of F-modules (i.e. the induced map $H(X) \to H(M)$ is an isomorphism), then for any N, $H(X \otimes_F N) \cong \mathrm{Tor}^F(M,N)$ canonically.

Observe that, if $H(F)$ is piecewise noetherian, then:

(1.2.1) The Hilbert "series" $\Sigma_i \mathrm{lenght}_{H_o(F)} H_i(F)t^i$ has integer coefficients if and only if $H_o(F)$ is an artinian ring; and

(1.2.2) If $\varepsilon: F \to \ell$ is an augmentation, the Poincaré series

$$P_F(t) = \sum_i \dim_\ell \mathrm{Tor}_i^F(\ell,\ell)t^i$$

has integer coefficients.

(1.3) If $\varepsilon_X: X \to \ell$ is a quasi-isomorphism of F-modules, with $X^{\#}$ $F^{\#}$-free, let $\mathbb{1} \in X_o$ be an element such that $\varepsilon_X(\mathbb{1}) = 1 \in \ell$. If $z \in IZ(F)$ is a cycle with augmentation zero, then $z\mathbb{1} \in IZ(X)$, hence there exists an $y \in IX$ such that $dy = z$. The assignment

$$[z] \to [y \otimes 1] \in X \otimes_F \ell$$

gives rise to a degree one map of $H_o(F)$-modules

$$\sigma^F: IH(F) \to \mathrm{Tor}^F(k,k),$$

called the suspension. It is well-defined, and natural: for details cf. e.g. [GM, (3.6)].

(1.4.1) If $h_1,\dots,h_n$ are classes in $IH(F)$, their Massey product is defined if there exist $a_{ij} \in IF$ $(1 \le i \le j \le n)$ with $da_{i,i} = 0$, $[a_{i,i}] = h_i$, $da_{ij} = \sum_{v=1}^{j} \bar{a}_{iv}a_{vj}$ (where $\bar{a} = (-1)^{\deg(a)+1}$, and $(i,j) \neq (1,n)$). Then $\sum_{v=1}^{n} \bar{a}_{1v}a_{vn}$ is a cycle in IF, and all classes defined by such cycles form the Massey product $\langle h_1,\dots,h_n\rangle \subset IH(F)$.

(1.4.2) Let $B = \{h_\alpha\}$ be a subset (finite of infinite) of $IH(F)$. It is said to admit a trivial Massey operation (t.M.o.), if there exists a function γ from the disjoint union $\amalg_{i=1}^{\infty} B^i$ to IF, such that

$$\gamma(h_\alpha) = z_\alpha \in IZ(F), \text{ with } [z_\alpha] = h_\alpha; \text{ and}$$
$$d_\gamma(h_{\alpha_1},\dots,h_{\alpha_n}) = \sum_{v=1}^{n} \overline{\gamma(h_{\alpha_1},\dots,h_{\alpha_v})}\cdot\gamma(h_{\alpha_{v+1}},\dots,h_{\alpha_n}).$$

(1.4.3) Ordinary Massey products have been generalized to operations on matrices in [Ma], to which we refer for definitions. We shall only need to know the set of all matric Massey products in a $H_o(F)$-submodule in $IH(F)$, denoted by $MH(F)$, and the following result:

(1.4.4) [GM, (5.12)]. For a piecewise noetherian F, there is equality:

$$MH(F) = \mathrm{Ker}\,\sigma^F.$$

(1.5) The DG algebras F and F' are said to be (homology) equivalent, if there exists a chain of quasi-isomorphisms

$$F \leftarrow G^{(1)} \to F^{(1)} \leftarrow G^{(2)} \to \dots \to F^{(n)} \leftarrow G^{(n)} \to F'.$$

In the augmented case, one furthermore requires all these algebras to map to the same field ℓ, and the quasi-isomorphisms to induce the identity on ℓ.

2. GOLOD ALGEBRAS VIA HOMOLOGY

For an augmented algebra F, denote by $G_F(t)$ its Golod "series":

$$G_F(t) = (1 - t \sum_{i \geq 0} \text{length}_{F_0} (IF)_i t^i)^{-1}$$

where the right-hand side denotes the "series"

$$1 + t \Sigma \text{ length } (IF)_i t^i + [t \Sigma \text{ length } (IF)_i t^i]^2 + \dots .$$

(2.0) In this section, $\varepsilon: F \to \ell$ denotes a piecewise noetherian augmented DG algebra, such that $H_0(F)$ contains some field.

(2.1) Lemma. For any such F, the inequality

$$P_F(t) \leqslant G_{H(F)}(t)$$

holds.

Proof. Localizing at $\text{Ker}(\varepsilon: F_0 \to \ell)$ does not change the left-hand side, and can only decrease the right-hand one, hence we can assume F_0 local with maximal ideal $\underline{m}$. Completing in the $\underline{m}$-adic topology leaves both sides unchanged, hence we shall moreover assume F_0 complete. In this context we have:

(2.1.1) [Av_4]. If F is an augmented DG algebra with F_0 complete and local, then it is equivalent to a supplemented DG ℓ-algebra F' (i.e. $\ell \subset F'_0$, and $\varepsilon': F'_0 \to \ell$ is the identity on ℓ), with F'_0 complete and local.

Both sides of the inequality being preserved by equivalences, we can assume F supplemented, and replace length $H_0(F)$ by $\dim_\ell$. With $\overline{B}$ denoting the reduced bar-construction of Eilenberg and MacLane [Ca], one has

(2.1.2) $\text{Tor}^F(\ell,\ell) = H\overline{B}(F)$.

Filtering $\overline{B}(F)$ by "the number of bars", one obtains an (Eilenberg-Moore) spectral sequence, converging to $\text{Tor}^F(\ell,\ell)$, such that

(2.1.3) $E^1_{p,q} = (\overline{B}H(F))_{p,q} = (sIH(F)^{\otimes p})_q$

(2.1.4) $E^2_{p,q} = \text{Tor}^{H(F)}_{p,q}(\ell,\ell)$

Now the required inequality of "series" follows from:

$$P_F(t) = \sum_{i \geq 0} (\sum_{p+q=i} \dim_\ell E^\infty_{p,q})t^i \leqslant \sum_{i \geq 0} (\sum_{p+q=i} \dim_\ell E^1_{p,q})t^i = G_{H(F)}(t) .$$

(2.2) Definition. An algebra F as in (2.0) is called a Golod algebra, if the upper bound is reached in the previous lemma, i.e. if $P_F(t) = G_{H(F)}(t)$.

(For a comparison with Levin's definition of a Golod algebra, cf. the end of this section.)

(2.3) Theorem. Let F satisfy the conditions of (2.0). Then the following are equivalent:

(1) F is Golod;

(2) $\operatorname{Ker} \sigma^F = 0$;

(3) $MH(F) = 0$;

(4) $H(F)$ has trivial Massey products, i.e. for every $n \geq 2$ and every set $h_1,\ldots,h_n$ $(h_i \in \mathrm{I}H(F))$, the Massey product $\langle h_1,\ldots,h_n\rangle$ is defined and contains only zero;

(4') there exists a set of generators $B = \{h_\alpha\}$ of $\mathrm{I}H(F)$ over $H_o(F)$, such that every Massey product $\langle h_{\alpha_1},\ldots,h_{\alpha_n}\rangle$ is defined for all $n > 0$;

(5) every set of elements of $\mathrm{I}H(F)$ admits a trivial Massey operation;

(5') some set of generators of $\mathrm{I}H(F)$ over $H_o(F)$ admits a t.M.o.;

(6)(i) $(\operatorname{Ker} \varepsilon)\cdot \mathrm{I}H(F) = 0$;

(ii) $\mathrm{I}H(F)$ has an ℓ-basis $\{h_\alpha\}_{\alpha\in A}$ which admits a t.M.o. γ;

(iii) taking a free F_o-module V with basis $\{v_\alpha\}_{\alpha\in A}$, and the surjective degree zero map $V \to \mathrm{I}H(F)$ $(v_\alpha \to h_\alpha)$, set:

$$d(1 \otimes sv_{\alpha_1} \otimes \ldots \otimes sv_{\alpha_n}) = \sum_{i=1} (h_{\alpha_1},\ldots,h_{\alpha_i}) \otimes sv_{\alpha_{i+1}} \otimes \ldots \otimes sv_{\alpha_n}$$

and extend this map to $X = F \otimes_{F_o} T(sV)$ by requiring additivity and $d(f \otimes v) = df \otimes v - \bar{f} \otimes dv$ for $v = sv_{\alpha_1} \otimes \ldots \otimes sv_{\alpha_n}$ in the tensor algebra $T(sV)$; then d is a differential on X ;

(iv) the natural augmentation $\varepsilon(f \otimes v) = \varepsilon(f)\varepsilon(v)$ induces an isomorphism $H(\varepsilon)\colon H(X) \xrightarrow{\sim} \ell$;

(6') there exists a DG F-module X with the following properties:

(i) $X^{\#} \simeq F^{\#} \otimes_{F_o} T(sV)$, where V is a free graded F_o-module equipped with a surjective F_o-linear map $V \to \mathrm{I}H(F)$, which induces an isomorphism $V \otimes_{F_o} \ell \simeq \mathrm{I}H(F)$;

(ii) $dX \subset (\mathrm{I}F)X$;

(iii) the augmentation $\varepsilon(f \otimes v) = \varepsilon(f)\varepsilon(v)$ induces an isomorphism $H(X) \xrightarrow{\sim} \ell$.

(7) $(\operatorname{Ker} \varepsilon)\, \mathrm{I}H(F) = 0$, and F is equivalent to the trivial extension $\ell \ltimes \mathrm{I}H(F)$;

(7') F is equivalent to the trivial extension of ℓ by some ℓ-vector space.

Note. If W is an ℓ vector space, $\ell \ltimes W = \ell \oplus W$ as vector spaces, ℓ is a subring, $W^2 = 0$, and $dW = 0$.

Proof. (1) $\Rightarrow$ (2). Condition (1) clearly is invariant under homology equivalences, and so is (2) by [Ma, Theorem 1.5]. Hence we can, as in the proof of Lemma (2.1), assume F contains ℓ . Now the equality (1) implies that $E^1 = E^\infty$ in the spectral sequence (2.1.3). However, since σ^F is the map induced in homology by $\mathrm{I}F \ni x \mapsto x \in B_{1,*}(F)$, this implies σ^F is injective.

(2) $\leftrightarrow$ (3) is Gugenheim and May's theorem, quoted in (1.4.4).

(3) $\Rightarrow$ (4). Massey products of 2 elements are (up to sign) ordinary products, hence are always defined, and by the assumption are trivial. Inductively, one can assume any product of $< n$ elements is defined and contains only zero. In particular, this holds for $\langle h_i,\dots,h_j\rangle$ for $1 \leq j-i \leq n-2$. According to [Ma, Lemma 1.3 and Definition 1.2], $\langle h_1,\dots,h_n\rangle$ is defined, and the equality $MH(F) = 0$ implies it contains only zero.

(4) $\Rightarrow$ (5). Let $B = \{h_\alpha\}$ be a set of classes in $IH(F)$, and set $\gamma(h_\alpha) = z_\alpha$ for an arbitrary choice of the cycle z_α in its class h_α. By induction, one assumes γ already defined on $\amalg_{i=1}^{n-1} B^i$. Then by (1.4.1) the class $z_{\alpha_1,\dots,\alpha_n} = \Sigma\,\overline{\gamma(h_{\alpha_1},\dots,h_{\alpha_i})}$ $\times\,\gamma(h_{\alpha_{i+1}},\dots,h_{\alpha_n})$ belongs to $\langle h_{\alpha_1},\dots,h_{\alpha_n}\rangle$, hence by the assumption it is a boundary. It follows that for each sequence $(\alpha_1,\dots,\alpha_n)$ one can choose $\gamma(h_{\alpha_1},\dots,h_{\alpha_n}) \in IF$, which bounds to $z_{\alpha_1,\dots,\alpha_n}$, hence one can extend the t.M.o. γ to $\amalg_{i=1}^{n} B^i$. The same argument shows that (4') implies (5'), while (4') and (5') follow trivially from (4) and (5) respectively.

(5') $\Rightarrow$ (6). Let $B = \{h_\alpha\}$ be the system of generators on which a t.M.o. γ is defined, and denote by $h_1,\dots,h_m$ those of degree zero. Since $IH(F)_o = \mathrm{Ker}(H_o(F) \to \ell)$ is a maximal ideal $\underline{\bar{n}}$ of $H_o(F)$, the fact that $\gamma(h_i)\gamma(h_\alpha) = d\gamma(h_i,h_\alpha)$ for $i = 1,\dots,m$ and all α, shows that $\underline{\bar{n}}\,IH(F) = 0$. One can now cut B down to a basis of $IM(F)$: the restriction of γ still is a t.M.o. It is now clear how to choose that map $V \to IM(F)$, and the basis $\{v_\alpha\}$.

The fact that $d^2 = 0$ on X is verified through an instant computation, which makes use of the definition (1.4.2) of a t.M.o. To show that $H(X) = \ell$, note that the inclusion of F in X leads to an exact sequence

$$0 \to F \to X \to X \otimes_{F_o} sV \to 0$$

where the tensor product is of DG modules, with $d(sV) = 0$. It yields an exact sequence

$$\dots \to \bigoplus_{i+j=n+1} H_i(X) \otimes_{F_o} V_{j-1} \xrightarrow{\partial_{n+1}} H_n(F) \to H_n(X) \to \bigoplus_{i+j=n} H_i(X) \otimes_{F_o} V_{j-1} \to \dots .$$

Obviously, $H_o(X) = \ell$, and $\partial_{n+1}(1 \otimes v_\alpha) = h_\alpha$, hence ∂_n is surjective for all n. However, working back from the sequence

$$0 \to H_1(X) \to H_o(X) \otimes V_o \xrightarrow{\delta_1} H_1(F) \to 0$$
$$H_o(X) \otimes V_o = \ell \otimes H_1(F)$$

one easily sees that this implies $H_i(X) = 0$ for $i \geq 1$.

(6) $\Rightarrow$ (6') is trivial. To deduce (1) from (6'), note that $\mathrm{Tor}^F(\ell,\ell) \simeq H(X \otimes_F \ell)$ (cf. (1.2)), while the assumptions on X imply:

$$H(X \otimes_F \ell) = X \otimes_F \ell = T_\ell(sV \otimes_{F_o} \ell) = T_\ell(sIH(F)) ,$$

whence the equality of power series.

(7) ⇒ (7') needs no proof, while (7') implies (1) because the equality to be established is invariant under homology equivalences, and one has isomorphisms of vector spaces:

$$\mathrm{Tor}^{\ell \ltimes W}(\ell,\ell) = H(\overline{B}(\ell \ltimes W)) = \overline{B}(\ell \ltimes W) \simeq T(sW) .$$

In order to complete the proof we show:

(4) ⇒ (7). By the preceding, one can assume $\ell \subset F$, and one knows $IH(F)$ is a piecewise finite-dimensional ℓ-vector space with $(IH(F))^2 = 0$; set $W = IH(F)$. Furthermore, let $(sW)^\vee$ be the graded ℓ-dual of sW, let $\mathbb{L}(sW)^\vee$ be the free graded Lie algebra on $(sW)^\vee$ and let G denote the algebra of alternating cochains on $s^{-1}\mathbb{L}(sW)^\vee$; $G_r = \bigoplus_{r=q-n} G^n_q$ where G^n_q is the set of degree $q-n$ linear functionals f on $(s\mathbb{L}(sW)^\vee)^{\otimes n}$, for which

$$f(sv_1 \otimes \dots \otimes sv_n) = \varepsilon(\sigma) f(sv_{\sigma(1)},\dots,sv_{\sigma(n)}), \quad v_i \in \mathbb{L}(sW)^\vee, \quad \Sigma \deg sv_i = -q ,$$

with $\varepsilon(\sigma)$ standing for the usual ("Koszul") sign involved in a permutation of homogeneous symbols. Note that, by construction, G is a free skew-commutative ℓ-algebra on the vector space $(s\mathbb{L}(sW)^\vee)^\vee = s^{-1}(\mathbb{L}(sW)^\vee)^\vee$. Furthermore, G has a differential defined by

$$df(sv_1,\dots,sv_n) = \sum_{i<j} \pm f(s[v_i,v_j], sv_1,\dots,\hat{sv_i},\dots,\hat{sv_j},\dots,sv_n) .$$

(For the correct signs and a more detailed description, we refer the reader to the excellent discussion in [Ta, Chapter I].)

Clearly, the construction of G can be performed starting from any (graded) Lie algebra, and its homology $H^n_q(G)$ is, by a definition (appropriately stretched for the occasion), the (bi-) graded cohomology of L. In particular, for the free algebra $L = \mathbb{L}_\ell(sW)^\vee$, $H^0_*(G) = \ell$, $H^1_*(G) \simeq \mathrm{Der}_\ell(\mathbb{L}(sW)^\vee,\ell) \simeq (sW)^{\vee\vee} = sW$, and $H^i_*(G) = 0$ for $i > 1$. It follows the total homology $H_*(G) = \bigoplus_{g-n=*} H^n_q(G)$ is isomorphic as an algebra to $\ell \ltimes W$. Furthermore, this isomorphism is easily seen to be induced by the projection $G \to \ell \ltimes W$, which extends the surjection $(s\mathbb{L}(sW)^\vee)^\vee \to (s(sW)^\vee)^\vee = W$ obtained by dualizing the canonical inclusion $(sW)^\vee \hookrightarrow \mathbb{L}(sW)^\vee$.

On the other hand dualizing the inclusion of $\mathbb{L}(sW)^\vee$ into $T(sW)^\vee$, one gets a surjection $j\colon T(sW) = (T(sW)^\vee)^\vee \to (\mathbb{L}(sW)^\vee)^\vee$. Thus, a basis $\{w_\alpha\}$ of W defines a system of generators $\{u_{\alpha_1\dots\alpha_n} = s^{-1}j(sw_{\alpha_1},\dots,sw_{\alpha_n})\}$ of $(\mathbb{L}(sW)^\vee)^\vee$, hence of the algebra G. By the remarks above, the map which sends $u_\alpha \in G$ to $w_\alpha \in W$, and $u_{\alpha_1,\dots,\alpha_n}$ $(n \geq 2)$ to zero, is a quasi-isomorphism. Also, a straightforward computation [Ta, (1.4.2)], shows the differential d of G can be expressed by the formula:

$$du_{\alpha_1\dots\alpha_n} = -\sum_{i=1}^{n} \bar{u}_{\alpha_1\dots\alpha_i} u_{\alpha_{i+1}\dots\alpha_n} .$$

Assuming (4), it is now easy to construct a quasi-isomorphism $g\colon G \to F$, which will establish (7). In fact, choose for each α a cycle z_α in IF which maps to

w_α under $IZ(F) \to IH(F)^\sim \to W$, and set $g(u_\alpha) = z_\alpha$. By induction, one can assume g defined on the subalgebra $G^{<n}$ of G, generated by the $\{u_{\alpha_1 \dots \alpha_m} | m < n\}$. Choose now a linearly independent set among the $u_{\alpha_1 \dots \alpha_n}$. For any one among them, the formula above shows the $a_{ij} = u_{\alpha_i \dots \alpha_j}$ $(1 \leq i < j \leq n)$ form a defining system for the Massey product $\langle -u_{\alpha_1}, \dots, u_{\alpha_n} \rangle$, hence

$$g(\Sigma \bar{u}_{\alpha_1 \dots \alpha_i} u_{\alpha_{i+1} \dots \alpha_n}) = \langle [z_{\alpha_1}], \dots, [z_{\alpha_n}] \rangle = dy_{\alpha_1 \dots \alpha_n}$$

for some $y_{\alpha_1 \dots \alpha_n} \in F$, so that $g(u_{\alpha_1 \dots \alpha_n}) = y_{\alpha_1 \dots \alpha_n}$ extends g to $G^{<(n+1)}$.

(2.4). Remarks. In the literature dealing with homology of local rings, e.g. [Go], [Gu], [$Lev_{1,2}$], [$Av_{1,2}$], [Ro], the goal has been to connect the trivial Massey products structure of $H(F)$, for some special choice of F, to the homology of the ring F_o. This has led to the interference of delicate questions from linear algebra (e.g: proving a certain complex is minimal) with the essentially quite simple homological problems arising in a context with trivial Massey products. The point of view we have taken is that the trivial Massey structure of F should be analyzed in terms intrinsic to the algebra. This said, no claim of originality is made for the arguments presented above. In particular, the equivalence of (1) with (4), (5) and (6) (and their companion versions) is due to Golod [Go] who introduced the construction of the resolution X in (6) (with F the Koszul complex of a local ring). The proof of the acyclicity of X is borrowed from Levin [Lev_2]. The introduction of the Gugenheim-May theory in the guise of (2) and (3) was done in [Av_1]. The very useful condition (7) seems to have been overlooked in algebra, but well known to topologists. I am grateful to Steve Halperin for pointing it out to me, with a somewhat different proof.

(2.5.1). Remarks. The term "Golod algebra" has been employed in the notes [Lev_2] in a non-equivalent context, to denote a DG algebra F such that:

(i) F_o is a noetherian local ring with maximal ideal $\underline{m}$;

(ii) for every $i \geq 0$, F_i is a free F_o-module of finite rank;

(iii) $dF \subset \underline{m} F$;

(iv) every set of elements of $IH(F)$ admits a t.M.o. with values in $\underline{m} F$.

Thus, if $H_o(F)$ contains a field, (2.3.5) shows our definition (2.2) is much more general than Levin's. However, it is not for generality's sake that we have adopted the new framework. The point is that the equicharacteristic assumption will come in for free for the DG algebras which arise in the main applications of Theorem (2.3) (cf. Sections 4 and 5), while in the same context it will be particularly difficult to verify the last part of condition (iv) above.

(2.5.2). A Golod algebra F which is a ring (i.e. $F = F_o$) has, by (2.3.7), to be of the form $\ell \ltimes W$ for some finite-dimensional ℓ-vector space. This is a very particular instance of the "Golod rings" of local algebra: how they fit in the present approach is made explicit in (5.4).

3. GOLOD ALGEBRAS VIA HOMOTOPY

An algebra F is called a Γ-algebra, if it has a system of divided powers $\{\gamma^i\}_{i\geq 0}$, defined on elements x of even positive degree, and satisfying the usual axioms: e.g. [Ca, Exposé 7]; if moreover F is a DG algebra whose differential satisfies $d\gamma^i(x) = (dx)\gamma^{i-1}(x)$, it is called a DG Γ-algebra.

Next we give a very brief account of the results of $[Av_4]$ (partly announced in $[Av_3]$), which are needed in order to go on with the exposition in this section.

(3.1). For any piecewise noetherian augmented DG Γ-algebra $F \to \ell$, $\mathrm{Tor}^F(\ell,\ell)$ has a natural structure of Hopf Γ-algebra. If moreover, $\ell \subset F$, then this coincides with the Hopf Γ-algebra structure on $H\overline{B}(F)$, where $\overline{B}(F)$ has the shuffle product, diagonal, and divided powers, defined in the work of Eilenberg-MacLane and Cartan (cf. [Ca]). Write $I^{(2)}$ for the ideal generated by I^2 and all $\gamma^i(x)$, when x is of even positive degree and $i \geq 2$. Then the ℓ-vector space

$$(I\mathrm{Tor}^F(k,k)/I^{(2)}\ \mathrm{Tor}^F(k,k))^\vee$$

is a graded Lie algebra, called the homotopy Lie algebra of F, and denoted $\pi^*(F)$. If $F \leftarrow G \to F'$ are quasi-isomorphisms of DG algebras (augmented to ℓ), and if both F and F' are DG Γ-algebras, then $\pi^*(F) \simeq \pi^*(F')$ in a natural way (it is important to note that G need not be a Γ-algebra).

(3.2). For the rest of this section, $\varepsilon: F \to \ell$ denotes a piecewise notherian augmented DG Γ-algebra, such that $H_o(F)$ contains a field.

Recall that for a graded Lie algebra L, the graded Lie subalgebra, generated by all $[a,b]$ $(a, b \in L)$ is denoted by $[L,L]$ (and called the commutator ideal); in characteristic 2, we include among the generators of $[L,L]$ also all $\kappa(a)$, where κ is the quadratic operator $L^{2i+1} \to L^{4i+2}$, which is part of the structure of L (cf. $[Av_3]$, $[Av_4]$ for details).

(3.3). Lemma. The suspension σ^F of (1.3) defines a natural degree zero map of graded ℓ-vector spaces:

$$\tau_F: \pi^*(F)/[\pi^*(F),\ \pi^*(F)] \to \mathrm{Hom}_{F_o}(sIH(F),\ell)\ .$$

Moreover, $\mathrm{Im}\ \tau_F = \mathrm{Im}(\sigma^F)^\vee$.

Proof. Compose the degree -1 isomorphism $sIH(F) \cong IH(F)$ with the suspension σ^F: $IH(F) \to I\mathrm{Tor}^F(\ell,\ell)$ (deg $\sigma^F = 1$) and follow this by the projection

$$\mathrm{ITor}^F(\ell,\ell) \to \mathrm{ITor}^F(\ell,\ell)/I^{(2)}\mathrm{Tor}^F(\ell,\ell)\ .$$

Since $\mathrm{Hom}_{F_o}(\ ,\ell) = \mathrm{Hom}_\ell(\ ,\ell)$ for vector spaces, dualization of this composition yields a degree zero map $\tilde{\tau}\colon \pi^*(F) \to \mathrm{Hom}_{F_o}(sIH(F),\ell)$. It remains to show $\mathrm{Ker}\ \tilde{\tau} \supset$ $\supset [\pi^*(F),\pi^*(F)]$, which can also be written as $[\pi^*(F),\pi^*(F)]\cdot \mathrm{Im}\ \sigma^F = 0$. Since commutators and images of the quadratic operator are decomposable in the universal envelope $(\mathrm{Tor}^F(\ell,\ell))^\vee$ of $\pi^*(F)$, it suffices to show $(\mathrm{ITor}^F(\ell,\ell)^\vee)^2\cdot \mathrm{Im}\ \sigma^F = 0$. This is equivalent to the inclusion of $\mathrm{Im}\ \sigma^F$ in the ℓ-dual of the indecomposables of the graded algebra $\mathrm{Tor}^F(\ell,\ell)^\vee$, which is canonically identified with the primitives of $\mathrm{Tor}^F(\ell,\ell)$. We shall show $\sigma^F[z]$ is primitive for any $z \in IH(F)$. As at the beginning of the proof of (2.1), one can for this purpose replace F by a supplemented DG ℓ-algebra F', with $H(F') = H(F) \otimes_{F_o} \hat{F}_o$, and $\mathrm{Tor}^{F'}(\ell,\ell) = \mathrm{Tor}^F(\ell,\ell)$, both equalities being provided by series of DG algebra maps. However, with the identification $\mathrm{Tor}^{F'}(\ell,\ell) = H\overline{B}(F')$, and for $z' \in IZ_*(F')$, $\sigma^{F'}[z'] = [x]$, where x is the element z' in $\overline{B}_{1,*}(F) \simeq IF$. The definition of the diagonal $\overline{\Delta}$ of $\overline{B}(F')$ shows x is primitive, whence the claim.

For the final statement, it suffices to note that since $\pi^*(F) \hookrightarrow (\mathrm{Tor}^F(\ell,\ell))^\vee$ is the inclusion of the Lie algebra into its universal envelope, it induces a canonical isomorphism of graded vector spaces

$$\pi^*(F)/[\pi^*(F),\pi^*(F)] \simeq (\mathrm{ITor}^F(\ell,\ell)^\vee)/(\mathrm{ITor}^F(\ell,\ell)^\vee)^2\ .$$

(3.4). *Theorem. For a DG Γ-algebra F, which satisfies the conditions of (3.2), the following are equialent:*

(1) F *is a Golod algebra;*

(2) $IH(F)$ *is a vector space (through the augmentation* $\varepsilon\colon F_o \to \ell$), *and there is an isomorphism of graded Lie algebras;*

$$\phi\colon \pi^*(F) \xrightarrow{\simeq} \mathbb{L}(sIH(F))^\vee\ (= L)$$

where $\mathbb{L}$ *denotes the free Lie algebra functor; furthermore* ϕ *induces the vertical map in the following commutative diagram:*

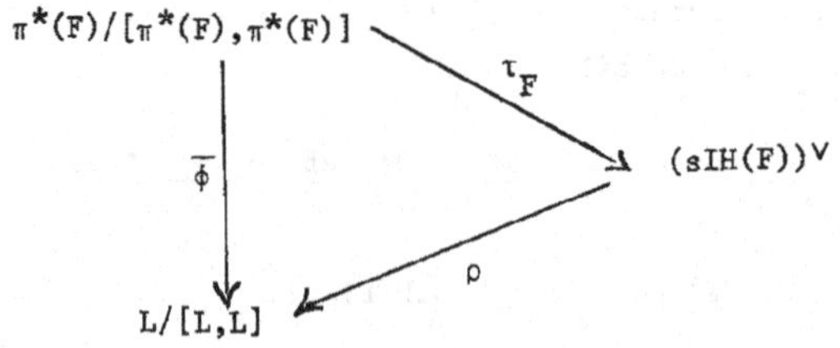

in which ρ *is the canonical isomorphism;*

(3) τ_F *is surjective;*

(4) $\pi^*(F)$ *is a free Lie algebra.*

Proof. (1) $\Rightarrow$ (2). It is easily seen that $\pi^*(F)$ does not change when one passes from F to its completion $\hat{F}$ in the $\mathrm{Ker}\,(F_o \to \ell)$-adic topology. Furthermore, the homology equivalence (2.1.1), which links F to a supplemented ℓ-algebra F', can be achieved by using only homomorphisms of DG Γ-algebras (for details cf. [Av_4]). Hence, by the remarks at the beginning of this section, F can be replaced by a supplemented DG Γ-algebra F' over ℓ. Since the trivial extension $\ell \ltimes W$ (cf. (2.3), Note)) is in a natural way a DG Γ-algebra (with $\gamma^i(x) = 0$ for every x of even positive degree and every $i \geq 2$), one can further replace F' by $k \ltimes W$. But now all the claims of (2) are obvious.

It is clear that (2) implies both (3) and (4). Noting that (3) is equivalent to $\mathrm{Ker}\,\sigma^F = 0$, (3) implies (1) by Theorem (2.3).

Now we assume (4) and shall prove that $H(F)$ is a Golod algebra. Consider first the homomorphism $f\colon F \to \overline{F} = H_o(F)$ of DG Γ-algebras, and the induced homomorphism $f^*\colon \pi^*(\overline{F}) \to \pi^*(F)$ of graded Lie algebras over L. Since $\pi^1(F)$ is naturally isomorphic to $(\underline{\overline{n}}/\underline{\overline{n}}^2)^\vee$, where $\underline{\overline{n}} = \mathrm{Ker}(\overline{F} \to \ell)$ (cf. [Av_4]), f^1 is an isomorphism. Subalgebras of free Lie algebras being free [Lem, Proposition A 1.10], it follows that $\pi^1(F)$ generates a free Lie subalgebra in $\pi^*(F)$, hence $\pi^1(\overline{F})$ generates a free subalgebra L of $\pi^*(\overline{F})$.

The ring $\overline{F}$ being equicharacteristic, we can after localization and completion assume it is the homomorphic image of the formal power series ring $\ell[[X_1,\ldots,X_n]]$ by an ideal $\underline{a}$ minimally generated by $a_1,\ldots,a_r$ with $a_i \in \underline{n}^2$, $\underline{n} = (X_1,\ldots,X_n)$. Hence there exist $\overline{a}^h_{ij} \in \ell$, such that

$$a_h - \sum_{i\le i\le j\le n} \overline{a}^h_{ij} X_i X_j \in \underline{n}^3, \quad h = 1,2,\ldots,r .$$

We now recall an equality established by Sjödin [Sj, p. 8]:

$$\dim_\ell \mathrm{Ext}^2_{\overline{F}}(\ell,\ell) = \dim_\ell (\mathrm{Ext}^1_{\overline{F}}(\ell,\ell))^2 + r - \mathrm{rank}\,(\overline{a}^h_{ij})$$

where $\mathrm{Ext}_{\overline{F}}(\ell,\ell)$ is considered with the Yoneda product, which (up to signs) coincides with the product on $(\mathrm{Tor}^{\overline{F}}(\ell,\ell))^\vee$ induced by the diagonal of $\mathrm{Tor}^{\overline{F}}(\ell,\ell)$. Since $(\mathrm{Tor}^{\overline{F}}(\ell,\ell))^\vee$ is the universal envelope of $\pi^*(\overline{F})$, the freeness of L yields

$$\dim_\ell (\mathrm{Ext}^1_{\overline{F}}(\ell,\ell))^2 = n^2 .$$

On the other hand, it is well known that

$$\dim_\ell (\mathrm{Tor}^{\overline{F}}_2(\ell,\ell)) = \binom{n}{2} + r .$$

The three equalities produce the expression

$$\mathrm{rank}\,(\overline{a}^h_{ij}) = \binom{n+1}{2} .$$

But this means $\dim_\ell (\underline{a} + \underline{n}^2)/\underline{n}^3 = \binom{n+1}{2} = \dim_\ell \underline{n}^2/\underline{n}^3$, whence $\underline{a} + \underline{n}^3 = \underline{n}^2$, and finally $\underline{a} = \underline{n}^2$ by Nakayama. As remarked by the referee, a shorter proof of this equality can be obtained by using some results of Löfwall [Lö].)

Summing up, we have proved $\bar{n}^2 = 0$ for the maximal ideal $\bar{n}$ of $H_o(F)$. Since, for any i, $H_i(F)$ is a finitely-generated $H_o(F)$-module, it has finite length over the (local) ring $H_o(F)$. In particular, with $\hat{F}$ as in the proof of (1) $\Rightarrow$ (2), $H(\hat{F}) = H(F) \otimes_{F_o} \hat{F}_o = H(F)$, hence by the arguments given there we can replace F by a supplemented DG Γ-algebra over ℓ. In particular, we can consider the (non-commutative) DG ℓ-algebra $(\bar{B}(F))^\vee$, whose product is the dual of the diagonal in the bar-construction of F over ℓ. As a vector space, $(\bar{B}(F))^\vee$ is the direct product $\Pi_{i\geq 0}(IF^{\otimes i})^\vee$, and clearly $\Pi_{i\geq p}(IF^{\otimes i})^\vee$ is a filtration by DG ideals. In the corresponding spectral sequence one has

$$E_0^{p,*} = (IF^{\otimes p})^\vee ;$$
$$E_1^{p,*} = (IH(F)^{\otimes p})^\vee .$$

(Note that for the second equality the already established finite length of $H_i(F)$ over $H_o(F)$, which implies $\dim_\ell H_i(F) < \infty$ for all i, is of crucial importance.) We have now a first quadrant spectral sequence $\{E_r^{**}\}$ of bigraded DG algebras, with differential $d_r\colon E^{p,q} \to E_r^{p+r,q-r+1}$, and first term defined by the isomorphism of algebras

$$E_1^{**} = T(E_1^{1,*})$$

with $E_1^{1,*} = (IH(F))^\vee$. For standard reasons, this sequence converges in the strong(est) sense to $H(\bar{B}(F))^\vee = (\mathrm{Tor}^F(\ell,\ell))^\vee$.

Next we set to show $E_1^{1,*} = E_\infty^{1,*}$. Assuming inequality, choose the smallest q for which $E_1^{1,q} \neq E^{1,q}$. The pattern of the action of d_r guarantees that $E_r^{1,q} = E_\infty^{1,q}$ for $r > q+1$. Hence there is a largest integer r, for which $d_r E_r^{1,q} \neq 0$. Denote by C^{**} the subspace of $E_r^{1,*} (\subset E_1^{1,*})$, defined by the conditions

$$C^{1,s} = \begin{cases} E_r^{1,s} & \text{for } s < q ; \\ \operatorname{Ker} d_r^{1,q} & \text{for } s = q ; \\ 0 & \text{for } s > q . \end{cases}$$

By the choice of C, its elements are permanent cycles in the spectral sequence, i.e. they are in the image of the map (= edge homomorphism)

$$\nu\colon (IH\bar{B}(F)_*)^\vee \twoheadrightarrow E_\infty^{1,*} \hookrightarrow E_1^{1,*} .$$

Since composed with the isomorphism $E_1^{1,*} \simeq (IH(F)_*)^\vee$, ν gives up to a degree shift, the dual of the suspension σ^F, Lemma (3.3) shows that a basis of C can be lifted to elements of $\pi^*(F)$, which form a basis for the generators of degree $\leq q+1$ of this Lie algebra. By our assumption and [Lem, Proposition A 1.10], they generate a free Lie subalgebra L' of $\pi^*(F)$, hence $(\mathrm{Tor}^F(\ell,\ell))^\vee = U\pi^*(F) \supset UL'$, which is the tensor algebra T on the graded vector space associated to C. Moreover, the inclusion is an equality in degrees $\leq q+1$, while in degree $q+2$ it yields

$$\dim H^{q+2}(\overline{B}(F)) = \dim E_\infty^{1,q+1} + \dim T^{q+2}.$$

Let now D denote the bigraded subalgebra of E_r^{**}, generated by C. We claim that $dE_r^{1,q} \subset D$. This is clear when $r \geq 2$, since in this case $dE^{1,q} \subset E^{p',q'}$ with $q' < q$, and by the definition of D one has $E^{**} = D^{**}$ in these dimensions. If $r = 1$ and $a \in E^{1,q}$, write

$$da = \Sigma\, c_i' \otimes c_i'' + b$$

with $b \in E_1^{\geq 2,*}$, $c_i'' \in E_*^{1,q}$, $c_i' \in E_*^{1,0} = (\overline{n})^\vee$, and furthermore choosing the c_i' linearly independent. Since $\overline{n}^2 = 0$, the formula for the differential of $\overline{B}(F)$ shows $dc_i' = 0$, hence

$$\Sigma\, c_i' \otimes dc_i'' = d(da+b) = db \in E_1^{\geq 3,*},$$

Comparing filtration degrees, one sees that $\Sigma\, c_i' \otimes dc_i'' = 0$, hence $dc_i'' = 0$, i.e. $c_i'' \in \operatorname{Ker} d_1^{1,q} = C^{1,q}$.

With T as above, consider the map $\alpha\colon T \to E_{r+1}^{*,*}$, defined by the inclusion $C \subset R_{r+1}$. The choice of r shows that $E_r^{*,*} = E_\infty^{*,*}$ in total degree $\leq q+1$, and in the same degrees α is an isomorphism. On the other hand, it guarantees the existence of some $a \in E_r^{1,q}$ with $d_r(a) \neq 0$, which produces a relation of degree $q+2$ among the generators of degree $\leq q+1$ of $E_{r+1}^{*,*}$, hence in degree $q+2$ $\operatorname{Ker}(T \to E_{r+1}) \neq 0$. This means that

$$\dim H^{q+2}\overline{B}(F) = \dim E_\infty^{1,q+1} + \sum_{\substack{s+t=q+2\\ s\geq 2}} \dim E_{r+1}^{s,t} < \dim E_\infty^{1,q+1} + \dim T^{q+2}$$

which contradicts the equality established earlier.

Having proved that $E_1^{1,*} = E_\infty^{1,*}$, it follows immediately that $E_1 = E_\infty$ since $E_1^{1,*}$ forms a system of generators of the ℓ-algebra E^1. Now we have, in view of the ℓ-finiteness of $H_i(F)$:

$$\sum_i \dim(H^i\overline{B}(F))t^i = \sum_i \Big(\sum_{p+q=i} \dim E_\infty^{p,q}\Big)t^i = \Big(1 - \sum_i \dim IH_i(F)t^{i+1}\Big)^{-1}$$

which is the definition (2.2) of a Golod algebra.

(3.5). Remark. Theorem (3.4) looks (and is) similar to results in [Av_2; in particular: Theorem (1.4) and Corollary (1.6)]. The crucial difference, which makes the results (and most of the arguments) of the previous paper unusable in the present context, is the basic assumption [Av_2, (1.3)] that F is a subalgebra of the minimal DG Γ-algebra resolution of F_o. Indeed, it is precisely this condition on $F = S \otimes_R X$ (see below) that we want to deduce from the properties of $\operatorname{Tor}^R(S,K)$.

4. LOCAL HOMOMORPHISMS

In this section $f: (R,\underline{m},k) \to (S,\underline{n},\ell)$ denotes a homomorphism of local (noetherian and commutative) rings, such that $f(\underline{m}) \subset \underline{n}$. Furthermore, X denotes a free DG Γ-algebra over R, such that X_i has finite R-rank for any i, and $H_*(X) \xrightarrow{\sim} k$ by a natural augmentation $\varepsilon (= \varepsilon_X)$, commuting with that of $R (\to R/\underline{m} = k)$.

The DG Γ-algebra $F = S \otimes_R X$, augmented to ℓ by $(S \to S/\underline{n} = \ell) \otimes \varepsilon$, is called <u>the (homotopy) fibre of</u> f, and the canonical inclusion $S \to S \otimes_R X$ is denoted by g.

The fundamental importance of the fibre in the study of the homology of the map f is given by the next result proved in $[Av_4]$:

(4.1) There is a natural exact sequence of graded Lie algebras over ℓ:

$$(4.1.1) \quad \ell \otimes_k \pi^*(R) \xleftarrow{f^*} \pi^*(S) \xleftarrow{g^*} \pi^*(F) \xleftarrow{\delta} \underline{C}(\mathrm{Coker}\, f^*) \longleftarrow 0$$

where for any piecewise finite-dimensional graded vector space W, $\underline{C}^*(W) = \pi^*(\Gamma_\ell W^\vee)$, with Γ_ℓ denoting the free Γ-algebra of $W^\vee$ over ℓ.

Moreover, $\mathrm{Im}\,\delta$ is central in $\pi^*(F)$.

If furthermore the flat dimension $\mathrm{fd}_R S$ is finite, then Coker f^* is concentrated in odd degrees, its dimension is $\leq \mathrm{fd}_R S + \mathrm{edim}\,(S/\underline{m}S)$, and $\underline{C}^i(\mathrm{Coker}\, f) \simeq \mathrm{Coker}(f^{i-1})$.

(Here and below we use the notation $\mathrm{edim}\, R = \dim_k(\underline{m}/\underline{m}^2))$.

(4.2). <u>Lemma</u>. <u>For any local homomorphism, there is an inequality of power "series":</u>

$$P_S(t) \preccurlyeq P_R(t)\cdot G_S^R(t)$$

<u>where</u> $G_S^R(t) = (1 - \Sigma\, \mathrm{length}_S(\mathrm{ITor}_i^R(S,k))t^{i+1})^{-1}$ <u>is the Golod "series" of</u> $H(F)$.

<u>Proof</u>. Setting $e_i(\ ) = \dim \pi^i(\ )$, $a_i = \dim\, (\mathrm{Coker}\, f^i)$, $b_i = \dim\, C^i(\mathrm{Coker}\, f^*)$, the exact sequence (4.1) yields

$$(4.2.1) \quad e_i(S) + a_i + b_i = e_i(R) + e_i(F).$$

Recall that for the universal envelope of a graded Lie algebra L, the PBW theorem gives the equality of formal power series

$$\Sigma \dim\, (UL)^i t^i = \frac{(1+t)^{\dim L_1}(1+t^3)^{\dim L_3}\cdots}{(1-t^2)^{\dim L_2}(1-t^4)^{\dim L_4}\cdots}.$$

Write $A(t)$ for the Hilbert series of the free Γ-algebra on Coker f^*, and $B(t)$ for that of the universal envelope of $\underline{C}(\mathrm{Coker}\, f^*)$. One now has:

$$\begin{aligned}(4.2.2). \quad P_S(t) &\preccurlyeq P_S(t)A(t)B(t) && \text{because } A(t) \succcurlyeq 0,\ B(t) \succcurlyeq 0;\\ &= P_R(t)P_F(t) && \text{by (4.2.1)};\\ &\preccurlyeq P_R(t)G_S^R(t) && \text{by (2.1)}.\end{aligned}$$

It should be emphasized that any two constructions of a fibre for f produce quasi-isomorphic DG Γ-algebras, so that in particular the homology structure of F is uniquely defined (note also that $H(F) = \mathrm{Tor}^R(S,k))$. In view of this, the next property is intrinsic to f , i.e. it does not depend on the choice of F .

(4.3). Definition. A local homomorphism f is called a Golod homomorphism if F is a Golod algebra (equivalently: if $\mathrm{Tor}^R(S,k)$ has trivial Massey products).

(A comparison with earlier definitions is relegated to (4.7) below.)

(4.4). Exceptional homomorphisms. Suppose $\mathrm{length}_S(\mathrm{ITor}^R(S,\ell)) = 1$. This means there exists an i , such that

$(4.4)_i$. $\mathrm{ITor}^R(S,k)$ is concentrated in degree i and isomorphic to ℓ .

In this case the Massey products of $\mathrm{Tor}^R(S,k)$ are (trivially) trivial hence such a map is necessarily Golod. However, it presents some deviations from the usual pattern, hence we have to give a closer look at this simple situation.

First of all note, that since the vanishing of Tor_1 implies, by the local criterion of flatness, the triviality of all higher Tor-s, in (4.4.1) $i = 0$, or $i = 1$.

When $i = 0$, S is R-flat, hence $F = S \otimes_R X \to S/\underline{m}S$ is a quasi-isomorphism, so that $\pi^*(F) \simeq \pi^*(S/\underline{m}S)$. Moreover, $S/\underline{m}S \simeq \ell \ltimes \ell$, hence $\pi^*(S/\underline{m}S)$ is the free Lie algebra on a single generator of degree 1 (these are very special cases of (3.4)). Accordingly, $\pi^n(F) = 0$ for $n \neq 1, 2$, with $\pi^1 \simeq \ell \simeq \pi^2$. According to (4.1) the following conditions become equivalent:

(a) $\mathrm{Ker}\ g^* \neq 0$;

(b) $\mathrm{Coker}\ f^* \neq 0$;

(c) $\mathrm{Coker}\ f^1 \neq 0$.

Dualizing over ℓ , one can express the last inequality in the easily checked form:

$$(4.4.2) \quad 0 \neq \mathrm{Ker}\ (f_1 : \ell \otimes_k \underline{m}/\underline{m}^2 \to \underline{n}/\underline{n}^2) .$$

Moreover, note that in this case $1 \leq \dim_\ell \mathrm{Ker}\ f_1 = \dim_\ell \mathrm{Coker}\ f^1 = \dim_\ell \underline{C}^2(\mathrm{Coker}\ f^*) \leq$ $\leq \dim_\ell \pi^2(F) = 1$, hence equalities hold throughout. Since by construction $\underline{C}^1(\mathrm{coker}\ f^*)$ $= 0$, the homotopy sequence from (4.1) reduces to the following statement:

(4.4.3). Assume both conditions (4.4) and (4.4.2) hold, and let A and B denote (abelian) one-dimensional Lie algebras, concentrated in degree 1 . Then there is an exact sequence of Lie algebras

$$0 \leftarrow A \leftarrow \ell \otimes_k \pi^*(R) \xleftarrow{f^*} \pi^*(S) \leftarrow B \leftarrow 0 .$$

In particular, $P_S(t) = P_R(t)$.

We now treat the case $i = 1$. According to (3.4), in this case $\pi^*(F)$ is free, generated by a single element of degree 2 , hence $\dim_\ell \pi^n(F) = 0$ for $n \neq 2$, $\pi^2(F) \simeq \ell$. As before, (4.1) shows $\mathrm{Ker}\ g^* \neq 0 \iff \mathrm{Coker}\ f^* \neq 0 \iff \mathrm{Coker}\ f^1 \neq 0$, hence in this context one obtains:

(4.4.4). Assume both conditions $(4.4)_1$ and (4.4.2) hold, and let A denote a one-dimensional abelian Lie algebra, concentrated in degree 1. Then there is an exact sequence of Lie algebras

$$0 \leftarrow A \leftarrow \ell \otimes_k \pi^*(R) \xleftarrow{f^*} \pi^*(S) \leftarrow 0 .$$

In particular, $P_S(t) = P_R(t)(1+t)^{-1}$.

We shall call a homomorphism exceptional, if $\text{length}_S \operatorname{Tor}^R_*(S,k) = 2$ and (4.4.2) holds. To vindicate the terminology, we note the following statements, whose easy proofs are left to the reader.

(4.5). Examples. (a) The projection $R \to R/(b)$, with $b \in \underline{m}$ a nonzero divisor not in $\underline{m}^2$, is an exceptional homomorphism with $(\operatorname{ITor}^R(S,k))_i = 0$ for $i \neq 1$.

(b) Let $X^2 + aX + b$ be a polynomial in $R[X]$ with $a \in \underline{m}$, $b \in \underline{m}$, $b \notin \underline{m}^2$. Then the composition

$$R \hookrightarrow R[X] \to R[X]/(X^2 + aX + b)$$

is an exceptional homomorphism with $(\operatorname{ITor}^R(S,k))_i = 0$ for $i \neq 0$.

(c) If S is a finitely generated R-module via f, and the residue field extension $k \subset \ell$ is trivial, then every exceptional homomorphism is obtained either as in (a) or as in (b) above.

A homomorphism which is not exceptional is said to be standard.

We have now come to the main result:

(4.6). Theorem. The following are equivalent, for a local homomorphism f with fibre F:

(1) f is Golod and standard;

(2) $P_S(t) = P_R(t) G^R_S(t)$;

(3) (i) $S/\underline{m}S \simeq \ell \ltimes N$, where N is an ℓ-vector space of dimension $\operatorname{edim} S - \operatorname{edim} R$;

(ii) the natural S-module structure on $V = \operatorname{ITor}^R(S,k)$ satisfies $\underline{n}V = 0$, hence V is in a canonical way a piecewise finitely dimensional graded vector space over ℓ;

(iii) $\pi^*(F)$ is the free Lie algebra on $V^\vee$, and the sequence

$$0 \leftarrow \ell \otimes_k \pi^*(R) \xleftarrow{f^*} \pi^*(S) \xleftarrow{g^*} \mathbf{L}(V) \leftarrow 0$$

is exact.

Proof. (1) $\Rightarrow$ (3). The fibre F being a Golod algebra, assertions (i) and (ii) follow from Theorem (3.4). When $\text{length}_S \operatorname{ITor}^R(S,k) \neq 1$, the algebra $\pi^*(F)$ is either trivial or free non-abelian. In both cases, it has a trivial center, hence $\operatorname{Ker} g^* = 0$ by (4.1) and this implies $\operatorname{Coker} f^* = 0$ by the construction of $\underline{C}^*$. It remains to prove g^* is injective assuming $\text{length}_S \operatorname{ITor}^R(S,k) = 1$. However, the discussion in (4.4) showed in this case $\operatorname{Ker} g^* \neq 0$ if and only if f is special, hence we are through.

(3) ⇒ (2) is trivial.

(2) ⇒ (1). The power series equality (2) shows $G_S^R(t)$ has (non-negative) integer coefficients. Using this and comparing (2) with (4.2.2), one gets the equalities:

$$P_S(t) = P_S(t)A(t)B(t) = P_R(t)P_F(t) = P_R(t)G_S^R(t)$$

which imply $P_F(t) = G_S^R(t)$ and $A(t)B(t) = 1$. By the definition of $G_S^R(t)$ in (4.1), and by Theorem (2.3), the first equality means F is a Golod algebra. On the other hand, A(t) and B(t) having non-negative coefficients, the second equality is possible only with $A(t) = 1 = B(t)$. In view of the definition of these series, this means f^* is surjective, hence (4.4.3) and (4.4.4) show f cannot be exceptional.

(4.7). Remarks. Golod homomorphisms were used implicitly by Golod [Go] for regular R, and by Gulliksen [Gu] in a more general context. Their explicit introduction, for surjective maps, is due to Levin, who started a systematic study in [Lev_1]. In [Av_2] the notion was extended to maps which induce the identity on residue fields. In particular, it was proved there that either set of conditions characterizes such homomorphisms:

$P_S(t) = P_R(t)G_S^R(t)$ and $\underline{n}\,\mathrm{ITor}^R(S,k) = 0$;

$\mathrm{Tor}^R(S,k)$ has trivial Massey products and f^* is surjective.

That for a surjective f one recovers the original notion was shown in [Lev_2], where several alternate characterizations can also be found.

It follows from Theorem (4.6) above, that the *a priori* much broader notion of Golod homomorphism adopted in (4.3) coincides, once exceptional maps are dropped, with the earlier concepts, whenever these are defined.

The structural result for Golod maps, contained in (4.6.3) is known in the residually trivial case from [Av_1] and [Lö].

All the papers just quoted provide numerous examples.

5. FINAL REMARKS

(5.1). Let $f: R \to S$ be a homomorphism of DG Γ-algebras, augmented to k and to ℓ respectively. In this case one can still define the fibre $F = S \otimes_R X$, and (4.1) holds with the change from flat dimension to formal dimension, i.e. setting $fd_R S = \{\max d \,|\, H_d(F) \neq 0\}$: (details are given in [Av_4]; when R and S are rings, the formal and flat dimensions coincide).

The definition given in (4.3) for a Golod homomorphism continues to make sense in the DG Γ-algebra setup, so we use it to define Golod maps there.

What does change somewhat is the discussion in (4.4) of the exceptional homomorphisms: with the local criterion of flatness unavailable, condition (4.4) can be satisfied for every $i \geq 0$. However, it is still easily shown that, for any i, $\pi^*(F)$ is the free Lie algebra generated by a single element of degree $(i+1)$, and the analysis of the exceptional cases, although slightly more involved, is not hard. We shall not pursue it further at this point.

(5.2). Example. For a DG Γ-algebra R, with R_o local, denote by $\tilde{R}$ the one obtained by adjoining a (minimal) set of degree 1 variables, which kill the generators of $m = \mathrm{Ker}(H_o(R) \to \ell)$. (Thus, when $R = R_o$ is a local ring, $\tilde{R}$ is "its" Koszul complex). In any case, $\tilde{R}$ is uniquely defined up to isomorphism, and it is shown in [Av_3], [Av_4] that the inclusion $R \hookrightarrow \tilde{R}$ induces the exact sequence

$$0 \leftarrow \pi^1(R) \leftarrow \pi^*(R) \leftarrow \pi^*(\tilde{R}) \leftarrow 0 ,$$

hence $\pi^*(\tilde{R}) = \pi^{\geq 2}(R)$, in a canonical way.

Suppose now $f\colon R \to S$ is a map, for which f^1 is bijective (e.g. $S = R/J$ for some ideal with $J_o \subset \underline{m}^2$). Then $S \otimes_R \tilde{R}$ is a good choice for $\tilde{S}$, and playing with the naturality of the sequence (4.1), one sees that $\pi^*(F) \cong \pi^*(\tilde{F})$, with $\tilde{f}\colon \tilde{R} \to \tilde{S}$ being the canonical map and $\tilde{F}$ denoting its fibre. Hence:

(5.2.1). f is Golod if and only if $\tilde{f}$ is.

This observation can be useful in some computations. For example, consider the spectral sequence with

$$(5.2.2).\quad E^2_{p,q} = \mathrm{Tor}^{H(\tilde{S})}_{p,q}(\ell,\ell) \Rightarrow \mathrm{Tor}^{\tilde{S}}_{p+q}(\ell,\ell).$$

It yields the formal power series inequality $P_{\tilde{S}}(t) \preccurlyeq P_{H(\tilde{S})}(t)$, which becomes an equality precisely when the sequence degenerates. Similarly, the spectral sequence

$$(5.2.3).\quad E^2_{p,q} = \mathrm{Tor}^{H(\tilde{R})}_{p,q}(H(\tilde{S}),\ell) \Rightarrow \mathrm{Tor}^{\tilde{R}}_{p+q}(\tilde{S},\ell) \quad \text{produces}$$

the inequality $G^{H(\tilde{R})}_{H(\tilde{S})} \succcurlyeq G^{\tilde{R}}_{\tilde{S}}$. Combining this with (4.2) one sees that:

$$\begin{array}{ccc} P_{H(\tilde{R})}(t)G^{H(\tilde{R})}_{H(\tilde{S})}(t) & \succcurlyeq & P_{\tilde{R}}(t)G^{\tilde{R}}_{\tilde{S}}(t) \\ \curlyvee & & \curlyvee \\ P_{H(\tilde{S})}(t) & \succcurlyeq & P_{\tilde{S}}(t) \end{array}$$

Since f is not exceptional, these inequalities can be used in conjunction with (4.6) and (5.2.1) to obtain the following sufficient condition for Golod homomorphisms:

(5.2.4). If $H(\tilde{f})$ is a Golod homomorphism, and the spectral sequence (5.2.2) degenerates, then f is Golod.

The preceding statement has been proved in [Ba, Lemma 5] under the additional assumption that the spectral sequence (5.2.2) for $\tilde{R}$ also degenerates. However, using the inequalities above, this is seen to be a consequence of the hypotheses in (5.2.4), which also imply the degeneracy of the sequence in (5.2.3). I am grateful to the referee for drawing my attention to Backelin's result.

(5.3). The inequality of Lemma (4.2), although giving a natural upper bound on the Poincaré series of S, is of little practical interest when $S/\underline{m}S$ is not an artinian ring. However, passing as above from F to $\tilde{F}$, one gets the inequality

$$P_S(t) \leqslant P_R(t) \cdot \frac{(1+t)^{\operatorname{edim} H_o(F)}}{1 - \sum_{i=1}^{\infty} \dim H_i(\tilde{F}) t^{i+1}}$$

which has the advantage of involving only series with integer coefficients.

(5.4). The preceding formula also brings us back to the origins of the entire circle of ideas considered in this paper. Indeed, when R is a regular local ring of the same embedding dimension as S, and f is surjective, $\tilde{S} = F = \tilde{F}$ is the Koszul complex of S, and the formula reduces to the well-known Serre inequality:

$$P_S(t) \leqslant \frac{(1+t)^{\operatorname{edim} S}}{1 - \sum_{i \geq 1}^{\infty} \dim H_i(\tilde{S}) t^{i+1}} .$$

The upper bound is reached precisely when S is a Golod ring [Go], and it is clear that this condition is equivalent to $\tilde{S}$ being a Golod algebra in the sense of definition (2.2).

(5.5). The results of this paper completely characterize those homomorphisms of local rings whose fibre has a free homotopy Lie algebra.

At the other end of the spectrum, one has the abelian Lie algebras, and it is interesting to ask for them the corresponding question. The answer here turns out to be much more subtle.

On the one hand, such homomorphisms exist. In fact, it follows from (4.1) that any f with f^* injective has this property. Examples of such maps can be found in [Lev_3], where Levin studies them under the name "large hombmorphisms".

On the other hand, under restriction that $fd_R S < \infty$, $\pi^*(F)$ can be abelian (or, for that matter, nilpotent), only if f is a complete intersection map: this follows from joint work with Halperin, which is being prepared for publication.

REFERENCES

[Av_1] L.L. Avramov, Small homomorphisms of local rings. J. Algebra 50 (1978), 400-453

[Av_2] L.L. Avramov, Free Lie subalgebras of the cohomology of local rings. Trans. Amer. Math. Soc. 270 (1982), 589-608.

[Av_3] L.L. Avramov, Local algebra and rational homotopy, in Proceedings of the conference "Méthodes d'algèbre homotopique en topologie", In Homotopie Algébrique et Algèbre Locale, Astérisque 113/114 (1984), 15-43.

[Av_4] L.L. Avramov, Homotopy Lie algebras for commutative rings and DG algebras. To appear.

[Ba] J. Backelin, Golod attached rings with few relations (III). Reports, Dept. of Math., Univ. of Stockholm, No. 13, 1983.

[Ca] H. Cartan, Algèbres d'Eilenberg-MacLane (Séminaire ENS, 1954-1955, Exposés 2 à 11), in Oeuvres, Volume III 1309-1394, Springer-Verlag, Berlin, 1979.

[Go] E.S. Golod, On the homologies of certain local rings. Dokl. Akad. Nauk SSSR 144 (1962), 479-482 (in Russian); English translation: Soviet Math. Dokl. 3 (1962), 745-748.

[Gu] T.H. Gulliksen, Massey operations and the homology of certain local rings. J. Algebra 22 (1972), 223-232.

[GM] V.K.A.M. Gugenheim and J.P. May, On the theory and applications of differential torsion products. Memoirs Amer. Math. Soc. 142 (1974).

[Lem] J.-M. Lemaire, Algèbres connexes et homologie des éspaces de lacets. Lecture Notes in Math. 422, Springer-Verlag, Berlin, 1974.

[Lev_1] G. Levin, Local rings and Golod homomorphisms. J. Algebra 37 (1975), 266-289.

[Lev_2] G. Levin, Lectures on Golod homomorphisms. Reports, Dept. Math., Univ. of Stockholm, No. 15, 1976.

[Lev_3] G. Levin, Large homomorphisms of local rings. Math. Scand. 46 (1980), 209-215.

[Lö] C. Löfwall, On the subalgebra generated by the one-dimensional elements of the Yoneda Ext-algebra. These proceedings.

[Ma] J.P. May, Matric Massey products. J. Algebra 12 (1969), 533-568.

[Mo] J.C. Moore, Algèbre homologique et homologie des éspaces classifiants. Séminaire H. Cartan, ENS 1959-1960, Exposé 7, Secrétariat Math., Paris, 1961.

[Ro] J.-E. Roos, Homology of loop spaces and of local rings. 18th Scandinavian Congress of Mathematics, Proceedings 1980, 441-468, Birkhäuser, Basel, 1981.

[Ta] D. Tanré, Homotopie rationnelle: Modèles de Chen, Quillen, Sullivan. Lecture Notes in Math. 1025, Springer-Verlag, Berlin, 1983.

L.L. Avramov
Institute for Algebraic Meditation
Department of Mathematics
University of Toronto
Toronto, Canada M5S 1A1

and

Institute of Mathematics
University of Sofia
ul. "Akad. G. Bončev" Bl. 8
1113 Sofia, Bulgaria

ON THE RATES OF GROWTH OF THE HOMOLOGIES OF VERONESE SUBRINGS

JÖRGEN BACKELIN

Let R be an (associative, non-negatively graded) connected algebra, generated by R_1, over a field k. In [5] Ralf Fröberg and I proved that if the homogeneous minimal relations of R appear only in degrees $\leq r < \infty$, then similar limitations hold for the Veronese subrings $R^{(d)}$ of R ($d = 2,3,\ldots$), with r decreasing as d increases till eventually all minimal relations are quadratic. (This generalizes a result of D. Mumford: [13, thm 1].) Likewise (generalizing a result of S. Barcanescu and N. Manolache: [6, thm 2.1]) we proved that if R is a Koszul algebra, then so are all $R^{(d)}$. (See [5, prop 3 and thm 4].) In this note both these results are generalized, by proving that if $\mathrm{Tor}_p^R(k,k)$ is concentrated in degrees $\leqq c(p-1)+1$ for $p = 2,\ldots,n$ (as in figure 1 below), then a similar assertion holds for $R^{(d)}$, with a new rate of growth $\lceil c/d \rceil$ replacing c. Eventually we get the rate 1. (The precise general result is given in theorem 3 below.)

The "rate of growth"-result is particulary interesting in the case where *all* p are concerned (theorem 1). In case R is commutative and finitely generated, we also get a "limit algebra" type of result, namely that $R^{(d)}$ is a Koszul algebra for $d \gg 0$ (theorem 2).

Figure 1.

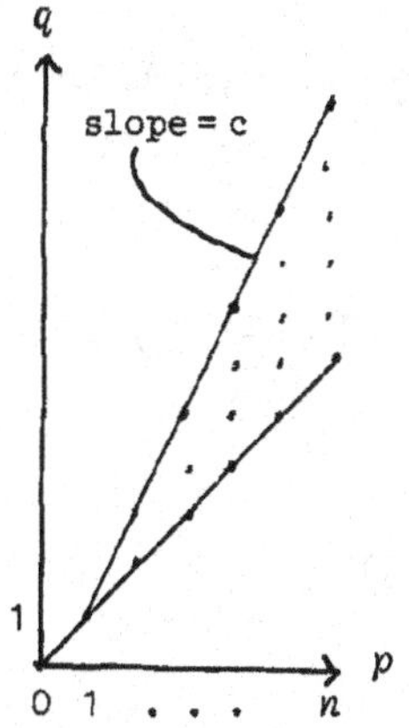

$\mathrm{Tor}_{p,q}^R(k,k) \neq 0$ only in dotted area

In section 1 the main results are given and are proved to follow from theorem 3. The main ideas of the proof of theorem 3 are outlined in section 2, and the details fill the succeeding three sections. In the last section possible improvements of the results are discussed.

1. DEFINITIONS AND MAIN RESULTS.

In this article, k, V, I, P and R will denote a fixed field, a fixed vector space over k, the augmentation ideal in the (naturally graded) tensor algebra $T(V)$ ($= k \oplus V \oplus V \otimes V \oplus \ldots$), a fixed homogeneous two-sided $T(V)$-ideal contained in I^2, and the graded residue class ring $T(V)/P$, respectively (*).

d will denote an integer such that $d \geq 2$.

The *Veronese subring* $R^{(d)} = \coprod_{j\geq 0} R_j^{(d)}$ is defined by

$R_j^{(d)} = R_{jd}$ for $j = 0,1,2,\ldots$.

k is a graded R-module (by the augmentation $R \twoheadrightarrow R/\bar{I} \tilde{\to} k$), whence the homology spaces

$\mathrm{Tor}_p^R(k,k)$, $p = 0,1,2,\ldots$

are graded:

$$\mathrm{Tor}_p^R(k,k) = \coprod_{q\geq 0} \mathrm{Tor}_{p,q}^R(k,k) .$$

If $\mathrm{Tor}_p^R(k,k)$ is finite-dimensional as a k-vector space for all p, then the *Poincaré-Betti series* of R is defined to be

$$P_R(x) = \sum_{p\geq 0} x^p \dim_k \mathrm{Tor}_p^R(k,k) .$$

$\lfloor c \rfloor$ denotes the integer part of c, while $\lceil c \rceil$ ($= -\lfloor -c \rfloor$) denotes the smallest integer not less than c.

Roughly following D. Anick ([1, section 4]), put

$$q_0(p,R) = \sup\{q \mid \mathrm{Tor}_{p,q}^R(k,k) \neq 0 \text{ or } q=p\} ;$$

(*) V is *not* assumed necessarily to be finite-dimensional. Sometimes in what follows results are quoted and applied generally, though in the sources finite dimension was assumed. In these cases, the proofs of the quoted results make no use of this assumption.

the extra case $q=p$ will only affect $q_0(p,R)$ if the global homological dimension $\mathrm{gldim}\, R < p$, since by e.g. [9, first inequality in lemma 2, formula (9)]

$\mathrm{Tor}^R_{p,q}(k,k) = 0$ for $q < p$.

Clearly $p \leq q_0(p,R) \leq \infty$.

Formally, let the *rate of growth* of the homology of R be

$\mathrm{rate}(R) = \sup\{ (q_0(p,R)-1)/(p-1) \mid p = 2,3,4,\ldots \}$.
Clearly $1 \leq \mathrm{rate}(R) \leq \infty$. R is called a *(homogeneous) Koszul algebra* or a *Fröberg ring* if $\mathrm{rate}(R) = 1$. (This generalizes the ordinary definition, where in addition V is assumed to be finite-dimensional; cf. e.g. [5, 1.16.].)

In several interesting cases $\mathrm{rate}(R) < \infty$, e.g. if $\dim_k V < \infty$ and R is commutative ([1, thm 4.2]), or if $q_0(2,R) < \infty$ and R fulfills a certain distributivity condition (cf. [3, prop 2.5 (b)]). In these cases, theorem 1 and its corollary apply.

THEOREM 1. - *Assume that* $\mathrm{rate}(R) = c < \infty$, *and that* $d \geq 2$ *is an integer. Then*

$$\mathrm{rate}(R^{(d)}) \leq \lceil c/d \rceil .$$

COROLLARY. - *If* $d \geq \mathrm{rate}(R)$, *then* $R^{(d)}$ *is a Koszul algebra.*

As is well known, commutative finitely generated Koszul algebras have rational Poincaré-Betti series; hence and from the results mentioned above we get

THEOREM 2. - *If* R *is* (as above and) *commutative and finitely generated, then* $R^{(d)}$ *is a Koszul algebra for* $d \gg 0$. *In particular,* $P_{R^{(d)}}(x)$ *is rational (for those d).*

Theorem 1 follows easily from

THEOREM 3. - *Assume that for some integers* j *and* n $(j \geq n \geq 2)$ *we have*

(1) $q_0(p,R) - 1 \leq \lfloor j/(n-1) \rfloor (p-1)d$, $p = 2,\ldots,n$.

Then

(2) $q_0(n,R^{(d)}) - 1 \leq j$.

For, assume that $\mathrm{rate}(R) = c < \infty$ and fix an $n \geq 2$. We want to prove that $q_0(n,R)-1 \leq \lceil c/d \rceil (n-1)$. However, if $j = \lceil c/d \rceil (n-1)$, then $\lfloor j/(n-1) \rfloor = \lceil c/d \rceil \geq c/d$, whence for $p = 2,\ldots,n$

$q_0(p,R)-1 \leq c(p-1) \leq \lfloor j/(n-1) \rfloor (p-1)d$,

whence indeed by (2) $q_0(p,R^{(d)})-1 \leq j$.

2. FUNDAMENTAL IDEAS OF THE PROOF OF THEOREM 3.

The following lemma is proved by Govorov in a more general form ([9, lemma 1]). (He however assumes but does not use that $\dim_k V < \infty$.)

LEMMA 1. - *For* $S = T(V)/Q$, *where* Q *is* *any* *proper homogeneous 2-sided* T(V)-*ideal, we have the following isomorphisms of graded* k-*vector spaces:*

$$\mathrm{Tor}^S_{2m+1}(k,k) \xrightarrow{\sim} (IQ^m \cap Q^m I)/(IQ^m I + Q^{m+1}) \quad (m = 0,1,2,\ldots), \quad \text{and}$$

$$\mathrm{Tor}^S_{2m}(k,k) \xrightarrow{\sim} (IQ^{m-1} I \cap Q^m)/(IQ^m + Q^m I) \quad (m = 1,2,\ldots) .$$

Note that we do not have to assume that $Q \subseteq I^2$. Thus, we may apply lemma 1 not only for R but also for $R^{(d)}$, if we make the following natural identifications:

(3) $R^{(d)} = T(V^{(d)})/P^{(d)}$,

where $V^{(d)} = I_d$, and where $P^{(d)}_i = P_{id} \subseteq I_{id} = I^{(d)}_i$ (for all i).

Hence we may rewrite the assumptions (1) in theorem 3 as follows:

For $i \geq 2 + \lfloor j/(n-1) \rfloor (p-1)d$ and $p = 2r+1$ odd

(4) $(IP^r)_i \cap (P^r I)_i = (IP^r I)_i + (P^{r+1})_i$,

while for $i \geq 2 + \lfloor j/(n-1) \rfloor (p-1)d$ and $p = 2r$ even

(5) $(IP^{r-1} I)_i \cap (P^r)_i = (IP^r)_i + (P^r I)_i$.

Equally, the conclusion (2) may be rewritten thus, if $n = 2m+1$ is odd:

For $t \geq 2+j$

(6) $(I^{(d)}(P^{(d)})^m)_t \cap ((P^{(d)})^m I^{(d)})_t = (I^{(d)}(P^{(d)})^m I^{(d)})_t + ((P^{(d)})^{m+1})_t$;

and thus, if $n = 2m$ is even:

For $t \geq 2+j$

(7) $(I^{(d)}(P^{(d)})^{m-1} I^{(d)})_t \cap ((P^{(d)})^m)_t = (I^{(d)}(P^{(d)})^m)_t + ((P^{(d)})^m I^{(d)})_t$.

Clearly it is sufficient to prove (6) and (7) for $t = j+2$.

By the identifications we may regard (6) and (7) as statements concerning subspaces of $I_{(j+2)d}$. To be concrete, we have e.g.

$$(P^m)_i = \sum_b P_{b_1} P_{b_2} \cdot\cdot\cdot P_{b_m} \quad (\text{sum over } b = (b_1,\ldots,b_m) \text{ such that } \sum_s b_s = i),$$

while by (3)

$$(P^{(d)})^m_{j+2} = \sum_b P_{b_1} P_{b_2} \cdot\cdot\cdot P_{b_m} \quad (\text{sum over } b = (b_1,\ldots,b_m) \text{ such that}$$
$$\sum_s b_s = (j+2)d \textit{ and } \text{that } d \mid b_s \text{ for all } s).$$

Thus $((P^{(d)})^m)_{j+2} \subseteq (P^m)_{(j+2)d}$, and more precisely a term $P_{b_1} \ldots P_{b_m}$ in $(P^m)_{(j+2)d}$ is one of the terms defining $((P^{(d)})^m)_{j+2}$ if and only if there is no "disallowed border" $a_k = \sum_{s=1}^{k} b_s$, not divisible by d, "between" the factors P_{b_k} and $P_{b_{k+1}}$ for some $k \in \{1,\ldots,m-1\}$; cf. the figure below.

Figure 2.

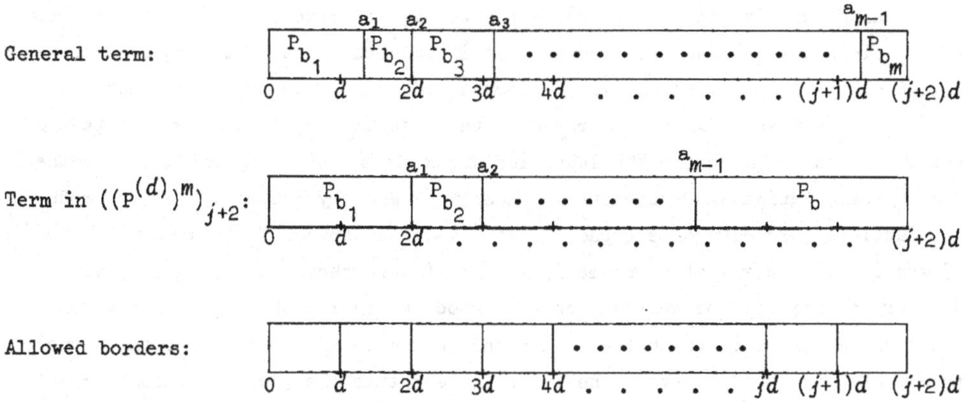

The situation is parallel for $(I^{(d)}(P^{(d)})^m)_{j+2}$ $(\subseteq (IP^m)_{(j+2)d})$, $((P^{(d)})^m I^{(d)})_{j+2}$ $(\subseteq (P^m I)_{(j+2)d})$, and $(I^{(d)}(P^{(d)})^m I^{(d)})_{j+2}$ $(\subseteq (IP^m I)_{(j+2)d})$. Thus, assuming $n = 2m+1$ odd and applying (4) for $p = n$, we get

$$(I^{(d)}(P^{(d)})^m)_{j+2} \cap ((P^{(d)})^m I^{(d)})_{j+2} \; (\subseteq (IP^m)_{(j+2)d} \cap (P^m I)_{(j+2)d}) \subseteq$$

$$\subseteq (IP^m I)_{(j+2)d} + (P^{m+1})_{(j+2)d},$$

and we want to prove that the assumptions imply that the inclusion remains true even if we omit the terms with "forbidden borders" on the right side.

In order to achieve this we "forbid" these borders one by one. The order in which the borders are forbidden is determined, in the first place by their residues modulo d, and in the second place by their own order. Thus the next border to be forbidden in the situation of figure 3 is $gd+e-1$.

Figure 3.

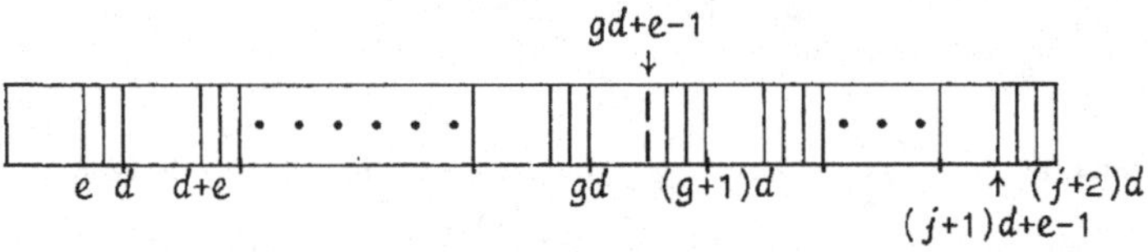

(The sum of the terms in P^m (IP^m, etc.) whose borders are compatible with figure 3 will be denoted $(P^m)^{d,d,e}_{(j+2)d,g-1}$ ($(IP^m)^{d,d,e}_{(j+2)d,g-1}$, etc.); cf. definitions (22) to (25).)

In order to eliminate this border, we use the conditions (4) and (5) and some technical generalizations of these. I.e., we have to make induction with respect to the claim that (4) and (5) remain true if on both sides of the equalities we forbid all borders having too small residues modulo d. This is the content of lemma 2 in section 3, which is proved in that section, up to the crucial part of the induction step. Before the induction hypotheses can be applied, some rather lengthy calculations must be done. Then only elementary properties of the modular lattices $(\{W\},+,\cap)$ of subspaces of the k-vector spaces I_i (given in section 4) are used. I fear that these manipulations (since they take too much place and are very technical, due to the lack of a good theory for the logical interrelations of different distributive conditions) may make the reader miss the actual points of the proof. However, the induction and thus the proof of theorem 3 is completed in section 5.

3. A TECHNICAL GENERALIZATION.

DEFINITIONS. - For any positive integers e, i and f, such that $e \leq d \geq f < i$, we define

(8) $(IP^0)^{f,d,e}_i = (P^0I)^{f,d,e}_i = (IP^0I)^{f,d,e}_i = I_i$;

(9) $(P^1)^{f,d,e}_i = P_i$;

and for $m \geq 1$

(10) $(P^{m+1})_i^{f,d,e} = \sum_a P_{a_1} P_{a_2-a_1} \cdots P_{a_m-a_{m-1}} P_{i-a_m}$;

(11) $(IP^m)_i^{f,d,e} = \sum_a I_{a_1} P_{a_2-a_1} \cdots P_{a_m-a_{m-1}} P_{i-a_m}$;

(12) $(P^m I)_i^{f,d,e} = \sum_a P_{a_1} P_{a_2-a_1} \cdots P_{a_m-a_{m-1}} I_{i-a_m}$; and (for $m \geq 2$)

(13) $(IP^{m-1}I)_i^{f,d,e} = \sum_a I_{a_1} P_{a_2-a_1} \cdots P_{a_m-a_{m-1}} I_{i-a_m}$;

where in each case the sum is taken over all m-tuples of integers $a = (a_1, \ldots, a_m)$, such that

$0 < a_1 < \ldots < a_m < i$, and that

$a_r - f \not\equiv 1, 2, \ldots, e-1 \pmod d \quad (r = 1, \ldots, m)$.

For $X = P^m, IP^m, P^m I$ or $IP^m I$ we have $X_i^{f,d,e} \subseteq X_i$, but in general some of the borders we want to get rid of are forbidden in $X_i^{f,d,e}$. In fact,

(14) $X_{(j+2)d}^{d,d,d} = X_{j+2}^{(d)}$, while

(15) $X_i^{f,d,1} = X_i$

(independently of the value of f). Thus, theorem 3 is a special case of the following lemma:

LEMMA 2. - *Let n, d, e, i and f be positive integers such that*

$d \geq 2$;

$e \leq d \geq f$;

$j := \lfloor (i-f-1)/d \rfloor \geq n-1$; *and* (if $n > 1$)

(16) $q_0(p,R) - 1 \leq \lfloor j/(n-1) \rfloor (p-1)d$ *for* $p = 1, \ldots, n$.

Then, if $n = 2m+1$ is odd we have

(17) $(IP^m)_i^{f,d,e} \cap (P^m I)_i^{f,d,e} = (IP^m I)_i^{f,d,e} + (P^{m+1})_i^{f,d,e}$,

and if $n = 2m$ is even we have

(18) $(IP^{m-1}I)_i^{f,d,e} \cap (P^m)_i^{f,d,e} = (IP^m)_i^{f,d,e} + (P^m I)_i^{f,d,e}$.

(Note that condition (16) always holds for $p=1$.)

The lemma is proved by induction, first on n, then on e, and then on i.

Let $L(n',e',i',f')$ be the claim that the implications in the lemma are true whenever $(n,e,i,f)=(n',e',i',f')$.

$L(1,e',i',f')$ follows directly from the definitions (8) and (9). Thus we may henceforth assume that

(19) *$n\geq 2$ and $L(n',e',i',f')$ for $n'=1,\ldots,n-1$* (and for all positive integers e', i' and f').

If $e=1$, then we note that if the assumptions of the lemma are fulfilled, we in particular have $q_0(n,R)<2+\lfloor j/(n-1)\rfloor(n-1)d\leq 2+jd\leq i$, whence

$$\mathrm{Tor}^R_{n,i}(k,k)=0\ ,$$

whence by lemma 1 and by (15) the conclusion of lemma 2 holds. Thus we may henceforth assume that

(20) *$e\geq 2$, and $L(n,e',i',f')$ for $e'=1,\ldots,e-1$.*

If $i=1$, then (since $j<0<n-1$) the assumptions of the lemma are not fulfilled, whence indeed $L(n,e,1,f')$. Thus we may henceforth assume that

(21) *the assumptions of* lemma 2 *are fulfilled, and $L(n,e,i',f')$ for $i'=1,\ldots,i-1$.*

From now on (n,e,i,f) is a fixed set of integers fulfilling (19), (20) and (21). When $L(n,e,i,f)$ is proved, then so is lemma 2.

DEFINITIONS. - For any integer h such that $0<f+hd+e<i+d$, and for any $m\geq 1$, let

$$(22)\quad (P^{m+1})^{f,d,e}_{i,h}=\sum_a P_{a_1}P_{a_2-a_1}\cdots P_{a_m-a_{m-1}}P_{i-a_m}\ ,$$

$$(23)\quad (IP^{m})^{f,d,e}_{i,h}=\sum_a I_{a_1}P_{a_2-a_1}\cdots P_{a_m-a_{m-1}}P_{i-a_m}\ ,$$

$$(24)\quad (P^{m}I)^{f,d,e}_{i,h}=\sum_a P_{a_1}P_{a_2-a_1}\cdots P_{a_m-a_{m-1}}I_{i-a_m}\ ,\quad \text{and (for } m\geq 2)$$

$$(25)\quad (IP^{m-1}I)^{f,d,e}_{i,h}=\sum_a I_{a_1}P_{a_2-a_1}\cdots P_{a_m-a_{m-1}}I_{i-a_m}\ ;$$

where in each case the sum is taken over all m-tuples of integers $a=(a_1,\ldots,a_m)$ such that

$$(26)\quad \begin{cases} 0 < a_1 < \ldots < a_m < i\ , \\ a_r - f \not\equiv 1,\ldots,e-2 \pmod{d} \quad (r = 1,\ldots,m)\ , \text{ and} \\ a_r - f \not\equiv e-1 \pmod{d} \text{ for } r \text{ such that } a_r < f+hd\ . \end{cases}$$

Likewise, if in addition $f+hd+e-1 < i$, let

$$(27)\quad '(P^{m+1})^{f,d,e}_{i,h} = \sum_{a} P_{a_1} P_{a_2-a_1} \cdots P_{i-a_m} \subseteq (P^{m+1})^{f,d,e}_{i,h}\ , \text{ et cetera,}$$

where the sums are taken over a such that (26) is fulfilled and that in addition for some r $(1 \leq r \leq m)$ we have

$$(28)\quad a_r - f = hd+e-1\ .$$

Thus, putting

$$h_1 := \lfloor (d-f-e+1)/d \rfloor \text{ and}$$

$$h_2 := \lfloor (i+d-f-e)/d \rfloor\ ,$$

and taking $X=P^m$, $=IP^m$, $=P^mI$, or $=IP^mI$, (for some relevant m,) we have:

$(X)^{f,d,e}_{i,h}$ is defined iff $h = h_1,\ldots,h_2$;

$'(X)^{f,d,e}_{i,h}$ is defined iff $h = h_1,\ldots,h_2-1$;

$$(29)\quad (X)^{f,d,e-1}_{i} = (X)^{f,d,e}_{i,h_1}$$

$$(30)\quad (X)^{f,d,e}_{i,h} = {}'(X)^{f,d,e}_{i,h} + (X)^{f,d,e}_{i,h+1} \quad \text{for } h = h_1,\ldots,h_2-1\ ; \text{ and}$$

$$(31)\quad (X)^{f,d,e}_{i,h_2} = (X)^{f,d,e}_{i}\ .$$

Assume henceforth that $n = 2m+1$ is odd; then we have to prove (17). (The even case is handled similarly.)

By the definitions, by $L(n,e-1,i,f)$ and by (29)

$$(IP^m)^{f,d,e}_{i} \cap (P^mI)^{f,d,e}_{i} \subseteq (IP^m)^{f,d,e-1}_{i} \cap (P^mI)^{f,d,e-1}_{i} =$$

$$= (IP^mI)^{f,d,e-1}_{i} + (P^{m+1})^{f,d,e-1}_{i} = (IP^mI)^{f,d,e}_{i,h_1} + (P^{m+1})^{f,d,e}_{i,h_1}\ ,$$

whence

$$(32)\quad (IP^m)^{f,d,e}_{i} \cap (P^mI)^{f,d,e}_{i} = (IP^m)^{f,d,e}_{i} \cap (P^mI)^{f,d,e}_{i} \cap \Big((IP^mI)^{f,d,e}_{i,h_1} + (P^{m+1})^{f,d,e}_{i,h_1} \Big).$$

In section 5 it is proved that for $h = h_1,\ldots,h_2-1$

$$(33)\quad \begin{aligned}&(IP^m)_i^{f,d,e} \cap (P^m I)_i^{f,d,e} \cap \left((IP^m I)_{i,h}^{f,d,e} + (P^{m+1})_{i,h}^{f,d,e}\right) = \\ &= (IP^m)_i^{f,d,e} \cap (P^m I)_i^{f,d,e} \cap \left((IP^m I)_{i,h+1}^{f,d,e} + (P^{m+1})_{i,h+1}^{f,d,e}\right).\end{aligned}$$

By (32), (33) and (31)

$$(IP^m)_i^{f,d,e} \cap (P^m I)_i^{f,d,e} = (IP^m)_i^{f,d,e} \cap (P^m I)_i^{f,d,e} \cap ((IP^m I)_i^{f,d,e} + (P^{m+1})_i^{f,d,e})$$

whence indeed

$$(IP^m)_i^{f,d,e} \cap (P^m I)_i^{f,d,e} \subseteq (IP^m I)_i^{f,d,e} + (P^{m+1})_i^{f,d,e}.$$

Since the opposit inclusion in (17) is trivial, (17) indeed follows. Thus theorem 3 indeed is reduced to (33) (and to a similar omitted even counterpart).

4. SOME LATTICE-THEORETICAL PREPARATIONS.

For a general reference on lattice theory, see e.g. [10].

Recall that in general a lattice $(L,\vee,\wedge)$ is called *modular* if

(34) $x\wedge(y\vee z) = (x\wedge y)\vee z$ for all $x,y,z \in L$ such that $x \geq z$ (the *modular law*)

or, equivalently, if

(35) $x\wedge(y\vee z) = x\wedge(y\vee(z\wedge(x\vee y)))$ for all $x,y,z \in L$ (the *shearing property*).

In the rest of this article, all lattices considered are assumed to be modular.

It is well known that for any k-vector space W, ({subspaces of W},+,∩) forms a (modular) lattice; in particular, we may take $W=I_{i'}$ $(0 \leq i' \leq i)$.

We shall also be concerned with some well-known relations between elements in the lattices of subspaces of W', W", and $W'\otimes_k W''$, for two k-spaces W' and W":

LEMMA 3. – *Let* $V_1',\dots,V_r'$ *and* $V_1'',\dots,V_s''$ *be subspaces of* W' *and of* W", *respectively. Then we have*

(36) $V_1'\otimes W'' \cap W'\otimes V_1'' = V_1'\otimes V_1''$, *and*

(37) $\left(\sum_{t=1}^{r} V_t'\right)\otimes\left(\sum_{u=1}^{s} V_u''\right) = \sum_t\sum_u (V_t'\otimes V_u'')$.

We also need

LEMMA 4. - *Assume that* $(L,\vee,\wedge)$ *is a modular lattice, that* s *is a positive integer, that* $a_1,\ldots,a_s,b_1,\ldots,b_s \in L$, *and that* $a_1 \leq a_2 \leq \ldots \leq a_s$ *and* $b_1 \geq b_2 \geq \ldots \geq b_s$. *Then*

$$(38) \quad \bigvee_{t=1}^{s}(a_t \wedge b_t) = b_1 \wedge \Big(\bigwedge_{t=1}^{s-1}(a_t \vee b_{t+1})\Big) \wedge a_s \quad .$$

Proof by induction on s and by the modular law (34).
(If $s \geq 2$ then $\wedge(\text{rest of factors}) \wedge (a_{s-1} \vee b_s) \wedge a_s =$
$= \wedge(\text{rest of factors}) \wedge (a_{s-1} \vee (b_s \wedge a_s)) = (\wedge(\text{rest of factors}) \wedge a_{s-1}) \vee (a_s \wedge b_s)$.)

COROLLARY. - *Assume that* $t \geq 1$, *that* $a_1,\ldots,a_{2t},b_1,\ldots,b_{2t} \in L$ *and that* $a_1 \leq \ldots \leq a_{2t}$ *and* $b_1 \geq \ldots \geq b_{2t}$. *Then*

$$(39) \quad \Big(\bigvee_{u=1}^{t}(a_{2u-1} \wedge b_{2u-1})\Big) \wedge \Big(\bigvee_{u=1}^{t}(a_{2u} \wedge b_{2u})\Big) = \bigvee_{u=1}^{2t-1}(a_u \wedge b_{u+1}) \quad .$$

For, $(\bigvee_u (a_{2u-1} \wedge b_{2u-1})) \wedge (\bigvee_u (a_{2u} \wedge b_{2u})) =$

$$= b_1 \wedge \Big(\bigwedge_{u=1}^{t-1}(a_{2u-1} \vee b_{2u+1})\Big) \wedge a_{2t-1} \wedge b_2 \wedge \Big(\bigwedge_{u=1}^{t-1}(a_{2u} \vee b_{2u+2})\Big) \wedge a_{2t} =$$

$$= b_2 \wedge \Big(\bigwedge_{u=1}^{2t-2}(a_u \vee b_{u+2})\Big) \wedge a_{2t-1} = \bigvee_u (a_u \wedge b_{u+1}) \quad .$$

5. PROOF OF (33).

By (30) and by the shearing property (35),

$$(IP^m)_i^{f,d,e} \cap (P^m I)_i^{f,d,e} \cap ((IP^m I)_{i,h}^{f,d,e} + (P^{m+1})_{i,h}^{f,d,e}) =$$

$$= (IP^m)_i^{f,d,e} \cap (P^m I)_i^{f,d,e} \cap ((IP^m I)_{i,h+1}^{f,d,e} + (P^{m+1})_{i,h+1}^{f,d,e} +' (IP^m I)_{i,h}^{f,d,e} +' (P^{m+1})_{i,h}^{f,d,e}) =$$

$$= (IP^m)_i^{f,d,e} \cap (P^m I)_i^{f,d,e} \cap ((IP^m I)_{i,h+1}^{f,d,e} + (P^{m+1})_{i,h+1}^{f,d,e} +$$

$$+(('(IP^m I)_{i,h}^{f,d,e} +' (P^{m+1})_{i,h}^{f,d,e}) \cap (((IP^m)_i^{f,d,e} \cap (P^m I)_i^{f,d,e}) + (IP^m I)_{i,h+1}^{f,d,e} + (P^{m+1})_{i,h+1}^{f,d,e}))).$$

Thus (33) follows from

$$(40)\quad \begin{cases} ('(IP^{m}I)^{f,d,e}_{i,h} + '(P^{m+1})^{f,d,e}_{i,h}) \cap \\ \cap (((IP^{m})^{f,d,e}_{i} \cap (P^{m}I)^{f,d,e}_{i}) + (IP^{m}I)^{f,d,e}_{i,h+1} + (P^{m+1})^{f,d,e}_{i,h+1}) \subseteq \\ \subseteq (IP^{m}I)^{f,d,e}_{i,h+1} + (P^{m+1})^{f,d,e}_{i,h+1} \ . \end{cases}$$

(40) may be regarded as an inequality concerning elements in the lattice of all subspaces to $I_i = I_{i'} \otimes I_{i''}$, where

$$i' := f+hd+e-1$$

the border to be eliminated, and where

$$i'' := i - i' \ .$$

Also, let

$$f'' := \ d-e+1 \ .$$

The idea of the proof of (40) is to reformulate the inequality in terms of subspaces v_*, x_*, u_* and w_* of $I_{i'}$ and $I_{i''}$, and then to apply the inductive hypotheses on these.

Each term $I_{a_1}P_{a_2-a_1}\cdots P_{a_{m+1}-a_m}I_{i-a_{m+1}}$ in the sum which defines $'(IP^{m}I)^{f,d,e}_{i,h}$ by (28) has a border $a_r = i'$ and thus is contained in $(IP^{r-1})^{f,d,e}_{i'}(P^{m-r+1}I)^{f'',d,e-1}_{i''}$ for some r such that $1 \leq r \leq m+1$. Hence and similarly we get

$$(41)\quad \begin{aligned} '(IP^{m}I)^{f,d,e}_{i,h} &= \sum_{r=1}^{m+1} (IP^{r-1})^{f,d,e}_{i'}(P^{m-r+1}I)^{f'',d,e-1}_{i''} = \\ &= \sum_{r=0}^{m} (IP^{r})^{f,d,e}_{i'}(P^{m-r}I)^{f'',d,e-1}_{i''} \ , \end{aligned}$$

and

$$(42)\quad '(P^{m+1})^{f,d,e}_{i,h} = \sum_{r=1}^{m} (P^{r})^{f,d,e}_{i'}(P^{m-r+1})^{f'',d,e-1}_{i''} \ .$$

For $r = 0,\ldots,m$, let us write

(43) $u_{2r+1} = (P^{m-r}I)^{f'',d,e-1}_{i''}$, and

(44) $v_{2r+1} = (IP^{r})^{f,d,e}_{i'}$;

and for $r = 1,\dots,m$, let us write

(45) $u_{2r} = (P^{m+1-r})^{f'',d,e-1}_{i''}$, and

(46) $v_{2r} = (P^{r})^{f,d,e}_{i'}$.

Then (recalling that $n = 2m+1$ and by lemma 3) we may summarize (41) and (42) thus:

(47) $'(IP^{m}I)^{f,d,e}_{i,h} + '(P^{m+1})^{f,d,e}_{i,h} = \sum_{s=1}^{n} v_s u_s = \sum_{s=1}^{n} I_{i'} u_s \cap v_s I_{i''}$.

Next, let $T = P_{a_1} P_{a_2-a_1} \dots P_{i-a_m}$ be a term in the sum defining $(P^{m+1})^{f,d,e}_{i,h+1}$. Since i' is not a border in T, either there is an r (with $1 \leq r \leq m-1$) such that $a_r < i' < a_{r+1}$, or $0 < i' < a_1$, or $a_m < i' < i$. Thus either

$T \subseteq P_{a_1} \dots P_{a_r - a_{r-1}} I_{a_{r+1}-a_r} P_{a_{r+2}-a_{r+1}} \dots P_{i-a_m} \subseteq (P^{r}I)^{f,d,e}_{i'} (IP^{m-r})^{f'',d,e-1}_{i''}$, or

$T \subseteq I_{a_1} P_{a_2-a_1} \dots P_{i-a_m} \subseteq (P^{0}I)^{f,d,e}_{i'} (IP^{m})^{f'',d,e-1}_{i''}$, or

$T \subseteq P_{a_1} \dots P_{a_m - a_{m-1}} I_{i-a_m} \subseteq (P^{m}I)^{f,d,e}_{i'} (IP^{0})^{f'',d,e-1}_{i''}$.

Hence

(48) $(P^{m+1})^{f,d,e}_{i,h+1} \subseteq \sum_{r=0}^{m} (P^{r}I)^{f,d,e}_{i'} (IP^{m-r})^{f'',d,e-1}_{i''}$.

Similarly $(IP^{m}I)^{f,d,e}_{i,h+1} \subseteq$

$\subseteq \sum_{r=1}^{m} (IP^{r-1}I)^{f,d,e}_{i'} (IP^{m-r}I)^{f'',d,e-1}_{i''} + I_{i'} (IP^{m}I)^{f'',d,e-1}_{i''} + (IP^{m}I)^{f,d,e}_{i'} I_{i''}$. Since

$I_{i'} (IP^{m}I)^{f'',d,e-1}_{i''} \subseteq (IP^{0}I)^{f,d,e}_{i'} (IP^{m-1}I)^{f'',d,e-1}_{i''}$ and

$(IP^{m}I)^{f,d,e}_{i'} I_{i''} \subseteq (IP^{m-1}I)^{f,d,e}_{i'} (IP^{0}I)^{f'',d,e-1}_{i''}$, we have

(49) $(IP^{m}I)^{f,d,e}_{i,h+1} \subseteq \sum_{r=1}^{m} (IP^{r-1}I)^{f,d,e}_{i'} (IP^{m-r}I)^{f'',d,e-1}_{i''}$.

For $r = 0,\dots,m$, let us write

$$(50)\quad w_{2r+1} = (IP^{m-r})_{i''}^{f'',d,e-1} \text{ , and}$$

$$(51)\quad x_{2r+1} = (P^{r}I)_{i'}^{f,d,e} \;;$$

and for $r = 1,\dots,m$, let us write

$$(52)\quad w_{2r} = (IP^{m-r}I)_{i''}^{f'',d,e-1} \text{ , and}$$

$$(53)\quad x_{2r} = (IP^{r-1}I)_{i'}^{f,d,e} \;.$$

Then we may summarize (48) and (49) thus:

$$(54)\quad (IP^{m}I)_{i,h+1}^{f,d,e}+(P^{m+1})_{i,h+1}^{f,d,e} = \sum_{t=1}^{n} x_t w_t = \sum_{t=1}^{n} I_{i'}w_t \cap x_t I_{i''} \;.$$

Also note that

$$(55)\quad u_1 \subseteq u_2 \subseteq \cdots \subseteq u_n \;,$$

$$(56)\quad v_1 \supseteq v_2 \supseteq \cdots \supseteq v_n \;,$$

$$(57)\quad w_1 \subseteq w_2 \subseteq \cdots \subseteq w_n \text{ , and}$$

$$(58)\quad x_1 \supseteq x_2 \supseteq \cdots \supseteq x_n \;,$$

and correspondingly for the $I_{i'}u_s$, $v_s I_{i''}$, $I_{i'}w_t$ and $x_t I_{i''}$.

Furthermore, by similar arguments

$$(IP^{m})_{i}^{f,d,e} \subseteq \sum_{r=1}^{m} x_{2r}w_{2r+1} = \sum_{r=1}^{m} I_{i'}w_{2r+1}\cap x_{2r}I_{i''} \quad \text{and}$$

$$(P^{m}I)_{i}^{f,d,e} \subseteq \sum_{r=1}^{m} x_{2r-1}w_{2r} = \sum_{r=1}^{m} I_{i'}w_{2r}\cap x_{2r-1}I_{i''} \quad ,$$

whence by the corollary to lemma 4

$$(59)\quad (IP^{m})_{i}^{f,d,e}\cap(P^{m}I)_{i}^{f,d,e} \subseteq \sum_{r=2}^{2m} x_r I_{i''}\cap I_{i'}w_r = \sum_{r=2}^{2m} x_r w_r \;.$$

Thus by (47), (54) and (59)

$$(60)\quad (\text{right side of the inequality (40)}) \subseteq \left(\sum_{s=1}^{n} v_s u_s\right)\cap\left(\sum_{t=1}^{n} x_t w_t\right) .$$

By lemma 3

$$(61)\quad (\sum_s v_s u_s) \cap (\sum_t x_t w_t) = \sum_s \sum_t (v_s \cap x_t)(u_s \cap w_t) .$$

Below I prove that for each s and t

$$(62)\quad (v_s \cap x_t)(u_s \cap w_t) \subseteq (IP^m I)_{i,h+1}^{f,d,e} + (P^{m+1})_{i,h+1}^{f,d,e} .$$

(60), (61) and (62) imply (40), indeed.

Proof of (62): Fix s and t. We get three cases and two subcases in each case, depending on whether $s<t$, $s>t$ or $s=t$, and whether s is odd or even, respectively.

i) If $s<t$, then $(v_s \cap x_t)(u_s \cap w_t) \subseteq x_t u_s \subseteq x_{s+1} u_s$; and:

ia) for $s=2r+1$ odd $x_{s+1}u_s = x_{2(r+1)}u_{2r+1} = (IP^r I)_{i'}^{f,d,e}(P^{m-r}I)_{i''}^{f'',d,e-1}$.

Now, if $I_{a_1}P_{a_2-a_1}\dots P_{a_{r+1}-a_r}I_{i'-a_{r+1}}$ is one of the terms defining $(IP^r I)_{i'}^{f,d,e}$, and if $P_{a_{r+2}-i'}P_{a_{r+3}-a_{r+2}}\dots P_{a_{m+1}-a_m}I_{i-a_{m+1}}$ is one of the terms defining $(P^{m-r}I)_{i''}^{f'',d,e-1}$, then

$$I_{a_1}P_{a_2-a_1}\dots I_{i'-a_{r+1}}P_{a_{r+2}-i'}\dots I_{i-a_{m+1}} \subseteq$$

$$\subseteq I_{a_1}P_{a_2-a_1}\dots P_{a_{r+2}-a_{r+1}}\dots I_{i-a_{m+1}} ,$$ which is one of the terms defining $(IP^m I)_{i,h+1}^{f,d,e}$. Thus (and by lemma 3) indeed

$$(IP^r I)_{i'}^{f,d,e}(P^{m-r}I)_{i''}^{f'',d,e-1} \subseteq (IP^m I)_{i,h+1}^{f,d,e} .$$

Similarly,

ib) for $s=2r$ even

$$x_{s+1}u_s = x_{2r+1}u_{2r} = (P^r I)_{i'}^{f,d,e}(P^{m+1-r})_{i''}^{f'',d,e-1} \subseteq (P^{m+1})_{i,h+1}^{f,d,e} .$$

ii) If $s>t$, then $(v_s \cap x_t)(u_s \cap w_t) \subseteq v_s w_t \subseteq v_{t+1} w_t$; proceed as in case i).

Thus we have finally exhausted the elementary properties of the involved lattices, and reduced the proof to a case where the inductive hypotheses of lemma 2 may be applied:

iii) Assume that $s = t$.

In this case some different subdivisions are necessary.

Let $s'' := n-s+1$.

Note that $h = \lfloor (i'-f-1)/d \rfloor$, and that $h'' := j-h-1 = \lfloor (i''-f''-1)/d \rfloor$. Furthermore we have

CLAIM. - *If* $2 \leq s \leq n-1$, *then*

$$\lfloor j/(n-1) \rfloor \leq \max(\lfloor h/(s-1) \rfloor, \lfloor h''/(s''-1) \rfloor).$$

PROOF. - Note that $(s-1)+(s''-1) = (n-1)$, while $h+h'' = j-1$.

If $j \not\equiv 0 (\mathrm{mod}\ n-1)$, then $\lfloor j/(n-1) \rfloor = \lfloor (j-1)/(n-1) \rfloor$ whence the claim follows from the fact that $(j-1)/(n-1) = (h+h'')/(s-1+s''-1) \leq \max(h/(s-1), h''/(s''-1))$.

Assume that $j = \ell(n-1) \equiv 0 (\mathrm{mod}\ n-1)$. Then $h+h'' = \ell(s-1)+\ell(s''-1)-1$, and since both h and h'' are integers, we must have either $h \geq \ell(s-1)$ or $h'' \geq \ell(s''-1)$. In either case the claim follows.

Thus we get the following six subcases:

iiia) $2 \leq s \leq n-1$ and $\lfloor j/(n-1) \rfloor \leq \lfloor h/(s-1) \rfloor$,
iiib) $2 \leq s \leq n-1$ and $\lfloor j/(n-1) \rfloor \leq \lfloor h''/(s''-1) \rfloor$,
iiic) $s = n$ and $h = j$,
iiid) $s = n$ and $h \leq j-1$,
iiie) $s = 1$ and $h = -1$, and
iiif) $s = 1$ and $h \geq 0$.

In case iiia) we have

$$q_0(p,R)-1 \leq \lfloor j/(n-1) \rfloor (p-1)d \leq \lfloor h/(s-1) \rfloor (p-1)d \quad \text{for } p=1,\dots,s;$$

and by the inductive hypothesis (19) we have $L(s,e,i',f)$.

Thus lemma 2 applies: If $s = 2r+1$ is odd, then by (17)

$$v_s \cap x_s = (IP^r)_{i'}^{f,d,e} \cap (P^r I)_{i'}^{f,d,e} = (IP^r I)_{i'}^{f,d,e} + (P^{r+1})_{i'}^{f,d,e} = v_{s+1} + x_{s+1}.$$

Similarly $v_s \cap x_s = v_{s+1} + x_{s+1}$ by (18) if s is even. Thus in either case (and by lemma 3)

$$(v_s \cap x_s)(u_s \cap w_s) = (v_{s+1} + x_{s+1})(u_s \cap w_s) = v_{s+1}(u_s \cap w_s) + x_{s+1}(u_s \cap v_s) \subseteq$$
$$\subseteq v_{s+1} w_s + x_{s+1} u_s \subseteq (IP^m I)_{i,h+1}^{f,d,e} + (P^{m+1})_{i,h+1}^{f,d,e}, \text{ indeed.}$$

Similarly, in case iiib) we have

$$q_0(p,R)-1 \le \lfloor h''/(s''-1)\rfloor(p-1)d \text{ for } p=1,\dots,s'',$$

whence by $L(s'',e-1,i'',f'')$

$$u_s \cap w_s = u_{s-1}+w_{s-1}\ ,$$

whence

$$(v_s \cap x_s)(u_s \cap w_s) \subseteq v_s w_{s-1}+x_s u_{s-1} \subseteq (IP^m I)^{f,d,e}_{i,h+1}+(P^{m+1})^{f,d,e}_{i,h+1}\ .$$

In case iiic) we must have $h_2 = j+1$, and by (16) and $L(n,e,i',f)$ (cf. (21)) we have

$$(v_n \cap x_n)(u_n \cap w_n) = ((IP^m)^{f,d,e}_{i'} \cap (P^m I)^{f,d,e}_{i'})(I_{i''} \cap I_{i''}) =$$
$$=(IP^m I)^{f,d,e}_{i'} I_{i''} + (P^{m+1})^{f,d,e}_{i'} I_{i''} \subseteq (IP^m I)^{f,d,e}_{i} + (P^{m+1})^{f,d,e}_{i}$$
$$(\ = (IP^m I)^{f,d,e}_{i,j+1} + (P^{m+1})^{f,d,e}_{i,j+1}\ )\ .$$

Similarly, in case iiie) we have $h'' = j$ and $i' = f-d+e-1 < f$, and by (16) and $L(n,e-1,i'',f'')$ (cf. (20)) we have

$$(v_1 \cap x_1)(u_1 \cap w_1) = (I_{i'} \cap I_{i'})((P^m I)^{f'',d,e-1}_{i''} \cap (IP^m)^{f'',d,e-1}_{i''}) =$$
$$= I_{i'}(IP^m I)^{f'',d,e-1}_{i''} + I_{i'}(P^{m+1})^{f'',d,e-1}_{i''} \subseteq (IP^m I)^{f,d,e}_{i,0} + (P^{m+1})^{f,d,e}_{i,0}\ .$$

In case iiid), like in case iiic) we have

$$(v_n \cap x_n)(u_n \cap w_n) = ((IP^m)^{f,d,e}_{i'} \cap (P^m I)^{f,d,e}_{i'})I_{i''} \subseteq (IP^m)^{f,d,e}_{i'} I_{i''}\ .$$

Since $h \le j-1$, we have an "allowed border" $f+jd > f+hd+e-1 = i'$, and $I_{i''} = I_{f+jd-i'} I_{i-f-jd}$. Thus, if $I_{a_1} P_{a_2-a_1} \dots P_{i'-a_m}$ is a term in the sum defining $(IP^m)^{f,d,e}_{i'}$, then $I_{a_1} \dots P_{i'-a_m} I_{i''} = I_{a_1} \dots P_{i'-a_m} I_{f+jd-i'} I_{i-f-jd} \subseteq$

$$\subseteq I_{a_1} \dots P_{f+jd-a_m} I_{i-f-jd} \subseteq (IP^m I)^{f,d,e}_{i} \subseteq (IP^m I)^{f,d,e}_{i,h+1}\ .$$

Case iiif), finally, is handled analogously.

Thus indeed (62) and hence (40), (33), lemma 2 and theorem 3 are proved.

6. POSSIBLE IMPROVEMENTS?

Some of the conditions (1) are slightly stronger than necessary in order to ensure the consequence (2) claimed in theorem 3. E.g., for $n=4$, by a refinement of the CLAIM in the preceeding section, if

$$\begin{cases} q_0(2,R)-1 \le \lfloor j/3 \rfloor d \ , \\ q_0(3,R)-1 \le \lfloor (2j+2)/3 \rfloor d \ , \text{ and} \\ q_0(4,R)-1 \le jd \ , \end{cases}$$

then indeed $q_0(4,R^{(d)})-1 \le j$.

I do not know what the best possible conditions (in general) are, but I suspect that there are but minor improvements of (1). One reason for this is the following example, where actually R has global homological dimension 2.

(If V has a basis $(T_1,T_2,\ldots)$, then we write $k\langle T_1,T_2,\ldots\rangle = T(V)$.)

EXAMPLE. - Let $R = k\langle T_1,T_2\rangle/(T_1T_2^c)$ for some positive integer c. Then $q_0(2,R)-1 = c$, while for all other p $q_0(p,R)-1 = p-1$. However, if $d \ge 2$ and $c' = \lceil c/d \rceil$, then $T(V^{(d)}) = k\langle T_1^d,\ T_1^{d-1}T_2, T_1^{d-2}T_2T_1,\ldots,T_2^d\rangle$ (2^d "variables") and $P^{(d)}$ is minimally generated by some monomials in these variables, including $T_2^{d-1}T_1\cdot(T_2^d)^{c'-1}\cdot T_2^{d-1}T_1$, whence by [2, thm 2] gldim $R^{(d)} = \infty$, and indeed $q_0(n,R^{(d)})-1 = c'(n-1)$ for all n.

Thus for $p=2$ there can be no greater improvement of (1) than replacing $\lceil c/d \rceil$ by $\lfloor 1+(c-1)/d \rfloor$ (which amounts to nothing if c is an integer).

In particular, proposition 3 a) in [5] can not be improved. However, there we also gave a roughly "twice as good" result for the case where R is *commutative*, namely that then

(63) $q_0(2,R)-1 \le c \Rightarrow q_0(2,R^{(d)})-1 \le \max(1, \lfloor c/d \rfloor)$;

in particular,

(64) $q_0(2,R^{(d)})-1 = 1$ if $d \ge \frac{1}{2}(\lfloor c \rfloor+1)$.

It is conceivable that in this case (1) in general (and thus the corollary to theorem 1) may be improved similarly.

Let us finally discuss how Anick's theorem ([1, thm 4.2]) on the finiteness of rate(R) for R finitely generated and commutative might be improved.

Let $n = \text{edim}(R) = \dim_k V < \infty$, and let $r = q_0(2,R)$. (Since R is commutative and $n < \infty$

we have $r<\infty$.) By analysing two proofs of Anick's theorem we get

CLAIM. - *There is an integer* $C=C(n,r)$ *(only depending on* n *and* r*) such that* $\mathrm{rate}(R)\le C$;

and we get some ideas on how we might get some good estimates for C.

Anick's proof goes roughly as follows:

1. Choose an ordered basis $(X_1,\dots,X_n)$ for V. Then $R\simeq\tilde{R}/\mathfrak{a}$ for some ideal $\mathfrak{a}\subset\tilde{R}:=k[X_1,\dots,X_n]$ (the commutative polynomial ring). By a classical method of Macaulay, this uniquely determines an associated ideal $\mathfrak{b}\subset\tilde{R}$, generated by monomials in $X_1,\dots,X_n$, and thus an associated "monomial ring" $A=\tilde{R}/\mathfrak{b}$ to R. Anick constructs a spectral sequence (see [1, lemma 4.1])

$$\mathrm{Tor}^A_{*,*}(k,k)\Rightarrow\mathrm{Tor}^R_{*,*}(k,k)\quad,$$

whence in particular $\mathrm{Tor}^R_{p,q}(k,k)\ne 0\Rightarrow\mathrm{Tor}^A_{p,q}(k,k)\ne 0$; thus

$$\mathrm{rate}(R)\le\mathrm{rate}(A)\ .$$

2. Since by [4] the *double Poincaré-Betti series*

$$P_A(x,y)=\sum_p\sum_q x^p y^q\dim_k\mathrm{Tor}^A_{p,q}(k,k)$$

is rational, and indeed $P_A(x,y)=(1+xy)^n/(1+\sum_{p,q>1}a_{p,q}x^py^q)$, where the integers $a_{p,q}$ are 0 for almost all (p,q), we have

$$\mathrm{rate}(A)\le\max(1\,;(q-1)/(p-1)\ :\ a_{p,q}\ne 0)<\infty\ .$$

L.L. Avramov has given an alternative proof, using the bigraded version of the Serre inequality ([8, formula (1)]):

(65) $P_R(x,y)\le(1+xy)^n/(1-\sum_{p,q>0}c_{p,q}x^{p+1}y^q)$ (term-wise inequality) .

where $c_{p,q}=\dim_k\mathrm{Tor}^{\tilde{R}}_{p,q}(R,k)$ ($=\dim_k(H_pK^R)_q$, where K^R is the Koszul complex of R), whence (similary to step 2 above)

(66) $\mathrm{rate}(R)\le\max(1\,;(q-1)/p:c_{p,q}\ne 0)<\infty$.

The Avramov approach may be combined with a result deduced in the proof of [7, prop 1], namely

(67) *There is an* $N=N(n,r)$ *such that* $c_{p,q}=0$ *for all* $q>N$;

in order to yield

(68) $\mathrm{rate}(R) \leq \max(r-1,(N(n,r)-1)/2) < \infty$.

Thus the claim is proved. However, the (implicitly) given boundaries for $N(n,r)$ in [7] are ridicuously large even for small n and r. In fact, Fröberg, Gulliksen and Löfwall use the limits calculated by G. Hermann in [11] in order to find an $N' = N'(n,r)$ which bounds the q:s such that $c_{1,q} \neq 0$; and then (roughly) bounds the q:s such that $c_{2,q} \neq 0$ by $N'(n,N'(n,r))$, et cetera. Since already the Hermann bound (even as improved by D. Lazard in [12]) is very large, $N(n,r)$ is quite out of (effective) control.

There is however possibly a more promising approach. Note, to begin with, that the existence proof in Anick's step 2 may be replaced by

CLAIM. - *If* B *is a "monomial ring", then*

(69) $\mathrm{rate}(B) = q_0(2,B)-1$.

I have proved this (unpublished) by a rather lengthy argument, involving a generalization of the Taylor resolution (cf. [14]). For our discussion, however, a weaker but short Avramov-type argument will suffice:

The Taylor resolution of B is an $\tilde{R}$-free resolution

$$\ldots \to F_p \to F_{p-1} \to \ldots \to F_0 \to B \to 0 ,$$

where each F_p is generated (as a graded $\tilde{R}$-module) by elements of degrees $\leq$ $\leq pq_0(2,B)$. Tensoring with k and taking homology, we get

$$c_{p,q} = 0 \quad \text{for } q > pq_0(2,B) ,$$

whence by (66)

$$\mathrm{rate}(B) < q_0(2,B) .$$

Thus, by Anick's argument

$$\mathrm{rate}(R) < q_0(2,A) \quad (\text{and actually } \mathrm{rate}(R) \leq q_0(2,A)-1)$$

and it is "enough" to find a bound $B = B(n,r)$ for
$\min(q_0(2,A) : A \text{ is an associated monomial ideal to } k' \otimes_k R \text{ for some field } k' \supseteq k)$.
As far as I know, the existence of such a bound is unproved, but (as Fröberg has suggested) it is plausible that a Hermann-type argument would yield one. One might hope that in fact there is an estimate of the type suggested by Lazard in [12]; i.e. that $B(n,r) \approx n(r-1)$.

Note added in proof:

D. Anick and R. Fröberg have remarked that indeed (in the notation from above) $B(n,r) < \infty$. One way to prove this is the following:

By the results in [7], for fixed n and r there are only finitely many possible Hilbert series.

Furthermore, there are only finitely many non-isomorphic "monomial rings" with a given fixed Hilbert series.

The latter statement, and indeed a direct bound on the highest degree of a relation, may be based on the ideas in the proof of Macaulay's theorem on feasible Hilbert series. If we adopt the terminology of the proof of this theorem given in R. Stanley, *Hilbert functions of graded algebras*, Adv. in Math. 28 (1978), 57-83 (Theorem 2.2), we see that if $(h_i)_{i\geq 0}$ is a given Hilbert series, if M is the corresponding 'order ideal of monomials' (taking lexicographically first monomials in the appropriate amounts), and if j is the maximal degree of relations in the "monomial ring" whose non-killed monomials are exactly M, then

$h_{i+1} = h_i^{\langle i \rangle}$ for all $i \geq j$, whence no (commutative) algebra with the Hilbert series (h_i) can have relations in degrees $> j$.

Note however that this gives a [7]-type, not a Hermann-type of estimate for $B(n,r)$, whence it does not directly yield any improvements on the bounds $C(n,r)$ for the rates of growth.

Also note that David Bayer and Michael Stillman (*On the complexity of computing syzygies*, manuscript, March 1985; the authors are at Columbia University and at University of Chicago, respectively) have shown that the true bound for $q_0(2,R)$ is high - of "double exponential" type - like G. Hermann's own bound. Thus $C(n,r)$ is also at least "doubly exponential".

REFERENCES.

1 D.J. Anick, *On the homology of associative algebras*, Trans. Amer. Math. Soc. (to appear).

2 J. Backelin, *La série de Poincaré-Betti d'une algebre graduée de type fini à une relation est rationelle*, C.R. Acad. Sc. Paris 287 (1978), 843-846.

3 J. Backelin, *A distributiveness property of augmented algebras, and some related homological results*, part of thesis, Stockholm, 1982.

4 J. Backelin, *Les anneaux locaux à relations monomiales ont des séries de Poincaré-Betti rationnelles*, C. R. Acad. Sc. Paris 295 (1982), Série I, 607-610.

5 J. Backelin and R. Fröberg, *Koszul algebras, Veronese subrings and rings with linear relations*, Rev. Roumaine Math. Pures Appl. 30 (1985), 85-97.

6 S. Barcanescu and N. Manolache, *Betti numbers of Segre-Veronese singularities*, Re. Roumaine Math. Pures Appl. 26 (1981), 549-565.

7 R. Fröberg, T.H. Gulliksen and C. Löfwall, *Flat families of local, artinian algebras with an infinite number of Poincaré series*, these proceedings.

8 E. S. Golod, О гомологиях некоторых локальных колец, Doklady Akademii Nauk SSSR 144, 1962, 479-482.

9 V.E. Govorov, Размерность и кратность градуированних алгебр, Sibirsk Math. J. 14 (1973), 1200-1206. (Translation: p. 840-845.)

10 G. Grätzer, *General lattice theory*, Birkhäuser Verlag, Basel und Stuttgart, 1978.

11 G. Hermann, *Die Frage der endlich vielen Schritte in der Theorie der Polynomideale*, Math. Ann. 95 (1926), 736-788.

12 D. Lazard, *Algèbre linéaire sur* $K[X_1,\ldots,X_n]$, *et élimination*, Bull Soc. Math. France 105 (1977), 165-198.

13 D. Mumford, *Varieties defined by quadratic equations*, in Questions on algebraic varieties, edited by E. Marchionna, C.I.M.E. III Ciclo, Varenna, 1969, Roma 1970, 29-100.

14 D. Taylor, *Ideals generated by monomials in an R-sequence*, Ph. D. Thesis, University of Chicago, 1966.

Department of Mathematics
University of Stockholm
Box 6701
S-113 85 STOCKHOLM
(SWEDEN)

WHEN IS THE DOUBLE YONEDA EXT-ALGEBRA OF A LOCAL NOETHERIAN RING AGAIN NOETHERIAN ?

by

Jörgen BACKELIN and Jan-Erik ROOS

§ 0. INTRODUCTION.

Let $(R,\underline{m})$ be a local commutative noetherian ring with maximal ideal $\underline{m}$ and residue field $k = R/\underline{m}$. Consider the graded vector space

(1) $$\mathrm{Ext}_R^*(k,k) = \bigoplus_{i\geq 0} \mathrm{Ext}_R^i(k,k)$$

which is a graded algebra under the Yoneda product [13],[19],[24]:

$$\mathrm{Ext}_R^i(k,k) \otimes \mathrm{Ext}_R^j(k,k) \longrightarrow \mathrm{Ext}_R^{i+j}(k,k) .$$

Indeed, $\mathrm{Ext}_R^*(k,k)$ is even a graded, cocommutative Hopf algebra over k [13],[24]. Now iterate this Ext construction and form

(2) $$\mathrm{Ext}^*_{\mathrm{Ext}_R^*(k,k)}(k,k)$$

which is again a graded algebra (but not a Hopf algebra) under the Yoneda product. Note that (2) has an extra grading, coming from the grading in $\mathrm{Ext}_R^*(k,k)$. Note also that while $\mathrm{Ext}_R^*(k,k)$ is highly non-commutative in general (indeed, if $\mathrm{Ext}_R^*(k,k)$ is graded commutative, it follows from [5] that R is a local complete intersection of a special type), it is always true that (2) is commutative in the bigraded sense, i.e. if x and y have bidegrees (r,s) and (r',s'), then $xy = (-1)^{rr'+ss'}yx$. This follows from the fact that $E = \mathrm{Ext}_R^*(k,k)$ is a Hopf algebra, and the fact that in this case two "natural" products on $\mathrm{Ext}_E^*(k,k)$ coincide (cf. [26], page 15-13).

It is very rare that $E = \mathrm{Ext}_R^*(k,k)$ is noetherian (left noetherian is equivalent to right noetherian, since E is a Hopf algebra). Indeed, in these proceedings, Bøgvad and Halperin [7] prove that E is noetherian precisely when R is a local complete intersection (and then E is even generated as an algebra by a finite number of generators of degree 1 and 2).

It happens more often that

(3) $$\mathrm{Ext}^*_{\mathrm{Ext}_R^*(k,k)}(k,k) \text{ is noetherian}$$

and when this happens we can deduce interesting results about $\mathrm{Ext}_R^*(k,k)$ and R. Thus, for instance, we will prove in § 1 that the following assertions follow from (3)

(i) $P_R(Z) = \sum_{i\geq 0} \dim_k(\mathrm{Ext}_R^i(k,k))\cdot Z^i = \sum_{i\geq 0} \dim_k(\mathrm{Tor}_i^R(k,k))\cdot Z^i$ is a rational function of Z.

More precisely, if (3) holds, then there is a finite set of algebra generators $\{\xi_i\}_{i=1}^n$ of (2) with bidegree$(\xi_i) = (m_i,n_i)$ (contrary to the convention in [26] we let both degrees take non-negative values) <u>and</u>

$$P_R(Z) = \prod_{i=1}^{n}{}' (1-(-1)^{m_i} Z^{n_i})/\mathrm{pol}(Z)$$

<u>where</u> $'$ <u>means that we take the product over those</u> i:s <u>such that</u> $m_i^2+n_i^2$ <u>is even</u>, char $k \neq 2$, <u>and where</u> pol(Z) <u>is a polynomial in</u> Z. [In particular, if the generators of the algebra (2) are of bidegree (1,1), it follows that $P_R(Z) = (1+Z)^n/\mathrm{pol}(Z)$.]

(ii) $\mathrm{Ext}_R^*(k,k)$ <u>is finitely generated as an algebra</u>.

(iii) <u>When we present</u>, <u>using</u> (ii), $\mathrm{Ext}_R^*(k,k)$ <u>as a quotient of a finitely generated free graded algebra</u>, <u>it is also true that the</u> (twosided) <u>kernel ideal is generated</u> (as a twosided ideal) <u>by a finite set of generators</u>. [We say that $\mathrm{Ext}_R^*(k,k)$ is a finitely presented algebra.]

(iv) <u>Furthermore</u>, <u>there are only a finite number of relations between the relations in</u> (iii) <u>etc</u>.

The first example where (ii) is false is given in [24], p.314 , using ideas from J.-M. Lemaire's thesis [15]. Using similar ideas, one can construct examples $(R,\underline{m})$, where (ii) is true, but (iii) is violated etc. Later Anick [1] gave many examples $(R,\underline{m})$, where $P_R(Z)$ is non rational, thus violating (i).

Thus there are many examples where (3) is not satisfied. However, in [23], one of us proved that <u>if</u> $(R,\underline{m})$ is a Golod ring (examples: $R = \tilde{R}/\tilde{\underline{m}}^s$, $s \geq 1$, where $(\tilde{R},\tilde{\underline{m}})$ is a regular local noetherian ring, or $R = \tilde{R}/x\cdot\underline{a}$, where $x \in \tilde{\underline{m}}$ and $\underline{a}$ is a proper ideal of $\tilde{R}$...), then (3) holds, and so does even the following more precise assertion [(3) follows from (4), if we take M = R]:

<u>If</u> M <u>is a finitely generated</u> R-<u>module</u>, <u>then the</u> (left) $\mathrm{Ext}^*_{\mathrm{Ext}_R^*(k,k)}(k,k)$-<u>module</u>

(4) (*) $$\mathrm{Ext}^*_{\mathrm{Ext}_R^*(k,k)}(\mathrm{Ext}_R^*(M,k),k)$$

<u>is noetherian</u>.

Note that here the Yoneda product

$$\mathrm{Ext}_R^i(k,k) \otimes \mathrm{Ext}_R^j(M,k) \longrightarrow \mathrm{Ext}_R^{i+j}(M,k)$$

gives $\mathrm{Ext}_R^*(M,k)$ the structure of a left $\mathrm{Ext}_R^*(k,k)$-module, and that in a similar way (*) becomes a left $\mathrm{Ext}^*_{\mathrm{Ext}_R^*(k,k)}(k,k)$-module. Using a reasoning, similar to that in [23], it follows that if R comes from a local complete intersection S by a Golod map S ——> R, then (4) and (<u>à fortiori</u>) (3) are true. For a complete proof of this, <u>cf</u>. § 2 below. In particular, (3) and (4) are true if R is a local complete intersection, a fact which can also be deduced from [12].

In [6], one of us proved among other things, that if k is a field and R is the quotient of a formal power series ring: $R = k[[X_1,\dots,X_n]]/(M_1,\dots,M_r)$, where the M_i:s are monomials in the X_j:s, then $P_R(Z)$ is rational of the form $(1+Z)^n/\mathrm{pol}(Z)$. It was therefore very natural to ask whether the stronger statement (3) could hold for these rings with monomial relations. This is indeed the case, and that is one of the main results of the present paper (Theorem 5 of § 3). The proof is a combination of the ideas from our papers [6] and [23]. It follows from the theory of § 1 below, combined with results of C. Jacobsson [14], that the stronger statement (4) can not hold in general for these rings with monomial relations. The fact that (3) does hold for these rings has already been used by Anick [2] in his study of the homology of certain loop spaces. Indeed he proves in [2] that $H_*(\Omega X,\mathbb{Q})$ is finitely presented if X is a generalized fat wedge. The paper by us, cited as [4] in [2], has been incorporated in the present paper. We study commutative and not anti-commutative R:s here, but the methods are the same.

The plan of the present paper is as follows: In § 1 we deduce general consequences of the validity of (3) and (4) for a local ring. In § 2 and § 3 we prove the main theorems, using general results about the (co)homology spectral sequences of extensions of Hopf algebras. Finally, in § 4 some open problems are mentioned.

§ 1. THE DOUBLE Ext-ALGEBRA AND THE DOUBLE Ext-MODULES.

THEOREM 1.- Let $(R,\underline{m})$ be a local, commutative noetherian ring such that $\mathrm{Ext}^*_{\mathrm{Ext}^*_R(k,k)}(k,k)$ is a (bigraded) noetherian ring. Then:

(a) $\mathrm{Ext}^*_{\mathrm{Ext}^*_R(k,k)}(k,k)$ has a finite set of algebra generators $\{\xi_i\}_{i=1}^n$ with bidegree$(\xi_i)=(m_i,n_i)$, and we have the following formula for the corresponding double Hilbert series:

$$(5)\quad \sum_{p,q\geq 0} \dim_k\left(\mathrm{Ext}^{p,q}_{\mathrm{Ext}^*_R(k,k)}(k,k)\right)\cdot X^pY^q = \frac{\mathrm{Pol}(X,Y)}{\prod'_{1\leq i\leq n}(1-X^{m_i}Y^{n_i})}$$

where $'$ means that we can restrict the product in (5) to those i:s such that $m_i^2+n_i^2$ is even, if char$(k)\neq 2$, and where $P(X,Y)$ is a polynomial in X,Y with integral coefficients.

(b) Assume furthermore, that M is a finitely generated R-module such that $\mathrm{Ext}^*_{\mathrm{Ext}^*_R(k,k)}(\mathrm{Ext}^*_R(M,k),k)$ is a (bigraded) finitely generated (thus noetherian) module over $\mathrm{Ext}^*_{\mathrm{Ext}^*_R(k,k)}(k,k)$. Then we have the following formula for the corresponding double Hilbert series:

$$(6)\quad \sum_{p,q\geq 0} \dim_k\left(\mathrm{Ext}^{p,q}_{\mathrm{Ext}^*_R(k,k)}(\mathrm{Ext}^*_R(M,k),k)\right)\cdot X^pY^q = \frac{\mathrm{Pol}_M(X,Y)}{\prod'_{1\leq i\leq n}(1-X^{m_i}Y^{n_i})}$$

where the numerator is a polynomial in X,Y (integral coefficients), depending on M, and where the denominator is the same as in (5).

(c) When (a) is satisfied it follows also that the algebra $\mathrm{Ext}_R^*(k,k)$ has a finite number of generators, that all relations between a minimal set of generators are generated by finitely many relations (we say that $\mathrm{Ext}_R^*(k,k)$ is finitely presented), that there are only "finitely many relations between these relations" etc. Furthermore, in this case (notations as in (5)):

$$(7)\qquad P_R(Z) = \sum_{i\geq 0} \dim_k(\mathrm{Tor}_i^R(k,k))\cdot Z^i = \frac{\prod'_{1\leq i\leq n}(1-(-1)^{m_i}Z^{n_i})}{\mathrm{Pol}(-1,Z)}$$

(d) When the stronger condition (b) is satisfied, it follows also that the left $\mathrm{Ext}_R^*(k,k)$-module $\mathrm{Ext}_R^*(M,k)$ is finitely generated, even finitely presented, and even that it has a minimal, free $\mathrm{Ext}_R^*(k,k)$-resolution, having a finite number of generators in each step. Furthermore, in this case (notations as in (5) and (6)):

$$(8)\qquad P_R^M(Z) = \sum_{i\geq 0} \dim_k(\mathrm{Tor}_i^R(M,k))\cdot Z^i = \frac{\mathrm{Pol}_M(-1,Z)}{\mathrm{Pol}(-1,Z)} .$$

PROOF: We will need the general observation [15], Chapitre I, that if A is a graded connected algebra over k, then a minimal set of graded generators of A corresponds to a basis of the graded vector space $\mathrm{Tor}_1^A(k,k) = I(A)/I(A)^2$, where I(A) is the augmentation ideal of A. In the same way, a minimal set of graded relations for A corresponds to a basis of the graded vector space $\mathrm{Tor}_2^A(k,k)$, etc. Furthermore, the graded vector spaces $\mathrm{Ext}_A^i(k,k)$ are dual to the $\mathrm{Tor}_i^A(k,k)$. We will use this below for $A = \mathrm{Ext}_R^*(k,k)$. But let us first recall that we have supposed that the bigraded algebra $B = \mathrm{Ext}_A^*(k,k)$ is noetherian. Thus $I(B)/I(B)^2$ is finite-dimensional and therefore B has a finite set of generators $\{\xi_i\}_{i=1}^n$ as in (a) of Theorem 1. Since B is the Ext-algebra of the Hopf algebra A, it follows from [26], Théorème 1, p.15-13, that $\xi_i\xi_j = (-1)^{m_im_j+n_in_j}\xi_j\xi_i$. We will now use the following bigraded variant of Theorem 11.1 in [3]:

LEMMA.- Let B be a bigraded connected algebra over a field k. Suppose that B has a finite set of algebra generators $\{\xi_i\}_{1\leq i\leq n}$ such that bidegree$(\xi_i) = (m_i, n_i)$ and

$$(*)\qquad \xi_i\xi_j = (-1)^{m_im_j+n_in_j}\,\xi_i\xi_j$$

Let V be a finitely generated bigraded module over B, and let

$$V(X,Y) = \sum_{p,q\geq 0} \dim_k(V_{p,q})\cdot X^pY^q$$

be the two-variable Hilbert series of V. Then

$$V(X,Y) = \frac{\mathrm{Pol}(X,Y)}{\prod'_{1\leq i\leq n}(1-X^{m_i}Y^{n_i})}$$

where Pol(X,Y) is a polynomial in X,Y (integral coefficients) and where ′ means that we can restrict ourselves to take the product for those i:s such that $m_i^2+n_i^2$ is even (if char(k)≠2).

SKETCH OF THE PROOF OF THE LEMMA: We will use induction over the number n of generators ξ_i of B. If n=0, then V is a finite-dimensional bigraded vector space and V(X,Y) is a polynomial. Suppose n > 0. The ξ_i:s "almost" commute and therefore B is noetherian. Furthermore, if char(k)≠2 and $m_i^2+n_i^2$ is odd, it follows from (*) that $\xi_i^2=0$. We then have an exact sequence

$$0 \longrightarrow \xi_i V \longrightarrow V \longrightarrow V/\xi_i V \longrightarrow 0$$

and $\xi_i V$ and $V/\xi_i V$ are finitely generated and bigraded modules over $B/(\xi_i)$ which has n-1 generators. Thus

$$V(X,Y) = (\xi_i V)(X,Y) + (V/\xi_i V)(X,Y)$$

and we can use the induction and ignore such an i. For a ξ_j such that $m_j^2+n_j^2$ is even, we consider, as in [3] the bigraded exact sequence:

$$(9) \quad 0 \longrightarrow K_{p,q} \longrightarrow V_{p,q} \xrightarrow{\cdot\xi_j} V_{p+m_j,q+n_j} \longrightarrow L_{p+m_j,q+n_j} \longrightarrow 0$$

where the finitely generated $B/(\xi_j)$-modules K and L are defined by (9). The alternating sum of the dimensions of the vector spaces in (9) is 0. Multiplying this sum by $X^{p+m_j}Y^{q+n_j}$ and summing over p,q we obtain that $(1-X^{m_j}Y^{n_j})V(X,Y)$ can be expressed as $L(X,Y) + X^{m_j}Y^{n_j}K(X,Y)$ + a polynomial in X,Y . The result now follows from the inductive hypothesis.

Now both (a) and (b) of Theorem 1 follow immediately from this lemma. Using a minor variant of Proposition A.2.4 of the Appendix of Lemaire′s thesis [15] (cf. also [11]) we can now also deduce the formulae (7) and (8) of (c) and (d) from the formulae (5) and (6). Indeed, if N is any graded module over a connected graded algebra Λ overk (assume that Λ and N are finite-dimensional in each degree), then

$$(10) \qquad N(Z) = \Lambda(Z)P_N(-1,Z)$$

Here N(Z) and Λ(Z) are the Hilbert series of N and Λ and

$$P_N(X,Y) = \sum_{p,q\geq 0} \dim_k(\mathrm{Tor}^{\Lambda}_{p,q}(k,N))\cdot X^pY^q$$

The proof of (10) follows (as in [15]) from taking a minimal Λ-free resolution of N and by taking alternating sums of dimensions in each degree of corresponding bigraded modules. Apply now (10) first to $\Lambda = \mathrm{Ext}^*_R(k,k)$ and N = k (this gives (7) as a consequence of (5)) and then to $\Lambda = \mathrm{Ext}^*_R(k,k)$ and $N = \mathrm{Ext}^*_R(M,k)$ (this gives (8) as a consequence of (6), using (7)).[We have used that the vector space duals of the $\mathrm{Tor}^{\Lambda}_{p,q}(k,N)$ are the $\mathrm{Ext}^{p,q}_{\Lambda}(N,k)$.] The remaining assertions of (c) and (d) now follow from the explicit formulae (5) and (6) , which give that for each fixed p, the graded vector spaces $\mathrm{Tor}_{p,*}^{\mathrm{Ext}^*_R(k,k)}(k,k)$ and $\mathrm{Tor}_{p,*}^{\mathrm{Ext}^*_R(k,k)}(\mathrm{Ext}^*_R(M,k),k)$ have total finite dimension. Thus Theorem 1 is completely proved.

§ 2. SOME LOCAL RINGS WHOSE DOUBLE Ext-ALGEBRA IS NOETHERIAN.

The following theorem contains Theorem 2 of [23] and gives a generalization that in a sense is the best possible one (for the definition of e.g. Golod maps, we refer e.g. to § 3 of [25]):

THEOREM 2.- a) Let $\tilde{R}$ be a local complete intersection, $\tilde{R} \longrightarrow R_1$ a Golod map and M a finitely generated R_1-module. Then

$$\mathrm{Ext}^*_{\mathrm{Ext}^*_{R_1}(k,k)}(\mathrm{Ext}^*_{R_1}(M,k),k) \text{ is a noetherian } \mathrm{Ext}^*_{\mathrm{Ext}^*_{R_1}(k,k)}(k,k)\text{-module}.$$

In particular $\mathrm{Ext}^*_{\mathrm{Ext}^*_{R_1}(k,k)}(k,k)$ is a noetherian algebra (take $M = R_1$).

b) Suppose further that $R_1 \longrightarrow R_2$ is a second Golod map. Then $\mathrm{Ext}^*_{\mathrm{Ext}^*_{R_2}(k,k)}(k,k)$ is again noetherian.

c) If R_1 is as in a), then $\mathrm{Ext}^*_{R_1}(k,k)$ is a (graded) coherent algebra [25].

Remark.- Theorem 2 is the best possible in the following sense:

i) There exist local (commutative, noetherian) rings R_3 that come by three Golod maps from a local complete intersection (even a regular ring) $\tilde{R}$:

$$\tilde{R} \xrightarrow{\text{Golod}} R_1 \xrightarrow{\text{Golod}} R_2 \xrightarrow{\text{Golod}} R_3$$

such that $\mathrm{Ext}^*_{\mathrm{Ext}^*_{R_3}(k,k)}(k,k)$ is not noetherian.

ii) We also have a situation ($\tilde{R}$ regular): $\tilde{R} \xrightarrow{\text{Golod}} R_1 \xrightarrow{\text{Golod}} R_2$ and a finitely generated module M over R_2 such that:

$$\mathrm{Ext}^*_{\mathrm{Ext}^*_{R_2}(k,k)}(\mathrm{Ext}^*_{R_2}(M,k),k) \text{ is not a noetherian } \mathrm{Ext}^*_{\mathrm{Ext}^*_{R_2}(k,k)}(k,k)\text{-module.}$$

Indeed, if the modules in i)-ii) were always noetherian, it would follow from Theorem of § 1 (formulae (7) and (8)) , that $P_{R_3}(Z)$ and $P^M_{R_2}(Z)$ were always rational. But in [14], Theorem 2.1 (and Corollaries 2.2 and 2.3) C. Jacobsson gives examples where this is not the case.(In these examples R_2 is even a ring with quadratic monomial relations!)

In order to prove Theorem 2 we will now first make a general analysis of the behaviour of double Ext-algebras with respect to a Golod map. Thus let $R \longrightarrow S$ be any Golod map ([25], § 3) and recall (loc.cit.) that we have an "exact sequence of Hopf algebras":

$$(11) \qquad k \longrightarrow T \longrightarrow \mathrm{Ext}^*_S(k,k) \longrightarrow \mathrm{Ext}^*_R(k,k) \longrightarrow k$$

where T is the free associative graded algebra on the graded vector space $\{ \mathrm{Ext}^{i-1}_R(S,k) \}_{i\geq 2}$. Thus the global homological dimension of T is 1 and the diagonal is the natural one. We want to compare the ring theoretical properties of the algebras $Ext_S^*(k,k)$ and $Ext_R^*(k,k)$ (in particular their Ext-algebras) and for this (and also for later) purpose(s) it is convenient to study first a very general situation of extensions of graded, connected, cocommutative Hopf algebras:

(12) $$k \longrightarrow A \xrightarrow{i} B \xrightarrow{j} C \longrightarrow k$$

Recall [22],[13],[9] that we say that (12) is an exact sequence of Hopf algebras if $B \xrightarrow{j} C$ is onto, and if i is the kernel of j in the category of coalgebras, i.e. i induces an isomorphism $A \xrightarrow{\simeq} k \,\square_C B$, where $\square$ is the cotensor product of coalgebras [20]. Then it follows that we have also an algebra isomorphism $k \otimes_A B \xrightarrow{\simeq} C$ (and $B \otimes_A k \xrightarrow{\simeq} C$). Now a cocommutative Hopf algebra can be considered as a group in the category of cocommutative coalgebras. It is therefore very natural to expect that corresponding to the extension (12) there should be a Hochschild-Serre spectral sequence, relating the (co)homology of B with the cohomology of A and C. This is indeed the case, and the spectral sequence has already been studied in [15],[23],[22],[21] and [18]. Our aim now is just to state all these results in a form that is the most convenient one for our applications in this and the next section: Consider the ring map $B \xrightarrow{j} C$ in (12) and let V be a graded left B-module. Consider the change of rings spectral sequence associated to this situation ([8], Chap.XVI):

(13) $$E_2^{p,q} = Ext_C^p(Tor_q^B(C,V),k) \Rightarrow Ext_B^n(V,k)$$

(of course there are many other variants of this spectral sequence).
Since $C = k \otimes_A B$ and B is flat (even free) as a left A-module ([20], prop. 4.4) it follows that

(14) $$Tor_q^B(C,V) = Tor_q^B(k \otimes_A B,V) \cong Tor_q^A(k,V)$$

which, combined with (13) gives the following "Hochschild-Serre spectral sequence":

(15) $$E_2^{p,q} = Ext_C^p(Tor_q^A(k,V),k) \Rightarrow Ext_B^n(V,k)$$

Note that we have suppressed the extra grading coming from the grading in A,B,C and V in (13)-(15). On what follows we will need the following more precise results about the operations of Yoneda Ext-algebras on the spectral sequences (13) and (15). The proofs of these results can be found in Ming [21]. First recall that the Yoneda product makes $Ext_B^*(V,k)$ into a left module over $Ext_B^*(k,k)$. Since $B \xrightarrow{j} C$ induces a ring map $Ext_C^*(k,k) \longrightarrow Ext_B^*(k,k)$, it follows that $Ext_B^*(V,k)$ is also a left $Ext_C^*(k,k)$-module. In the spectral sequence (15) (or (13)...) the E_2-term $Ext_C^*(Tor_q^A(k,V),k)$ is also a left $Ext_C^*(k,k)$-module by means of the Yoneda product. How is the spectral sequence (15) (or (13)...) related to these $Ext_C^*(k,k)$-module structures? Here is the answer (Ming [21] Theorem 2.7, p. 238):
There are structure maps:

$$Ext_C^s(k,k) \otimes E_r^{p,q} \xrightarrow{\Theta_r} E_r^{p+s,q}$$

such that:

i) Θ_2 is exactly the Yoneda product

$$\mathrm{Ext}_C^s(k,k)\otimes \mathrm{Ext}_C^p(\mathrm{Tor}_q^A(k,V),k) \longrightarrow \mathrm{Ext}_C^{p+s}(\mathrm{Tor}_q^A(k,V),k)$$

ii) The differential d_r of (15) is a left graded $\mathrm{Ext}_C^*(k,k)$-module homomorphism, and Θ_{r+1} is induced from Θ_r by passing to cohomology.

iii) Let $\{ F^p\mathrm{Ext}_B^n(V,k) \}_{p\geq 0}$ be the decreasing filtration of $\mathrm{Ext}_B^n(V,k)$, corresponding to the spectral sequence (15), and let ρ_p be the natural quotient map defined by:

$$0 \longrightarrow F^{p+1}\mathrm{Ext}_B^n(V,k) \longrightarrow F^p\mathrm{Ext}_B^n(V,k) \xrightarrow{\ \rho_p\ } E_\infty^{p,n-p} \longrightarrow 0$$

With these notations the following diagram is commutative:

$$\begin{array}{ccc} \mathrm{Ext}_C^s(k,k)\otimes F^p\mathrm{Ext}_B^n(V,k) & \xrightarrow{\ Y'\ } & F^{p+s}\mathrm{Ext}_B^{n+s}(V,k)\\ \Big\downarrow {\scriptstyle \mathrm{Id}\otimes \mathrm{inclusion}} & & \Big\downarrow {\scriptstyle \mathrm{inclusion}}\\ \mathrm{Ext}_C^s(k,k)\otimes E_\infty^{p,n-p} & \xrightarrow{\ \Theta_\infty\ } & E_\infty^{p+s,n-p} \end{array}$$

Here Y' is induced by the Yoneda product and the natural algebra map $\mathrm{Ext}_C^*(k,k) \xrightarrow{j^*} \mathrm{Ext}_B^*(k,k)$, so that the following diagram is commutative:

$$\begin{array}{ccccc} \mathrm{Ext}_C^s(k,k)\otimes \mathrm{Ext}_B^n(V,k) & \xrightarrow{\ j^s\otimes \mathrm{Id}\ } & \mathrm{Ext}_B^s(k,k)\otimes \mathrm{Ext}_B^n(V,k) & \xrightarrow{\ \mathrm{Yoneda}\ } & \mathrm{Ext}_B^{n+s}(V,k)\\ \Big\uparrow {\scriptstyle \mathrm{Id}\otimes \mathrm{inclusion}} & & & & \Big\uparrow {\scriptstyle \mathrm{inclusion}}\\ \mathrm{Ext}_C^s(k,k)\otimes F^p\mathrm{Ext}_B^n(V,k) & & \xrightarrow{\qquad Y'\qquad} & & F^{p+s}\mathrm{Ext}_B^{n+s}(V,k) \end{array}$$

Summing up, one can say that the spectral sequence (15) is compatible with all the left $\mathrm{Ext}_C^*(k,k)$-module structures in sight. We will use this in some special cases:

THEOREM 3.- <u>Let</u>

$$k \longrightarrow A \longrightarrow B \longrightarrow C \longrightarrow k$$

<u>be an extension of cocommutative graded connected Hopf algebras such that</u> A <u>is is a free algebra</u> (i.e. gldim A = 1). <u>Then, for each graded left</u> B-<u>module</u> V, <u>we have an exact sequence of left</u> $\mathrm{Ext}_C^*(k,k)$-<u>modules</u>:

$$(16) \ldots \to \mathrm{Ext}_C^{*-2}(\mathrm{Tor}_1^B(C,V),k) \to \mathrm{Ext}_C^*(C\otimes_B V,k) \to \mathrm{Ext}_B^*(V,k) \to \mathrm{Ext}_C^{*-1}(\mathrm{Tor}_1^B(C,V),k) \to \ldots$$

PROOF: The isomorphism (14) gives that the $E_2^{p,q}$ of (13) (or (15)) are zero for $q > 1$, if gldim A = 1. Therefore, in this case the spectral sequence (13) degenerates into a long exact sequence (16). The assertions about the left $\mathrm{Ext}_C^*(k,k)$-module structure are just reformulations of Ming's results, quoted above, in this special case.

Remark.- Theorem 3 can be applied to the Hopf algebra extension (11), coming from a Golod map $R \longrightarrow S$. In this case $A = T$ = the free associative graded algebra on the graded vector space $\{ \mathrm{Ext}_R^{i-1}(S,k) \}_{i \geq 2}$ [This graded vector space will be henceforth be denoted by $s^{-1}\overline{\mathrm{Ext}_R^*}(S,k)$, i.e. the "suspension" of the elements of degree > 0 in $\mathrm{Ext}_R^*(S,k)$.], $B = \mathrm{Ext}_S^*(k,k)$ and $C = \mathrm{Ext}_R^*(k,k)$. Furthermore, for $V = k$, the isomorphism (14) becomes:

$$\mathrm{Tor}_1^B(C,k) \simeq \mathrm{Tor}_1^A(k,k) \simeq s^{-1}\overline{\mathrm{Ext}_R^*}(S,k) \qquad (17)$$

and here the left $C = \mathrm{Ext}_R^*(k,k)$-module structure on $\mathrm{Tor}_1^B(C,k)$ corresponds to the left $\mathrm{Ext}_R^*(k,k)$-module structure on $s^{-1}\overline{\mathrm{Ext}_R^*}(S,k)$, defined by Yoneda product [18]. Therefore we have:

COROLLARY.- *Let* $R \xrightarrow{\varphi} S$ *be a Golod map. Then we have a long exact sequence of left* $\mathrm{Ext}^*_{\mathrm{Ext}_R^*(k,k)}(k,k)$-*modules:*

$$\cdots \longrightarrow \mathrm{Ext}^{*-2}_{\mathrm{Ext}_R^*(k,k)}(s^{-1}\overline{\mathrm{Ext}_R^*}(S,k),k) \longrightarrow \mathrm{Ext}^*_{\mathrm{Ext}_R^*(k,k)}(k,k) \xrightarrow{\varphi^{**}} \qquad (18)$$

$$\longrightarrow \mathrm{Ext}^*_{\mathrm{Ext}_S^*(k,k)}(k,k) \longrightarrow \mathrm{Ext}^{*-1}_{\mathrm{Ext}_R^*(k,k)}(s^{-1}\overline{\mathrm{Ext}_R^*}(S,k),k) \longrightarrow \cdots$$

where a l l *the module structures are defined by Yoneda products and by the algebra map* φ^{**}.

Here is another application of the spectral sequence (13) (or (15))and the Ming theory:

THEOREM 4.- *Let*

$$k \longrightarrow A \longrightarrow B \longrightarrow C \longrightarrow k \qquad (19)$$

be an extension of cocommutative graded connected Hopf algebras such that gldim $A = N < \infty$. *Suppose that* V *is a graded left* B-*module such that the* $\mathrm{Ext}_C^*(\mathrm{Tor}_t^B(C,V),k)$ *are noetherian* (coherent) $\mathrm{Ext}_C^*(k,k)$-*modules for* $0 \leq t \leq N$ (module structure defined by the Yoneda product). *Then* $\mathrm{Ext}_B^*(V,k)$ *is a noetherian* (coherent) $\mathrm{Ext}_C^*(k,k)$-*module, via the algebra map* $\mathrm{Ext}_C^*(k,k) \longrightarrow \mathrm{Ext}_B^*(k,k)$ *and the Yoneda product.*

Remark.- Recall that a (graded) finitely generated module is called (graded) coherent if every finitely generated (graded) submodule is finitely presented. It follows that extensions of coherent modules are coherent, and that the kernels, cokernels and images of module maps between coherent modules are again coherent. A ring is called coherent if it is coherent as a module over itself. For more details, cf. e.g. [25] and the literature cited there.

PROOF OF THEOREM 4: We consider the spectral sequence (13), whose $E_2^{p,q}$-term is $\mathrm{Ext}_C^p(\mathrm{Tor}_q^B(C,V),k)$. Since gldim $A = N < \infty$, it follows from (14) that $E_2^{p,q} = 0$ for $q > N$. From the Ming theory (i)-(iii) above, it now follows that for each t $(0 \leq t \leq N)$, the $\{ F^{i-t}\mathrm{Ext}_B^i(V,k) \}_{i\geq 0}$ (put $F^s = F^o$ if $s < 0$) are sub-$\mathrm{Ext}_C^*(k,k)$-modules of $\mathrm{Ext}_B^*(V,k)$. Denote these submodules by $F^{*-t}\mathrm{Ext}_B^*(V,k)$. Then we have $F^{*-t}\mathrm{Ext}_B^*(V,k)/F^{*-(t-1)}\mathrm{Ext}_B^*(V,k) = E_\infty^{*-t,t}$ for $0\leq t\leq N$. (Note that $F^{*-N}\mathrm{Ext}_B^* = \mathrm{Ext}_B^*$ and that $F^{*+1}\mathrm{Ext}_B^* = 0$.) It is therefore sufficient to prove that each $E_\infty^{*-t,t}$ $(0\leq t\leq N)$ is a noetherian (coherent) $\mathrm{Ext}_C^*(k,k)$-module. To prove this, it is sufficient to prove that each $E_2^{*-t,t}$ (note that $E_{N+2}^{*-t,t} = E_{N+3}^{*-t,t} = \ldots = E_\infty^{*-t,t}$) is a noetherian (coherent) $\mathrm{Ext}_C^*(k,k)$-module. But this is exactly the hypothesis of Theorem 4 (modulo a translation of degrees), and the Theorem 4 is proved.

Here are some applications of Theorem 4 (and Theorem 3):

COROLLARY 1.- Let $B = \mathrm{Ext}_R^*(k,k)$ where R is a local complete intersection. Let V be a finitely generated graded left B-module. Then $\mathrm{Ext}_B^*(V,k)$ is a noetherian left $\mathrm{Ext}_B^*(k,k)$-module. (In particular $\mathrm{Ext}_B^*(k,k)$ is a noetherian (bi)graded algebra.)

PROOF: It is known (cf. e.g. [13] or [4]) that B sits in the middle of an exact sequence (19), where C is an exterior algebra (on a finite number of generators of degree 1) and where A is a commutative polynomial algebra on a finite number of generators of degree 2. Thus gldim $A < \infty$, and we are in the situation of Theorem 4. We start by proving that the $\mathrm{Ext}_C^*(k,k)$-modules $\mathrm{Ext}_C^*(\mathrm{Tor}_t^B(C,V),k)$ are noetherian. But since C is an exterior algebra on finitely many generators of degree 1, it follows that $\mathrm{Ext}_C^*(k,k)$ is a bigraded commutative polynomial algebra on finitely many variables of bidegree (1,1). Therefore $\mathrm{Ext}_C^*(k,k)$ is noetherian. Every graded C-module L that is of finite length (or finitely generated - that is the same thing here) over C is an iterated extension of a finite number of trivial C-modules k (placed in different degrees). The exact sequence of $\mathrm{Ext}_C^*(\ ,k)$, therefore gives that $\mathrm{Ext}_C^*(L,k)$ is a noetherian $\mathrm{Ext}_C^*(k,k)$-module. Now V is a finitely generated B-module and therefore also a finitely generated A-module (this is a general assertion about exact sequences (19), where C is finite-dimensional). Therefore $\mathrm{Tor}_t^B(C,V) \simeq \mathrm{Tor}_t^A(k,V)$ is a finite-dimensional vector space, so that $\mathrm{Tor}_t^B(C,V)$ is a C-module of finite length, and therefore, by what we have just proved, $\mathrm{Ext}_C^*(\mathrm{Tor}_t^B(C,V),k)$ is noetherian over $\mathrm{Ext}_C^*(k,k)$. Now Theorem 4 implies that $\mathrm{Ext}_B^*(V,k)$ is a noetherian bigraded module over $\mathrm{Ext}_C^*(k,k)$ via the ring map $\mathrm{Ext}_C^*(k,k) \longrightarrow \mathrm{Ext}_B^*(k,k)$ and the Yoneda product. But this assertion is even stronger than the assertion in Corollary 1 which is therefore proved.
Remark.- The fact, just proved, that $\mathrm{Ext}_B^*(V,k)$ is a finitely generated bigraded module over $\mathrm{Ext}_C^*(k,k)$, implies (Use the Lemma of § 11) that the double Hilbert series of $\mathrm{Ext}_B^*(V,k)$ is equal to $\mathrm{Pol}_V(X,Y)/(1-XY)^N$.

COROLLARY 2.- Let R be a local complete intersection, and let $R \xrightarrow{j} S$ be a Golod map. Let V be a left finitely presented graded $Ext_S^*(k,k)$-module. Then $Ext^*_{Ext_S^*(k,k)}(V,k)$ is a noetherian left $Ext^*_{Ext_R^*(k,k)}(k,k)$-module, via the natural ring map $Ext^*_{Ext_R^*(k,k)}(k,k) \xrightarrow{j^{**}} Ext^*_{Ext_S^*(k,k)}(k,k)$ and the Yoneda product.

PROOF: We have an exact sequence of Hopf algebras:

$$(20) \quad k \longrightarrow T \longrightarrow Ext_S^*(k,k) \longrightarrow Ext_R^*(k,k) \longrightarrow k$$

where T is a free algebra. Apply Theorem 3 to this sequence (20)! Using at some places the short notation $B = Ext_S^*(k,k)$ and $C = Ext_R^*(k,k)$ we obtain (a part of (16)) the following exact sequence of $Ext^*_{Ext_R^*(k,k)}(k,k)$-modules:

$$(21) \quad \cdots \longrightarrow Ext^*_{Ext_R^*(k,k)}(C\otimes_B V,k) \longrightarrow Ext^*_{Ext_S^*(k,k)}(V,k) \longrightarrow Ext^{*-1}_{Ext_R^*(k,k)}(Tor_1^B(C,V),k) \longrightarrow .$$

Now V is a finitely presented left B-module and therefore we have an exact sequence

$$(22) \qquad 0 \longrightarrow W \longrightarrow F \longrightarrow V \longrightarrow 0$$

of left B-modules, where F is a finitely generated and free B-module and where W is a finitely generated B-module. Tensor (22) with C! We obtain an exact sequence of left C-modules:

$$(23) \quad 0 \longrightarrow Tor_1^B(C,V) \longrightarrow C\otimes_B W \longrightarrow C\otimes_B F \longrightarrow C\otimes_B V \longrightarrow 0$$

Here $C\otimes_B F$ is a finitely generated free C-module. Therefore $V_o = C\otimes_B V$ is also a finitely generated C-module. I claim that $V_1 = Tor_1^B(C,V)$ is also a finitely generated C-module. Since W is a finitely generated B-module, it follows as before that $C\otimes_B W$ is a finitely generated C-module. But since $C = Ext_R^*(k,k)$ is noetherian (R is a local complete intersection) we have that $C\otimes_B W$ is noetherian, and therefore its submodule $Tor_1^B(C,V)$ (use(23)!) is finitely generated. Applying Corollary 1 to the finitely generated $Ext_R^*(k,k)$-modules V_o and V_1, we obtain that $Ext^*_{Ext_R^*(k,k)}(V_i,k)$ (i=0,1) are both noetherian $Ext^*_{Ext_R^*(k,k)}(k,k)$-modules, and therefore the middle term of (21) is also so, and the Corollary 2 is proved.

COROLLARY 3.- Using the notations and hypotheses of Corollary 2, we have that $Ext_S^*(k,k)$ is a (graded) coherent algebra. (Left coherence and right coherence are equivalent since $Ext_S^*(k,k)$ is a Hopf algebra.) Furthermore, for each finitely generated S-module M, we have that the left $Ext_S^*(k,k)$-module $Ext_S^*(M,k)$ is coherent.

PROOF: Put $B = Ext_S^*(k,k)$. Recall that we proved in Corollary 2, that if V was a finitely presented left B-module [i.e. if $\dim_k(Tor_i^B(k,V)) < \infty$, $0\le i\le 1$, or , equivalently, if the dual vector spaces $Ext_B^i(V,k)$ had finite dimension for $0\le i\le 1$], then $Ext_B^*(V,k)$ was a finitely generated (bi)graded $Ext^*_{Ext_R^*(k,k)}(k,k)$-module. But this last (bigraded) algebra has a finite number of generators, since R is a local complete intersection. From this one easily sees that $\dim_k \mathrm{Ext}_B^i(V,k) < \infty$ for all $i \geq 2$ too, so that $\mathrm{Ext}_S^*(k,k)$ is coherent [25]. The last part of Corollary 3 now follows from Theorem 1 of [23], and therefore Corollary 3 is completely proved.

Remark.- Taking M = S in Corollary 3, we obtain that k is a coherent (thus finitely presented) $\mathrm{Ext}_S^*(k,k)$-module. Therefore Corollary 2 for V = k shows in particular that $\mathrm{Ext}^*_{\mathrm{Ext}_S^*(k,k)}(k,k)$ is a finitely generated (bigraded) algebra. This will be applied in:

COROLLARY 4.- Let S be a local ring that comes from a local complete intersection by a Golod map, and let $S \xrightarrow{\varphi} S'$ be a second Golod map. Then $\mathrm{Ext}^*_{\mathrm{Ext}_{S'}^*(k,k)}(k,k)$ is a noetherian (bigraded) algebra.

PROOF: Apply the middle part of the exact sequence (18) of the Corollary of Theorem 3 to the Golod map $S \xrightarrow{\varphi} S'$. We obtain an exact sequence of $\mathrm{Ext}^*_{\mathrm{Ext}_S^*(k,k)}(k,k)$-modules

$$(24) \ldots \longrightarrow \mathrm{Ext}^*_{\mathrm{Ext}_S^*(k,k)}(k,k) \xrightarrow{\varphi^{**}} \mathrm{Ext}^*_{\mathrm{Ext}_{S'}^*(k,k)}(k,k) \longrightarrow \mathrm{Ext}^{*-1}_{\mathrm{Ext}_S^*(k,k)}(s^{-1}\overline{\mathrm{Ext}_S^*}(S',k),k) \longrightarrow$$

Now according to Corollary 3 and the Remark following it, both $\mathrm{Ext}_S^*(S',k)$ and k are coherent $\mathrm{Ext}_S^*(k,k)$-modules. Thus $s^{-1}\overline{\mathrm{Ext}_S^*}(S',k)$ is also a coherent $\mathrm{Ext}_S^*(k,k)$-module, and therefore Corollary 2 implies that $\mathrm{Ext}^*_{\mathrm{Ext}_S^*(k,k)}(s^{-1}\overline{\mathrm{Ext}_S^*}(S',k),k)$ is a finitely generated $\mathrm{Ext}^*_{\mathrm{Ext}_S^*(k,k)}(k,k)$-module. This last ring is noetherian (Remark following Corollary 3) and therefore $\mathrm{Ext}^*_{\mathrm{Ext}_{S'}^*(k,k)}(k,k)$ sits between two noetherian modules in the exact sequence (24). Thus $\mathrm{Ext}^*_{\mathrm{Ext}_{S'}^*(k,k)}(k,k)$ is a noetherian $\mathrm{Ext}^*_{\mathrm{Ext}_S^*(k,k)}(k,k)$-module, and à fortiori Corollary 4 is proved.

PROOF OF THEOREM 2: This is now immediate: a) and c) follow from Corollary 2 and Corollary 3 of Theorem 4 and b) follows from Corollary 4 of the same Theorem 4.

§ 3. RINGS WITH MONOMIAL RELATIONS.

THEOREM 5.- Let k be a field, $k[X_1,\ldots,X_n]$ the (commutative) polynomial ring in n variables, let $M_1,\ldots,M_r$ be monomials in the X_i:s, and let $R = k[X_1,\ldots,X_n]/(M_1,\ldots,M_r)$ Then $\mathrm{Ext}^*_{\mathrm{Ext}_R^*(k,k)}(k,k)$ is a (bigraded commutative) noetherian ring.

Remark 1.- The R of Theorem 5 is not local in general, but it has the same Ext-algebra as the corresponding local ring $k[[X_1,\ldots,X_n]]/(M_1,\ldots,M_r)$, and so the preceding theory can be applied to R and related rings.

Remark 2.- Since each variable X_i defines a grading on R, it follows that R is n-graded and that $\mathrm{Ext}^*_{\mathrm{Ext}^*_R(k,k)}(k,k)$ is (n+2)-graded. In the course of the proof of Theorem 5 we will obtain a more precise result about how the finite set of generators of $\mathrm{Ext}^*_{\mathrm{Ext}^*_R(k,k)}(k,k)$ can be chosen.

PROOF OF THEOREM 5: Consider first the case where all the M_i:s are squarefree. Fix a t, $1 \leq t \leq n$, and consider

$$S_t = R/(X_t) = k[X_1,\ldots,\hat{X}_t,\ldots,X_n]/(\text{those } M_j\text{:s, where } X_t \text{ does not occur})$$

(^ means that the corresponding variable is omitted). For those M_i:s, where X_t <u>does</u> occur, we write $M_i = X_t M_i'$ (note that there is no X_t in M_i', since all M_s:s are squarefree). Let $\underline{a}_t$ be the ideal in S_t, generated by the images of these M_i'. Then $R = S_t[X_t]/\underline{a}_t X_t S_t[X_t]$. Of course we can suppose that all M_i:s have degree ≥ 2, and then $\underline{a}_t$ is generated by elements of degree ≥ 1. Writing for simplicity $S = S_t$, $\underline{a} = \underline{a}_t$ and $X = X_t$, we therefore have a Golod map:

$$S[X] \xrightarrow{\ \varphi\ } \frac{S[X]}{\underline{a}\cdot X\cdot S[X]} = R \tag{25}$$

(of course (25) is also a Golod map for more general S:s and $\underline{a}$:s). We now apply the Corollary of Theorem 3 to the Golod map (25), and we obtain from (18) the following long exact sequence of $\mathrm{Ext}^*_{\mathrm{Ext}^*_{S[X]}(k,k)}(k,k)$-modules:

$$\cdots\to \mathrm{Ext}^*_{\mathrm{Ext}^*_{S[X]}(k,k)}(k,k) \xrightarrow{\varphi^{**}} \mathrm{Ext}^*_{\mathrm{Ext}^*_R(k,k)}(k,k) \to \mathrm{Ext}^{*-1}_{\mathrm{Ext}^*_{S[X]}(k,k)}(s^{-1}\overline{\mathrm{Ext}^*_{S[X]}(R,k)},k) \to \mathrm{Ext}^{*+1}_{\mathrm{Ext}^*_{S[X]}(k,k)}(k,k) \to \cdots \tag{26}$$

But $\mathrm{Ext}^*_{S[X]}(k,k) = \mathrm{Ext}^*_S(k,k) \otimes_k E(T)$, where $E(T)$ is the exterior algebra on one variable T of degree 1 $(T^2 = 0)$. Therefore

$$\mathrm{Ext}^*_{\mathrm{Ext}^*_{S[X]}(k,k)}(k,k) = \mathrm{Ext}^*_{\mathrm{Ext}^*_S(k,k)}(k,k) \otimes_k \mathrm{Ext}^*_{E(T)}(k,k) \tag{27}$$

and $\mathrm{Ext}^*_{E(T)}(k,k) = k[V]$, i.e. the commutative polynomial ring on one variable V of bidegree (1,1). Thus (26) is an exact sequence of graded modules over $\mathrm{Ext}^*_{\mathrm{Ext}^*_S(k,k)}(k,k) \otimes_k k[V]$. In particular (26) inherits a grading from X and the operations of V are compatible with this grading.We claim that V operates on the four modules surrounding $\mathrm{Ext}^*_{\mathrm{Ext}^*_R(k,k)}(k,k)$ in (26), as it does on (27), i.e.

1) multiplication by V is a monomorphism,

2) each element of X-degree $\upsilon > 1$ is a multiple by V of an element of the module of X-degree $\upsilon-1$.

Of course this is clear for $\mathrm{Ext}^*_{\mathrm{Ext}^*_{S[X]}(k,k)}(k,k)$ and $\mathrm{Ext}^{*+1}_{\mathrm{Ext}^*_{S[X]}(k,k)}(k,k)$, in view of (27) (we may even take $\upsilon \geq 1$ in (27)). But, since

$$\overline{\mathrm{Ext}}^*_{S[X]}\left(\frac{S[X]}{X\cdot\underline{a}\cdot S[X]},k\right) \simeq s^{-1}\mathrm{Ext}^*_{S[X]}(X\cdot\underline{a}\cdot S[X],k) \simeq s^{-1}\left(\mathrm{Ext}^*_S(\underline{a},k) \otimes_k \mathrm{Hom}_{k[X]}(X\cdot k[X],k)\right),$$

this is also true for $\mathrm{Ext}^{*-1}_{\mathrm{Ext}^*_{S[X]}(k,k)}(s^{-1}\overline{\mathrm{Ext}}^*_{S[X]}(R,k),k)$, and $\mathrm{Ext}^{*-2}(s^{-1}\ldots$

Now it follows easily, using 1) and 2) and the exact sequence (26), that if we put $v = \varphi^{**}(V)$, then every element ξ of X-degree $\upsilon > 1$ in $\mathrm{Ext}^*_{\mathrm{Ext}^*_R(k,k)}(k,k)$ is of the form $\xi = v\cdot\xi'$, where ξ' has X-degree $\upsilon-1$. Furthermore ξ' is unique. (Use the 5-lemma!)

Returning to the old notations $X_t = X$, $S_t = S$ and $\underline{a}_t = \underline{a}$, introducing the notation $v_t = v$, and using the previous result for all t $(1 \leq t \leq n)$ we obtain that each (n+2)-multihomogeneous element $\xi \in \mathrm{Ext}^*_{\mathrm{Ext}^*_R(k,k)}(k,k)$ is a linear combination of products of v_i:s [each v_i has (n+2)-multidegree $(1,1,0..,1,...,0)$, where the last 1 is at the $(i+2)^{nd}$ place] with <u>elements whose last</u> n <u>multidegrees</u> $(a_1,\ldots,a_n)$ <u>verify</u> $0 \leq a_i \leq 1$ $(1 \leq i \leq n)$. But the linear space, spanned by these last elements is finite-dimensional. This last assertion is an easy consequence of the classical result - proved by means of the bar resolution - that if B is a non negatively graded, connected algebra over k, then the graded vector spaces $\mathrm{Tor}^B_p(k,k)$ can contain no nonzero elements of degree < p. Thus $\mathrm{Ext}^*_{\mathrm{Ext}^*_R(k,k)}(k,k)$ has a finite number of generators, if the M_j:s are squarefree. Note that if one of the variables X_j:s is absent in all M_i:s, then we do <u>not</u> need any extra elements above with last n multidegrees $(a_1,\ldots,a_{j-1},1,a_{j+1},\ldots,a_n)$.

We now pass to the general case, when some of the M_j:s might contain squares of the X_i:s (we assume, of course, that all $\deg(M_j) \geq 2$). First of all it is clear that there are still canonical elements $v_i \in \mathrm{Ext}^1_{\mathrm{Ext}^*_R(k,k)}(k,k)$ (corresponding to X_i) of multidegree $(1,1,0,\ldots,1,\ldots,0)$ -the last 1 is at the $(i+2)^{nd}$ place (just use the interpretation of Ext^1). Let m_i = the maximal exponent of X_i among the M_j:s. If all $m_i \leq 1$, then we are in the preceding situation. Assume therefore $m_1 > 1$. Using a procedure, independently due to Fröberg and Weyman ([10], p. 30), we will reduce ourselves to a ring with lower m_1. We can assume (renumber the M:s if necessary!) that $M_1 = X_1^{i_1}M_1'$, ... , $M_k = X_1^{i_k}M_k'$, $i_1 > 0$, ..., $i_k > 0$, but that X_1 does not occur in M_{k+1}, ... , M_r. Now introduce a new variable X_o and put

$$R' = k[X_o,X_1,\ldots,X_n]/(X_oX_1^{i_1-1}M_1', \ldots, X_oX_1^{i_k-1}M_k', M_{k+1}, \ldots, M_r)$$

Now X_o-X_1 is a non-zero divisor in R' and $R'/(X_o-X_1) \simeq R$. Thus we have a "large" map

$R' \xrightarrow{j} R$ in the sense of Levin [17]. It follows that the Hopf algebra map $\mathrm{Ext}^*_R(k,k) \longrightarrow \mathrm{Ext}^*_{R'}(k,k)$ is a monomorphism, and that (*cf.* *loc. cit.* p. 212) we have an isomorphism of left $\mathrm{Ext}^*_{R'}(k,k)$-modules:

(28) $$\mathrm{Ext}^*_{R'}(k,k) \otimes_{\mathrm{Ext}^*_R(k,k)} k \xrightarrow{\sim} \mathrm{Ext}^*_{R'}(R,k)$$

Now $\mathrm{Ext}^i_{R'}(R,k) = 0$, $i > 1$ and $\mathrm{Ext}^i_{R'}(R,k) \simeq k$, $0 \leq i \leq 1$. Therefore we have an exact sequence of $\mathrm{Ext}^*_{R'}(k,k)$-modules (with trivial operations on k and $s^{-1}k$):

$$0 \longrightarrow s^{-1}k \longrightarrow \mathrm{Ext}^*_{R'}(k,k) \xrightarrow{\pi} k \longrightarrow 0$$

which gives rise to a long exact sequence of $\mathrm{Ext}^*_{\mathrm{Ext}^*_{R'}(k,k)}(k,k)$-modules:

(29) $$\cdots \to \mathrm{Ext}^*_{\mathrm{Ext}^*_{R'}(k,k)}(k,k) \xrightarrow{\pi^*} \mathrm{Ext}^*_{\mathrm{Ext}^*_{R'}(k,k)}(\mathrm{Ext}^*_{R'}(R,k),k) \longrightarrow \mathrm{Ext}^*_{\mathrm{Ext}^*_{R'}(k,k)}(s^{-1}k,k) \xrightarrow{\delta} \mathrm{Ext}^{*+1}_{\mathrm{Ext}^*_{R'}(k,k)}(k,k) \longrightarrow \cdots$$

In view of the formula (28) and the fact that $\mathrm{Ext}^*_{R'}(k,k)$ is $\mathrm{Ext}^*_R(k,k)$-free, we obtain that the map π^* can be identified with the ring map

(30) $$\mathrm{Ext}^*_{\mathrm{Ext}^*_{R'}(k,k)}(k,k) \xrightarrow{j^{**}} \mathrm{Ext}^*_{\mathrm{Ext}^*_R(k,k)}(k,k)$$

Suppose now, inductively, that $\mathrm{Ext}^*_{\mathrm{Ext}^*_{R'}(k,k)}(k,k)$ is noetherian. It then follows from (29) that $\mathrm{Ext}^*_{\mathrm{Ext}^*_{R'}(k,k)}(\mathrm{Ext}^*_R(R',k),k)$ is a *noetherian* $\mathrm{Ext}^*_{\mathrm{Ext}^*_{R'}(k,k)}(k,k)$-module, since it is an extension of two such modules. Using the identification of π^* with j^{**}, we now obtain that $\mathrm{Ext}^*_{\mathrm{Ext}^*_R(k,k)}(k,k)$ is a noetherian $\mathrm{Ext}^*_{\mathrm{Ext}^*_{R'}(k,k)}(k,k)$-module, and therefore, *à fortiori*, it is a noetherian ring. *Thus Theorem* 5 *is proved!*

However, we wish to continue and obtain a more precise result about where the generators of $\mathrm{Ext}^*_{\mathrm{Ext}^*_R(k,k)}(k,k)$ are situated. Here is the result we are aiming at:

THEOREM 5'.- *Let* $R = k[X_1,\ldots,X_n]/(M_1,\ldots,M_r)$, *where the* M_j:s *are monomials in the* X_i:s (of degree ≥ 2) , *let* v_i *be the element of multidegree* $(1,1,0,\ldots,1,\ldots,0)$ *in* $\mathrm{Ext}^1_{\mathrm{Ext}^*_R(k,k)}(k,k)$ (corresponding to X_i and defined in general above) *and let* m_i *be the maximal exponent of* X_i *in the* M_j:s. *Then:*

a) *The* $(n+2)$-*multigraded algebra* $\mathrm{Ext}^*_{\mathrm{Ext}^*_R(k,k)}(k,k)$ *is generated by* $v_1,\ldots,v_n$ *and* *a* (necessarily) *finite number of elements having last multidegrees* $(a_1,\ldots,a_n)$, *with* $0 \leq a_i \leq m_i$ $(1 \leq i \leq n)$.

b) Multiplication by v_i in $\mathrm{Ext}^*_{\mathrm{Ext}^*_R(k,k)}(k,k)$ gives an isomorphism of the space of elements of degree $(s,t,b_1,\ldots,b_i,\ldots,b_n)$ onto the space of elements of degree $(s+1,t+1,b_1,\ldots,b_i+1,\ldots,b_n)$, provided $b_i \geq m_i$. Therefore the $(n+2)$-variable Hilbert series of our $(n+2)$-graded double Ext-algebra is:

$p(Z_1,Z_2,Y_1,\ldots,Y_n)/\prod_{i=1}^{n}(1-Z_1Z_2Y_i)$, where $p(Z_1,Z_2,Y_1,\ldots,Y_n)$ is a polynomial in n+2 variables with non-negative integral coefficients, where furthermore for any monomial $Z_1^{i_1}Z_2^{i_2}Y_1^{j_1}\ldots Y_n^{j_n}$ with non-vanishing coefficient we have $j_1 \leq m_1,\ldots,j_n \leq m_n$ and $\max(i_1,i_2) \leq \sum_{i=1}^{n} j_i$.

PROOF OF THEOREM 5′: The theorem follows already from the proof of Theorem 5 if all $m_i \leq 1$. Let us assume $m_1 > 1$ etc. as above and consider $R' \longrightarrow R = R'/(X_0 - X_1)$.

Clearly R' is $(n+1)$-graded by grades $(b_0,b_1,\ldots,b_n)$ (we use X_0 too, to define the grading) and the ring map $R' \longrightarrow R$ is n-multihomogeneous if we use the n-grading $(b_0+b_1,b_2,\ldots,b_n)$ on R'. Those elements in $\mathrm{Ext}^1_{\mathrm{Ext}^*_{R'}(k,k)}(k,k)$ that are analogous to the v_j:s for R will be denoted by v_i' $(0 \leq i \leq n)$. The long exact sequence (29) is $(n+2)$-multihomogeneous, and it is easily seen that δ is the multiplication by a non-zero scalar multiple of $v_0' - v_1'$. If we could prove that $v_0' - v_1'$ were a non-zero-divisor in $\mathrm{Ext}^*_{\mathrm{Ext}^*_{R'}(k,k)}(k,k)$, it would follow from (29) that we have an $(n+2)$-graded exact sequence of $\mathrm{Ext}^*_{\mathrm{Ext}^*_{R'}(k,k)}(k,k)$-modules:

$$(31)\quad 0 \longrightarrow \mathrm{Ext}^{*-1}_{\mathrm{Ext}^*_{R'}(k,k)}(s^{-1}k,k) \xrightarrow{v_0' - v_1'} \mathrm{Ext}^*_{\mathrm{Ext}^*_{R'}(k,k)}(k,k) \longrightarrow \mathrm{Ext}^*_{\mathrm{Ext}^*_R(k,k)}(k,k) \longrightarrow 0$$

and thus all assertions in Theorem 5′ would follow by induction from (31) (recall that $m_0' = 1$, $m_1' = m_1 - 1$, $m_i' = m_i$, $i > 1$).

We therefore now endthe proof of Theorem 5′ by proving that $v_0' - v_1'$ is a non-zero-divisor. Assume the contrary, i.e. that there is a non-zero $c \in \mathrm{Ext}^*_{\mathrm{Ext}^*_{R'}(k,k)}(k,k)$ such that

$$(32)\qquad (v_0' - v_1')\cdot c = 0$$

We decompose c into homogeneous components with respect to the degree (denoted by $\deg_0(\;)$) defined by X_0 : $c = \sum_{j=0}^{N} c_j$ $(c_N \neq 0)$. Thus $\deg_0(c_j) = j$, $\deg_0(v_0')=1$, $\deg(v_1') = 0$ and it now follows by taking the $\deg_0$-component of degree N+1 of (32) that $v_0'\cdot c_N = 0$. But $m_0'=1$, and if $N > 0$, it would follow from the inductive hypothesis (about multiplication by v_0') that $c_N = 0$, which is impossible. Therefore $N = 0$, i.e. $c = c_0$, and now (32) gives $v_1'\cdot c = 0$ (take components of $\deg_0$-degree 0 in (32)). But this is an equality of elements, whose $\deg_0$-grading is zero. Now use the following trick of multihomogeneous algebra: Let R_0' be the subring of R', where

$\deg_o = 0$. We have (recall the notations in the proof of Theorem 5 !):
$R'_o = k[X_1,\dots,X_n]/(M_{k+1},\dots,M_r)$, where the $M_{k+1},\dots,M_r$ do not contain X_1, so that $R'_o \simeq A[X_1]$, where $A = k[X_2,\dots,X_n]/(M_{k+1},\dots,M_r)$. Clearly $\mathrm{Ext}^*_{\mathrm{Ext}^*_{R'_o}(k,k)}(k,k)$ = that part of $\mathrm{Ext}^*_{\mathrm{Ext}^*_{R'}(k,k)}(k,k)$, where $\deg_o$ is zero. Since the equality $v'_1 \cdot c = 0$ takes place in this last part, and since $R'_o = A[X_1] \Rightarrow \mathrm{Ext}^*_{\mathrm{Ext}^*_{R'_o}(k,k)}(k,k) = \mathrm{Ext}^*_{\mathrm{Ext}^*_A(k,k)}(k,k)[v'_1]$ (v'_1 is a polynomial variable), it follows that $c = 0$, which is a contradiction and the Theorem 5′ is completely proved.

Remark.- We had to work rather hard to get that δ was a monomorphism or, equivalently, that the ring map (30) was an epimorphism. There are reasons for that. Indeed, in general, if (R',m') is a local commutative noetherian ring, $x' \in \underline{m}' \setminus (\underline{m}')^2$ a non-zero-divisor, then $R' \xrightarrow{j} R'/(x') = R$ is still large [17], but it is not true that j^{**} is an epimorphism or, equivalently that

(33) $$\mathrm{Tor}_*^{\mathrm{Ext}^*_R(k,k)}(k,k) \longrightarrow \mathrm{Tor}_*^{\mathrm{Ext}^*_{R'}(k,k)}(k,k)$$

is a monomorphism. Here is a counterexample, due to Clas Löfwall and reproduced here with his permission:

$$R' = k[[X,Y,Z]]/(XZ - Y^3) \quad , \quad R = R'/(X) = k[[Y,Z]]/(Y^3) \qquad (\text{thus } x' = X).$$

Clearly X is a non-zerodivisor in R'. Furthermore, both R' and R are local complete intersections, and their Ext-algebras are generated by elements of degree 1 and elements of degree 1 and 2 respectively (cf.e.g. [27]). Consider the inclusion:

(34) $$\mathrm{Ext}^*_R(k,k) \longrightarrow \mathrm{Ext}^*_{R'}(k,k)$$

and take an indecomposable generator τ of degree 2 of $\mathrm{Ext}^*_R(k,k)$. The image of τ under (34) must be a decomposable element, since $\mathrm{Ext}^*_{R'}(k,k)$ is generated by elements of degree 1. Thus, already on the Tor_1 -level, the map (33) is not a monomorphism.

§ 4. FINAL REMARKS. OPEN PROBLEMS.

It might be interesting to try to find other classes of local commutative noetherian rings R, for which

(35) $$\mathrm{Ext}^*_{\mathrm{Ext}^*_R(k,k)}(\mathrm{Ext}^*_R(M,k),k) \text{ is a noetherian } \mathrm{Ext}^*_{\mathrm{Ext}^*_R(k,k)}(k,k)\text{-module for all finitely generated modules } M$$

PROBLEM 1.- Could we "classify" those rings that satisfy (35) ?

It is also possible to study the right $\mathrm{Ext}^*_R(k,k)$-modules $\mathrm{Ext}^*_R(k,N)$ (one of us did so in [23] and Lescot also did so in [16]. This was applied to "Bass series" $I^N(Z) = \sum_{i \geq 0} \dim_k(\mathrm{Ext}^i_R(k,N))\cdot Z^i$ in [23] and [16]. In particular, Lescot has proved [16] that the Bass series $I^R(Z)$ is rational for any ring R with monomial relations as in § 3. He has proved this as a consequence of a very general theorem about rationality of $P_R^M(Z)$:s for multigraded M:s over such R:s, and indeed he has rationality of the multigraded version of P_R^M (n+2 variables) too. There is probably a "double-Ext" - version of this, corresponding to the theory of § 3.

Finally it should be remarked that the rings studied in § 3 contain the "Stanley-Reisner" rings (or "face" rings [28]) associated to a finite simplicial complex Δ .

PROBLEM 2.- Give a combinatorial-geometrical interpretation of the coefficients in the rational function of n+2 variables in Theorem 5′ for the case when R is the Stanley-Reisner ring associated to a finite simplicial complex Δ.

BIBLIOGRAPHY

[1] ANICK, D.J., A counterexample to a conjecture of Serre, Ann. Math., 115, 1982, 1-33. Correction: Ann. Math., 116, 1983, 661.

[2] ANICK, D.J., Connections between Yoneda and Pontrjagin algebras, Lecture Notes in Mathematics, 1051, 1984, 331-350, Springer-Verlag, Berlin, Heidelberg, New York and Tokyo.

[3] ATIYAH, M.F. and MACDONALD, I.G., Introduction to Commutative Algebra, Addison-Wesley, Reading, Mass., 1969.

[4] AVRAMOV, L., Local algebra and rational homotopy, Astérisque, 113-114, 1984, 15-43.

[5] AVRAMOV, L., Differential graded models for local rings, RIMS Kokyuroku, 446,1981, 80-88, Kyoto Research Institute for Mathematical Sciences, Kyoto, Japan.

[6] BACKELIN, J., Les anneaux locaux à relations monomiales ont des séries de Poincaré-Betti rationnelles, Comptes rendus Acad. Sc. Paris, 295, Série I, 1982, 607-610.

[7] BØGVAD,R. and HALPERIN, S., On a conjecture of Roos, These Proceedings.

[8] CARTAN, H. and EILENBERG, S., Homological Algebra, Princeton Univ. Press, Princeton, 1956.

[9] COHEN, F.R., MOORE, J.C. and NEISENDORFER, J.A., Torsion in homotopy groups, Ann. Math., 109, 1979, 121-168.

[10] FRÖBERG, R., A study of graded extremal rings and of monomial rings, Math. Scand., 51, 1982, 22-34.

[11] GOVOROV, V.E., Dimension and multiplicity of graded algebras, Siberian Math. J., 14, 1973, 840-845.

[12] GULLIKSEN, T.H., A change of ring theorem with applications to Poincaré series and intersection multiplicity, Math. Scand. 34, 1974, 167-183.

[13] GULLIKSEN, T.H. and LEVIN, G., Homology of local rings, Queen's Papers in Pure Appl. Math., n° 20, Queen's Univ., Kingston, Ontario, 1969.

[14] JACOBSSON, C., Finitely presented graded Lie algebras and homomorphisms of local rings, J. Pure Appl. Algebra, 38, 1985, 243-253.

[15] LEMAIRE, J.-M., Algèbres connexes et homologie des espaces de lacets, Lecture Notes in Mathematics, 422, 1974, Springer-Verlag, Berlin, Heidelberg, New York.

[16] LESCOT, Thèse, Caen 1985 and letter from J. LESCOT to J.-E. ROOS, June 14, 1985.

[17] LEVIN, G., Large homomorphisms of local rings, Math. Scand., 46, 1980, 209-215.

[18] LEVIN, G., Finitely generated Ext algebras, Math. Scand., 49, 1981, 161-180.

[19] MACLANE, S., Homology, Springer-Verlag, Berlin, Heidelberg, New York, 1963.

[20] MILNOR, J. and MOORE, J., On the structure of Hopf algebras, Ann. Math., 81, 1965 211-264.

[21] MING, R., Yoneda products in the Cartan-Eilenberg change of rings spectral sequence with applications to $BP_*(BO(n))$, Trans. Amer. Math. Soc., 219, 1976, 235-252.

[22] MOORE, J.C. and SMITH, L., Hopf algebras and multiplicative fibrations I-II, Amer. J. Math., 90, 1968, 752-780 and 1113-1150.

[23] ROOS, J.-E., Sur l'algèbre Ext de Yoneda d'un anneau local de Golod, Comptes rendus Acad. Sc. Paris, 286, série A, 1978, 9-12.

[24] ROOS, J.-E., Relations between the Poincaré-Betti series of loop spaces and local rings, Lecture Notes in Mathematics, 740, 1979, 285-322, Springer-Verlag, Berlin, Heidelberg, New York.

[25] ROOS, J.-E., On the use of graded Lie algebras in the theory of local rings, London Math. Soc. Lecture Notes Series, 72, 1982, 204-230, Cambridge University Press, Cambridge.

[26] Séminaire H. CARTAN, 11e année 1958/59, Invariant de Hopf et opérations cohomologiques sécondaires, Paris, Secr. Math., 11 rue Pierre Curie, Paris 5,1959. (Has also been published by Benjamin, New York.)

[27] SJÖDIN,G., A set of generators for $Ext_R(k,k)$, Math. Scand. 38, 1976, 1-12.

[28] STANLEY, R.P., Combinatorics and Commutative Algebra,Progress in Mathematics, vol. 41, 1983, Birkhäuser, Boston, Basel, Stuttgart.

Department of Mathematics
University of Stockholm
Box 6701
S-113 85 STOCKHOLM (SWEDEN)

ON A CONJECTURE OF ROOS

by

Rikard Bøgvad and Stephen Halperin

1. Introduction. In this paper we prove the following two theorems:

Theorem A: Let R be a local commutative ring whose Yoneda Ext-algebra is noetherian. Then R is a complete intersection.

Theorem B: Let S be a 1-connected finite CW complex and suppose the Pontrjagin algebra $H_*(\Omega S;\mathbb{Q})$ is noetherian. Then $\pi_*(S)\otimes\mathbb{Q}$ vanishes in all but finitely many degrees. (S is elliptic - cf. [8].)

Theorem A was a question of Roos [14]. Theorem B is its translation to topology via the standard dictionary ([3],[4]), and was posed by Roos in [13].

The main tool in the proof is Sullivan's notion ([12],[15]) of minimal models, defined by him for the study of topological spaces, and adapted by Avramov ([3]) for the study of local rings.

The first key ingredient is the notion of "category" of a minimal model. This was introduced by Felix-Halperin in [7] for Sullivan models, and shown to coincide with the classical definition of Lusternik-Schnirelmann category. Here we adapt it to Avramov's models as well (in sec. 2).

The second ingredient is the fact that a minimal model determines a graded Lie algebra whose universal enveloping algebra is closely related to the Ext-algebra or Pontryagin algebra ([3],[1]). We recall the necessary facts in sec. 3.

In sec. 4 we combine these ingredients to prove a single theorem about models of which both Theorems A and B are corollaries. In sec. 5 we deduce a theorem on graded Lie algebras.

We thank L. Avramov and C. Löfwall for many helpful discussions.

2. The category of a minimal model. Let $X=\oplus_{p\in\mathbb{Z}}X_p$ be a graded vector space over a field $\mathbf{k}$ (possibly of characteristic >0). If $x\in X_p$ we say $\deg x=p$ and $|x|=|p|$. For $n\geq 0$ we put $X(n)=\oplus_{|p|\geq n}X_p$.

By ΛX we shall mean the tensor product of the exterior algebra on X_{odd} with the symmetric algebra on X_{even}. Then $\Lambda X=\oplus_{p\geq 0}\Lambda^p X$; where where $\Lambda^p X=X\wedge\ldots\wedge X$ (p factors).

We shall use minimal model to mean a DGA of the form $(\Lambda X,d)$ in which

(i) $X=X_{>0}$ or $X=X_{<0}$, and each X_p has finite dimension.

(ii) d is a derivation of degree-1, and $d^2=0$.

(iii) There is a homogeneous basis, $\{x_i\}$, of X with $|x_{i-1}|\leq|x_i|$ and such that $dx_i \varepsilon \Lambda\{x_1,\ldots,x_{i-1}\}$.

(iv) $\mathrm{Im} d \subset \Lambda^{\geq 2}X$.

The basis $\{x_i\}$ will be called a <u>KS basis</u> for $(\Lambda X,d)$.

Given such a model and some $m\geq 1$, the projection $\zeta_m:\Lambda X\rightarrow \Lambda X/\Lambda^{>m}X$ endows the quotient with a differential $\bar{d}$. It is straight forward to factor ζ_m as the composite of homomorphisms

(2.1) $(\Lambda X,d)\xrightarrow{i}(\Lambda X\otimes\Lambda Y,D)\xrightarrow{\rho}(\Lambda X/\Lambda^{>m}X,\bar{d})$,

where Y is a graded space and

(i) $Y=Y_{>0}$ (resp. $Y=Y_{<0}$) if $X=X_{>0}$ (resp. $X=X_{<0}$).

(ii) i is the obvious inclusion; D is a degree-1 derivation; $D^2=0$.

(iii) Y admits a homogeneous basis $\{y_i\}$ with $Dy_i \varepsilon \Lambda X\otimes\Lambda\{y_1,\ldots,y_{i-1}\}$.

(iv) $H(\rho)$ is an isomorphism.

Following [7] we make the

2.2 <u>Definition</u>. The <u>category</u>, $\mathrm{cat}(\Lambda X,d)$, of a minimal model is the least m (possibly ∞) such that there is a sequence (2.1) admitting a morphism

$$r : (\Lambda X\otimes\Lambda Y,D) \rightarrow (\Lambda X,d)$$

with $ri=id$.

2.3 <u>Lemma</u>. (i) If $\mathrm{cat}(\Lambda X,d)\leq m$ then $H_+(\Lambda X)\cdot\ldots\cdot H_+(\Lambda X)$ (m+1 factors)=0, and any cycle in $\Lambda^{>m}X$ is a boundary.

(ii) If $H_i(\Lambda X,d)=0$ for $|i|>n$ then $\mathrm{cat}(\Lambda X,d)\leq n$.

<u>Proof</u>. (i) is immediate. For (ii) we note that there is a d stable acyclic ideal $I\subset\Lambda X$ such that $I_i=(\Lambda X)_i$ for $|i|>n$. The projection $\pi:\Lambda X\rightarrow\Lambda X/I$ then factors through ζ_n. Decomposing ζ_n as in (2.1) we factor π as

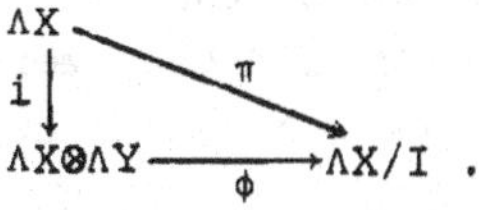

Because $H(\pi)$ is an isomorphism and π is surjective an induction argument on the basis y_i of Y shows one can lift ϕ through ρ to get $r : \Lambda X\otimes\Lambda Y\rightarrow\Lambda X$.

□

For this paper we need an elementary version of the mapping theorem [7; Theorem 5.1]. The proof of [7] needs modification because of the possibility that char $\mathbb{k}>0$. To state the result we note that if $(\Lambda X,d)$ is a minimal model and if we divide by the ideal generated by elements $x\varepsilon X$ with $|x|<p$ then we obtain a quotient minimal model

$(\Lambda X(p),\bar{d})$. The result we need is

2.4 Proposition. $\mathrm{cat}(\Lambda X,d) \geq \mathrm{cat}(\Lambda X(p),\bar{d})$ for all p.

Proof. Let Z be the span of the x_i $(i\geq 2)$, where $\{x_i\}_{i\geq 1}$ is a KS basis for $(\Lambda X,d)$. Thus $\Lambda X=\Lambda(x_1)\otimes\Lambda Z$ and dividing by x_1 gives a minimal model $(\Lambda Z,\bar{d})$. It is clearly sufficient to prove $\mathrm{cat}(\Lambda X,d) \geq \mathrm{cat}(\Lambda Z,\bar{d})$.

Let (u) be a one dimensional vector space with basis u and $\deg u = \deg x_1+1$. Let $\Gamma(u)$ denote the exterior algebra on u if deg u is odd; otherwise $\Gamma(u)$ is the graded algebra with basis $\{\gamma^p u\}_{p\geq 0}$ such that $\gamma^0 u=1$, $\gamma^1 u=u$, $\gamma^p u\cdot\gamma^q u=\binom{p+q}{q}\gamma^{p+q}u$ and $\deg\gamma^p u=p\deg u$. Extend $(\Lambda X,d)$ to a DGA $(\Lambda X\otimes\Gamma(u),\delta)$ by setting $\delta(\gamma^p u)=x_1\otimes\gamma^{p-1}u$. The projection $(\Lambda X,d)\rightarrow(\Lambda Z,\bar{d})$ factors as

$$(\Lambda X,d) \rightarrow (\Lambda X\otimes\Gamma(u),\delta)\xrightarrow{\phi}(\Lambda Z,\bar{d})$$

with $\phi(\gamma^p u)=0$, $p>0$. Because $(\Lambda(x_1)\otimes\Gamma(u),\delta)$ is acyclic, $H(\phi)$ is an isomorphism.

Now suppose $\mathrm{cat}(\Lambda X,d)=m$. Then for a suitable factorization (2.1) of ζ_m we get morphisms

$$\begin{array}{ccccc}(\Lambda X,d) & \xrightarrow{i} & (\Lambda X\otimes\Lambda Y,D) & \xrightarrow{r} & (\Lambda X,d)\\ & & \simeq\big\downarrow\rho & & \\ & & (\Lambda X/\Lambda^{>m}X,\bar{d}) & & ,\end{array}$$

with $H(\rho)$ an isomorphism and $ri=id$. Put $A=\Lambda X/\Lambda^{>m}X$ and apply $\otimes_{\Lambda x_1}(\Lambda x_1\otimes\Gamma(u),\delta)$ to get

$$\begin{array}{ccccc}(\Lambda X\otimes\Gamma(u),\delta) & \xrightarrow{i'} & (\Lambda X\otimes\Lambda Y\otimes\Gamma(u),D') & \xrightarrow{r'} & (\Lambda X\otimes\Gamma(u),\delta)\\ & & \big\downarrow\rho' & & \\ & & A\otimes\Gamma(u),\delta'), & & \end{array} \tag{2.5}$$

Clearly $r'i'=id$. Let $S^p=\bigoplus_{j=0}^{p}(\gamma^j u)$. A simple induction on p then shows that each restriction $(\Lambda X\otimes\Lambda Y\otimes S^p,D')\rightarrow(A\otimes S^p,\delta')$ of ρ' gives an isomorphism in homology. Hence $H(\rho')$ is an isomorphism.

Let $I\subset A\otimes\Gamma(u)$ be the ideal generated by Z if $|x_1|$ is odd, and by Z, x_1^2 and $x_1 u$ if $|x_1|$ is even. In either case condition (iv) in the definition of minimal models implies that $\delta'(I)\subset I$. Moreover, division by I gives a DGA with homology $\mathbb{k}$ in degree zero and zero in the other degrees. Thus the inclusion $I\oplus\mathbb{k}\rightarrow A\otimes\Gamma(u)$ induces an isomorphism in homology.

Similarly if we set $J=(\rho')^{-1}(I)$ and $K=(i')^{-1}(J)$ we obtain that

$$J\oplus\mathbb{k}\rightarrow\Lambda X\otimes\Lambda Y\otimes\Gamma(u) \quad\text{and}\quad K\oplus\mathbb{k}\rightarrow\Lambda X\otimes\Gamma(u)$$

are homology isomorphisms.

In particular, the composite $\psi : K\oplus\mathbb{k}\rightarrow\Lambda X\otimes\Gamma(u)\xrightarrow{\phi}\Lambda Z$ is a surjective homology isomorphism. By induction on a KS basis of ΛZ we obtain a morphism $\sigma : \Lambda Z\rightarrow K\oplus\mathbb{k}$ such that $\psi\sigma=id$ (and hence $H(\sigma)$ is an isomorphism). Thus (2.5) yields the DGA diagram

$$\begin{array}{ccccc} \Lambda Z & \xrightarrow{i\sigma} & J\oplus k & \xrightarrow{\phi r'} & \Lambda Z \\ & & \downarrow{\scriptstyle \rho''} & & \\ & & I\oplus k & & \end{array},$$

ρ'' the restriction of ρ'. Again $H(\rho'')$ is an isomorphism, ρ'' is surjective, and $(\phi r')\circ(i\sigma)=\mathrm{id}$.

But by construction $I\cdot I\cdot\ldots\cdot I$ (m+1 factors)=0. Thus $\rho'' i'\sigma$ factors through the projection $\Lambda Z\to\Lambda Z/\Lambda^{>m}Z$. If this is decomposed as $\Lambda Z\to\Lambda Z\otimes\Lambda W\to\Lambda Z/\Lambda^{>m}Z$ as in (2.1) then the induced map $\Lambda Z\otimes\Lambda W\to I\oplus k$ lifts through ρ'' to a DGA morphism $\Lambda Z\otimes\Lambda W\to J\oplus k$. Composing with $\phi r'$ yields the desired retraction $\Lambda Z\otimes\Lambda W\to\Lambda Z$.

□

Next observe that the proof of [8; Theorem 2.1] applies verbatim in our context to imply

2.6 Proposition. If $(\Lambda X,d)$ is a minimal model with $\mathrm{cat}(\Lambda X,d)<\infty$ and $\dim X=\infty$, then there is a sequence of odd integers n_i such that $\dim X_{n_i}\to\infty$ as $i\to\infty$.

There follows an observation of Felix-Halperin-Thomas (unpublished).

2.7 Corollary. If $(\Lambda X,d)$ is a minimal model with $\mathrm{cat}(\Lambda X,d)<\infty$ and $\dim X=\infty$, then for some p, $H_*(\Lambda X(p),\bar{d})$ is non-zero in infinitely many degrees.

Proof: Put $m=\mathrm{cat}(\Lambda X,d)$. By (2.6) we may choose k so that k is odd and $\dim X_k\geq m$. Let $z_1,\ldots,z_r$ $(r\geq m)$ be a basis of X_k.

Now set $p=|k|+1$. Suppose $H_*=H_*(\Lambda X(p),\bar{d})$ is zero in only finitely many degrees and choose n so $H_n\neq 0$ and $H_\ell=0$ for $|\ell|>|n|$. If $\alpha\neq 0$ in H_n is represented by $\Omega\in[\Lambda X(p)]_n$ then $z_1\wedge\ldots\wedge z_r\wedge\Omega$ is a cycle in ΛX which is not a boundary. This contradicts (2.3 (i)).

□

3. The Lie algebra of a minimal model. Let $(\Lambda X,d)$ be a minimal model. Then $d=d_2+d_3+\ldots$ where the d_i are derivations mapping X to $\Lambda^i X$. In particular, $(\Lambda X,d_2)$ is itself a minimal model. The gradation $\Lambda X=\bigoplus_p\Lambda^p X$ is inherited by $H(\Lambda X,d_2)$ which thus becomes a bigraded algebra $\bigoplus_{p,q} H^p_q(\Lambda X,d_2)$.

On the other hand (cf. [3],[1]) $(\Lambda X,d_2)$ determines a graded Lie algebra $L=\bigoplus_i L_i$ as follows. As a graded space L is given by $L_i=\mathrm{Hom}(X_{-i-1};k)$. Thus the elements of $\Lambda^2 X$ may be interpreted as bilinear functions in L via

$$\langle x\wedge y;\alpha,\beta\rangle = \langle x,\beta\rangle\langle y,\alpha\rangle + (-1)^{|y|\,|x|}\langle x,\alpha\rangle\langle y,\beta\rangle, x,y\in X, \alpha,\beta\in L.$$

The Lie bracket in L is then defined by

$$\langle x;[\alpha,\beta]\rangle = (-1)^{|\beta|}\langle d_2x;\alpha,\beta\rangle. \qquad (3.1)$$

By [3; Theorem 4.2], $(L,[\ ,\ ])$ is a graded Lie algebra: it is called the homotopy Lie algebra of $(\Lambda X,d)$.

The grading of L induces a grading in the universal enveloping algebra, UL. In particular, there is a resolution

$$\longrightarrow M_*(i) \xrightarrow{d_i} M_*(i-1) \to \cdots \to \mathbb{k}$$

of the UL-module $\mathbb{k}$ in which each $M_*(i)$ is a free graded UL-module, and d_i is homogeneous of degree zero. This induces in turn a gradation in each $\mathrm{Tor}_i^{UL}(\mathbb{k},\mathbb{k})$, which we write as $\mathrm{Tor}_i^{UL}(\mathbb{k},\mathbb{k}) = \oplus_p \mathrm{Tor}_{i,p}^{UL}(\mathbb{k},\mathbb{k})$.

3.2 Proposition. If L is the homotopy Lie algebra of a minimal model $(\Lambda X,d)$, then

$$H_q^p(\Lambda X,d_2) = \mathrm{Hom}(\mathrm{Tor}_{p,-p-q}^{UL}(\mathbb{k},\mathbb{k});\mathbb{k})$$

Proof: When L is ungraded this is essentially [6; Theorem 7.1, p. 280]. The proof here is a modification.

Denote by A the graded dual of ΛX : $A_p^q = \mathrm{Hom}(\Lambda^p X)_q;\mathbb{k})$. Let ∂ be the graded dual ([16]) of d_2, and for $x \in X$ let $i(x):A_p \to A_{p-1}$ be the graded dual of left multiplication by x. Then $i(x)$ vanishes in A^q if $|x| > |q|$. Thus if $x_\lambda \in X$, $\alpha_\lambda \in L$ are dual bases we can define $D : UL\otimes A_p \to UL\otimes A_{p-1}$ by

$$D(z\otimes a) = (-1)^{|z|} z\otimes\partial a - \sum_\lambda (-1)^{|z||x_\lambda|} \alpha_\lambda \cdot z\otimes i(x_\lambda)a.$$

We shall show that

$$\longrightarrow UL\otimes A_p \xrightarrow{D} UL\otimes A_{p-1} \xrightarrow{D} \cdots \xrightarrow{D} UL \xrightarrow{\varepsilon} k \tag{3.3}$$

is a resolution. The proposition is then immediate.

A straightforward computation shows that $D^2=0$. Now write (3.3) in the form $\varepsilon : (UL\otimes A,D) \to \mathbb{k}$. Denote by $U_i \subset UL$ the linear span of words of the form $\beta_1 \cdot \ldots \cdot \beta_s$ $(\beta_j \in L,\ s \le i)$. Filter $UL\otimes A$ by the increasing filtration F_*:

$$F_i(UL\otimes A_p) = U_{i-p}\otimes A_p.$$

Finally, notice that $\Delta : x \mapsto x\otimes 1+1\otimes x$ makes ΛX into a coalgebra; the dual algebra structure in A is the free divided powers algebra, $\Gamma(A_1)$, on the graded space A_1. Thus the Poincaré-Birkoff-Witt theorem identifies the associated graded algebra (for the filtration, F_*) of $UL\otimes A$ as $\Lambda L\otimes\Gamma(A_1)$.

But by definition, $A_1^q = L_{-q-1}$. The filtration F_* is stable with respect to D and the associated differential, δ, in $\Lambda L\otimes\Gamma(A_1)$ is the unique derivation such that $\delta_1(L)=0$ and $\delta_1(\gamma^p a) = a\otimes\gamma^{p-1}a$, $a \in A_1$. In particular, $H(\Lambda L\otimes\Gamma(A_1),\delta) = \mathbb{k}$ and so (3.3) is exact.

□

4. The proofs of Theorems A and B. We shall deduce both from

4.1 Theorem. Let $(\Lambda X,d)$ be a minimal model with homotopy Lie algebra L such that

(i) $\mathrm{cat}(\Lambda X,d)$ is finite, and

(ii) UL is noetherian.

Then X is finite dimensional.

Proof: We suppose dimX infinite, and deduce a contradiction. First observe that both (i) and (ii) also hold for the quotient models $(\Lambda X(k),\bar{d})$, because of Proposition (2.4) and the fact that the corresponding Lie algebra is a subalgebra of L. In view of Corollary 2.7 we may thus suppose that $H(\Lambda X,d)$ has infinite dimension.

Now filter $(\Lambda X,d)$ by the ideals $\Lambda^{\geq p}X$ to produce a spectral sequence convergent to $H(\Lambda X,d)$. In view of Proposition 3.2 the columns of the E^2-term can be identified with the graded duals of the spaces $\mathrm{Tor}^{UL}_{p,*}(\Bbbk,\Bbbk)$. Since UL is noetherian, each column has finite total dimension.

But Lemma 2.3 (i) shows that (since $\mathrm{cat}(\Lambda X,d)=m<\infty$) only the first m+1 columns contribute to E^∞. Hence E^∞ (and so $H(\Lambda X,d)$) has finite total dimension. This is the desired contradiction.

□

Proof of Theorem A: Let K^R be the Koszul complex of R. In [3] Avramov associates with K^R a minimal model $(\Lambda X,d)$ over the residue field of R such that $H(\Lambda X,d)=H(K^R)$. Because $H_i(K^R)=0$ unless $0\leq i\leq$(embedding dimension - depth)(R), Lemma 2.3 (ii) shows that $\mathrm{cat}(\Lambda X,d)$ is finite.

On the other hand [3; 1.1, 1.2, 3.1 and 4.2] if L is the homotopy Lie algebra of $(\Lambda X,d)$ then L is a sub Lie algebra of a graded Lie algebra E for which UE is the Ext-algebra; moreover dim(E/L) is finite.

By hypothesis, UE is noetherian. Thus so is UL and so Theorem 4.1 implies that L (and hence E) is finite dimensional. Now a result of Gulliksen [10] shows that R is a complete intersection.

Proof of Theorem B: Let $(\Lambda X,d)$ be the Sullivan minimal model of S (over $\mathbb{Q}$); identify $X^i=X_{-i}$. Then $H(\Lambda X,d) = H(S;\mathbb{Q})$ is finite dimensional, so that $\mathrm{cat}(\Lambda X,d)$ is finite (Lemma 2.3 (i)).

On the other hand ([1]), the homotopy Lie algebra of $(\Lambda X,d)$ is given by $L = \pi_*(\Omega S)\otimes\mathbb{Q}$ with the Samelson product. Thus $UL = H_*(\Omega S;\mathbb{Q})$ with the Pontrjagin product, and so we have only to apply Theorem 4.1.

□

5. Graded Lie algebras. Let L be a graded Lie algebra such that $L_{<0}=0$ or $L_{>-2}=0$ and each L_i is finite dimensional (over some field $\Bbbk$, possibly of positive characteristic) - cf. [3]. Then it is easy to see that (3.1) defines a minimal model $(\Lambda X,d)$ with $d=d_2$. It is an easy exercise to deduce that $\mathrm{cat}(\Lambda X,d_2)$ is the greatest integer m for which $H^m_*(\Lambda X,d_2)$ is non-zero. Thus (cf. [8]) it follows from Proposition 3.2 that

$$\mathrm{cat}(\Lambda X,d_2) = \mathrm{gl.dim}\, UL.$$

Now Theorem 4.1 specializes to

Theorem C. If the graded Lie algebra L has a noetherian universal enveloping algebra of finite global dimension then L is finite dimensional and nilpotent.

□

(For the more or less pan-homological proof for groups see Bieri [5].)

Department of Mathematics,
Massachusetts Institute of Technology,
Cambridge, Mass 02139
U.S.A.

Department of Mathematics,
University of Toronto,
Toronto, Canada
M5S 1A1

Department of Mathematics,
University of Stockholm,
Box 6701
S-1 1385, Stockholm,
Sweden

References

[1] P. Andrews and M. Arkowitz, Sullivan's minimal models and higher order Whitehead products, Canad. J. Math. 30 (1978) 961-982.

[2] L.L. Avramov, Free Lie subalgebras of the cohomology of local rings, Trans. Amer. Math. Soc. 270 (1982), 589-608.

[3] L.L. Avramov, Local algebra and rational homotopy, in Homotopie Algébrique et Algebre Locale, Asterisque 113-114 (1984) 15-43.

[4] L. Avramov and S. Halperin, Through the looking glass: a dictionary between local algebra and rational homotopy, these proceedings.

[5] R. Bieri, Homological Dimension of Discrete Groups, Math. Notes, Queen Mary College, London (1976).

[6] H. Cartan and S. Eilenberg, Homological Algebra. Princeton University Press, Princeton, N.J. (1956).

[7] Y. Felix and S. Halperin, Rational L.S. category and its applications, Trans. Amer. Math. Soc. 273 (1982) 1-37.

[8] Y. Felix, S. Halperin and J.-C. Thomas, The homotopy Lie algebra for finite complexes, Publ. Math. I.H.E.S. 56 (1982) 387-410.

[9] Y. Felix and J.-C. Thomas, Characterization of spaces whose rational LS category is two, to appear Ill. J. Math.

[10] T.H. Gulliksen, A homological characterization of local complete intersections, Composition Math. 23 (1971) 251-255.

[11] T.H. Gulliksen and G. Levin, Homology of Local Rings, Queen's Papers in Pure and Applied Mathematics 20, Queen's University,

Kingston, Canada (1969).

[12] S. Halperin, Lectures on Minimal Models, Mem. de la Soc. Math. de France, 9/10, Paris (1984).

[13] J.-E. Roos, Relations between the Poincaré - Betti series of loop spaces and of local rings. Lect. Notes Math. 740, 285-322, Springer-Verlag, Berlin, 1979.

[14] J.-E. Roos, Homology of loop spaces and of local rings, in Proc. 18th Scand. Cong. of Math. (1980) 441-468, Progress in Mathematics 11, Birkhauser, Basel 1981.

[15] D. Sullivan, Infinitesimal computations in topology, Publ. Math. I.H.E.S. 47 (1978) 269-331.

[16] D. Tanré, Homotopie Rationnelle: Modèles de Chen, Quillen, Sullivan, Lect. Notes Math. 1025, Springer Verlag, Berlin, 1983.

Two examples of smooth projective varieties with non-zero Massey products.

Torsten Ekedahl

The purpose of this note is to try to convince the reader that there exist non-zero Massey products in the mod ℓ cohomology rings of some smooth projective varieties, in sharp contrast to the rational Massey products (cf. [1]). I will first give a non-simply connected example and then show that from it a simply connected example may be constructed. As this not is primarily directed to topologists I will spend some time on the algebro-geometric details.

In the following ℓ will be a fixed prime and we will consider only varieties over the complex numbers. (The interested reader will be able to see that we could work in any characteristic different from ℓ.)

1. Let us recall that if X is a topological space and $a,b,c \in H^i(X,\mathbb{Z}/\ell)$ have the property that ab=bc=0 then we define the triple Massey product $\langle a,b,c\rangle$ as follows: Choose (singular) cocycles $a',b',c' \in Z^i(X,\mathbb{Z}/\ell)$ lifting a, b resp. c and cochains $x,y \in C^{2i-1}(X,\mathbb{Z}/\ell)$ s.t. $dx = a'b'$ and $dy = b'c'$. Then $a'y+(-1)^{i+1}xc'$ is a cocycle of degree $3i-1$. Varying the choices made we see that we get a well defined element $\langle a,b,c\rangle$ in $H^{3i-1}(X,\mathbb{Z}/\ell)/aH^{2i-1}+H^{2i-1}\cdot c$, the triple Massey product. From the definition we immediately see that if $t \in H^j(X,\mathbb{Z}/\ell)$ then

(1.1) $\langle ta,tb,tc\rangle = t^3\langle a,b,c\rangle H^{3(i+j)-1}(X,\mathbb{Z}/\ell)/taH^{2(i+j)-1}+H^{2(i+j)-1}\cdot tc$

We also have an evident functorial property for maps between spaces.

Finally, I will assume known that there are spaces with non-zero Massey products for i=1.

2. Using the classification of principal fibrations it is clear that there is a fibration $K((\mathbb{Z}/\ell)^2,1) \longrightarrow T \xrightarrow{g} K((\mathbb{Z}/\ell)^3,1)$

such that if a triple $d,e,f \in H^1(Y,\mathbb{Z}/\ell)$ is represented by a map $f\colon Y \longrightarrow K((\mathbb{Z}/\ )^3,1)$ then f lifts to T iff de=ef=0. Hence, as there is some space with a non-zero degree 1 triple Massey product, the triple a,b,c of cohomology elements represented by g has a well defined and non-zero Massey product. The long exact homotopy sequence also shows that T is a $K(\pi,1)$ for some ℓ-group π of order ℓ^5.

<u>Lemma</u> 2.1 Any connected space X with $\pi_1 X \twoheadrightarrow \pi$ has a non-zero $\langle d,e,f\rangle \in H^2(X,\mathbb{Z}/\ell)/dH^1+H^1f$.

Indeed, making a classifying map $h\colon X \longrightarrow T$ into a fibration gives us a fibre F with $\pi_1 F = \{1\}$. In the Serre spectral sequences of this fibration $E_2^{i,j} = H^i(T,H^j(F,\mathbb{Z}/\ell)) \Rightarrow H^{i+j}(X,\mathbb{Z}/\ell)$, we therefore have $E_2^{i,1} = 0$ for all i. This shows that $h^1\colon H^1(T,\mathbb{Z}/\ell) \longrightarrow H^1(X,\mathbb{Z}/\ell)$ is an isomorphism and $h^2\colon H^2(T,\mathbb{Z}/\ell) \longrightarrow H^2(X,\mathbb{Z}/\ell)$ a monomorphism. Hence $\bar{h}^2\colon H^2(T,\mathbb{Z}/\ell)/aH^1+H^1c \longrightarrow H^2(X,\mathbb{Z}/\ell)/h^1(a)H^1+H^1h^1(c)$ is a monomorphism and thus $\langle h^1(a),h^1(b),h^1(c)\rangle = \bar{h}^2\langle a,b,c\rangle \neq 0$.

3. It is a general fact, due to Serre [4], that every finite group is the fundamental group of a smooth, connected and projective complex surface. Combining this with Lemma 2.1 we get our first example

<u>Theorem</u> 3.1 There exists a smooth, connected and projective surface X with $a,b,c\ H^1(X,\mathbb{Z}/\ell)$ such that ab=bc=0 and $\langle a,b,c\rangle \neq 0$.

4. The idea is now to use a blow-up to get a simply connected example. Consider any X as in Thm 3.1. It may be embedded in $\mathbb{P}^6$, complex projective 6-space (cf. [2:p. 173]). Choose such an embedding $j\colon X \hookrightarrow \mathbb{P}^6$ and blow up (cf. [3:II,8:24]) the subvariety X in $\mathbb{P}^6$ to get $\tilde{\mathbb{P}}^6$. We have the following situation

$$\begin{array}{ccc} \tilde{X} & \overset{i}{\hookrightarrow} & \tilde{\mathbb{P}}^6 \\ \downarrow{\scriptstyle \rho} & & \downarrow{\scriptstyle \rho'} \\ X & \overset{j}{\hookrightarrow} & \mathbb{P}^6 \end{array} \tag{4.1}$$

If we put $U := \mathbb{P}^6 \setminus X$ then $\rho'^{-1}(U) \longrightarrow U$ is an isomorphism and

$\rho: \tilde{X} \longrightarrow X$ is isomorphic to $\mathbb{P}(N) \longrightarrow X$, the projective bundle associated to N, the conormal bundle of X in $\mathbb{P}^6$.
(Note that, as is usually the case in algebraic geometry, $\mathbb{P}(N)$ is the bundle of hyperplanes and not lines in N.)

Theorem 4.2 $\tilde{\mathbb{P}}^6$ is simply connected and there are elements $a'', b'', c'' \in H^3(\tilde{\mathbb{P}}^6, \mathbb{Z}/\ell)$ s.t. $a''b''=b''c''=0$ and $\langle a'', b'', c''\rangle \neq 0$.

Proof: The first part is well-known. As $\mathbb{P}^6$ is simply connected it suffices to show that ρ' induces an isomorphism on fundamental groups. Consider the following diagram:

$$\begin{array}{ccc} \tilde{\mathbb{P}}^6 & \longleftrightarrow & \rho^{-1}(U) =: U' \\ \downarrow \rho' & & \downarrow \\ \mathbb{P}^6 & \longleftrightarrow & U \end{array}$$

As X has real codimension greater than 2 in $\mathbb{P}^6$ $\pi_1 P^6 = \pi_1 U = \pi_1 U'$ and as $\tilde{X}$ has real codimension 2 in $\tilde{\mathbb{P}}^6$ $\pi_1 U' \longrightarrow \pi_1 P^6$ is surjective. This immediately gives the desired conclusion.

To continue the proof let us recall some further properties of (4.1).

As $\tilde{X}$ is of (complex) codimension 1 in $\tilde{\mathbb{P}}^6$ its defining ideal I is a linebundle on $\tilde{\mathbb{P}}^6$. Its restriction to X is the tautological linebundle $\mathcal{O}_{P(N)}(1)$. Let $c_1 \in H^2(\tilde{X}, \mathbb{Z}/\ell)$ be the first Chern class of $I_{|\tilde{X}}$. Then, as N is of rank 4, $H^*(\tilde{X}, \mathbb{Z}/\ell)$ is, through ρ^*, a free $H^*(X, \mathbb{Z}/\ell)$-module on generators 1, c_1, c_1^2 and c_1^3. As a first step we now consider $a', b', c' := \rho^*a, \rho^*b$ resp. ρ^*c and we find that $\langle a', b', c'\rangle \neq 0$ because ρ^* is an isomorphism in degree 1 and a monomorphism in degree 2.

The next step is to push a', b' and c' into $\tilde{\mathbb{P}}^6$. As $\tilde{X}$ and $\tilde{\mathbb{P}}^6$ are canonically oriented (real) manifolds we have a Gysin-map $i_*: H^*(\tilde{X}, \mathbb{Z}/\ell) \longrightarrow H^*(\tilde{\mathbb{P}}^6, \mathbb{Z}/\ell)$ of degree $\dim_{\mathbb{R}} \tilde{\mathbb{P}}^6 - \dim_{\mathbb{R}} \tilde{X} = 2$. Apart from the projection formula $x \cdot i_* y = i_*(i^* x \cdot y)$ we will need a well-known formula for $i^* \circ i_*$. I will take the opportunity to present a simple proof of it in the next section. The element i_*

element $i_*(1)$ is the class of the subvariety $\tilde{X}$ in $H^*(\mathbb{P}^6, \mathbb{Z}/\ell)$. As $\tilde{X}$ is of codimension 1 this class equals the first Chern class of the associated linebundle which is I^{-1}. Hence $i^*i_*(1) = -c_1$ and (5.1) shows that

(4.2.1) $\quad i^*i_*(x) = -c_1 x \qquad x \in H^*(\tilde{X}, \mathbb{Z}/\ell)$

We are now fully prepared to construct a", b" and c". Put $a'', b'', c'' := i_*a', i_*b', i_*c'$. Let us verify that indeed $a''b'' = b''c'' = 0$. By the projection formula $a''b'' = i_*(i^*i_*(a')\cdot b')$ which by (4.2.1) equals $i_*(-c_1 a'b') = 0$ and similarly for $b''c''$. This implies that a'', b'', c'' is defined and it only remains to show that it is non-zero. To do this we may pull back to $\tilde{X}$. Using (1.1) and (4.2.1) we see that $i^*\langle a'', b'', c''\rangle = -c_1^3\langle a', b', c'\rangle$ $H^8(\tilde{X}, \mathbb{Z}/\ell)/c_1 a'H^5 + H^5 c_1 c'$. Assume that there are $x, y \in H^5(\tilde{X}, \mathbb{Z}/\ell)$ s.t. $-c_1^3\langle a', b', c'\rangle = c_1 a'x + c_1 c'y$. Writing $x = x_0 + c_1 x_1 + c_1^2 x_2$ and $y = y_0 + c_1 y_1 + c_1^2 y_2$ with $x_i, y_i \in H^*(X, \mathbb{Z}/\ell)$ and using the linear independence of 1, c_1, c_1^2 and c_1^3 we see that $x_0 = x_1 = y_0 = y_1 = 0$ and $-\langle a', b', c'\rangle = a'x_2 + c'y_2$ which contradicts the fact that $\langle a', b', c'\rangle \neq 0$.

5. I will end by giving the promised short proof of the following proposition.

<u>Proposition</u> 5.1 Let $i: Y \hookrightarrow X$ be a closed immersion of smooth varieties. Then $i^*i_*(x) = i^*i_*(1)\cdot x$ for $x \in H^*(Y, R)$, R a commutative ring.

Indeed, we have two morphisms i^*i_*, $i^*i_*(1)\cdot(-) : i_*R \longrightarrow i_*R[2\operatorname{codim} Y]$ in $D(X,R)$ and it suffices to show that they are equal. The projection formula shows that their composites with the natural mapping $R \longrightarrow i_*R$ are equal. It is thus sufficient to show that the mapping a: $\operatorname{Hom}_{D(X,R)}(i_*R, i_*R$

a: $\operatorname{Hom}_{D(X,R)}(i_*R, i_*R[2\operatorname{codim} Y]) \longrightarrow$
$\operatorname{Hom}_{D(X,R)}(R, i_*R[2\operatorname{codim} Y])$ is injective. The mapping a is clearly equal to the natural mapping $R^0\Gamma_Y(i_*R[2\operatorname{codim} Y])$

$\longrightarrow R^0\Gamma(i_* R[2\text{codim } Y])$ from hypercohomology with support in Y to ordinary hypercohomology. As $i_* R[2\text{codim } Y]$ has support in Y, a is in fact an isomorphism.

[1]. P. Deligne, P. Griffiths, J. Morgan and D. Sullivan: Real homotopy theory of Kähler manifolds. Inv. math. 29, p. 245-274 (1975)

[2]. P. Griffiths and J. Harris: Principles of algebraic geometry. Wiley-Interscience 1978

[3]. R. Hartshorne: Algebraic geometry. Springer Verlag 1977

[4]. J.-P. Serre: Sur la topologie des variétés algébriques en caractéristique p. Symposium de Topologica Algebraica, Mexico (1956)

Department of Mathematics
University of Stockholm
Box 6701
S-113 85 STOCKHOLM
(SWEDEN)

THE RADICAL OF $\pi_*(\Omega S) \otimes \mathbb{Q}$

by

Yves Felix, Stephen Halperin, Daniel Tanré & Jean-Claude Thomas

Let S be a simply connected topological space whose rational homology is finite dimensional in each degree. This is then true for the rational homotopy as well, so that the graded Lie algebra $\pi_*(\Omega S) \otimes \mathbb{Q}$ (Samelson product) is also finite dimensional in each degree. It can be identified with the Lie algebra of primitive elements in $H_*(\Omega S;\mathbb{Q})$, and is called the <u>rational homotopy Lie algebra of</u> S.

Denote by $\mathrm{cat}_0(S)$ its rational Lusternik-Schnirelmann category [F-H]. It coincides with the classical L-S category of the localization $S_{\mathbb{Q}}$ [T] when S is a CW complex. The restriction $\mathrm{cat}_0(S) < \infty$ imposes serious conditions on the rational homotopy Lie algebra [FHT]; in particular, with rare exceptions it will be infinite dimensional.

On the other hand, in general there is little nontrivial Lie theory for infinite dimensional Lie algebras. For example, the sum of solvable ideals may not be solvable. By contrast we have

<u>Theorem 1</u>. Suppose S is simply connected with finite rational homology in each degree and assume that $\mathrm{cat}_0(S)$ is finite. The sum, R, of all the solvable ideals in $\pi_*(\Omega S) \otimes \mathbb{Q}$ is then solvable.

<u>Definition</u>. R is called the <u>radical</u> of $\pi_*(\Omega S) \otimes \mathbb{Q}$.

Theorem 1 is an immediate consequence of Theorem 2 below. Recall that the solv length of a solvable Lie algebra is the length of its derived series.

<u>Theorem 2</u>. Let S be as in Theorem 1 with $\mathrm{cat}_0(S) \leq m$. Suppose I is a graded ideal in $\pi_*(\Omega S) \otimes \mathbb{Q}$. The following conditions are then equivalent:

(i) Either $\dim \pi_*(\Omega S) \otimes \mathbb{Q}$ is finite, or for each k

$$\sum_{i=k}^{2k-1} \dim I_{2i} < m .$$

(ii) I is solvable.

When these conditions hold then

$$\text{solv length } (I) < 4 + 2\log_2 m .$$

<u>Corollary</u>. If I is solvable then every finitely generated subalgebra of I_{even} is

finite dimensional.

Proof of the corollary. Let $\alpha_1,\ldots,\alpha_t \in I$ have even degrees and consider a sequence of elements

$$\gamma_\nu = [\alpha_{i_\nu}[\alpha_{i_{\nu-1}}[\ldots[\alpha_{i_2},\alpha_{i_1}]\ldots] \qquad 1 \leq i_\nu \leq t .$$

If $\ell = \max \deg \alpha_i$ then the degrees of the γ_ν lie in an interval of the form $[k,2k-2]$ if $\ell m < \nu \leq (\ell+1)m$. It follows from (i) that $\gamma_\nu = 0$, $\nu \geq (k+1)m$. □

Proof of Theorem 2. Suppose first that (i) fails for I. Theorem 2.2 of [FHT] then asserts that either (i) fails for I' or that for some $\alpha \in I$ of even degree and some $\beta \in L_{odd}$, $(ad\alpha)^i\beta \neq 0$ all i. In the latter case (i) must still fail for I' as follows from Proposition 8.1 of [FHT]. By induction (i) fails for each term of the derived series, and I is not solvable.

Conversely, assume that the $(k+1)$st term, $I^{(k+1)}$, of the derived series is nonzero for some $k+1 \geq 4 + 2\log_2 m$. There are then homogeneous elements $\alpha, \beta \in I^{(k)}$ such that $\deg \alpha \leq \deg \beta$ and $[\alpha,\beta] \neq 0$.

We may, moreover suppose that for each $1 \leq j \leq k$ we are given 2^j elements

$$\alpha(\varepsilon_1,\ldots,\varepsilon_j) \in I^{(k-j)}, \qquad \varepsilon_\gamma = 0 \text{ or } 1,$$

for which

$$\alpha = [\alpha(0),\alpha(1)] \quad \text{and}$$

$$\alpha(\varepsilon_1,\ldots,\varepsilon_j) = [\alpha(\varepsilon_1,\ldots,\varepsilon_j,0),\ \alpha(\varepsilon_1,\ldots,\varepsilon_j,1)] .$$

Now note that L is necessarily infinite dimensional, since if L were of finite dimension we would have ([F-H]):

$$\dim L_{odd} \leq \dim L_{even} \leq m .$$

But for some ordering $\alpha_1,\ldots,\alpha_{2^k}$ of the $\alpha(\varepsilon_1,\ldots,\varepsilon_k)$ we have

$$[\alpha_i,[\alpha_{i+1},\ldots[\alpha_{2^k},\beta]]\ldots] \neq 0, \qquad 1 \leq i \leq 2^k . \tag{1}$$

This provides 2^k linearly independent elements of L, which contradicts $k > 1 + \log_2(m)$. Hence L is infinite dimensional.

Next observe that if at least $2m$ elements of the $\alpha(\varepsilon_1,\ldots,\varepsilon_k)$ have odd degree then the sequence (1) above contains at least m even elements whose degrees are even and lie in the interval $[\deg\beta+1,\ \deg\beta+\deg\alpha]$. This violates (i).

On the other hand, suppose at most $2m-1$ of the $\alpha(\varepsilon_1,\ldots,\varepsilon_k)$ have odd degree. Choose the least integer n such that $2^n \geq 2m$. Then amongst the 2^n elements $\alpha(\varepsilon_1,\ldots,\varepsilon_n)$ at least one (say γ) is built up from elements $\alpha(\varepsilon_1,\ldots,\varepsilon_k)$ all of which have even degree.

As above we may expand γ to a sequence of independent elements of the form

$$[\gamma_i, \gamma_{i+1}, \ldots [\gamma_{2^{k-n-1}}, \gamma']] \ldots] , \qquad 1 \leq i \leq 2^{k-n-1} ,$$

all of which have even degrees in the interval $[\deg \gamma' + 2, 2\deg \gamma']$. Since $2^{n-1} < 2m$, $2^{k-n-1} > 2^k/8m \geq 8m^2/8m = m$ and so (i) is still violated.

REFERENCES

[B] N. Bourbaki, Groupes et algèbres de Lie - Chap. I.

[F-H] Y. Felix et S. Halperin, L.S. category and its applications. Trans. AMS 273 (1982), 1-38.

[F-H-T] Y. Felix, S. Halperin et J.C. Thomas, The homotopy Lie algebra for finite complexes. Publ. I.H.E.S. n° 56 (1982), 387-410.

[T] G. Toomer, Lusternik-Schnirelmann category and the Milnor-Moore spectral sequence. Math. Z. 138 (1974), 123-143.

Y. Felix
Université de Louvain-la-Neuve
2, Chemin du Cyclotron
1348 - Louvain-la-Neuve, Belgique

S. Halperin
Department of Mathematics
University of Toronto
Toronto, Canada M5S 1A1

D. Tanré et J.-C. Thomas
Université de Lille I
U.E.R. de Mathématiques Pures et Appliquées
59655 - Villeneuve d'Ascq Cedex, France

SUR L'OPERATION D'HOLONOMIE RATIONNELLE

Y. FELIX[*] et J.C. THOMAS[**]

Etant donné une fibration (au sens de Hurewicz) $F \xrightarrow{j} E \xrightarrow{p} B$, la propriété de relèvement des homotopies définit (à homotopie près) l'opération (bien connue) du H-espace homotopiquement associatif ΩB sur l'espace topologique F, notée

$$\nu : \Omega B \times F \longrightarrow F.$$

Dans le cas particulier de la fibration canonique $\Omega B \to PB \to B$, ν est simplement la multiplication des lacets.

Pour chaque anneau R, commutatif unitaire, ν induit une opération à gauche de l'anneau de Pontryagin $H_*(\Omega B;R)$ sur le R-module gradué $H_*(F;R)$

$$\nu_* : H_*(\Omega B;R) \otimes H_*(F;R) \longrightarrow H_*(F;R).$$

Par référence à la géométrie différentielle, nous appelons ν une *représentation d'holonomie de la fibration*. En effet, si la fibration est un fibré différentiable localement trivial, de base paracompacte, la donnée d'une connexion de Ehresmann fournit une application $u_k : \Omega_k(B) \to G$ où G désigne un sous-groupe de difféomorphismes de F et $\Omega_k(B)$ l'espace des lacets de classe C^k par morceaux. Les inclusions $G \hookrightarrow \mathrm{Diff}(F) \hookrightarrow \mathrm{Homéo}(F) \hookrightarrow \mathrm{aut}(F)$, où $\mathrm{aut}(F)$ désigne le monoïde des "self equivalences" de F, fournissent alors le morphisme ν par adjonction.

L'opération ν est dite triviale (resp. $\mathbb{Q}$-triviale, resp. H-triviale), si ν est homotope[***], en tant qu'application continue, à la projection $\pi_2 : \Omega B \times F \to F$ (resp. $\nu \underset{\mathbb{Q}}{\sim} \pi_2$, resp. $H_*(\nu,\mathbb{Q}) = H_*(\pi_2,\mathbb{Q})$).

(*) Chercheur qualifié F.N.R.S.

(**) E.R.A. au C.N.R.S. n° 07 590

(***) Nous ne demandons pas que les applications $\nu_t : \Omega B \times F \to F$, définies par l'homotopie, soient des opérations de ΩB pour tout t.

L'étude de l'opération ν a été abordée notamment par Stasheff et Ganéa qui démontrent :

- [St] ν est une application fortement homotopiquement multiplicative (S.H.M.) ;
- [Ga,2] ν est triviale si et seulement si l'injection $j : F \to E$ est un monomorphisme dans la catégorie homotopique.

Dans cet article, nous nous intéressons principalement à l'opération d'holonomie rationnelle

$$\nu_* : H_*(\Omega B;\mathbb{Q}) \otimes H_*(F;\mathbb{Q}) \longrightarrow H_*(F;\mathbb{Q}).$$

Les espaces sont supposés connexes par arcs et ayant le type d'homotopie de C.W. complexes de type fini. La base B sera supposée 1-connexe. Sauf mention explicite, l'homologie et la cohomologie des espaces sont supposées à coefficients rationnels. La notation H_+ désigne l'homologie réduite.

D'après les théorèmes de Milnor-Moore et Cartan-Serre, $H_*(\Omega B)$ s'identifie à l'algèbre enveloppante de l'algèbre de Lie d'homotopie $\pi_*(\Omega B) \otimes \mathbb{Q}$. Donc une représentation du premier équivaut à une représentation du second. En particulier, si θ_ν note la représentation de $\pi_*(\Omega B) \otimes \mathbb{Q}$ dans $H_*(F)$, alors une représentation $\theta_\nu^\#$ de $\pi_*(\Omega B) \otimes \mathbb{Q}$ dans $H^*(F)$ est définie par :

$$\theta_\nu^\#(u)(a) = (-1)^{\deg u.\deg a+1}\theta_\nu(u)^*(a) .$$

Elle induit une représentation $\tau : H_*(\Omega B) \otimes H^*(F) \to H^*(F)$.
Nous verrons au § 3.2 que l'action $\theta_\nu^\#$ s'obtient par des dérivations.

A l'aide de la théorie des modèles de Sullivan, nous donnons au § 3 une description calculatoire de ν^* et de τ. Ceci nous permet de démontrer les résultats énoncés ci-après.

D'après Bogvad ([Bo]), si $\mathrm{cat}_0(B) < \infty$, alors $\dim \pi_*(\Omega B) \otimes \mathbb{Q} < \infty$ si et seulement si $H_*(\Omega B)$ est une algèbre noetherienne. Notre premier théorème généralise ce résultat :

Théorème 4.1.- Si $\pi_*(\Omega B) \otimes \mathbb{Q}$ *et* $H^*(E)$ *sont des espaces vectoriels de dimension finie, et si* B *et* F *sont* 1*-connexes, alors*

i) $H_*(F)$ *est un* $H_*(\Omega B)$*-module noetherien ;*

ii) *La série de Poincaré de* $H_*(F)$ *est une fraction rationnelle de la forme* $P(t)/\prod_i (1-t^{2i})^{\alpha_{2i}}$ *avec* $P(t) \in \mathbb{Z}[t]$ *et* $\alpha_i = \dim \pi_i(\Omega B) \otimes \mathbb{Q}$.

Dans la correspondance topologie-algèbre locale [A.H], ce résultat correspond à un théorème de Gulliksen ([Gu,2]).

Une question surgit alors : si $\dim \pi_*(\Omega B) \otimes \mathbb{Q}$ est infini, sous quelles conditions $H_*(F)$ est-il un $H_*(\Omega B)$-module finiment engendré ? Si B est un bouquet de sphères, la seule condition $\dim H^*(E) < \infty$ suffit (§ 8). Plus généralement, cette condition suffit-elle ?

L'intérêt de cette question est illustrée par le théorème suivant :

Théorème 4.6.- Si le connectant de la suite longue d'homotopie rationnelle d'une fibration est nul et si $H_*(F)$ *est un* $H_*(\Omega B)$*-module finiment engendré, alors l'algèbre de cohomologie* $H^*(F)$ *est de nilpotence finie.*

Dans de nombreux cas, $H_+(F)$ est un $H_*(\Omega B)$-module libre. En effet,

Théorème 5.1.- Soit $i : A \to X$ *une cofibration fermée et* F *la fibre homotopique de la projection* $X \to X/A$, *alors* $H_+(F)$ *est un* $H_*(\Omega(X/A))$*-module libre engendré par* $H_+(A)$.

Ce résultat permet de retrouver l'homologie des espaces $\Omega\Sigma A$, $\Omega A * \Omega A$. Il permet aussi le calcul de l'homologie des espaces de Ganéa et des produits réduits relatifs de James.

A l'antipode des actions libres, nous trouvons les actions triviales introduites par Ganéa (cf. plus haut). Leur étude apporte des précisions sur la structure fibrée.

Rappellons qu'un 0-cône est un point et qu'un n-cône est obtenu en rattachant un bouquet de cellules (non nécessairement toutes de même dimension) à un (n-1)-cône.

<u>*Proposition.*</u>- *Si* B *est un* 1*-cône, alors une fibration de base* B *est homotopiquement triviale si l'opération d'holonomie est triviale.*

La suite de Wang qui est définie quand B est un 1-cône se généralise dans certains cas :

<u>*Proposition*</u> *6.3.- Si* B *est un* 2*-cône et si l'opération d'holonomie dans la fibration* $F \xrightarrow{j} E \xrightarrow{p} B$ *est triviale, alors il existe une longue suite exacte*

$$\longrightarrow H^{n-1}(F) \xrightarrow{\Delta} (H_+(B) \otimes H(F))^n \longrightarrow H^n(E) \xrightarrow{j^*} H^n(F) \xrightarrow{\Delta} \cdots$$

Si on considère l'opération ν_* en homologie (à coefficients dans un anneau R), on voit que

$$\nu_*(H_+(\Omega B;R) \otimes H_*(F;R)) \subset \mathrm{Ker}(j_* : H_*(F;R) \longrightarrow H_*(E;R))$$

puisque $j\nu \sim j\pi_2$, où π_2 désigne la seconde projection $\Omega B \times F \to F$.

Il en résulte (pour le cas ii) utiliser [Go]).

<u>*Proposition.*</u>-

i) Toute fibration T.N.C.Z. rel R (i.e. $j_* : H_*(F;R) \to H_*(E;R)$ *est injective) a une holonomie* H*-triviale.*

ii) Si la fibre d'une fibration est une variété fermée dont la caractéristique d'Euler-Poincaré est non nulle, alors la classe fondamentale n'appartient pas à $\nu_*(H_+(\Omega B;R) \otimes H_*(F;R))$.

A la fibration $F \xrightarrow{j} E \xrightarrow{p} B$ est associée classiquement ([E.M]) une suite spectrale

$$E_2 = \mathrm{Ext}_{H^*(B)}(H^*(E),\mathbb{Q}) \Longrightarrow H_*(F)$$

appelée suite spectrale d'Eilenberg-Moore. Or la multiplication de Yoneda fait de $\mathrm{Ext}_{H^*(B)}(\mathbb{Q},\mathbb{Q})$ une algèbre et de $\mathrm{Ext}_{H^*(B)}(H^*(E),\mathbb{Q})$ un $\mathrm{Ext}_{H^*(B)}(\mathbb{Q},\mathbb{Q})$-module.

Théorème 7.1.- Si l'application p *est formalisable* ([L.S.]), *alors les formalisations de* B *et de* E *induisent des isomorphismes*

$$\mathrm{Ext}_{H^*(B)}(\mathbb{Q},\mathbb{Q}) \cong H_*(\Omega B;\mathbb{Q})$$

et

$$\mathrm{Ext}_{H^*(B)}(H^*(E),\mathbb{Q}) \cong H_*(F;\mathbb{Q}).$$

Le premier est un isomorphisme d'algèbres, le second un isomorphisme de modules sur ces algèbres.

En particulier, les suites spectrales d'Eilenberg-Moore dégénèrent ([V]).

A titre d'exemple, rappelons que toute application holomorphe entre variétés kälhériennes est formalisable ([D.G.M.S.]).

Des théorèmes 4.1 et 7.1, on tire immédiatement un analogue d'un théorème de Levin ([Le]).

Corollaire.- Si H *est une* $\mathbb{Q}$*-algèbre graduée* 1*-connexe et* H' *une* H*-algèbre graduée, et si* $\mathrm{Ext}_H(\mathbb{Q},\mathbb{Q})$ *est une algèbre noethérienne (pour le produit de Yoneda), alors* $\mathrm{Ext}_H^+(H',\mathbb{Q})$ *est un* $\mathrm{Ext}_H(\mathbb{Q},\mathbb{Q})$*-module noetherien.*

De même, des théorèmes 5.2 et 7.1, on tire une version graduée d'un résultat de Roos ([R]).

Corollaire 7.2.- Soit H *une* $\mathbb{Q}$*-algèbre graduée* 1*-connexe. Si* $(H^+)^n \neq 0$ *et* $(H^+)^{n+1} = 0$*, alors* $\mathrm{Ext}_H^+(H/_{(H^+)^n},\mathbb{Q})$ *est un* $\mathrm{Ext}_H(\mathbb{Q},\mathbb{Q})$*-module libre engendré par* $\mathrm{Ext}_H^1(H/_{(H^+)^n},\mathbb{Q})$.

La représentation d'holonomie fournit finalement une suite spectrale, appelée suite spectrale d'holonomie qui généralise la suite spectrale de Milnor-

Moore d'un espace [Q].

<u>Théorème</u> *8.1.- Pour chaque fibration, il existe une suite spectrale du premier quadrant vérifiant*

$$E_2 = \mathrm{Ext}_{H_*(\Omega B)}(\mathbb{Q}, H^*(F)) \Longrightarrow H^*(E).$$

Si B *a le type d'homotopie rationnelle d'une suspension, cette suite spectrale dégénère au terme* E_2 *et on a un isomorphisme d'espaces vectoriels gradués*

$$\mathrm{Ext}_{H_*(\Omega B)}(\mathbb{Q}, H^*(F)) \cong H^*(E).$$

Le texte s'organise comme suit :

1. *Définition de l'opération d'holonomie.*
2. *Quelques points d'homotopie rationnelle.*
3. *Calcul de l'opération d'holonomie rationnelle.*
4. *Opération d'holonomie noethérienne.*
5. *Opération d'holonomie libre.*
6. *Opération d'holonomie triviale.*
7. *Holonomie et suite spectrale d'Eilenberg-Moore.*
8. *Suite spectrale d'holonomie.*

§ 1 - <u>*Définition de l'opération d'holonomie.*</u>

1.1.- Soit $F \to E \to B$ une fibration. Supposons B pointé en b_o. Notons PB l'espace des chemins de B d'origine b_o. Désignons par ℓ : $F \to PB \times_B E$ l'application envoyant f sur (c_{b_o}, f) où c_{b_o} désigne le chemin constant en b_o. L'injection canonique $\Omega B \times F \to PB \times_B E$ se factorise à homotopie près par F fournissant ainsi un morphisme $\nu : \Omega B \times F \to F$ appelé *opération d'holonomie* de la fibration

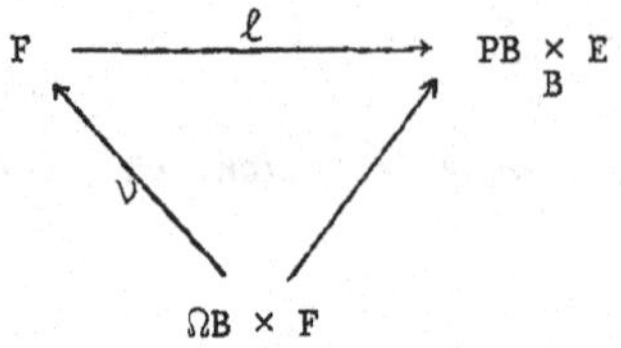

Proposition ([W]).- *L'opération d'holonomie* ν *est une opération du H-espace homotopiquement associatif* $(\Omega B,\mu)$ *sur l'espace* F.

Cette action est s.h.m. ([St]).

1.2.- Exemples.

(1) Si $f : E \to B$ est une application continue et $F_f \to E_f \to B$ la fibration homotopique associée $(E_f = B^I \times_B E)$, l'opération d'holonomie $\nu : \Omega B \times F_f \to F_f$ est définie par

$$\nu(\omega,(\gamma,e)) = (\omega \circ \gamma, e).$$

(2) La fibration canonique $\Omega B \to PB \to B$ étant la fibration associée à l'injection du point de base b_o dans B, l'opération d'holonomie ν est simplement la composition dans l'espace des lacets.

(3) L'opération d'holonomie γ dans la fibration des lacets libres $\Omega B \xrightarrow{j} B^{S^1} \xrightarrow{p} B$ $(p(\omega) = \omega(0) = \omega(1))$ est la conjugaison des lacets

$$\nu(\omega,\omega') = \omega \circ \omega' \circ \omega^{-1}.$$

(4) Le connectant δ de la suite de Baratt

$$\Omega E \xrightarrow{\Omega p} \Omega B \xrightarrow{\delta} F \xrightarrow{j} E \xrightarrow{p} B$$

est par construction la restrictiton de ν à $\Omega B \times \{f_o\}$.

δ est compatible avec les opérations de ΩB sur ΩB et sur F.

(5) D'après Ganea [Ga 1], la fibre homotopique F de l'inclusion $S \vee T \to S \times T$ a le type d'homotopie du joint $\Omega(S) * \Omega(T)$. On vérifie alors que l'opération d'holonomie est fournie par l'opération diagonale

$$(\Omega(S) \times \Omega(T)) \times (\Omega S * \Omega T) \to \Omega S * \Omega T.$$

§ 2 - *Quelques points d'homotopie rationnelle.*

Tous les espaces considérés sont supposés connexes par arcs et du type d'homotopie faible d'un C.W.-complexe de type fini. On notera $H_*(S)$ et $H^*(S)$ les espaces vectoriels $H_*(S;\mathbb{Q})$ et $H^*(S;\mathbb{Q})$. L'outil principal dans la suite est la théorie des modèles minimaux. Nous rappelons ici quelques définitions et propriétés élémentaires. Pour plus de détails, le lecteur est invité à se reporter à [Su, Ha, Ta].

2.1.- *Le modèle minimal de Sullivan.*

Tous les espaces vectoriels et toutes les algèbres sont supposées définies sur le corps $\mathbb{Q}$.

Une *algèbre différentielle graduée commutative* (*a.d.g.c.*) (A,d_A) est une algèbre graduée commutative $(xy = (-1)^{\deg x.\deg y}yx)$ munie d'une dérivation d_A de degré $+1$ et de carré nul. Elle est dite *libre* si A est le produit tensoriel d'une algèbre symétrique sur un espace vectoriel Y concentré en degrés pairs par une algèbre extérieure sur un espace vectoriel Z concentré en degrés impairs : on note $A = \Lambda X$, X étant la somme directe $Y \oplus Z$. Notons alors $\Lambda^i X$ l'espace vectoriel des mots de longueur i en X.

(A,d_A) est dite *minimale* si $A = \Lambda X$, si $d_A(X) \subset \Lambda^{\geq 2}X$ et s'il existe une base $(x_\alpha)_{\alpha\in A}$ de X indexée par un ensemble bien ordonné tel que $d_A(x_\alpha) \subset \Lambda X_{<\alpha}$.

Un morphisme d'a.d.g.c. $\varphi : (A,d_A) \to (B,d_B)$ est appelé un *quasi-isomorphisme* si $\varphi^* : H^*(A,d_A) \to H^*(B,d_B)$ est un isomorphisme.

Si (A,d_A) est une a.d.g.c. vérifiant $H^0(A,d_A) = \mathbb{Q}$, il existe une *unique* (à *isomorphisme* près) a.d.g.c. minimale $(\Lambda X,d)$ munie d'un quasi-isomorphisme $\varphi_A : (\Lambda X,d) \to (A,d_A)$. $(\Lambda X,d)$ s'appelle le *modèle minimal* de (A,d_A).

Le foncteur PL-formes construit par Sullivan ([Su]), noté A_{PL}, associe à chaque espace S une a.d.g.c. $(A_{PL}(S),d_S)$. Le modèle minimal de $(A_{PL}(S),d_S)$ est par définition le modèle minimal de S.

2.2.- *L'algèbre de Lie d'un modèle minimal.*

Soit L une algèbre de Lie graduée connexe (a.l.g) de type fini, alors le complexe de Koszul ([Ta]) $(C^*(L),d)$ est l'a.d.g.c. $(\Lambda X,d)$ 1-connexe définie par

(i) $X = \mathrm{Hom}(sL,\mathbb{Q})$

(ii) $d : X \longrightarrow \Lambda^2 X$

(iii) $<dx;su,sv> = (-1)^{\deg v}<x,s[u,v]>$.

C'est un modèle minimal.

Par contre, si $(\Lambda X,d)$ est un modèle minimal, posons $L_p = \mathrm{Hom}(s(\Lambda^+X/\Lambda^{\geq 2}X)^{P+1},\mathbb{Q})$. Les éléments u de L définissent des fonctions $su : \Lambda^+X \to \mathbb{Q}$ par la formule $<b,su> = (-1)^{\deg u}<sb,u>$.

L'espace $\Lambda^2 sL$ s'interprète alors comme $\mathrm{Hom}(\Lambda^{\geq 2}X/\Lambda^{\geq 3}X;\mathbb{Q})$ et la formule

$$<d\Phi;su \wedge sv> = <d\Phi;su,sv> = (-1)^{\deg v}<\Phi,s[u,v]>, \quad \Phi \in \Lambda^+X,$$

définit une structure d'algèbre de Lie graduée sur L.

L s'appelle alors *l'algèbre de Lie* de $(\Lambda X,d)$.

Décomposons d sous la forme $d = d_2 + d_3 + \ldots$ où d_i désigne la dérivation vérifiant $d_i(X) \subset \Lambda^i X$, alors $(\Lambda X,d_2)$ est une a.d.g.c. ; $(\Lambda X,d)$ et $(\Lambda X,d_2)$ ont même algèbre de Lie L et $(\Lambda X,d_2) \cong C^*(L)$.

Théorème ([Su], [A.A.]).- *L'algèbre de Lie* L *du modèle minimal d'un espace 1-connexe est isomorphe à l'algèbre de Lie d'homotopie rationnelle* $\pi_*(\Omega S) \otimes \mathbb{Q}$ *de* S.

2.3.- *K.S.-extensions.*

Soient (A,d_A) et (B,d_B) des a.d.g.c. telles que $H^o(A) = H^o(B) = \mathbb{Q}$ et (A,d_A) est augmentée par ε_A sur $\mathbb{Q}$. Dans ([Ha]), S. Halperin montre que à chaque homomorphisme d'a.d.g.c. $f : (A,d_A) \to (B,d_B)$ est associé un diagramme commutatif :

$$\begin{array}{ccccc}
(A,d_A) & \xrightarrow{\quad f \quad} & (B,d_B) & & \\
 & {\scriptstyle i}\searrow & \uparrow {\scriptstyle g}\ \simeq & & \\
 & & (A \otimes \Lambda X, d) & \xrightarrow[\varepsilon_A \otimes id]{\qquad} & (\Lambda X, \bar{d})
\end{array}$$

où : 1) g est un quasi-isomorphisme ;

2) i est l'injection canonique ;

3) Il existe une base $(x_\alpha)_{\alpha\in\kappa}$ de X et un bon ordre sur κ tel que $dx_\alpha \in A \otimes \Lambda X_{<\alpha}$.

Un tel diagramme s'appelle un K.S. (Koszul-Sullivan)-modèle de f. Si en outre $\alpha < \beta \Longrightarrow |x_\alpha| \leq |x_\beta|$, la K.S.-extension est dite minimale. Elle est alors unique à isomorphisme près ([Ha]).

L'utilité de la construction réside dans le théorème suivant dû à Grivel dans le cas B 1-connexe ([Gr]) et à S. Halperin dans le cas général ([Ha]).

Théorème.- Si $\xi : F \longrightarrow E \xrightarrow{p} B$ *est une fibration où* $\pi_1(B)$ *opère de façon nilpotente sur* $H^*(F)$, *et si* $f = A_{PL}(p)$, *alors* $(\Lambda X, \bar{d})$ *est un modèle minimal de* F.

§ 3 - *Calcul de l'opération d'holonomie rationnelle.*

Dans ce paragraphe, nous supposerons toujours que $\xi : F \xrightarrow{j} E \xrightarrow{p} B$ est une fibration avec B un C.W.-complexe 1-connexe de type fini et E, F connexes par arcs. Nous calculons tout d'abord le modèle minimal de $\nu : \Omega B \times F \to F$. Nous explicitons ensuite l'opération $\tau : H_*(\Omega B) \otimes H^*(F) \to H^*(F)$ définie dans l'introduction et montrons que pour $\alpha \in \pi_*(\Omega B) \otimes \mathbb{Q}$ on a : $\tau(\alpha) \in \mathrm{Der}\, H^*(F)$.

3.1.- *Le modèle minimal de* $\nu : \Omega B \times F \to F$.

Traduisons en homotopie rationnelle la construction de ν faite au § 1.1.

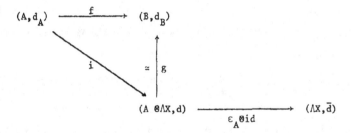

Désignons par $(A,d) \xrightarrow{i} (A \otimes \Lambda Y,d) \xrightarrow{k} (\Lambda Y,\bar{d})$ un K.S.-modèle de la fibration ξ. Formons le K.S.-modèle acyclique $(A \otimes \Lambda\bar{X},D)$ de la fibration $PB \to B$. Il résulte alors de ([Ha], § 2 0) que $(A \otimes \Lambda\bar{X} \otimes \Lambda Y,D) \underset{\text{déf}}{=\!=} (A \otimes \Lambda\bar{X}) \underset{A}{\otimes} (A \otimes \Lambda V)$ est un modèle de $PB \underset{B}{\times} E$. De plus, les inclusions de F et de $\Omega B \times F$ dans $PB \underset{B}{\times} E$ ont pour modèles les projections $q_1 : (A \otimes \Lambda\bar{X} \otimes \Lambda Y,D) \to (\Lambda Y,\bar{d})$ et $q_2 : (A \otimes \Lambda\bar{X} \otimes \Lambda Y,D) \to (\Lambda\bar{X},0) \otimes (\Lambda Y,\bar{d})$. Notons σ une section du quasi-isomorphisme q_1. Un modèle ϕ de ν est alors donné par la composée $q_2 \circ \sigma$.

3.2.- *La représentation* $\tau : H_*(\Omega B) \otimes H^*(F) \longrightarrow H^*(F)$.

Pour chaque a.d.g.c. (A,d_A), désignons par $\mathrm{Der}_o(A)$ l'espace vectoriel des dérivations de A de degré 0 vérifiant $\theta d_A + d_A\theta = 0$ et, si $p > 0$, par $\mathrm{Der}_p(A)$ l'espace des dérivations de degré $-p$.

La somme directe $\mathrm{Der}(A) = \bigoplus_{p\geq 0} \mathrm{Der}_p(A)$, munie du crochet $[\theta,\theta'] = \theta.\theta' - (-1)^{\deg\theta.\deg\theta'}\theta'.\theta$ et de la différentielle $\partial\theta = [d_A,\theta]$ est une a.l.d.g. .

Considérons la K.S.-extension $(\Lambda X,d) \xrightarrow{i} (\Lambda X \otimes \Lambda Y,d) \xrightarrow{k} (\Lambda Y,\bar{d})$ modèle de la fibration ξ, et choisissons une base $(b_\alpha)_{\alpha\in A}$ de Λ^+X homogène pour la bigraduation $X = \bigoplus_{p,q} (\Lambda^p X)^q$. La différentielle d définit des éléments θ^α de $\mathrm{Der}(\Lambda Y)$ par la formule

$$d|_{\Lambda Y} = 1 \otimes \bar{d} + \sum_{\alpha\in A} b_\alpha \otimes \theta^\alpha .$$

De la relation $d^2 = 0$, on tire alors :

$$(*) \qquad 0 = \sum_\alpha db_\alpha \otimes \theta^\alpha + \sum_\alpha (-)^{|b_\alpha|} b_\alpha \otimes \partial\theta^\alpha + \frac{1}{2} \sum_{\alpha,\beta} (-1)^{|b_\alpha|} b_\alpha b_\beta \otimes [\theta^\beta,\theta^\alpha] .$$

Notons L l'algèbre de Lie de $(\Lambda X,d)$ et $\psi : L \to \mathrm{Der}(\Lambda Y)$ l'application linéaire définie par

$$\psi(u) = \sum_\alpha \langle b_\alpha, su\rangle \theta^\alpha .$$

La première étape de la construction de τ repose sur le lemme suivant :

Lemme.-

1) $\partial\psi = 0$;

2) L'application $\psi_* : L \to H_*(\mathrm{Der}(\Lambda Y),\partial)$ *induite par* ψ *est un homomorphisme d'algèbres de Lie.*

Démonstration du lemme :

1) La formule (*) montre que si $b_\alpha \in \Lambda^1 X$, alors $\partial\theta^\alpha = 0$, comme $\langle\Lambda^{\geq 2}X, su\rangle = 0$, il est clair que $\partial\psi = 0$.

2) Rappelons [Ta] que

$$\langle dx, su, sv\rangle = (-1)^{\deg v}\langle x, s[u,v]\rangle.$$

Un simple calcul donne alors le résultat. ∎

Composons maintenant ψ_* avec le morphisme canonique $H_*(\mathrm{Der}(\Lambda Y),\partial) \to \mathrm{Der}(H^*(\Lambda Y,\bar{d}))$. Nous obtenons ainsi une représentation $\hat{\tau}$ de L dans $H^*(\Lambda Y,\bar{d})$ par des dérivations. Ceci se prolonge en une opération de l'algèbre enveloppante UL :

$$\hat{\tau} : UL \otimes H^*(\Lambda Y,\bar{d}) \longrightarrow H^*(\Lambda Y,\bar{d}).$$

Théorème 3.2.- Le morphisme $\hat{\tau}$ *coïncide avec le morphisme* $\tau : H_*(\Omega B;\mathbb{Q}) \otimes H^*(F) \to H^*(F)$ *défini dans l'introduction. En particulier,* τ *restreint à* $\pi_*(\Omega B) \otimes \mathbb{Q}$ *définit une action de cette algèbre de Lie par dérivations dans* $H^*(F)$.

Démonstration du théorème 3.2 : Puisque l'algèbre de Hopf $H_*(\Omega B)$ est primitivement engendrée par $\pi_*(\Omega B) \otimes \mathbb{Q} \cong L$ et puisque les opérations τ et $\hat{\tau}$ sont naturelles, il suffit de faire la démonstration lorsque B est une sphère

a) Cas d'une sphère impaire.

Dans ce cas, la K.S.-extension de la fibration prend la forme

$(\Lambda b,0) \xrightarrow{i} (\Lambda b \otimes \Lambda Y,d) \xrightarrow{k} (\Lambda Y,\bar{d})$, deg b impair, avec $d = 1 \otimes \bar{d} + b \otimes \theta$.

Pour construire l'action ν^*, on construit tout d'abord une section σ du quasi-isomorphisme $q_1 : (\Lambda b \otimes \Lambda \bar{b} \otimes \Lambda Y,D) \to (\Lambda Y,\bar{d})$ $(D\bar{b} = b)$ par la formule

$$\sigma(\alpha) = \sum_{n\geq 0} \frac{(-1)^n}{n!} \bar{b}^n \otimes \theta^n(\alpha).$$

On effectue ensuite la composition

$$\omega : \Lambda Y \xrightarrow[\sigma]{} \Lambda b \otimes \Lambda \bar{b} \otimes \Lambda Y \xrightarrow[q_2]{} \Lambda \bar{b} \otimes \Lambda Y.$$

La formule pour σ ne contenant pas de b, on peut identifier σ à ω.

Soit maintenant u dans L (algèbre de Lie de $(\Lambda b,0)$) l'élément défini par $\langle b,su\rangle = 1$. Puisque $D\bar{b} = b$, on écrit $\bar{b} = sb$ et on a alors $\langle \bar{b},u\rangle = 1$ et

$$\langle \tau(u \otimes [\Phi]),\alpha\rangle = -\langle \nu^*([\Phi]),u\otimes\alpha\rangle = -\langle \omega(\Phi),u\otimes\alpha\rangle = \theta(\Phi)(\alpha) = \langle \hat{\tau}(u \otimes [\Phi]),\alpha\rangle ,$$

avec Φ cocycle de $(\Lambda Y,\bar{d})$ et $\alpha \in H_*(F)$.

b) *Cas d'une sphère paire.*

Notons $(\Lambda b/_{b^2},0) \to (\Lambda b/_{b^2} \otimes \Lambda Y,d) \to (\Lambda Y,\bar{d})$ un K.S.-modèle de la fibration : $d = 1 \otimes \bar{d} + b \otimes \theta$. Soit $(\Lambda b/_{b^2} \otimes \Lambda(\bar{b},\bar{c}) \otimes \Lambda Y,D)$ l'extension de $(\Lambda b/_{b^2} \otimes \Lambda Y,d)$ définie par $D\bar{b} = b$, $D\bar{c} = -b\bar{b}$.

Un morphisme ω induisant ν^* est alors construit en composant la projection $q_2 : (\Lambda b/_{b^2} \otimes \Lambda(\bar{b},\bar{c}) \otimes \Lambda Y,D) \to (\Lambda(\bar{b},\bar{c}) \otimes \Lambda Y,D')$ avec une section σ de la projection q_1 sur $(\Lambda Y,\bar{d})$.

Une telle section σ est nécessairement de la forme $\sigma = id + \bar{b}\gamma_1 + b\gamma_2 + \bar{c}\gamma_3 + \gamma$ où les γ_i sont des endomorphismes linéaires de ΛY et où $\gamma : \Lambda Y \to (b \otimes \Lambda^+(\bar{b},\bar{c}) \otimes \Lambda Y) \oplus (\Lambda^{\geq 2}(\bar{b},\bar{c}) \otimes \Lambda Y)$. De l'équation $\sigma\bar{d} = D\sigma$, on déduit :

$$\theta + \gamma_1 = \gamma_2\bar{d} - \bar{d}\gamma_2.$$

En particulier, si $u \in L$ satisfait à $<b,su> = 1$, alors pour tout cocycle Φ de ΛY et tout α de $H_*(F)$, on a

$$<\tau(u \otimes [\Phi]),\alpha> = <u.[\Phi],\alpha>$$
$$= (-1)^{\deg u.\deg \Phi+1}<[\Phi],\nu_*(u \otimes \alpha)>$$
$$= (-1)^{\deg u.\deg \Phi+1}<\nu^*([\Phi]),u \otimes \alpha>$$
$$= (-1)^{\deg u.\deg \Phi+1}<\bar{b} \otimes \gamma_1\Phi,u \otimes \alpha>$$
$$= (-1)^{\deg u.\deg \Phi+1}(-1)^{\deg u.(\deg \Phi+1)}<\gamma_1\Phi,\alpha><\bar{b},u>$$
$$= <\theta(\Phi),\alpha> = <\hat{\tau}(u \otimes [\Phi]),\alpha> \quad \text{car } <\bar{b},u> = -1. \blacksquare$$

3.3.- Seconde construction de τ.

Notons de nouveau $(\Lambda X,d) \xrightarrow{i} (\Lambda X \otimes \Lambda Y,D) \to (\Lambda Y,\bar{d})$ de K.S.-modèle de la fibration ξ, et $Der_{\Lambda X}(\Lambda X \otimes \Lambda\bar{X}) \subset Der(\Lambda X \otimes \Lambda\bar{X})$ la sous-a.l.d.g. formée des dérivations s'annulant sur ΛX.

L'injection canonique $Der_{\Lambda X}(\Lambda X \otimes \Lambda\bar{X}) \to Der_{\Lambda X\otimes\Lambda Y}(\Lambda X \otimes \Lambda\bar{X} \otimes \Lambda Y)$ induit une action par dérivations Ψ de $H_*(Der_{\Lambda X}(\Lambda X \otimes \Lambda\bar{X}))$ sur $H^*(\Lambda X \otimes \Lambda\bar{X} \otimes \Lambda Y) \simeq H^*(\Lambda Y)$.

D'autre part, l'algèbre de Lie $Der_{\Lambda X}(\Lambda X \otimes \Lambda\bar{X})$ étant quasi-isomorphe au modèle de Quillen de B ([FHT]), on a un isomorphisme d'algèbres de Lie,

$$\rho_* : H_*(Der_{\Lambda X}(\Lambda X \otimes \Lambda\bar{X})) \xrightarrow{\simeq} L.$$

Ceci est donné explicitement par

$$<\bar{x},\rho_*([\theta])> = (-1)^{\deg \theta+1}\varepsilon\theta(\bar{x}), \quad \varepsilon \text{ désignant}$$

l'augmentation canonique de $\Lambda X \otimes \Lambda\bar{X}$.

Proposition 3.3.- *L'isomorphisme* ρ_* *identifie l'opération* Ψ *à l'opération* $\tau : \pi_*(\Omega B) \otimes H^*(F) \to H^*(F)$ *définie dans l'introduction.*

Démonstration : Il suffit de montrer que $\Psi = \hat{\tau}$ (l'opération de 3.2). Les opérations $\hat{\tau}$ et Ψ étant naturelles, il suffit de nouveau de faire la démonstration lorsque la base est une sphère.

a) Cas d'une sphère impaire.

Notons $(\Lambda b,0) \to (\Lambda b \otimes \Lambda Y,d) \to (\Lambda Y,\bar{b})$ un K.S.-modèle de la fibration : $d = 1 \otimes \bar{d} + b \otimes \theta$. Soit L l'algèbre de Lie de $(\Lambda b,0)$ et u dans L avec $\langle b,su\rangle = 1$. La dérivation d_u de $Der_{\Lambda b}(\Lambda b \otimes \Lambda\bar{b})$ définie par $d_u(\bar{b}) = -1$ est un cycle vérifiant $\rho_*([d_u]) = u$. Soit D_u son image dans $Der_{\Lambda b\otimes\Lambda Y}(\Lambda b\otimes\Lambda\bar{b}\otimes\Lambda Y)$.

Notons alors q_1 et σ la projection et section construites en 3.2.a.

$$\sigma(\Phi) = \sum_{n\geq 0} \frac{(-1)^n}{n!} \bar{b}^n \theta^n(\Phi).$$

$$\text{Alors } \Psi(u)([\Phi]) = [q_1 D_u \sigma(\Phi)] = [q_1 (\sum \frac{(-1)^n}{n!} \bar{b}^n \theta^{n+1}(\Phi))]$$
$$= [\theta(\Phi)] \overset{\text{déf}}{=} \hat{\tau}(u,[\Phi]).$$

b) Cas d'une sphère paire.

Notons $(\Lambda b/_{b^2},0) \to (\Lambda b/_{b^2} \otimes \Lambda Y,d) \to (\Lambda Y,\bar{d})$ un K.S.-modèle de la fibration $(d = 1 \otimes \bar{d} + b \otimes \theta)$ et L l'algèbre de Lie de $(\Lambda b/_{b^2},0)$.

Soit u dans L avec $\langle b,su\rangle = 1$. On reprend les notations de 3.2 b. En particulier, $\langle\bar{b},u\rangle = -1$. La dérivation d_u de $Der_{\Lambda b}(\Lambda b/_{b^2} \otimes \Lambda(\bar{b},\bar{c}))$ définie par $d_u(\bar{b}) = -1$ et $d_u(\bar{c}) = -\bar{b}$ est un cycle vérifiant $\rho_*([d_u]) = u$.

Notons q_1 et σ la projection canonique $\Lambda b/_{b^2} \otimes \Lambda(\bar{b},\bar{c}) \otimes \Lambda Y \to \Lambda Y$ et une section de celle-ci. Désignons par D_u l'image de d_u dans $Der_{\Lambda b/_{b^2}\otimes\Lambda Y}(\Lambda b/_{b^2} \otimes \Lambda(\bar{b},\bar{c}) \otimes \Lambda Y)$. Alors,

$$\Psi(u)([\Phi]) = [q_1 D_u \sigma(\Phi)]$$
$$= [q_1 D_u(\Phi + \bar{b}\gamma_1(\Phi) + b\gamma_2(\Phi) + \bar{c}\gamma_3(\Phi) + \gamma(\Phi)]$$
$$= -[\gamma_1(\Phi)] = [\theta(\Phi)] \overset{\text{déf}}{=} \hat{\tau}(u,[\Phi]). \quad \blacksquare$$

§ 4 - *Opération d'holonomie noethérienne.*

Théorème 4.1.- Soit $F \to E \to B$ *une fibration entre espaces 1-connexes. Si* $\dim \pi(B) \otimes \mathbb{Q} < \infty$ *et* $\dim H(E) < \infty$ *alors*

(1) $H_*(F)$ *est un* $H_*(\Omega B)$*-module noethérien ;*

(2) *La série de Poincaré de* $H_*(F)$ *est une fraction rationnelle de la forme* $P(t)/\pi(1-t^{2i})^{\dim \pi_{2i}(\Omega B) \otimes \mathbb{Q}}$ *où* $P(t)$ *désigne un polynôme à coefficients entiers.*

Avant d'entreprendre la démonstration de ce résultat, rappelons quelques définitions extraites de [Gu 1,2].

Soit $\pi(t) \in \mathbb{Z}[t]$ un polynôme vérifiant $\pi(0) = \pm 1$. Une série formelle $P(t)$ est dite π-rationnelle s'il existe un polynôme $R(t)$ dans $\mathbb{Z}[t]$ tel que $P(t) = R(t)/\pi(t)$.

Si H est un espace vectoriel gradué vérifiant $\dim H^i < \infty$ pour chaque i, on appelle *série de Hilbert* de H la série formelle

$$|H|(t) = \sum_{i\geq 0} \dim H^i . t^i .$$

Ainsi, lorsque H désigne la cohomologie d'un espace topologique S, $|H|(t)$ désigne la série de Poincaré $P_S(t)$ de S.

Soit $G = \bigoplus_{p\geq 0} G_p$ un anneau connexe gradué inférieurement. Un G-module gradué H est dit π-rationnel si

a) La série $|H|$ est définie ;

b) Pour tout sous G-module N de H, la série $|N|(t)$ est π-rationnelle.

Dans les lemmes 1, 2, 3 suivants, les modules sont supposés gradués inférieurement $(M = \sum_{p\geq 0} M_p)$. Le lemme 2 est le dual de l'énoncé de Gulliksen.

Lemme 1 [Gu 1].- *Soit* $0 \to H' \to H \to H'' \to 0$ *une suite exacte de* G*-modules gradués.* H *est noethérien et* π*-rationnel si et seulement si* H' *et* H'' *le sont.*

<u>*Lemme*</u> *2* |Gu 2|.- *Soit* $H' \xrightarrow{J} H' \xrightarrow{I} H$ *une suite exacte de* G-*modules gradués avec* I *de degré zéro tel que* J *soit la multiplication par un élément de* G *de degré strictement positif. Alors, si* H *est un* G-*module noethérien* π-*rationnel,* H' *est un* G-*module noethérien et* π'-*rationnel avec* $\pi'(t) = (1 - t^{\deg g})\pi(t)$.

<u>*Lemme*</u> *3.- Soit* L *une* $\mathbb{Q}$-*algèbre de Lie graduée et connexe de dimension finie, alors* UL *est un* UL-*module noethérien et* π_L-*rationnel avec*

$$\pi_L(t) = \prod_i (1 - t^{2i})^{\dim L_{2i}} .$$

<u>*Démonstration du lemme*</u> 3 : Procédons par récurrence sur la dimension de L. Le résultat est évident pour $\dim L = 0,1$. Supposons donc L de dimension $n > 0$ et soit $g \in L$ un élément de degré maximal, qui est donc dans le centre. Par l'hypothèse de récurrence $U(L/g)$ est un $U(L/g)$-module (et donc UL-module) noetherien et $\pi_{L/g}$-rationnel. Si g est impair, $UL.g \cong U(L/g)$ et la suite exacte de UL-modules $0 \to U(L).g \to U(L) \to U(L/g)$ avec le lemme 1 fournit le résultat. Si g est pair, le résultat provient du lemme 2 appliqué à la suite exacte de UL-modules $UL \xrightarrow{\times g} UL \longrightarrow UL/g$.

<u>*Démonstration du théorème*</u> *4.1.* : La suite spectrale de Serre de la fibration $\Omega B \to F \to E$ est une suite spectrale de $H_*(\Omega B)$-modules avec $E_2 = H_*(E) \otimes H_*(\Omega B)$, et convergente vers le $H_*(\Omega B)$-module $H_*(F)$.

Comme $H_*(E)$ est de dimension finie, le lemme 3 montre que E_2 et donc $H_*(F)$, est noetherien et π-rationnel. ∎

<u>*Corollaire*</u> *4.2.- Si* S *est un espace topologique vérifiant*

a) $\dim H^*(S;\mathbb{Q}) < \infty$;

b) $\pi_*(\Omega S) \otimes \mathbb{Q}$ *contient une sous-algèbre de Lie libre de codimension finie, alors la série de Poincaré de* ΩS *est rationnelle.*

Démonstration du corollaire 4.2. : Notons p le plus petit entier tel que $\pi_{>p}(\Omega S) \otimes \mathbb{Q}$ soit libre. Notons alors S_p le p^{e} étage de la tour de Postnikov de S. La fibre homotopique F de $\varphi : S \to S_p$ est un bouquet de sphères et donc :

$$\pi_*(\Omega F) \otimes \mathbb{Q} = \mathbb{L}(V) \quad \text{avec} \quad P(F) = t.|V| + 1.$$

D'après la formule de Hilton-Steer, $P(\Omega F) = (1 - (P(F) - 1))^{-1}$. Comme $P(\Omega S) = P(\Omega F).P(\Omega S_p)$, le théorème 4.1. montre que $P(\Omega S)$ est rationnel.

Remarque 4.3.- L'hypothèse $\dim \pi(B) \otimes \mathbb{Q} < \infty$ dans le théorème 4.1. est nécessaire. En effet, d'après le résultat de Bogvad ([Bo]), si B est un C.W. complexe fini, alors $H_*(\Omega B)$ est un anneau noethérien si et seulement si $\pi_*(B) \otimes \mathbb{Q}$ est de dimension finie. Il en résulte que dans la fibration $\Omega B \to PB \to B$, $H_*(\Omega B)$ ne peut être un $H_*(\Omega B)$-module noethérien que si $\pi_*(B) \otimes \mathbb{Q}$ est de dimension finie.

Question 4.4.- Si $F \to E \to B$ est une fibration entre espaces 1-connexes avec $\dim H^*(E) < \infty$, $H_*(F)$ est-il un $H_*(\Omega B)$-module finiment engendré ?

Exemple 4.5.- Si $F \to E \to B$ est une fibration de base un bouquet de sphères et si $H^*(E)$ est de dimension finie, alors $H_*(F)$ est un $H_*(\Omega B)$-module finiment engendré.

En effet, dans ce cas un K.S.-modèle de la fibration est de la forme $(H^*(B),0) \to (H^*(B) \otimes \Lambda Y,d) \to (\Lambda Y,\bar{d})$.

La différentielle d s'écrit donc de la forme

$$d = 1 \otimes \bar{d} + \sum_i b_i \otimes \theta^i ,$$

où b_i parcourt une base de $H^+(B)$. Si α est un cocycle de $(\Lambda Y,\bar{d})$ de degré plus élevé que la dimension cohomologique de E, alors dans $(H^*(B) \otimes \Lambda Y,d)$, $d\alpha \neq 0$. Il existe donc b_i avec $[\theta^i(\alpha)] \neq 0$.

Théorème 4.6.- Si $F \xrightarrow{j} E \xrightarrow{p} B$ *est une fibration de fibre et base 1-connexes, telle que* $H_*(F)$ *soit un* $H_*(\Omega B)$*-module finiment engendré, alors les conditions suivantes sont équivalentes :*

(1) *L'algèbre de cohomologie est de nilpotence finie.* ($(H^+)^n = 0$ *pour un certain* n).

(2) *L'application* $\psi_n : \pi_{n+1}(B) \otimes \mathbb{Q} \to H_n(F) \otimes \mathbb{Q}$ *(composée du connectant de la fibration avec l'homomorphisme d'Hurewicz) est nul pour tout* n *supérieur à un certain* q *et pour tout* n *pair.*

Démonstration : Soit $X^n \subset H^n(F)$ un sous-espace dual à $\mathrm{Im}\,\psi_n$. Dans ([Op]), Oprea démontre que le composé $\Lambda(\sum_n X^n) \to H^*(F) \xrightarrow{\delta^*} H^*(\Omega B)$ est un isomorphisme de $\Lambda(\sum_n X^n)$ sur $\mathrm{Im}\,\delta^*$. Il en résulte clairement que

1) $\Longrightarrow$ 2).

2) $\Longrightarrow$ 1). Notons $I_1 = \ker \delta^*$. Comme δ^* est un morphisme de $H_*(\Omega B)$-modules, I_1 est un $H_*(\Omega B)$-module ainsi que les idéaux I_n définis par $I_n = I_{n-1}.H^+(F)$. Notons $R = H_+(F)/\mathrm{Im}(\delta_* : H_*(\Omega B) \to H_*(F))$. R est un $H_*(\Omega B)$-module finiment engendré. Les sous $H_*(\Omega B)$-modules J_n de R formés des éléments orthogonaux à I_n forment une suite croissante. Il existe, d'autre part, un n_o pour lequel J_{n_o} contient tous les générateurs de R. J_{n_o} est donc R et $I_{n_o} = 0$.

Finalement, le résultat de Oprea montre que $\mathrm{Im}\,\delta^*$ est l'algèbre extérieure sur un nombre fini r de générateurs. On a donc

$$(H^+(F))^{r+1} \subset I_1 \quad \text{et} \quad (H^+(F))^{(r+1)n_o} = 0. \quad \blacksquare$$

§ 5 - *Opération d'holonomie libre.*

5.1.- Soit $f : X \to Y$ une application continue de cofibre $g : Y \to C_f$. Notons $h : F \to Y$ la fibre homotopique de g. Le composé gf étant homotopiquement trivial, il existe une application $k : X \to F$ avec $hk \sim f$. Prolongeons k en une application $k : \Omega C_f \times X \to F$ compatible avec les actions à gauche de ΩC_f sur $\Omega C_f \times X$ et sur F. La seconde projection $\Omega C_f \times X \to X$ est homotope à l'inclusion de la fibre homotopique de $g \circ f$ dans X.

Théorème 5.1.- *Avec les notations précédentes, si* Y *est 1-connexe et* X *connexe par arcs, l'application* k *induit un isomorphisme de* $H_*(\Omega C_f)$*-modules :* $H_*(\Omega C_f) \otimes H_+(X) \to H_+(F)$.

Démonstration : Choisissons un modèle surjectif de f : $A_Y \xrightarrow{\varphi} A_X$. Un modèle de C_f est alors fourni par l'a.d.g.c. $K\varphi = (\operatorname{Ker} \varphi \oplus \mathbb{Q})$ ([Ha]). Notons alors $(K\varphi \otimes \Lambda\bar{Z}, d)$ une K.S.-extension acyclique minimale. Dans ce cas, $\Lambda\bar{Z} = H^*(\Omega C_f)$. D'après ([Ha], § 20), $A_Y \underset{K\varphi}{\otimes} (K\varphi \otimes \Lambda\bar{Z}) \cong A_Y \otimes \Lambda\bar{Z}$ est un modèle pour F. La courte suite exacte $0 \to K\varphi \to A_Y \to A_X^+ \to 0$ d'espaces vectoriels différentiels induit par tensorisation la suite exacte

$$0 \to K\varphi \otimes \Lambda\bar{Z} \longrightarrow A_Y \otimes \Lambda\bar{Z} \xrightarrow{\omega} A_X^+ \otimes \Lambda\bar{Z} \to 0.$$

$H^+(K\varphi \otimes \Lambda\bar{Z})$ étant nul, $H^+(\omega)$ est un isomorphisme. D'autre part, ω étant le composé $A_Y \otimes \Lambda\bar{Z} \xrightarrow{\varphi\otimes 1} A_X \otimes \Lambda\bar{Z} \longrightarrow A_X^+ \otimes \Lambda\bar{Z}$ et $\varphi \otimes 1$ un modèle de k, le résultat s'en déduit aussitôt. ■

Corollaire : *Soit* $j : Y \to Z = Y \cup_\varphi (\bigvee_{i=1}^{r} e^{n_i+1})$ *une application continue, alors l'homologie réduite de la fibre homotopique de* j *est isomorphe comme* $H_*(\Omega Z)$*-module au module libre* $H_*(\Omega Z) \otimes (\bigoplus_{i=1}^{r} H_+(S^{n_i}))$.

Ceci généralise à toute cofibration le point b du théorème suivant de Halperin-Lemaire ([HL]).

Théorème.- Soit $j : X \to Y = X \cup_{\psi} (\vee_i e^i)$ *une application continue.*

a) $\pi_*(j)$ *est surjectif si et seulement si* $\mathrm{Ker}\ \pi_*(j)$ *est une algèbre de Lie libre* $\mathbb{L}(V)$.

b) *Dans ce cas,* V *est un* $H_*(\Omega Y)$*-module libre.*

5.2.- *Exemples.*

1) Les espaces G_n de Ganea. Soit X un espace topologique, désignons par G_o l'espace réduit à un point $*$ et $f_o : X \to G_o$ l'application constante. Si G_n et $f_n : X \to G_n$ sont définis, posons C_n la cofibre de f_n et G_{n+1} la fibre homotopique de l'inclusion $G_n \to C_n$. Le théorème 5.1. montre alors :

si X est 1-connexe, $H_+(G_n) \cong H_*(\Omega C_{n-1}) \otimes H_+(X)$.

2) Considérons l'injection $S^3 \vee S^3 \vee S^3 \xrightarrow{i} T(S^3,S^3,S^3)$, où $T(S^3,S^3,S^3)$ désigne le "fat-wedge" des trois sphères S^3. La fibre homotopique F de i admet comme 7-squelette dans une décomposition homologique rationnelle

$$(S^5 \vee S^5 \vee S^5) \vee (\bigvee_{1 \leq i \leq 9} S_i^7).$$

L'application $f : F \to S^3 \vee S^3 \vee S^3$ est définie comme suit : f envoie les sphères S^5 sur les représentants d'une base de $\pi_5(S^3 \vee S^3 \vee S^3) \otimes \mathbb{Q}$, les sphères S_i^7, $i \leq 8$, sur les représentants d'une base de $\pi_7(S^3 \vee S^3 \vee S^3) \otimes \mathbb{Q}$ et la sphère S_9^7 sur le point de base.

S_9^7 provient par la longue suite exacte d'homotopie de l'élément non nul de $\pi_8(T(S^3,S^3,S^3)) \otimes \mathbb{Q}$.

Néanmoins, $H_7(F) \cong H_2(\Omega T(S^3,S^3,S^3)) \otimes H_5(F)$. L'élément de $H_7(F)$ correspondant à S_9^7 est à la fois dans $H_2(\Omega T(S^3,S^3,S^3)).H_5(F)$ et dans $H_7(\Omega T(S^3,S^3,S^3)).H_o(F)$.

3) Considérons la fibration de Hopf généralisée

$$\Omega X * \Omega X \longrightarrow \Sigma\Omega X \longrightarrow X.$$

Si X est coformel et de catégorie de Lusternik-Schnirelmann 2, alors $H_+(\Omega X * \Omega X)$ est un $H_*(\Omega X)$-module libre.

5.3.- Désignons par F la fibre homotopique de la projection p :

$$(S^3_a \vee S^3_b) \bigcup_{[a,[a,b]]} e^8 \to S^3_b.$$

p n'est pas la cofibre d'une application. Néanmoins,
$H_+(F) \cong H_*(\Omega B) \otimes (u,v) = T(x) \otimes (u,v)$ avec $\deg x = 2$, $\deg u = 3$, $\deg v = 8$.
Une décomposition homologique rationnelle de F peut être décrite comme suit :

$$F = (\bigvee_{n\geq 0} S^{3+2n}_{a_n}) \bigcup_{[a_1,a_n]} (\bigvee_{n\geq 0} e^{8+2n}).$$

L'action de $H_*(\Omega S^3_b)$ sur $H_+(F)$ est définie comme suit :

$$x.S^{3+2n} = S^{3+2(n+1)}$$

$$x.e^{8+2n} = e^{8+2(n+1)}, \quad x \in H_2(\Omega S^3_b). \blacksquare$$

5.4.- *Problème.*

Déterminer des conditions plus générales que celles mentionnées dans le théorème 1 sous lesquelles $H_+(F)$ est un $H_*(\Omega B)$-module libre.

§ 6 - *Opération d'holonomie triviale.*

Soit $\xi : F \xrightarrow{j} E \xrightarrow{p} B$ une fibration de base B 1-connexe. Désignons par $\nu : \Omega B \times F \to F$ l'opération d'holonomie et par $\pi_2 : \Omega B \times F \to F$ la seconde projection. Notons alors $(\Lambda X,d) \to (\Lambda X \otimes \Lambda Y,d) \xrightarrow{\rho} (\Lambda Y,\bar{d})$ un K.S.-modèle minimal de ξ et L l'algèbre de Lie de $(\Lambda X,d)$.

Définition.- ν est triviale (resp. $\mathbb{Q}$-triviale, H-triviale) *si* $\nu \sim \pi_2$ (resp. $\nu \underset{\mathbb{Q}}{\sim} \pi_2$, $H_*(\nu) = H_*(\pi_2)$).

ν est donc H-triviale si, avec les notations du § 3.2, le morphisme $\hat{\tau} : L \to \mathrm{Der}\, H^*(F)$ est nul. Il en résulte que l'opération d'holonomie est toujours H-triviale dans une fibration ou la fibre F est une variété kälhérienne compacte 1-connexe, car dans ce cas $\mathrm{Der}_{<0} H^*(F) = 0$ ([Me]).

6.1.- *$\mathbb{Q}$-trivialité.*

Lemme.- Toute fibration à holonomie $\mathbb{Q}$-triviale de base une sphère est rationnellement triviale.

Démonstration : Toute fibration $F \to E \to S^n$ de base une sphère est entièrement définie par la classe d'homotopie de son morphisme d'embrayage $S^{n-1} \times F \xrightarrow{k} F$. La fibration est rationnellement triviale si k est rationnellement homotope à l'application seconde projection.

Si $\alpha : S^{n-1} \to \Omega S^n$ est l'adjoint de l'identité de S^n, alors $k \simeq \nu(\alpha \times id)$, d'où le lemme. ∎

Théorème.- Les propositions suivantes sont équivalentes :

1) *ν est $\mathbb{Q}$-triviale ;*
2) *Le morphisme d'algèbres de Lie* $\psi : L \to H_*(\mathrm{Der}(\Lambda Y))$ *défini en 3.2. est nul.*
3) *ξ admet un modèle pour lequel* $dY \subset \Lambda Y \otimes (\Lambda^{\geq 2} X \otimes \Lambda Y)$.

Démonstration :

1) ⟹ 2). Si γ est $\mathbb{Q}$-triviale, alors pour chaque $\alpha \in L$, $\alpha : S^n \to B$, la fibration image réciproque à une holonomie $\mathbb{Q}$-triviale. Elle est donc rationnellement triviale et la dérivation $\mathcal{V}(\alpha)$ est nulle.

2) ⟹ 3). Choisissons sur une base b_α de ΛX homogène pour la bigraduation $\Lambda X = \bigoplus_{p,q} (\Lambda^p X)^q$; d peut donc s'écrire $d = 1 \otimes \bar{d} + \sum_{\alpha \in A} b_\alpha \otimes \theta^\alpha$.

Il s'ensuit (voir *, § 3) que pour tout $b_\alpha \in \Lambda^1 X, \partial\theta^\alpha = 0$. Si maintenant $u_\beta \in L$ satisfait à $\langle b_\alpha, su_\beta\rangle = \delta_{\alpha\beta}$, alors $\mathcal{V}(u_\beta) = [\theta^\beta]$. Ceci est donc nul et $\theta^\beta = i_\beta \bar{d} - (-1)^{\deg i_\beta} \bar{d} i_\beta$ pour $i_\beta \in \mathrm{Der}(\Lambda Y)$. Définissons alors un morphisme $\gamma : \Lambda Y \to \Lambda X \otimes \Lambda Y$ en posant pour $y \in Y$, $\gamma(y) = y + \sum_{b_\alpha \in \Lambda X} b_\alpha i_\alpha(y)$ et en prolongeant multiplicativement. On a alors : $\gamma(z) - (z + \sum_\alpha b_\alpha i_\alpha(z)) \in \Lambda^{\geq 2} X \otimes \Lambda Y$ pour $z \in \Lambda Y$. Il suffit alors de remplacer $(\Lambda X \otimes \Lambda Y, D)$ par $[\Lambda X \otimes \Lambda(\gamma(Y)), D]$.

3) ⟹ 1). Désignons par $j : F \to E$ l'inclusion de la fibre. Les applications $j\nu$ et $j\pi_2$ étant homotopes, il suffit de voir que pour tout espace X et pour tout couple d'applications $f_1, f_2 : X \to F$ si $jf_1 \underset{\mathbb{Q}}{\sim} jf_2$, alors $f_1 \underset{\mathbb{Q}}{\sim} f_2$.

Pour chaque a.d.g.c. $(\Lambda X, d)$, notons $(\Lambda X, d)^I$ l'a.d.g.c. $(\Lambda X \otimes \Lambda\bar{X} \otimes \Lambda\hat{X}, D)$ où $Dx = dx$, $D\bar{x} = \hat{x}$ et $D\hat{x} = 0$.

Soient alors $g, h : (\Lambda Y, \bar{d}) \to (\Lambda Z, D)$ deux homomorphismes d'a.d.g.c. et $\Phi : (\Lambda X \otimes \Lambda Y, d)^I \to (\Lambda Z, D)$ une homotopie ([Ha]) entre $g\rho$ et $h\rho$. Rappelons que, par définition, on a $\Phi|_{X \otimes Y} = g\rho$ et $\Phi . e^{sd+ds}|_{X \otimes Y} = h\rho$, où s désigne la dérivation de degré -1 définie par :

$$s(x) = \bar{x}, \quad s(\bar{x}) = 0, \quad s(\hat{x}) = 0, \quad s(y) = \bar{y}, \quad s(\bar{y}) = 0, \quad s(\hat{y}) = 0.$$

Désignons par $\mu : (\Lambda X \otimes \Lambda Y)^I \to \Lambda\bar{X} \otimes (\Lambda Y)^I$ la projection définie par $\mu(x) = \mu(\hat{x}) = 0$. Nous montrons que Φ se factorise à travers μ en un morphisme Φ'. Par contre $\Lambda Y^I \to \Lambda X \otimes \Lambda Y^I$ est un morphisme d'adgc et nous montrons que le composé $\Lambda Y^I \to \Lambda\bar{X} \otimes \Lambda Y^I \to \Lambda Z$ est alors l'homotopie recherchée entre g et h.

a) Désignons par J l'idéal de $(\Lambda X \otimes \Lambda Y,d)^I$ engendré par $\Lambda^2 X$, $s(\Lambda^2 X)$ et $\hat{X}$. Un petit calcul montre que J est stable par s et par d.

b) La relation $\rho(x) = 0$ pour $x \in X$ montre que $\Phi|_X = 0$.

c) Pour tout $x \in X$, $\Phi(\hat{x}) = 0$. Ceci se démontre par récurrence le long d'une K.S. base $(x_\alpha)_{\alpha\in A}$ de $(\Lambda X,d)$. Supposons $\Phi(\hat{x}_\beta) = 0$ pour $\beta < \alpha$. Comme $(sd)(x_\alpha) \in I$, d'après (a) $(sd)^p(x_\alpha) \in I$ pour tout $p \geq 1$. On a donc, par hypothèse de récurrence $\Phi((sd)^p(x_\alpha)) = 0$, $\forall\, p \geq 1$. Il résulte alors de la formule [Ha]

$$h\rho(x_\alpha) = g\rho(x_\alpha) + \Phi(x_\alpha) + \sum_{p\geq 1} \frac{1}{p!} \Phi((sd)^p(x_\alpha))$$

que $\Phi(\hat{x}_\alpha) = 0$.

d) Puisque $J \subset$ l'idéal engendré par X et $\hat{X}$, $\Phi(J) = 0$.
La décomposition en somme directe $(\Lambda X \otimes \Lambda Y)^I = [(\Lambda X^I)^+ \otimes (\Lambda Y)^I] \oplus (\Lambda Y)^I$ permet d'écrire $d = d_1 + d_2$, avec pour tout y de Y :

$$d_1(y) \in \Lambda^{\geq 2}X \otimes \Lambda Y, \quad d_2(y) \in \Lambda Y, \quad d_1(\bar{y}) = d_1(\hat{y}) = d_2(\hat{y}) = 0, \quad d_2(\bar{y}) = \hat{y}.$$

En particulier, $\mathrm{Im}\, d_1 \subset J$.

Puisque J est stable par s, l'image de $\theta_1 = sd_1 + d_1 s$ est contenue dans J. Posons $\theta = sd+ds$ et $\theta_2 = sd_2+d_2 s$. On a alors $\mathrm{Im}(\theta^n - \theta_2^n) \subset J$ et donc $\Phi \circ e^{\theta_2} = \Phi \circ e^{\theta}$.

e) Soit $\Phi' : (\Lambda Y)^I \to \Lambda Z$ définie en restreignant Φ à la sous-algèbre $(\Lambda Y)^I$, alors $\Phi' d_2 = D\Phi'$ et il résulte de (d) que Φ' est une homotopie d'a.d.g.c. entre g et h. ∎

6.2.- Classifiants.

A chaque espace 1-connexe F est associé une fibration universelle $\xi_F : F \to EF \to BF$ ($BF = B\,\mathrm{aut}\,F$, [Go]). Les fibrations à fibre F et à base B

1-connexe sont alors classifiées par les classes d'homotopie d'applications continues de B dans le revêtement universel $\widetilde{BF}$ de BF.

Désignons par $(\Lambda Y,\bar{d})$ un modèle de F et par $\mathrm{Der}(\Lambda Y)$ la sous-a.l.d.g. de $\mathrm{Der}(\Lambda Y)$ définie par $\overline{\mathrm{Der}}_o = \partial \mathrm{Der}_1$ et $\overline{\mathrm{Der}}_p = \mathrm{Der}_p$ pour $p > 0$; $H_*(\overline{\mathrm{Der}}) = H_+(\mathrm{Der})$.

$\widetilde{BF} = \varinjlim_{\alpha} B_\alpha$, B_α parcourant les sous-complexes finis de $\widetilde{BF}$.

On a $\pi_*(\Omega\widetilde{BF}) = \varinjlim_{\alpha} \pi_*(\Omega B_\alpha)$. L'injection $B_\alpha \to \widetilde{BF}$ définit par pull back une fibration de fibre F sur B_α et donc (§ 3.1) un morphisme $\psi_\alpha : \pi_*(\Omega B_\alpha) \otimes \mathbb{Q} \to H_*(\overline{\mathrm{Der}}(\Lambda Y))$. La compatibilité des ψ_α avec les inclusions $B_\alpha \hookrightarrow B_\beta$ définit par passage à la limite un morphisme d'algèbres de Lie $\mu : \pi_*(\Omega\widetilde{BF}) \otimes \mathbb{Q} \to H_*(\overline{\mathrm{Der}}(\Lambda Y))$.

<u>*Théorème*</u> *6.2.- Avec les notations précédentes, on a*

1) Pour toute fibration ξ *de fibre* F *et de base* B *le diagramme suivant commute*

$$\begin{array}{ccc} \pi_*(\Omega\widetilde{BF}) \otimes \mathbb{Q} & \xrightarrow{\ \mu\ } & H_*(\overline{\mathrm{Der}}(\Lambda Y)) \\ & {\scriptstyle \pi_*(\Omega\Psi)} \nwarrow \quad \nearrow {\scriptstyle \psi} & \\ & \pi_*(\Omega B) \otimes \mathbb{Q} & \end{array}$$

où ψ *désigne le morphisme construit en 3.2 et* Ψ *l'application classifiante de la fibration* ξ.

2) μ *est un isomorphisme d'algèbres de Lie.*

<u>*Démonstration*</u> :

1) provient de la construction de μ et de la naturalité de ψ.

2) Soit $\alpha : S^n \to \Omega\widetilde{BF}$ avec $\mu(\alpha) = 0$.

L'application adjointe $\alpha' : S^{n+1} \to \widetilde{BF}$ définit une fibration ξ de base S^{n+1}. Comme $\psi(\alpha) = \mu\, \pi_*(\Omega\alpha') = 0$, ξ est rationnellement triviale (Lemme 6.1) et $\alpha = 0$. μ est donc injective.

La surjectivité de μ se voit comme suit : si $\theta \in Der_n(Y)$ satisfait $\partial\theta = 0$ et $[\theta] \neq 0$, θ induit une fibration ξ non rationnellement triviale de base S^{n+1} avec $[\theta] \in \mathrm{Im}\,\psi$. $[\theta]$ appartient donc aussi à $\mathrm{Im}\,\mu$. ■

Dans la fibration universelle $F \to E\tilde{F} \to B\tilde{F}$ l'opération d'holonomie est donc l'isomorphisme $H_*(\overline{Der(\Lambda Y)}) \to H_+(Der\ \Lambda Y))$. Si $\dim \pi_*(F) < \infty$, elle n'est jamais $\mathbb{Q}$-triviale (si $\pi_n(F) \otimes \mathbb{Q} \neq 0$ et $\pi_{>n}(F) \otimes \mathbb{Q} = 0$, on a $H_n(Der(\Lambda Y)) \neq 0$). Elle peut cependant être H-triviale : supposons, par exemple, que F soit une sphère paire, alors toute fibration de fibre F étant T.N.C.Z. ([Th]), il en est de même de la fibration universelle de fibre F. L'holonomie est dans ce cas H-triviale.

6.3.- *Fibrations de base un 2-cône.*

Proposition 3.- *Si* $\xi : F \xrightarrow{j} E \xrightarrow{p} B$ *est une fibration où l'opération d'holonomie est* $\mathbb{Q}$*-triviale et où* B *est un 2-cône, alors il existe une longue suite exacte en cohomologie :*

$$\cdots \to H^{n+1}(F) \xrightarrow{\Delta_{n+1}} [H^+(B) \otimes H^*(F)]^n \to H^n(E) \xrightarrow{j^n} H^n(F) \xrightarrow{n} \cdots$$

Démonstration : B étant un 2-cône, ξ admet ([FT]) un modèle du type $(A,d_A) \to (A \otimes \Lambda Y,d) \xrightarrow{q} (\Lambda Y,\bar{d})$ avec $A = \mathbb{Q} \otimes A_1 \otimes A_2$, $A_1.A_1 \subset A_2$, $A_1.A_2 = A_2.A_2 = 0$, $d(A_1) \subset A_2$, $d(A_2) = 0$. Dans le modèle minimal $\Lambda X \to A$, $\Lambda^{\geq 2}X$ est envoyé dans A_2. La $\mathbb{Q}$-trivialité de l'opération montre qu'on peut supposer $d(Y) \subset \Lambda Y \otimes (A_2 \otimes \Lambda Y)$. Considérons alors la courte suite exacte

$$0 \longrightarrow \mathrm{Ker}\, q \longrightarrow (A \otimes \Lambda Y,d) \longrightarrow (\Lambda Y,\bar{d}) \longrightarrow 0.$$

Comme $\mathrm{Ker}\, q = (A_1 \otimes A_2) \otimes \Lambda Y$ et $d|_{\mathrm{Ker}\, q} = d_A \otimes 1 = 1 \otimes \bar{d}$, on obtient la longue suite exacte

$$\cdots \to [H^+(B) \otimes H^*(F)]^n \longrightarrow H^n(E) \xrightarrow{j^*} H^n(F) \xrightarrow{\Delta} [H^+(B) \otimes H^*(F)]^{n+1} \to \cdots$$

où Δ est induit par $d_2 : Y \to A_2 \otimes Y$. ■

Remarque : Si $H^*(E)$ est de dimension finie, il en est de même de $H^*(B)$ et de $H^*(F)$. En effet, décomposons $H^+(B)$ sous la forme $H^+(B) = H_1(B) \oplus H_2(B)$ à partir de la décomposition $A^+ = A_1 \oplus A_2$. $H_1(B) \otimes H^*(F)$ s'injecte alors dans $H^*(E)$. Comme $\dim H^*(E) < \infty$ et $H_1(B) \neq 0$, on a $\dim H^*(F) < \infty$. Il en résulte que $\dim H^*(B) < \infty$.

Ceci soulève la question suivante :

Question : Si $F \xrightarrow{j} E \xrightarrow{p} B$ est une fibration avec holonomie $\mathbb{Q}$-triviale et $\dim H^*(E) < \infty$, a-t-on $\dim H^*(F) < \infty$?

§ 7 - *Opération d'holonomie et opération de Yoneda.*

7.1.- Pour toute fibration $F \xrightarrow{j} E \xrightarrow{p} B$ entre espaces 1-connexes, Eilenberg et Moore [E-M] ont construit une suite spectrale vérifiant

$$E_2 = \operatorname{Ext}_{H^*(B)}(H^*(E),\mathbb{Q}) \Rightarrow H_*(F;\mathbb{Q}).$$

Rappelons que l'application p est dit *formalisable* ([LS]) s'il existe un diagramme commutatif

$$\begin{array}{ccc} H^*(B;\mathbb{Q}) & \xrightarrow{p^*} & H^*(E;\mathbb{Q}) \\ \uparrow{\scriptstyle \psi_1} & & \uparrow{\scriptstyle \psi_2} \\ (\mathcal{M}_B,d_B) & \xrightarrow{\tilde{p}} & (\mathcal{M}_E,d_E) \end{array}$$

avec ψ_1 et ψ_2 des quasi-isomorphismes et $\tilde{p}$ un modèle minimal de p.

Si p est formalisable, alors, ([V]), la suite spectrale d'Eilenberg-Moore "collapse" au terme E_2 et l'on a

$$\mathrm{Ext}_{H^*(B)}(H^*(E),\mathbb{Q}) \cong H_*(F;\mathbb{Q}).$$

Sous cette même hypothèse de formalisabilité, B est formel et l'on a

$$\mathrm{Ext}_{H^*(B)}(\mathbb{Q},\mathbb{Q}) = H_*(\Omega B;\mathbb{Q}).$$

Théorème 7.1.- Si p *est formalisable, l'opération d'holonomie* $H_*(\Omega B) \otimes H_*(F) \xrightarrow{\nu} H_*(F)$ *coïncide, via les isomorphismes précédents, avec l'opération de Yoneda*

$$\mathrm{Ext}_{H^*(B)}(\mathbb{Q},\mathbb{Q}) \otimes \mathrm{Ext}_{H^*(B)}(H^*(E),\mathbb{Q}) \to \mathrm{Ext}_{H^*(B)}(H^*(E),\mathbb{Q}).$$

Démonstration : Si P est une résolution projective de $H^*(E)$ comme $H^*(B)$-module et si P' est une résolution projective de $\mathbb{Q}$ comme $H^*(B)$-module alors la coopération de Yoneda Δ sur les Tor est définie ([Le]) par la commutativité du diagramme suivant :

$$\begin{array}{ccc}
\mathrm{Tor}^{H^*(B)}(H^*(E),\mathbb{Q}) & \xrightarrow{\quad\Delta\quad} & \mathrm{Tor}^{H^*(B)}(\mathbb{Q},\mathbb{Q}) \otimes \mathrm{Tor}^{H^*(B)}(H^*(E),\mathbb{Q}) \\
\| & & \| \\
H^*(P \otimes_{H^*(B)} \mathbb{Q}) \xrightarrow[\sim]{1\otimes\varepsilon_*^{-1}} H^*(P \otimes_{H^*(B)} P') & \longrightarrow & H^*((P \otimes_{H^*(B)} \mathbb{Q}) \otimes_{\mathbb{Q}} (P' \otimes_{H^*(B)} \mathbb{Q}))
\end{array}$$

où ε désigne l'augmentation $P' \to \mathbb{Q}$.

Notons $(\Lambda X,d) \to (\Lambda X \otimes \Lambda Y,d) \to (\Lambda Y,\bar{d})$ un K.S. modèle de la fibration p et $(\Lambda X,d) \to (\Lambda X \otimes \Lambda\bar{X},D) \to (\Lambda\bar{X},0)$ est un K.S.-modèle de la fibration des chemins $PB \to B$.

Comme B est formel, il existe un quasi-isomorphisme $(\Lambda X,d) \xrightarrow{\psi} (H^*(B),0)$. $(H^*(B) \otimes \Lambda Y,d)$ et $(H^*(B) \otimes \Lambda\bar{X},D)$ sont alors des résolutions projectives P et P' de $H^*(E)$ et $\mathbb{Q}$ comme $H^*(B)$-modules. Le diagramme ci-dessus devient alors la traduction de la définition (3.1) de l'opération de l'holonomie. ∎

7.2.- *Corollaire* (version graduée d'un théorème de Roos ([R]).

Soit H *une* $\mathbb{Q}$-*algèbre graduée connexe. Si* $(H^+)^m \neq 0$ et $(H^+)^{m+1} = 0$, *alors* $\mathrm{Ext}_H^+(H/_{(H^+)^m},\mathbb{Q})$ *est un* $\mathrm{Ext}_H(\mathbb{Q},\mathbb{Q})$-*module libre engendré par* $\mathrm{Ext}_H^1(H/_{(H^+)^m},\mathbb{Q})$.

Démonstration : Notons $(H \oplus s(H^+)^m,d)$ l'a.d.g.c. définie par : $sh.sh' = 0$, $sh.h' = 0$, $dh = 0$, $dsh = h$.

La projection $q : (H \oplus s(H^+)^m) \to H/_{(H^+)^m}$ de noyau $(H^+)^m \oplus s(H^+)^m$ est un quasi-isomorphisme. La courte suite exacte :

$$0 \to (H,0) \to (H \oplus s(H^+)^m,d) \to (s(H^+)^m,0) \to 0$$

montre alors que $(H,0) \xrightarrow{i} (H/_{(H^+)^m},0)$ est un modèle de la cofibre d'une application. Ce modèle étant formel, l'homologie réduite de sa fibre homotopique est $\mathrm{Ext}_H^+(H/_{(H^+)^m},\mathbb{Q})$, qui est donc (§ 5) un $\mathrm{Ext}_H(\mathbb{Q},\mathbb{Q})$-module libre engendré par $s(H^+)^m \cong \mathrm{Ext}_H^1(H/_{(H^+)^m},\mathbb{Q})$. ∎

7.3.- *Corollaire*.- *Soit* H *une* $\mathbb{Q}$-*algèbre graduée connexe,* I *un idéal gradué de* H, *alors* $\mathrm{Ext}_I^+(H,\mathbb{Q})$ *est un* $\mathrm{Ext}_{\hat{I}}(\mathbb{Q},\mathbb{Q})$-*module libre engendré par* $H/_{\hat{I}}$, *où* $\hat{I} = I \oplus \mathbb{Q}$.

Démonstration : Ceci résulte du § 5 et de la courte suite exacte

$$0 \to I \to H \to H/_I \to 0.$$

§ 8 - *Suite spectrale d'holonomie d'une fibration.*

Soit $\xi : F \xrightarrow{j} E \xrightarrow{p} B$ une fibration $(\Lambda X,d) \xrightarrow{i} (\Lambda X \otimes \Lambda Y,d) \xrightarrow{k} (\Lambda Y,\bar{d})$ un K.S.-modèle minimal et L l'algèbre de Lie de $(\Lambda X,d)$.

En filtrant l'a.d.g.c. $(\Lambda X \otimes \Lambda Y,d)$ par la longueur des mots en X, on génère une suite spectrale du premier quadrant vérifiant $E_1^{p,q} = [\Lambda^p X \otimes H^*(\Lambda Y,\bar{d})]^{p+q}$ et convergeant vers $H^*(\Lambda X \otimes \Lambda Y,d)$. La différentielle $d_1 : E_1^{p,q} \to E_1^{p+1,q}$ s'écrit : $d_1 = d_L + d_\psi$ avec d_L la différentielle de $(\Lambda X,d_L) \cong C^*(L)$ et d_ψ définie par la représentation $\psi : L \to \mathrm{Der}\, H^*(\Lambda X,\bar{d})$ avec,

$$d_\psi(\alpha) = \sum_{i\in I} x_i \otimes u_i.\alpha \quad , \quad \alpha \in H^*(\Lambda Y,\bar{d}).$$

$(x_i)_{I\in I}$ désigne une base de X et $(u_i \in L)$ la base duale :

$$\langle x_i, su_j\rangle = \delta_{ij} .$$

Il résulte alors clairement de ([GHV], § 5.25) que

$$E_2^{p,q} \cong [H^q(L;H^*(F))]^{p+q} \cong \mathrm{Ext}_{UL}^{p,q}(\mathbb{Q},H^*(F)).$$

Nous pouvons donc énoncer :

<u>Proposition</u> 8.1.- *Il existe une suite spectrale du premier quadrant*

$$E_2^{p,q} = \mathrm{Ext}^{p,q}_{H_*(\Omega B)}(\mathbb{Q},H^*(F)) \implies H^*(E)$$

que nous appelons suite spectrale d'holonomie de la fibration ξ.

<u>Remarque</u> : La formule de dualité de Cartan-Eilenberg construit une suite spectrale duale :

$$E^2_{p,q} = \mathrm{Tor}^{H_*(\Omega B)}_{p,q}(\mathbb{Q},H_*(F)) \implies H_*(E).$$

<u>Corollaire</u> 1.- *Si* $F \longrightarrow E \longrightarrow B$ *est une fibration, et si* B *a le type d'homotopie rationnelle d'un bouquet de sphères, alors la suite spectrale d'holo-*

nomie dégénère au terme E_2 :

$$H_*(E) = \mathrm{Tor}^{H_*(\Omega B)}(H_*(F),\mathbb{Q}).$$

<u>*Corollaire*</u> *2* (Réciproque du théorème 4.1).- *Soit* $F \to E \to B$ *une fibration avec* B *un C.W.-complexe fini. Si* $H_*(F)$ *est un* $H_*(\Omega B)$*-module noethérien, alors* $H_*(E)$ *est de dimension finie.*

<u>*Démonstration*</u> : Notons $(\Lambda X,d)$ un modèle minimal de B. L'hypothèse de finitude sur B montre qu'il existe un quasi-isomorphisme $\Psi : (\Lambda X,d) \to (\Lambda X_{/I},\bar{d})$ où I est un idéal de ΛX contenant $\Lambda^{\geq r}X$ pour un certain r. E admet alors pour modèle $(\Lambda X_{/I} \otimes \Lambda Y,D) \underset{\text{déf}}{=\!=} (\Lambda X_{/I},\bar{d}) \otimes_{(\Lambda X,d)} (\Lambda X \otimes \Lambda Y,d)$ et le gradué associé à $H^*(E)$ vérifie $E^{\infty}_{p,*} = 0$ pour $p > r$.

D'autre part, $H_*(F)$ étant un $H_*(\Omega B)$-module noethérien, les espaces $\mathrm{Tor}_i^{H_*(\Omega B)}(H_*(F),\mathbb{Q})$ sont de dimension finie pour chaque i. Il en est donc de même de chaque $E^{\infty}_{i,*}$.

B I B L I O G R A P H I E

[A.A] ANDREWS P. and ARKOWITZ M. - Sullivan's minimal models and higher order Whitehead products. Can. J. of Math. 30, n° 5 (1978), 961-982.

[A.H] AVRAMOV L. and HALPERIN S. - Through the looking glass : A dictionary between rational homotopy theory and local algebra (These proceedings).

[Bo] BØGVAD R. - Graded Lie algebras in local algebra and rational homotopy. Thesis Stockholm (1983).

[DGMS] DELIGNE P., GRIFFITHS P., MORGAN J. and SULLIVAN D. - Real homotopy theory of Kähler manifolds. Invent. Math. 29 (1975), 245-274.

[E.M] EILENBERG S. and MOORE J.C. - Homology and fibrations I. Coalgebras, cotensor product and its derived functors.
Comment. Math. Helv. 40 (1966), 199-236.

[F.M] FELIX Y. and HALPERIN S. - Rational L.S. category and its applications.
Trans. A.M.S. 273 (1983), 1-37.

[F.H.T] FELIX Y., HALPERIN S. et THOMAS J.C. - Sur certaines algèbres de Lie de dérivations.
Ann. Inst. Fourier, 32, (1982), 143-150.

[F.T] FELIX Y. et THOMAS J.C. - Sur la structure des espaces de catégorie 2.
A paraître Ill. J. of Math.

[Ga-1] GANEA T. - A generalization of the homology and homotopy suspension.
Comment. Math. Helvet. 39 (1965), 295-322.

[Ga-2] GANEA T. - On monomorphisms in homotopy theory.
Topology, Vol. 6, (1967), 149-152.

[Go] GOTTLIEB D. - On fiber spaces and the evaluation map.
Ann. of Math. 87 (1968), 42-55.

[G.H.V] GREUB W., HALPERIN S. and VANSTONE R. - Connections, curvature and cohomology III.
Academic Press, 1976.

[Gr] GRIVEL P.P. - Formes différentielles et suites spectrales.
Ann. Inst. Fourier, 29 (1979), 17-37.

[Gu-1] GULLIKSEN T. - A change of ring theorem with applications to Poincaré series and intersection multiplicity.
Math. Scand. 34 (1974), 167-183.

[Gu-2] GULLIKSEN T. - On the Hilbert series of the homology of differentiel graded algebras.
Math. Scand. 46 (1980), 15-22.

[Ha] HALPERIN S. - Lectures on minimal models.
Mémoire de la S.M.F. n° 9/10 (1983).

[H.L] HALPERIN S. et LEMAIRE J.M. - Suites inertes dans les algèbres de Lie.
Preprint (1983), Nice. (To appear in Math. Scand.)

[L.S] LEMAIRE J.M. et SIGRIST F. - Sur les invariants d'homotopie rationnelle liés à la L.S. catégorie.
Comment. Math. Helv. 56 (1981), 103-122.

[Le] LEVIN G. - Finitely generated Ext-algebras.
Math. Scand. 49 (1981), 161-180.

[Me] MEIER W. - Some topological properties of Kähler manifolds and homogeneous spaces.
Math. Z. 183, (1983), 473-481.

[Op] OPREA J. - Infinite implications in rational homotopy theory.
To appear in Proceedings of A.M.S.

[Q] QUILLEN D. - Rational homotopy theory.
Ann. of Math. 90 (1969), 205-295.

[R] ROOS J.E. - Homology of loop spaces and local rings.
Proc. of the 18th scand. congress Math.
Aarhus (1980).(Progress in Mathematics, n° 11, Birhäuser, 1981.)

[St] STASHEFF J. - Parallel transport and classification of fibrations.
Lect. Notes in math. N° 428, (1974).

[Su] SULLIVAN D. - Infinitesimal computations in topology.
Publ. I.H.E.S. 47 (1977), 269-331.

[Ta] TANRÉ D. - Homotopie rationnelle : Modèles de Chen, Quillen, Sullivan.
Lect. Notes in Math. n° 1025 (1983), Springer Verlag.

[Th] THOMAS J.C. - Rational homotopy of Serre fibrations.
Ann. Inst. Fourier 31 (1978), 71-90.

[V] VIGUÉ M. - Réalisation de morphismes donnés en cohomologie et suite spectrale d'Eilenberg-Moore.
Trans. A.M.S. 265 (1981), 447-484.

[W] WHITEHEAD G. - Elements of homotopy theory.
Graduate texts in math. (1978), Springer Verlag.

Yves F E L I X
UNIVERSITE CATHOLIQUE DE LOUVAIN
1348 - LOUVAIN-LA-NEUVE
(Belgique)

Jean-Claude T H O M A S
UNIVERSITE DE LILLE I
59655 - VILLENEUVE D'ASCQ CEDEX
(France)

Flat families of local, artinian algebras with an infinite number of Poincaré series

by

Ralf Fröberg, Tor Gulliksen and Clas Löfwall.

Introduction. For a local ring (R,m,k) let $P_R(z)$ denote the Poincaré series $\sum_{i\geq 0}\dim_k \mathrm{Tor}_i^R(k,k)z^i$. The origin of the present work is a question how Poincaré series may vary in a flat family of local artinian k-algebras. In particular we were interested in knowing if such a family might have an infinite number of Poincaré series. We will show that this is indeed the case by exhibiting a one-parameter family $\{R_\lambda\}_{\lambda\in\mathbb{Q}}$ of local artinian $\mathbb{Q}$-algebras of length 85 such that the corresponding Poincaré series form an infinite set. We also get as a bonus an example of an augmented $\mathbb{Z}$-algebra A, free of rank 85 as $\mathbb{Z}$-module, such that A/p are local rings and $P_{A/p}(z)$ are different for all primes p, and also $\mathrm{Tor}^A(\mathbb{Z},\mathbb{Z})$ has p-torsion for all primes p.

It turns out that it is possible to construct families of local artinian $\mathbb{C}$-algebras whose Poincaré series vary quite vividly and depend on various algebraic and/or arithmetic properties of the parameters. For instance we show that there exists a family $\{R_\lambda\}_{\lambda\in\mathbb{C}}$ and a power series f(z) such that $P_{R_\lambda}(z)=f(z)$ if and only if λ is transcendent over $\mathbb{Q}$, and that there exists a family $\{S_\lambda\}_{\lambda\in\mathbb{Q}^2}$ such that the calculation of $P_{S_\lambda}(z)$ for all λ is equivalent to solving Fermat's equation $\lambda_1^n+\lambda_2^n=1$ for all n and $\lambda_1,\lambda_2\in\mathbb{Q}$.

It was natural for us to start looking at local k-algebras (R,m,k) with $m^3=0$. Let $\mathcal{C}$ be the class of such k-algebras and let $\mathcal{B}$ be the class of algebras of type $k\langle T_1,\dots,T_n\rangle/(g_1,\dots,g_s)$, where $k\langle T_1,\dots,T_n\rangle$ is the free associative (non-commutative) algebra and the g_i's are linear combinations of the elements T_i^2, $1\leq i\leq n$, and $T_iT_j+T_jT_i$,

$1 \leq i < j \leq n$. It follows from results of Löfwall [Lö] that if we can construct a family $\{B_\lambda\}$ in $\mathcal{B}$ with infinitely many Hilbert series $B_\lambda(z) = \sum_{i\geq 0} \dim_k (B_\lambda)_i z^i$, we get a family $\{A_\lambda\}$ in $\mathcal{C}$ with infinitely many Poincaré series. Each element $B \in \mathcal{B}$ is the universal enveloping algebra U(G) of a graded Lie algebra G. Anick and Löfwall-Roos, see [Lö-Ro], have a construction, which to any graded (non-commutative) algebra N gives a graded Lie algebra G, such that U(G)(z) is determined by N(z). If N is generated by elements of degree one and has relations of degree two only, then $U(G) \in \mathcal{B}$. Thus we have a construction available, which to any family of non-commutative algebras with generators of degree one and relations of degree two and with infinitely many Hilbert series gives a family of commutative local rings with $m^3 = 0$ and with infinitely many Poincaré series. For this reason we were lead to the study of non-commutative graded algebras.

Exhibiting a family in $\mathcal{B}$ with infinitely many Hilbert series also makes it possible for us to construct a family of topological spaces $\{X_\lambda\}$, in fact mapping cones of maps between wedges of spheres $\bigvee_1^{69} S^3 \xrightarrow{f_\lambda} \bigvee_1^{15} S^2$, with infinitely many series $\sum_{i\geq 0} (\dim_{\mathbb{Q}} \pi_i(X_\lambda) \otimes \mathbb{Q}) z^i$, π_i denoting homotopy groups.

1. Poincaré and Hilbert series of families of graded k-algebras.

A graded algebra will in this paper mean an algebra which has a presentation $k\langle T_1,\ldots,T_n\rangle/I$. Here $k\langle T_1,\ldots,T_n\rangle$ is the free associative (non-commutative) algebra in the variables $T_1,\ldots,T_n$ of degree one and I is a homogeneous two-sided ideal in $k\langle T_1,\ldots,T_n\rangle$, k a field. Of special interest to us will be the case when I is generated by elements of degree two. We call such algebras 2-related.

For a graded module $M = \bigoplus_{i\geq 0} M_i$ over a graded algebra we define the Hilbert series of M to be

$$M(z) = \sum_{i\geq 0} \dim_k M_i \cdot z^i.$$

The set of elements of positive degree in a graded module M will be denoted M^+.

For an augmented k-algebra A (or a local ring (A,m,k)) we define the Poincaré series of A to be

$$P_A(z) = \sum_{i\geq 0} \dim_k \mathrm{Tor}_i^A(k,k)\cdot z^i.$$

By a family of k-algebras $\{A_\lambda\}$, $\lambda = (\lambda_1,\ldots,\lambda_m) \in k^m$, we will mean a set of k-algebras together with a finitely presented k[X]-algebra A, $X = (X_1,\ldots,X_m)$, such that $A_\lambda = A/(X-\lambda)$ for all $\lambda \in k^m$. The family $\{A_\lambda\}$ is called a flat family if A is k[X]-flat. We call the family graded if A has a presentation $A = k[X]\langle Y\rangle/(f_1,\ldots,f_r)$, $Y = (Y_1,\ldots,Y_n)$, where the f_i's are homogeneous in Y. Thus all A_λ are graded k-algebras in a graded family $\{A_\lambda\}$. Finally we call the family commutative if A is commutative.

First we examine how Hilbert series $A_\lambda(z)$ may vary in a commutative graded family $\{A_\lambda\}$ of algebras.

Claim. In a commutative graded family there are only finitely many Hilbert series.

In fact a much more general statement is true as the following proposition shows. We note that in a graded family $\{A_\lambda\}$ there is a uniform bound for the number of generators and the degree of the relations, namely if A can be presented as a free algebra in n variables over k[X] with relations of degree $\leq d$ in Y, then each A_λ has a presentation $k\langle Y_1,\ldots,Y_n\rangle$ modulo forms of degree $\leq d$.

Proposition 1. Let n and d be fixed integers. There are only finitely many possibilities for A(z) when A belongs to the class of graded algebras of the form $k[X_1,\ldots,X_n]/(f_1,\ldots,f_r)$ where k is a field and the f_i's are forms of degree $\leq d$.

Proof. Let $B = k[X_1,\ldots,X_n]$ and let A be a graded factor ring of B. The syzygy theorem of Hilbert states that A has a minimal graded resolution

$$0 \to \bigoplus_{i=1}^{b_r} B[-n_{i,r}] \to \ldots \to \bigoplus_{i=1}^{b_1} B[-n_{i,1}] \to B \to A \to 0 \qquad (1)$$

for some $r \leq n$, where the brackets stand for a shift in degree,

$(B[-k])_d = B_{-k+d}$. To construct a step in this resolution is equivalent to solve some linear system of equations with coefficients which are forms in B. It is shown in [He] (also c.f. [Se] and [La]) that there is a bound M, only depending on n and the degrees of the coefficients in the system, such that all solutions can be generated by solutions of degree $\leq M$. Since the resolution is of length $\leq n$, this gives a bound $N = N(n,d)$ for all $n_{i,j}$. Since the degrees of the syzygies are uniformly bounded it follows that the number b_i of syzygies are uniformly bounded. In each fixed degree the resolution (1) is an exact sequence of vector spaces and thus their alternating sum of dimensions is zero. Taking generating functions we get the formula

$$A(z) = (1 - \sum_{i=1}^{b_1} z^{n_{i,1}} + \sum_{i=1}^{b_2} z^{n_{i,2}} - \ldots + (-1)^r \sum_{i=1}^{b_r} z^{n_{i,r}})/(1-z)^n$$

and hence we see that there are only finitely many possibilities for $A(z)$.

We are interested in the following property of a commutative graded family $\{A_\lambda\}$:

(P) The set $\{P_{A_\lambda}(z)\}$ of Poincaré series is finite.

We will show that there exist <u>flat</u> families of local graded artinian k-algebras not satisfying (P).

There is another, seemingly weaker, property for a commutative graded family $\{A_\lambda\}$:

(P') There is a number N such that, if $\dim_k \mathrm{Tor}_i^{A_{\lambda_1}}(k,k) = \dim_k \mathrm{Tor}_i^{A_{\lambda_2}}(k,k)$ for all $i \leq N$, then $P_{A_{\lambda_1}}(z) = P_{A_{\lambda_2}}(z)$.

In fact (P) is equivalent to (P') for a family $\{A_\lambda\}$. Of course (P) implies (P'). But if $\{A_\lambda\}$ is a graded family, then there are only finitely many possibilities for $\dim_k \mathrm{Tor}_i^{A_\lambda}(k,k)$ for fixed i. This follows by the same reasoning as in the proof of proposition 1: Let A be a graded k-algebra in n variables and with relations of degree $\leq d$. Constructing a step in a graded free A-resolution of k is equivalent to solving a system of linear equations over A. This can be lifted to B. By induction over i it follows from the theorem of Herrmann mentioned above that there are only finitely many possibilities for $\sum_{i=0}^{N} \dim_k \mathrm{Tor}_i^{A_\lambda}(k,k) z^i$. Hence (P') implies (P).

In the study of non-commutative families of graded algebras we will be interested in the following two properties for a family $\{B_\lambda\}$:

(H) The set $\{B_\lambda(z)\}$ of Hilbert series is finite.

(H') There is a number N such that, if $\dim_k(B_{\lambda_1})_i = \dim_k(B_{\lambda_2})_i$ for all $i \leq N$, then $B_{\lambda_1}(z) = B_{\lambda_2}(z)$.

The properties (H) and (H') are equivalent for a family $\{B_\lambda\}$. Of course (H) implies (H'). If n is a bound for $\dim_k(B_\lambda)_1$ for all λ, then $\dim_k(B_\lambda)_i \leq n^i$, thus there are only finitely many possibilities for $\sum_{i=0}^{N} \dim_k(B_\lambda)_i z^i$, hence (H') implies (H).

In next section we will show that (H) is not satisfied for all families of non-commutative algebras.

2. Hilbert series of non-commutative graded families.

In this section we give two methods of constructing graded algebras with badly varying Hilbert series.

Construction 1. Let A be a graded (non-commutative) algebra and let A_L and A_R be two graded vector subspaces of A^+. Let T be the coproduct of A with $k\langle a\rangle/(a^2)$. If $W = \{1\} \cup W^+$ is a graded k-basis for A, then $\{WaW^+aW^+\ldots aW^+aW\}$ is a graded k-basis for T. Let $I \subset T$ be the two-sided ideal generated by aA_L and A_Ra and let $\tilde{A} = T/I$. We note that if A is 2-related and $A_L, A_R \subset A_1$, then $\tilde{A}$ will be 2-related. Let W_1 (and W_3, respectively) be a graded basis for a complement to AA_R (and A_LA, respectively) in A and let W_2 be a graded basis for a complement to $A_LA + AA_R$ in A^+. Then a k-basis for $\tilde{A}$ is $W \cup \{W_1aW_2aW_2\ldots W_2aW_3\}$ and hence

$$\tilde{A}(z) = \sum_{n\geq 0} W_1(z)W_3(z)z^{n+1}(W_2(z))^n \quad + \quad A(z) =$$

$$= (A(z) - AA_R(z))(A(z) - A_LA(z))z[1 - z(A(z) - 1 - (A_LA+AA_R)(z))]^{-1} + A(z) \qquad (2)$$

Example 1. Let $A = k<b,c>/(bc-cb-\lambda c^2)$, $\lambda \in k$. As a k-vector space A_i is generated by $c^i, c^{i-1}b, \dots, cb^{i-1}, b^i$ hence $A(z) \leq (1-z)^{-2}$. In fact we have equality since $k<b,c>/(bc-cb) = k[b,c]$ is made to a cyclic left A-module by $b*c^ib^j = c^ib^{j+1} + i\lambda c^{i+1}b^j$ and $c*c^ib^j = c^{i+1}b^j$ and hence $A(z) \geq (1-z)^{-2}$.

Let $A_L = (c-b)k$ and $A_R = b\cdot k$, then

$$A_LA(z) = AA_R(z) = z(1-z)^{-2}$$

since, as is easily seen, $(c-b)c^i, (c-b)c^{i-1}b, \dots, (c-b)b^i$ (and $c^ib, c^{i-1}b\cdot b, \dots, b^i\cdot b$, respectively) are linearly independent. Finally we have

$$(A_LA+AA_R)_{i+1} = A_ib + (c-b)c^i\cdot k = A_ib + (1-i\lambda)c^{i+1}\cdot k.$$

Thus if $\lambda = 0$ or $\lambda^{-1} \notin \{1,2,\dots\}$ we have

$$(A_LA+AA_R)(z) = (1-z)^{-2} - 1.$$

If char k = 0 and $\lambda^{-1} = q \in \{1,2,\dots\}$ we have

$$(A_LA+AA_R)(z) = (1-z)^{-2} - 1 - z^{q+1}.$$

If char k = p and $\lambda^{-1} = q \in \{1,2,\dots,p-1\}$ we have

$$(A_LA+AA_R)(z) = (1-z)^{-2} - 1 - z^{q+1}(1-z^p)^{-1}.$$

Hence if $\lambda = 0$ or $\lambda^{-1} \notin \{1,2,\dots\}$, then $\tilde{A}_\lambda = k<a,b,c>/(a^2, bc-cb-\lambda c^2, ac-ab, ba)$ has Hilbert series (by formula (2))

$$\tilde{A}_\lambda(z) = (1-z)^{-2}(1+z).$$

If char k = 0 and $\lambda^{-1} = q \in \{1,2,\dots\}$, then

$$\tilde{A}_\lambda(z) = (1-z)^{-2}(1+z-z^{q+2})(1-z^{q+2})^{-1}.$$

Finally, if char k = p and $\lambda^{-1} = q \in \{1,2,\dots,p-1\}$, then

$$\tilde{A}_\lambda(z) = (1-z)^{-2}(1+z-z^p-z^{p+1}-z^{q+2})(1-z^p-z^{q+2})^{-1}.$$

If we replace k by $\mathbb{Z}$ and put $\lambda = 1$ we get a $\mathbb{Z}$-algebra

$$B = \mathbb{Z}<a,b,c>/(a^2, bc-cb-c^2, ac-ab, ba) \qquad (3)$$

such that the Hilbert series of B/pB are different for <u>all</u> primes p (and also different from the series of $B\otimes_{\mathbb{Z}}\mathbb{Q}$). This phenomenon can not occur in the commutative case according to proposition 1.

Example 2. Let $A = \mathbb{C}\langle b,c\rangle/(bc-\lambda cb)$, $\lambda \in \mathbb{C}$ and let $A_L = A_R = (c-b)\cdot\mathbb{C}$. This gives

$$\tilde{A}_\lambda = \mathbb{C}\langle a,b,c\rangle/(a^2, bc-\lambda cb, ba-ca, ab-ac).$$

It is possible to compute $A_LA(z)$, $AA_R(z)$ and $(A_LA+AA_R)(z)$ as in example 1 to get

$$\tilde{A}_\lambda(z) = (1+z-z^2)(1-z^2)^{-1}(1-z)^{-2}$$

if $\lambda^n \neq 1$ for all $n > 0$ and

$$\tilde{A}_\lambda(z) = (1+z-z^2-z^{n+2})(1-z^2-z^{n+2})^{-1}(1-z)^{-2}$$

if λ is an n'th primitive root of unity.

Construction 2. (This is an alternative to construction 1, it yields algebras with smaller Hilbert series but needs one more generator.) Let as before A be a graded algebra and A_L and A_R be subspaces of A^+. Let T' be the coproduct of $k\langle L\rangle/(L^2)$ and A and $k\langle R\rangle/(R^2)$ modulo the two-sided ideal $(*L, R*)$ where $*$ stands for anything of positive degree. As a graded vector space this algebra equals

$$k\langle L\rangle/(L^2)\otimes_k A\otimes_k k\langle R\rangle/(R^2).$$

Let I' be the two-sided ideal generated by LA_L and A_RR. Then $I' = LA_LA + AA_RR + L(A_LA+AA_R)R$ and if $\tilde{\tilde{A}} = T'/I'$ then

$$\tilde{\tilde{A}}(z) = A(z) + z(2A(z) - A_LA(z) - AA_R(z)) + z^2(A(z) - (A_LA+AA_R)(z)). \qquad (4)$$

We note that if A is 2-related and $A_L, A_R \subset A_1$ then $\tilde{\tilde{A}}$ will be 2-related.

To be able to give simple descriptions of the spaces A_LA, AA_R and A_LA+AA_R we will put restrictions on the algebra A. If B is an algebra and M a B-bi-module, the trivial extension $B\ltimes M$ is $B\oplus M$ as vector space and has multiplication $(b,m)(b',m') = (bb', bm'+mb')$. From now on we put

$$A = B\ltimes(V\otimes_k B),$$

where $B = k\langle S\rangle$ and S is a finite set of elements of degree one and V is a k-vector space of finite dimension. We make $V\otimes_k B$ a B-bi-module in the following way. For each $s\in S$ there is given a linear transformation $J_s\colon V \to V$. To each monomial $\beta = s_1 s_2 \ldots s_n$ in B we consider the composite map

$$J_\beta = J_{s_1} \circ J_{s_2} \circ \ldots \circ J_{s_n}.$$

For $\beta = 1$ we let J_β denote the identity map.

This defines V as a <u>non-graded</u> left B-module by $\beta\cdot v = J_\beta(v)$ and k-linear extension. For each $s\in S$ we define

$$s(v\otimes b) = sv\otimes sb$$

which extends as above to an operation of B to the left on $V\otimes_k B$. The operation of B to the right on $V\otimes_k B$ is the obvious

$$(v\otimes b)b' = v\otimes(bb').$$

These two operations are compatible. To define $V\otimes_k B$ as a <u>graded</u> B-bi-module we let the degree of $v\otimes b$ be $1+\deg b$, that is we consider V as concentrated in degree one. With this definition $V\otimes_k B$ indeed becomes graded, since

$$\deg(b_1(v\otimes b_2)) = \deg(b_1 v\otimes b_1 b_2) = \deg b_1 + \deg b_2 + 1 = \deg b_1 + \deg(v\otimes b_2)$$

and

$$\deg((v\otimes b_1)b_2) = \deg(v\otimes b_1 b_2) = \deg b_1 + \deg b_2 + 1 = \deg(v\otimes b_1) + \deg b_2 .$$

<u>Proposition 2</u>. Let E be a basis for V. The algebra $B\ltimes(V\otimes_k B)$ has the following presentation

$$H = k\langle S\cup E\rangle/(\{ee', se-J_s(e)s;\ e,e'\in E,\ s\in S\})$$

and hence $B\ltimes(V\otimes_k B)$ is 2-related.

<u>Proof</u>. First consider the surjection

$$p\colon k\langle S\cup E\rangle \twoheadrightarrow B\ltimes(V\otimes_k B)$$

defined by $p(s) = (s,0)$ for $s\in S$ and $p(e) = (0,e\otimes 1)$ for $e\in E$. Then

$$p(ee') = (0,e\otimes 1)(0,e'\otimes 1) = (0,0)$$

and

$$p(se-J_s(e)s) = p(s)p(e) - p(J_s(e))p(s) =$$
$$= (s,0)(0,e\otimes 1) - (0,J_s(e)\otimes 1)(s,0) =$$

$$= (0,s(e\otimes 1)) - (0,(J_s(e)\otimes 1)s) = (0,J_s(e)\otimes s) - (0,J_s(e)\otimes s) = 0.$$

Thus p factors through H and we get

$$H(z) \geq B\ltimes(V\otimes_k B)(z).$$

Using the relations $ee' = 0$ only, we see that words of type $w_1e_1w_2e_2\ldots w_re_rw_{r+1}$, where w_i is a word in S and $e_i \in E$ for all i, generate H. Using also the relations $se = J_s(e)s$ we see that it suffices to take words w_1 and ew_2, where w_1,w_2 are words in S and $e\in E$, to generate H. Hence

$$H(z) \leq B(z) + z\dim_k V\cdot B(z) = B\ltimes(V\otimes_k B)(z).$$

We have shown that

$$H(z) = B\ltimes(V\otimes_k B)(z)$$

which gives $H \simeq B\ltimes(V\otimes_k B)$ since $B\ltimes(V\otimes_k B)$ is a factor of H.

To be able to apply the construction 1 (or 2) we need to specify two subspaces A_L and A_R of A^+. We will choose these as subspaces of $V\subset A_1$ and call them V_L and V_R respectively. Let $\{\beta\}$ be the k-basis of monomials for B and let $B_\beta = \beta\cdot k$. We have

$$A_LA = V_L\otimes B = \bigoplus_\beta (V_L\otimes B_\beta)$$

and

$$AA_R = \mathrm{im}(B\otimes V_R \to V\otimes B) = \bigoplus_\beta J_\beta(V_R)\otimes_k B_\beta$$

whence

$$A_LA+AA_R = \bigoplus_\beta((J_\beta(V_R)+V_L)\otimes_k B_\beta).$$

Taking k-dimensions we get

$$A_LA(z) = z\cdot\dim_k V_L\cdot B(z) \tag{5}$$

$$AA_R(z) = \sum_\beta \dim_k J_\beta(V_R)\cdot z^{\deg\beta+1} \tag{6}$$

$$(A_LA+AA_R)(z) = \sum_\beta \dim_k(J_\beta(V_R)+V_L)\cdot z^{\deg\beta+1} \tag{7}$$

If J_s is an isomorphism for all $s\in S$ then (6) may be simplified to

$$AA_R(z) = z\cdot\dim_k V_R\cdot B(z) \tag{6'}$$

The formulas (5), (6) (or (6')) and (7) may be inserted in either construction 1 or 2, formulas (3) or (4), to get formulas for $\tilde{A}(z)$ or

$\tilde{\tilde{A}}(z)$ depending only (if J_s is iso for all s) on $\dim_k V_L$, $\dim_k V_R$, the number of elements in S and the crucial term

$$\sum_\beta \dim_k(J_\beta(V_R)+V_L)z^{1+\deg\beta}.$$

We have

$$\tilde{\tilde{A}}(z) = (1+z)^2B(z)(1+z\dim_k V) - z^2(\dim_k V_L\cdot B(z) + \sum_\beta \dim_k J_\beta(V_R)z^{\deg\beta}) - $$
$$- z^3(\sum_\beta \dim_k(J_\beta(V_R)+V_L)z^{\deg\beta}.$$

Suppose now $B = \mathbb{C}\langle s\rangle$, the free algebra over $\mathbb{C}$ on one generator of degree one, and suppose $V = \mathbb{C}^2$. Then there are two canonical forms for J_s, namely $\begin{pmatrix}\lambda_1 & 0\\ 0 & \lambda_2\end{pmatrix}$ and $\begin{pmatrix}\lambda & 0\\ 1 & \lambda\end{pmatrix}$. This motivates the first two examples below. The properties of the rings in examples 3 and 4 are similar to those in examples 1 and 2. The set of algebras obtained in examples 1 and 2 have one generator less, they can not however be obtained as $\tilde{A}$ where $A = B\ltimes(V\otimes_k B)$ for any B and V.

<u>Remark</u>. We note that the above reasoning is also valid in the more general case when B is a graded monoid algebra, that is $B = k\langle S\rangle$ modulo a set of monomials and differences of monomials of equal degree. If the monoid algebra is 2-related, then A will be 2-related. In particular we could have used $B = k[s_1,s_2,\dots,s_n]$ or $B = k[s_1,s_2,\dots,s_n]/(s_1^2,s_2^2,\dots,s_r^2)$. This of course imposes restrictions on the maps J_s.

<u>Example 3</u>. Let $B = \mathbb{C}\langle s\rangle$, $V = \mathbb{C}^2$, $V_R = (1,0)\cdot\mathbb{C}$, $V_L = (1,1)\cdot\mathbb{C}$ and $J_s = \begin{pmatrix}\lambda & 0\\ 1 & \lambda\end{pmatrix}$. Then $J_s^n\begin{pmatrix}1\\0\end{pmatrix} = \begin{pmatrix}\lambda^n & 0\\ n\lambda^{n-1} & \lambda^n\end{pmatrix}\begin{pmatrix}1\\0\end{pmatrix} = \begin{pmatrix}\lambda^n\\ n\lambda^{n-1}\end{pmatrix}$, hence $\dim_{\mathbb{C}}(J_s^n(V_R)+V_L) = 1$ if $\lambda = n > 0$ or $\lambda = 0$ and $n > 1$, and $\dim_{\mathbb{C}}(J_s^n(V_R)+V_L) = 2$ otherwise. Thus we have a family $\{\tilde{A}_\lambda\}$ with "generic" value of $\tilde{A}_\lambda(z)$ if $\lambda\notin\mathbb{N}$ and with $\tilde{A}_\lambda(z)$ different for all $\lambda\in\mathbb{N}$. (When $\lambda = 0$, J_s is <u>not</u> iso so the series $AA_R(z)$ has to be computed by means of (6).)

<u>Example 4</u>. Let $B = \mathbb{C}\langle s\rangle$, $V = \mathbb{C}^2$, $V_L = V_R = (1,1)\cdot\mathbb{C}$ and $J_s = \begin{pmatrix}\lambda & 0\\ 0 & 1\end{pmatrix}$.

Then $J_s^n\begin{pmatrix}1\\1\end{pmatrix} = \begin{pmatrix}\lambda^n & 0\\0 & 1\end{pmatrix}\begin{pmatrix}1\\1\end{pmatrix} = \begin{pmatrix}\lambda^n\\1\end{pmatrix}$. Hence $\dim_{\mathbb{C}}(J_s^n(V_R)+V_L) = 2$ if $\lambda^n \neq 1$ and $\dim_{\mathbb{C}}(J_s^n(V_R)+V_L) = 1$ if $\lambda^n = 1$. This gives a family $\{\tilde{A}_\lambda\}$ with "generic" value of $\tilde{A}_\lambda(z)$ if and only if λ is not a root of unity.

Example 5. Let $B = \mathbb{Q}<s>$, $V = \mathbb{Q}^3$, $V_R = (\lambda_1^3, \lambda_2^3, \lambda_3^3)\cdot\mathbb{Q}$, $V_L = (1,0,-1)\mathbb{Q} + (0,1,-1)\mathbb{Q}$ and let $J_s = \begin{pmatrix}\lambda_1 & 0 & 0\\ 0 & \lambda_2 & 0\\ 0 & 0 & \lambda_3\end{pmatrix}$. Then $J_s^n(V_R) = (\lambda_1^{n+3}, \lambda_2^{n+3}, -\lambda_3^{n+3})\cdot\mathbb{Q}$ and $\dim_{\mathbb{Q}}(J_s^n(V_R)+V_L) = 2$ if the determinant

$$\begin{vmatrix}\lambda_1^{n+3} & 1 & 0\\ \lambda_2^{n+3} & 0 & 1\\ -\lambda_3^{n+3} & -1 & -1\end{vmatrix} = \lambda_1^{n+3} + \lambda_2^{n+3} - \lambda_3^{n+3} = 0 \text{ and } \dim_{\mathbb{Q}}(J_s^n(V_R)+V_L) = 3 \text{ otherwise.}$$

Moreover $\dim_{\mathbb{Q}}(J_s^n(V_R)) = 1$ if $(\lambda_1,\lambda_2,\lambda_3) \neq (0,0,0)$. For $\lambda_3 = 1$ this gives exceptional values of $\tilde{A}_\lambda(z)$ if and only if $\lambda_1^{n+3} + \lambda_2^{n+3} = 1$ for some n. We have one value of $\tilde{A}_\lambda(z)$ for $\lambda = (\lambda_1,\lambda_2) = (1,0)$ or $(0,1)$, another value for $\lambda = (-1,0)$ or $(0,-1)$. The statement that $\tilde{A}_\lambda(z)$ is independent of λ for all other values of λ is equivalent to Fermat's last theorem.

Example 6. Let $\{\alpha_0=1,\alpha_1,\dots,\alpha_N\}$ be the set of monomials in $\{\lambda_1,\dots,\lambda_m\}$ of degree $\leq d$, let $S = \{s_0,\dots,s_N,s_0',\dots,s_N'\}$, let $B = \mathbb{C}<S>$, $V = \mathbb{C}^2$, $V_L = V_R = (1,0)\cdot\mathbb{C}$ and let $J_{s_i} = \begin{pmatrix}1 & 0\\ \alpha_i & 1\end{pmatrix}$ and $J_{s_i'} = \begin{pmatrix}1 & 0\\ -\alpha_i & 1\end{pmatrix}$ for $i = 0,1,\dots,N$. Let β be a monomial in S, then $J_\beta = \begin{pmatrix}1 & 0\\ l(\alpha) & 1\end{pmatrix}$ where $l(\alpha)$ is a linear combination of $\alpha_0,\dots,\alpha_N$ with coefficients in $\mathbb{Z}$. Any such linear combination can be achieved by appropriate choice of β. Thus we have exceptional values for $\tilde{A}_\lambda(z)$ if and only if $l(\alpha) = 0$, that is if and only if $(\lambda_1,\dots,\lambda_m)$ satisfies a polynomial equation of degree $\leq d$. If we restrict to $m = 1$ we get a family $\{\tilde{A}_\lambda\}$ with exceptional values of $\tilde{A}_\lambda(z)$ for algebraic numbers λ of degree $\leq d$, and if we restrict further and also let $d = 1$, we get exceptional values for $\lambda \in \mathbb{Q}$.

Example 7. Let $S = \{s_0, s_0', s_1, s_2, s_2', s_3, s_3', \ldots, s_m, s_m'\}$, let $B = \mathbb{C}\langle S\rangle$, $V = \mathbb{C}^2$, $V_L = V_R = (0,1)\cdot\mathbb{C}$ and let

$$J_{s_0} = \begin{pmatrix} 1 & 1 \\ 0 & 1 \end{pmatrix},\ J_{s_0'} = \begin{pmatrix} 1 & -1 \\ 0 & 1 \end{pmatrix},\ J_{s_i} = \begin{pmatrix} \lambda_i & 0 \\ 0 & 1 \end{pmatrix} \text{ for } i = 1,2,\ldots,m \text{ and } J_{s_i'} = \begin{pmatrix} \lambda_i^{-1} & 0 \\ 0 & 1 \end{pmatrix}$$

for $i = 2,3,\ldots,m$. If β is a monomial in S, then $J_\beta = \begin{pmatrix} * & p(\lambda) \\ 0 & 1 \end{pmatrix}$, where $p(\lambda)$ is a polynomial in $\{\lambda_1, \lambda_2, \ldots, \lambda_m, \lambda_2^{-1}, \lambda_3^{-1}, \ldots, \lambda_m^{-1}\}$ and we have an exceptional value of $\tilde{A}_\lambda(z)$ if $p(\lambda) = 0$.

Claim. We can get any polynomial $p(\lambda) \in \mathbb{Z}[\lambda_1, \ldots, \lambda_m]$ in this way.

Proof. First we see that we can get any $q(\lambda) \in \mathbb{Z}[\lambda_2, \ldots, \lambda_m]$ in this way. Let α be any monomial in $\lambda_2, \ldots, \lambda_m$, then

$$\begin{pmatrix} \alpha & 0 \\ 0 & 1 \end{pmatrix}\begin{pmatrix} 1 & \pm 1 \\ 0 & 1 \end{pmatrix}\begin{pmatrix} \alpha^{-1} & 0 \\ 0 & 1 \end{pmatrix} = \begin{pmatrix} 1 & \pm\alpha \\ 0 & 1 \end{pmatrix}.$$

Since $\begin{pmatrix} \alpha & 0 \\ 0 & 1 \end{pmatrix}$ is obtained as an appropriate product of the J_{s_i}'s, it follows that there is a β such that $J_\beta = \begin{pmatrix} 1 & \pm\alpha \\ 0 & 1 \end{pmatrix}$.

Also if $J_{\beta_1} = \begin{pmatrix} 1 & x \\ 0 & 1 \end{pmatrix}$ and $J_{\beta_2} = \begin{pmatrix} 1 & y \\ 0 & 1 \end{pmatrix}$ then $J_{\beta_1\beta_2} = \begin{pmatrix} 1 & x+y \\ 0 & 1 \end{pmatrix}$. Hence any matrix of the type $\begin{pmatrix} 1 & q(\lambda) \\ 0 & 1 \end{pmatrix}$, $q(\lambda) \in \mathbb{Z}[\lambda_2, \ldots, \lambda_m]$ is obtainable as a J_β. Suppose now

$$p(\lambda) = p_0(\lambda_2, \ldots, \lambda_m) + \lambda_1 p_1(\lambda_2, \ldots, \lambda_m) + \ldots + \lambda_1^d p_d(\lambda_2, \ldots, \lambda_m)$$

then

$$\begin{pmatrix} \lambda_1^d & p(\lambda) \\ 0 & 1 \end{pmatrix} = \begin{pmatrix} 1 & p_0 \\ 0 & 1 \end{pmatrix}\begin{pmatrix} \lambda_1 & 0 \\ 0 & 1 \end{pmatrix}\begin{pmatrix} 1 & p_1 \\ 0 & 1 \end{pmatrix}\cdots\begin{pmatrix} \lambda_1 & 0 \\ 0 & 1 \end{pmatrix}\begin{pmatrix} 1 & p_d \\ 0 & 1 \end{pmatrix}$$

and the claim is proved.

In the presentation of $\tilde{A}_\lambda$ which follows from proposition 2 the only relations containing $\lambda_2^{-1}, \ldots, \lambda_m^{-1}$ are the relations $s_i' e_1 = \lambda_i^{-1} e_1 s_i'$ for $i = 2,3,\ldots,m$. If we replace these relations with $\lambda_i s_i' e_1 = e_1 s_i'$, $i = 2,3,\ldots,m$, we get a family $\{\tilde{A}_\lambda'\}$ with exceptional values of $A_\lambda'(z)$ for all $\lambda = (\lambda_1, \ldots, \lambda_m)$ which satify some polynomial equation over $\mathbb{Z}$ at least if $\lambda_1\lambda_2\ldots\lambda_m \neq 0$. But if $\lambda_1\lambda_2\ldots\lambda_m = 0$ we get, as is easily checked, a value of $\dim_k(\tilde{A}_\lambda')_3$ which does not agree with the generric value. Hence the family $\{\tilde{A}'_\lambda\}$ has exceptional values of $A'_\lambda(z)$ if and only if $\lambda = (\lambda_1,\dots,\lambda_m)$ satisfies some polynomial equation over $\mathbb{Z}$. In particular when $m = 1$ we get a one-parameter family $\{\tilde{A}_\lambda\}_{\lambda\in\mathbb{C}}$ with exceptional values of $\tilde{A}_\lambda(z)$ if and only if λ is an algebraic number over $\mathbb{Q}$.

3. Families of graded Hopf algebras with infinitely many Hilbert series.

We now recall the construction in [Lö-Ro] mentioned in the introduction. Let $N = k\langle T_1,\dots,T_n\rangle/(g_1,\dots,g_t)$ be a 2-related algebra. Let $m_1,\dots,m_{n^2-t}$ be a set of monomials in $\{T_1,\dots,T_n\}$ of degree two whose images in N constitutes a k-basis for N_2. To the algebra N we define a graded Lie algebra G in the following way. The Lie algebra G is generated by a set

$$\{T_1,\dots,T_n,L_1,\dots,L_n,Y,R_1,\dots,R_n,Z_{m_1},\dots,Z_{m_{n^2-t}}\}$$

of variables of degree one and has the following relations

(1) $T_i^2 = 0,\ 1 \le i \le n$, and $[T_i,T_j] = 0,\ 1 \le i < j \le n$

(2) $[L_i,T_j] = [T_i,R_j],\ 1 \le i,j \le n$

(3) $\sum_{i,j} c_{ij}[L_i,T_j] = 0$ if and only if $\sum_{i,j} c_{ij}T_iT_j \in (g_1,\dots,g_t)$

(4) $[Y,T_i] = 0,\ 1 \le i \le n$

(5) $[Z,T_i] = 0,\ 1 \le i \le n$ and all $Z \in \{Z_{m_1},\dots,Z_{m_{n^2-t}}\}$

(6) $[L_i,R_j] = 0,\ 1 \le i,j \le n$

(7) $[L_i,Z] = 0,\ 1 \le i \le n$ and all $Z \in \{Z_{m_1},\dots,Z_{m_{n^2-t}}\}$

(8) $[Y,R_i] = 0,\ 1 \le i \le n$

(9) $[Y,Z_{m_i}] = [L_{j_i},T_{k_i}]$ if $m_i = T_{j_i}T_{k_i}$.

It follows from [Lö-Ro] that G is an extension of graded Lie algebras

$$0 \longrightarrow N^+_{ab} \longrightarrow G \longrightarrow F \longrightarrow 0.$$

Here N^+_{ab} denotes N^+ considered as an abelian Lie algebra and $F = F_1 \times F_2$ is a product of two free Lie algebras F_1 and F_2 generated in degree one. The generators for F_1 are $L_1,\dots,L_n$ and Y, and the generators for F_2 are

$R_1,\dots,R_n$ and $Z_{m_1},\dots,Z_{m_{n^2-t}}$. In particular

$$G(z) = N^+(z) + F(z).$$

For the enveloping algebras U(G), which are 2-related Hopf algebras and has a presentation as above (where the relations [A,B] is interpreted as AB+BA), this gives

$$U(G)(z) = U(N^+_{ab})(z)\cdot U(F)(z). \tag{8}$$

We have

$$U(F)(z) = (1-(n+1)z)^{-1}(1-(n+n^2-t)z)^{-1}$$

and if $N(z) = \sum_{i\geq 0} a_i z^i$, then

$$U(N^+_{ab})(z) = \prod_{i\geq 1}[(1+z^{2i-1})^{a_{2i-1}}/(1-z^{2i})^{a_{2i}}] \tag{9}$$

Let $\{\tilde{A}_\lambda\}$ (or $\{\tilde{\tilde{A}}_\lambda\}$) be any family from section 2. We use the construction above with $N = \tilde{A}_\lambda$. As is seen by inspection, $\dim_k(\tilde{A}_\lambda)_i$ are independent of λ for $i = 1,2$. Hence U(F)(z) is independent of λ. From (8) and (9) follows that $U(G)_{\lambda_1}(z) = U(G)_{\lambda_2}(z)$ if and only if $\tilde{A}_{\lambda_1}(z) = \tilde{A}_{\lambda_2}(z)$. Thus we have constructed for each family from section 2 a family of 2-related Hopf algebras with infinitely many Hilbert series.

Finally we list the relations in the case $\{\tilde{A}_\lambda\}$ is the family from example 1 in section 2. We get a family $\{U(G)_\lambda\}$, where

$$U(G)_\lambda = k\langle T_1,T_2,T_3,L_1,L_2,L_3,Y,R_1,R_2,R_3,Z_{12},Z_{22},Z_{31},Z_{32},Z_{33}\rangle/I$$

where I is generated by

(1) T_i^2, $1 \leq i \leq 3$ and $[T_i,T_j]$ $(=T_iT_j+T_jT_i)$ $1 \leq i < j \leq 3$

(2) $[L_i,T_j] - [T_i,R_j]$, $1 \leq i,j \leq 3$

(3) $[L_1,T_1]$, $[L_2,T_1]$, $[L_1,T_3] - [L_1,T_2]$, $[L_2,T_3] - [L_3,T_2] - \lambda[L_3,T_3]$

(4) $[Y,T_i]$, $1 \leq i \leq 3$

(5) $[Z,T_i]$, $1 \leq i \leq 3$ and all $Z \in \{Z_{12},Z_{22},Z_{31},Z_{32},Z_{33}\}$

(6) $[L_i,R_j]$, $1 \leq i,j \leq 3$

(7) $[L_i,Z]$, $1 \leq i \leq 3$ and all $Z \in \{Z_{12},Z_{22},Z_{31},Z_{32},Z_{33}\}$

(8) $[Y,R_i]$, $1 \leq i \leq 3$

(9) $[Y,Z_{12}]-[L_1,T_2], [Y,Z_{22}]-[L_2,T_2], [Y,Z_{31}]-[L_3,T_1], [Y,Z_{32}]-[L_3,T_2]$, $[Y,Z_{33}]-[L_3,T_3]$.

We have thus constructed a family $\{U(G)_\lambda\}$ of Hopf algebras with the following properties. We have $\dim_k(U(G)_\lambda)_i$ independent of λ for $i = 1,2$. If char k = 0 we have infinitely many values of $U(G)_\lambda(z)$. Replacing k with $\mathbb{Z}$ in the definition of $U(G)_\lambda$ above and restricting to $\lambda = 1$, we get a graded $\mathbb{Z}$-algebra C such that the Hopf algebras C/p have different values of C/p(z) for all primes p (also different from $C\otimes\mathbb{Q}(z)$).

Remark. The construction in [Lö-Ro] presumes char k = 0, but this is an unnecessary restriction.

4. Flat families of local k-algebras with infinitely many Poincaré series.

In this section we pass from Hopf algebras to commutative local rings. Let $\tilde{N}$ be the isomorphism classes of (not necessarily commutative) graded k-algebras N with $N_3 = 0$. Let A be the isomorphism classes of 2-related not necessarily commutative algebras. Then, quoting [Lö]:

There is a 1-1 correspondence between $\tilde{N}$ and A. If $N \in \tilde{N}$ corresponds to $A \in A$, then

$$P_N(z) = zA(z)/(1 + z - A(z)(1 - \dim_k N_1 \cdot z + \dim_k N_2 \cdot z^2)) \tag{10}$$

We restrict the class $\tilde{N}$ to the class C of commutative graded algebras C with $C_3 = 0$. Since artinian commutative graded k-algebras are local, all algebras in C are local. We get a 1-1 correspondence between C and B = the isomorphism classes of 2-related Hopf k-algebras, that is algebras of type

$k<T_1,\dots,T_n>$ modulo relations which are linear combinations of T_i^2, $1 \leq i \leq n$, and $[T_i,T_j] = T_iT_j + T_jT_i$, $1 \leq i < j \leq n$. Thus $\mathcal{B}$ contains the $U(G)$'s from section 3. From (10) it follows that if $B_1, B_2 \in \mathcal{B}$ and $B_1(z) \neq B_2(z)$, then $P_{C_1}(z) \neq P_{C_2}(z)$, where $C_i \in \mathcal{C}$ corresponds to B_i for $i = 1,2$. Thus for each family $\{\tilde{A}_\lambda\}$ (or $\{\tilde{\tilde{A}}_\lambda\}$) from section 2, we get a family $\{C_\lambda\}$ of commutative local rings with $m^3 = 0$ and with infinitely many Poincaré series. That the families $\{C_\lambda\}$ are flat follows from the following well-known lemma. For the convenience of the reader we supply a proof.

Lemma 3. Let C be a commutative Noetherian domain and let M be a finitely generated C-module. Then the following conditions are equivalent:

(i) M is C-flat.

(ii) $\dim_{C_p/pC_p} M_p/pM_p$ is independent of p for all $p \in \operatorname{Spec} C$.

(iii) If Q is the fraction field of C, then $\dim_{C_m/mC_m} M_m/mM_m = \dim_Q(M \otimes_C Q)$ for every maximal ideal m in C.

If the Jacobson radical of C is zero these conditions are equivalent to:

(iv) $\dim_{C_m/mC_m} M_m/mM_m$ is independent of m for every maximal ideal m in C.

Proof. (i) $\Rightarrow$ (ii): Let $p \in \operatorname{Spec} C$. A flat module over a local ring is free, hence M_p is free, $M_p = C_p^n$. This implies that $M \otimes_C Q = Q^n$, hence $n = \dim_{C_p/pC_p} M_p/pM_p$ is independent of p.

(ii) $\Rightarrow$ (iii) is trivial.

(iii) $\Rightarrow$ (i): Since M is C-flat if and only if M_m is C_m-flat for all maximal ideals m in C, we can assume C to be local. Suppose $n = \dim_{C_m/mC_m} M_m/mM_m = \dim_Q(M \otimes_C Q)$. Let $C^n \xrightarrow{f} M$ be a surjection which is an iso when tensored with C_m/mC_m. Then

$$0 \longrightarrow \operatorname{Ker} f \otimes_C Q \longrightarrow Q^n \xrightarrow{\bar{f}} M \otimes_C Q \longrightarrow 0$$

is exact since Q is C-flat. Since $\dim_Q Q^n = \dim_Q(M \otimes_C Q) = n$ we have that $\bar{f}$ is an iso and hence that $\operatorname{Ker} f \otimes_C Q = 0$. This implies that $\operatorname{Ker} f$ is a torsion C-module. But $\operatorname{Ker} f \subset C^n$, hence $\operatorname{Ker} f = 0$ and M is free.

Now suppose that the Jacobson radical is zero. We prove (iv)$\Rightarrow$(iii): Consider $U = \{p \in \operatorname{Spec} C\,;\ \dim_{C_p/pC_p} M_p/pM_p \leq \dim_Q(M \otimes_C Q)\}$. The sets $S_r = \bigcap_{C^r \xrightarrow{f} M} \{p;\ p \supset \operatorname{Ann}(\operatorname{Coker} f)\}$ are closed. Hence U, which is the complement of $S_{\dim_Q(M \otimes_C Q)}$, is open. Furthermore U is non-empty since $(0) \in U$. Since the intersection of all maximal ideals is zero, U must contain a maximal ideal m. Thus $\dim_{C_m/mC_m} M_m/mM_m \leq \dim_Q(M \otimes_C Q)$, hence the same inequality is true for all maximal ideals. The opposite inequality is trivial and hence we have equality.

Let $\{\tilde{A}_\lambda\}$ (or $\{\tilde{\tilde{A}}_\lambda\}$) be a family from section 2 and let $\{C_\lambda\}$ be the corresponding family of local rings. Since $\dim_k(\tilde{A}_\lambda)_i$ is independent of λ for $i = 1,2$ we have that $\dim_k(C_\lambda) = 1 + \dim_k(C_\lambda)_1 + \dim_k(C_\lambda)_2$ is independent of λ. To show that $\{C_\lambda\}$ is flat, we can assume that k is algebraically closed. To see this, let A be the finitely presented k[X]-algebra from the definition of a family. Then $A \otimes_k \hat{k}$ is $\hat{k}[X]$-flat if and only if A is k[X]-flat since $\hat{k}[X]$ is a faithfully flat k[X]-algebra, where $\hat{k}$ is the algebraic closure of k. If k is algebraically closed all maximal ideals of k[X] are of type $(X-\lambda)$, $\lambda \in k$. Hence (iv) in the lemma with $C = k[X]$ and $M = A$ shows that all our families are flat.

The material in this section supplies a proof of the following theorem. To get an explicit example we use the family $\{U(G)_\lambda\}$ from the end of section 3.

Theorem A. Consider the family $\{C_\lambda\}$ where

$$C_\lambda = k[T_1,T_2,T_3,L_1,L_2,L_3,Y,R_1,R_2,R_3,Z_{12},Z_{22},Z_{31},Z_{32},Z_{33}]/I_\lambda$$

and I_λ is generated by $(T,L,Y,R,Z)^3$ and

(1) L_iL_j and L_iY and Y^2, $1 \le i \le j \le 3$

(2) R_iR_j and R_iZ_I and Z_IZ_J, $1 \le i \le j \le 3$ and $I,J \in \{12,22,31,32,33\}$

(3) $L_1T_2+L_1T_3+YZ_{12}+T_1R_2+T_1R_3$, $L_2T_2+T_2R_2+YZ_{22}$, $L_3T_1+T_3R_1+YZ_{31}$,
$L_2T_3+T_2R_3+L_3T_2+T_3R_2+YZ_{32}$, $\lambda L_2T_3+\lambda T_2R_3+L_3T_3+T_3R_3+YZ_{33}$.

Then $\{C_\lambda\}$ is a flat family of local k-algebras of length 85. If $\operatorname{char} k = 0$, then $\{C_\lambda\}$ has infinitely many Poincaré series.

Theorem B. If we replace k with $\mathbb{Z}$ in the definition of C_λ above, and put $\lambda = 1$, we get a graded $\mathbb{Z}$-algebra D such that D is $\mathbb{Z}$-free of rank 85 and D/p are local rings with different Poincaré series for all primes p. (They are also all different from $P_{D\otimes\mathbb{Q}}(z)$.) Furthermore $\operatorname{Tor}^D(\mathbb{Z},\mathbb{Z})$ has p-torsion for each prime p.

Remark. Theorem B contrasts to [Mo, Cor.9] where it is shown that in some situations there can only be a finite number of Poincaré series.

The proof of the first statement of Theorem B follows from the beginning of this section using the last sentence of section 3. The last statement will follow from the lemma below. That we could draw this conclusion about p-torsion was observed by L. Avramov who also supplied a proof of the lemma.

Lemma 4. Let D be an augmented $\mathbb{Z}$-algebra, which is finitely generated as a $\mathbb{Z}$-module, and let p be a prime number. Then $\mathrm{Tor}^D(\mathbb{Z},\mathbb{Z})$ has no p-torsion if and only if for all i ,

$$\dim_{\mathbb{Z}/p\mathbb{Z}} \mathrm{Tor}_i^{D/pD}(\mathbb{Z}/p\mathbb{Z},\mathbb{Z}/p\mathbb{Z}) = \dim_{\mathbb{Q}} \mathrm{Tor}_i^{D\otimes\mathbb{Q}}(\mathbb{Q},\mathbb{Q}) .$$

Proof. Let $X \to \mathbb{Z}$ be a D-free resolution. Since D is $\mathbb{Z}$-free, $X \to \mathbb{Z}$ is a $\mathbb{Z}$-free resolution. Hence X/pX is acyclic and $X/pX \to \mathbb{Z}/p\mathbb{Z}$ is a D/pD - resolution. But

$$X/pX \otimes_{D/pD} \mathbb{Z}/p\mathbb{Z} = (X \otimes_D D/pD) \otimes_{D/pD} \mathbb{Z}/p\mathbb{Z} = X \otimes_D \mathbb{Z}/p\mathbb{Z} = (X \otimes_D \mathbb{Z}) \otimes_{\mathbb{Z}} \mathbb{Z}/p\mathbb{Z} .$$

Hence $\mathrm{Tor}^{D/pD}(\mathbb{Z}/p\mathbb{Z},\mathbb{Z}/p\mathbb{Z}) = H((X \otimes_D \mathbb{Z}) \otimes_{\mathbb{Z}} \mathbb{Z}/p\mathbb{Z})$. Moreover we have $\mathrm{Tor}^D(\mathbb{Z},\mathbb{Z}) = H(X \otimes_D \mathbb{Z})$. By the universal coefficient formula,

$$0 \to H_i(X\otimes_D\mathbb{Z})\otimes_{\mathbb{Z}}\mathbb{Z}/p\mathbb{Z} \to H_i((X\otimes_D\mathbb{Z})\otimes_{\mathbb{Z}}\mathbb{Z}/p\mathbb{Z}) \to \mathrm{Tor}_1^{\mathbb{Z}}(H_{i-1}(X\otimes_D\mathbb{Z}),\mathbb{Z}/p\mathbb{Z}) \to 0$$

is an exact sequence, and hence from above the following sequence is exact,

$$0 \to \mathrm{Tor}_i^D(\mathbb{Z},\mathbb{Z})\otimes_{\mathbb{Z}}\mathbb{Z}/p\mathbb{Z} \to \mathrm{Tor}_i^{D/pD}(\mathbb{Z}/p\mathbb{Z},\mathbb{Z}/p\mathbb{Z}) \to \mathrm{Tor}_1^{\mathbb{Z}}(\mathrm{Tor}_{i-1}^D(\mathbb{Z},\mathbb{Z}),\mathbb{Z}/p\mathbb{Z}) \to 0 .$$

Hence for all i ,

$$\dim_{\mathbb{Z}/p\mathbb{Z}}(\mathrm{Tor}_i^D(\mathbb{Z},\mathbb{Z}) \otimes_{\mathbb{Z}} \mathbb{Z}/p\mathbb{Z}) \leq \dim_{\mathbb{Z}/p\mathbb{Z}} \mathrm{Tor}_i^{D/pD}(\mathbb{Z}/p\mathbb{Z},\mathbb{Z}/p\mathbb{Z})$$

with equality for all i if and only if $\mathrm{Tor}^D(\mathbb{Z},\mathbb{Z})$ has no p-torsion. On the other hand, for all i ,

$$\dim_{\mathbb{Q}} \mathrm{Tor}_i^{D\otimes\mathbb{Q}}(\mathbb{Q},\mathbb{Q}) = \dim_{\mathbb{Q}}(\mathrm{Tor}_i^D(\mathbb{Z},\mathbb{Z}) \otimes_{\mathbb{Z}} \mathbb{Q}) \leq \dim_{\mathbb{Z}/p\mathbb{Z}}(\mathrm{Tor}_i^D(\mathbb{Z},\mathbb{Z}) \otimes_{\mathbb{Z}} \mathbb{Z}/p\mathbb{Z})$$

with equality for all i if and only if $\mathrm{Tor}^D(\mathbb{Z},\mathbb{Z})$ has no p-torsion. Putting this together we get,

$$\dim_{\mathbb{Q}} \mathrm{Tor}_i^{D\otimes\mathbb{Q}}(\mathbb{Q},\mathbb{Q}) \leq \dim_{\mathbb{Z}/p\mathbb{Z}}(\mathrm{Tor}_i^D(\mathbb{Z},\mathbb{Z}) \otimes_{\mathbb{Z}} \mathbb{Z}/p\mathbb{Z}) \leq \dim_{\mathbb{Z}/p\mathbb{Z}} \mathrm{Tor}_i^{D/pD}(\mathbb{Z}/p\mathbb{Z},\mathbb{Z}/p\mathbb{Z})$$

with equality in both places for all i if and only if $\mathrm{Tor}^D(\mathbb{Z},\mathbb{Z})$ has no p-torsion.

5. Flat families of non-commutative algebras with infinitely many Poincaré series.

We note that we can construct families with infinitely many

Poincaré series without the construction in section 3. However, these families will consist of non-commutative graded algebras.

Example 8. Let $B_\lambda = k\langle a,b,c\rangle/((ab+ac,\ b^2,\ bc+cb,\ bc+\lambda c^2,\ ca) + (a,b,c)^3)$. Then, in the correspondence of section 4, B_λ will correspond to $A_\lambda = k\langle a,b,c\rangle/(a^2,\ bc-cb-\lambda c^2,\ ac-ab,\ ba)$, which is the algebras from example 1 of section 2. Then (10) of section 4 gives

$$P_{B_\lambda}(z) = (1-3z)^{-1}$$

if $\lambda = 0$ or $\lambda^{-1} \notin \{1,2,\dots\}$. If char $k = 0$ and $\lambda^{-1} = q \in \{1,2,\dots\}$ we have

$$P_{B_\lambda}(z) = (1+z+z^{q+2})/(1-2z-3z^2-2z^{q+2}+5z^{q+3}-z^{q+4})$$

and if char $k = p$ and $\lambda^{-1} \in \{1,2,\dots,p-1\}$, then

$$P_{B_\lambda}(z) = (1+z-z^p-z^{p+1}-z^{q+2})/(1-2z-3z^2-z^{p+1}+2z^{p+2}+3z^{p+3}+z^{q+2}-2z^{q+3}+5z^{q+4}-z^{q+5}).$$

Example 9. A graded algebra A is called a Koszul algebra if $P_A(z)A(-z) = 1$. We can show that the family $\{\tilde{A}_\lambda\}$ above generically consists of Koszul algebras. More precisely $\tilde{A}_\lambda$ is a Koszul algebra if and only if $\lambda = 0$ or $\lambda^{-1} \notin \{1,2,\dots\}$.

6. A family of topological spaces with infinitely many Poincaré-Betti series.

The third homotopy group of a wedge of 2-spheres $\bigvee_{i=1}^{m} S_i^2$, is known to be a free abelian group on $m(m+1)/2$ generators (see e.g. [Gr-Mo, p. 240]). The generators are the Whitehead products $[s_i^2, s_j^2]$, $i \leq j$, where s^2 denotes the canonical element in $\pi_2(S^2)$. Thus elements in $\pi_3(\bigvee_1^m S_i^2)$ may be identified with elements of $\mathbb{Z}\langle T_1,\dots,T_m\rangle$ which are linear combinations of $T_iT_j + T_jT_i$, $i \leq j$. Let $f_1,\dots,f_r$ be elements of $\mathbb{Z}\langle T_1,\dots,T_n\rangle$ which are such linear combinations. Then $f_1,\dots,f_r$ gives rise to a continous map

$$f:\ \bigvee_{i=1}^{r} S_i^3 \longrightarrow \bigvee_{i=1}^{m} S_i^2.$$

let X denote the mapping cone of f. It is implicitely proved by Lemaire that there is a rational correspondence between the Hilbert series of

$\mathbb{Q}<T_1,\dots,T_m>/(f_1,\dots,f_r)$ and the homology algebra of the loop space X of X, which we denote by $H_*(\Omega X;\mathbb{Q})$ (see [Le, p. 64] and [Le', p. 117] and for an explicit formula [Ro, p. 450, formula (9)]). It is a well-known fact that $H_*(\Omega X;\mathbb{Q})$ is the enveloping algebra of the graded Lie algebra $\bigoplus_{i=1}^{\infty}(\pi_{i+1}(X)\otimes\mathbb{Q})$ defined by the Whitehead product. Hence if $a_i = \dim_{\mathbb{Q}}(\pi_{i+1}(X)\otimes\mathbb{Q})$ then

$$H_*(\Omega X;\mathbb{Q}) = \prod_{i=1}(1+z^{2i-1})^{a_{2i-1}}(1-z^{2i})^{-a_{2i}}.$$

We now choose the Hopf algebras $U(G)_\lambda$ from the end of section 3. From these we get a family $\{X_\lambda\}$ of topological spaces, namely X_λ is the mapping cone of a map $f_\lambda: \bigvee_1^{69} S^3 \longrightarrow \bigvee_1^{15} S^2$, with the property that the series $\sum_{i=1}^{\infty}\dim_{\mathbb{Q}}(\pi_{i+1}(X_\lambda)\otimes\mathbb{Q})z^i$ takes infinitely many values when λ varies in $\mathbb{Q}$.

Remark. During the work on this paper we have been informed that D. Anick independently has constructed a family of graded non-commutative algebras over $\mathbb{Q}$ with infinitely many Hilbert series. He also has a family which generically is of global dimension 2. This can be achived also by our methods. More precisely, consider the following slight variation of our example 1:

Let $D_\lambda = k<a,b,c,d>/(ab-ac, bc-cb-\lambda c^2, bd)$. Then D_λ has global dimension 2 if and only if $\lambda = 0$ or $\lambda^{-1} \notin \{1,2,\dots\}$.

References.

[He], Herrmann G., Die Frage der endlich vielen Schritte in der Theorie der Polynomideale, Math. Ann. 95 (1926), 736-788.

[La], Lazard D., Algèbre linéaire sur $K[X_1,\ldots,X_n]$, et élimination, Bull. Soc. Math. France 105 (1977) no. 2, 165-190.

[Le], Lemaire J.M., Algèbres connexes et homologies des Espaces des Lacets, Lect. Notes in Math. 422 (1979), Springer.

[Le´], Lemaire J.M., A finite complex whose rational homotopy is not finitely generated, Lect. Notes in Math. 196 (1971), 114-120, Springer.

[Lö], Löfwall C., Une algèbre nilpotente dont la série de Poincaré-Betti est non rationelle, C.R. Acad. Sc. Paris 288 (1979), sér. A, 327-330.

[Lö-Ro], Löfwall C. et Roos J.-E., Cohomologie des algèbres de Lie graduées et séries de Poincaré-Betti non rationelles, C.R. Acad. Sc. Paris 290 (1980), sér. A, 733-736.

[Mo], Mount K.R., Fitting´s invariants and a theorem of Grauert, Math. Ann. 203 (1973), 289-294.

[Ro], Roos J.-E., Homology of loop spaces and of local rings, Proc. 18th Scand. Congr. of Math., Progress in Math. 11 (1981), 441-468.

[Se], Seidenberg A., Construction in algebra, Trans. Am. Math. Soc. 197 (1974), 273-313.

Ralf Fröberg and Clas Löfwall
Department of Mathematics
University of Stockholm
Box 6701, S-113 85 STOCKHOLM (SWEDEN)

Tor Gulliksen
Department of Mathematics
University of Oslo
Blindern - OSLO 3 (NORWAY)

A NOTE ON INTERSECTION MULTIPLICITIES

by

Tor H. Gulliksen

Let $(R,\underline{m})$ denote a local (noetherian) ring and let M and N be finitely generated R-modules such that $M\otimes N$ has finite length. If R is regular we define the intersection multiplicity:

$$\chi^R(M,N) = \Sigma_i(-1)^i \text{length}(\text{Tor}_i^R(M,N))$$

as in [3]. If R is not regular we will define $\chi^R(M,N)$ as follows:

DEFINITION. Let $\hat{R}$ denote the $\underline{m}$-adic completion of R. We choose a surjective ring homomorphism $\tilde{R} \to \hat{R}$ where $\tilde{R}$ is a regular local ring of minimal dimension. Then we put

$$\chi^R(M,N) = \chi^{\tilde{R}}(\hat{M},\hat{N})$$

The purpose of this note is to show that $\chi^R(M,N)$ only depends on the R-modules M and N, and not on the choice of minimal regular "embedding" $\tilde{R} \to \hat{R}$. This will follow from the theorem below.

REMARK. If R is a local complete intersection, then $\chi^R(M,N)$ can be computed in terms of $\text{Tor}^R(M,N)$ in the following way. Consider the Poincaré series

$$P_R^{M,N}(t) = \sum_{i\geq 0} \text{length}(\text{Tor}_i^R(M,N))t^i.$$

It follows from corollary 4.2 in [1] that

$$\chi^R(M,N) = \lim_{t\to -1} P_R^{M,N}(t)(1-t^2)^{n-\dim R}$$

where n is the embedding dimension of R, i.e. the dimension of $\underline{m}/\underline{m}^2$ as a vectorspace over $R/\underline{m}$.

QUESTION. One may ask if it is possible to express $\chi^R(M,N)$ in terms of $\text{Tor}^R(M,N)$ also for rings R which are not local complete intersections.

THEOREM. Let C be an arbitrary local (noetherian) ring, and let M and N be C-modules of finite type such that $M\otimes N$ has finite length. Assume that $A \to C$ and $B \to C$ are surjective ring homomorphisms, A

and B being regular local rings of minimal dimension, that is the dimension of A and that of B equal the embedding dimension of C. Then we have

$$\chi^{A}(M,N) = \chi^{B}(M,N).$$

PROOF. We may assume that A, B and C are complete local rings. Hence so is the fiber-product $A \times_C B$. By Cohen's structure theorem $A \times_C B$ is a homomorphic image of a regular local ring R, thus we have a commutative diagram of surjective ring homomorphisms

$$\begin{array}{ccc} R & \longrightarrow & B \\ \downarrow & & \downarrow \\ A & \longrightarrow & C \end{array}$$

Put $\underline{a} := \mathrm{Ker}(R \to A)$ and $\underline{b} := \mathrm{Ker}(R \to B)$. Let $\underline{m}$ be the maximal ideal of R. Since A, B and R are all regular, the inclusions $\underline{a} \subset \underline{m}$ and $\underline{b} \subset \underline{m}$ give rise to injections

$$\underline{a}/\underline{ma} \to \underline{m}/\underline{m}^2$$
$$\underline{b}/\underline{mb} \to \underline{m}/\underline{m}^2$$

By means of these maps we will consider $\underline{a}/\underline{ma}$ and $\underline{b}/\underline{mb}$ as subspaces of $\underline{m}/\underline{m}^2$. Put

$$s = \dim R - \dim A$$

Since $\dim A = \dim B$ both $\underline{a}$ and $\underline{b}$ are minimally generated by s elements. Let $\partial_1,\dots,\partial_r$ $(r<s)$ be a basis for $\underline{a}/\underline{ma} \cap \underline{b}/\underline{mb}$. Let $a_1,\dots,a_r$ respectively $b_1,\dots,b_r$ be elements in $\underline{a}$ respectively $\underline{b}$ representing $\partial_1,\dots,\partial_r$. Now extend these two sequences to minimal sets of generators

$$a_1,\dots,a_r,\dots,a_s$$
$$b_1,\dots,b_r,\dots,b_s$$

for $\underline{a}$ and $\underline{b}$ respectively. For each i $(0<i<s)$ the elements

$$a_1,\dots,a_i,\ b_{i+1},\dots,b_s$$

represent linearly independent elements in $\underline{m}/\underline{m}^2$. Hence they are part of a regular system of parameters for R. Let $\underline{a}_i$ denote the ideal they generate and put

$$A_i := R/\underline{a}_i.$$

Then each A_i is a regular local ring. Observe that $A_0 = B$ and $A_s = A$. In the following let $1<i<s$. To prove the theorem it clarly suffices to prove

$$\chi^{A_{i-1}}(M,N) = \chi^{A_i}(M,N)$$

Here we will use a technique which was used in [2] for a similar purpose. To simplify the notation we put $P := A_{i-1}$ and $Q := A_i$. Let L be the ring $R/\underline{c}$ where $\underline{c}$ is the ideal generated by

$$a_1,\ldots,a_{i-1},a_i,b_i,b_{i+1},\ldots,b_s.$$

Observe that L need not be regualar. We have exact sequences:

$$0 \to P \xrightarrow{a_i} P \to L \to 0$$

$$0 \to Q \xrightarrow{b_i} Q \to L \to 0$$

where a_i and b_i denote multiplication by a_i and b_i respectively. From the sequences above we obtain standard spectral sequences

$$\mathrm{Tor}_p^L(M,\mathrm{Tor}_q^P(N,L)) \Rightarrow \mathrm{Tor}_{p+q}^P(M,N)$$

and

$$\mathrm{Tor}_p^L(M,\mathrm{Tor}_q^Q(N,L)) \Rightarrow \mathrm{Tor}_{p+q}^Q(M,N)$$

where $\mathrm{Tor}_q^P(N,L)$ and $\mathrm{Tor}_q^Q(N,L)$ equals N for $q = 0,1$ and equals zero for $q \neq 0,1$. Hence we obtain exact sequences

$$\ldots\mathrm{Tor}_i^L(M,N) \to \mathrm{Tor}_{i+1}^P(M,N) \to \mathrm{Tor}_{i+1}^L(M,N) \to \mathrm{Tor}_{i-1}^L(M,N) \to \ldots$$

$$\ldots\mathrm{Tor}_i^L(M,N) \to \mathrm{Tor}_{i+1}^Q(M,N) \to \mathrm{Tor}_{i+1}^L(M,N) \to \mathrm{Tor}_{i-1}^L(M,N) \to \ldots$$

from which it follows that

$$\chi^P(M,N) = \chi^Q(M,N).$$ □

REFERENCES

[1] T.H. Gulliksen, A change of ring theorem with applications to Poincaré series and intersection multiplicity, Math. Scand. 34 (1974) 167-183.

[2] M.-P. Malliavin - Brameret, Une remarque sur les anneaux locaux réguliers, Seminaire Dubreil - Pisot (Algèbre et Théorie des Nombres), 1970/1971 no. 13.

[3] J.P. Serre, Algèbre Locale Multiplicités, Lecture Notes in Mathematics, 11, Springer-Verlag, Berlin, 1965.

Department of Mathematics
University of Oslo
Blindern - Oslo 3
(NORWAY)

REDUCING THE POINCARÉ SERIES OF LOCAL RINGS TO THE CASE OF QUADRATIC RELATIONS

by
Tor H. Gulliksen

Let $(R,\underline{m},k)$ be a commutative, noetherian local ring and let M be a finitely generated R-module. By the Poincaré series of M we mean the formal power series

$$P_R^M = \sum_{p\geq 0} \dim_k \mathrm{Tor}_q^R(M,k)t^q$$

For a given M it is usually very difficult to compute P_R^M and many attempts have been made to reduce the general case to hopefully simpler cases. One such reduction appeared in [1, theorem 2] where computation of the Poincaré series of M was reduced to the case $M = k$, but over two different rings. More presicely we have the formula

$$P_{R(M)}^k = P_R^k \cdot (1-tP_R^M)^{-1}$$

where $R(M)$ is the trivial ring extension of R by M.

Later G. Levin made a beautiful reduction showing how the Poincaré series of a module can be expressed in terms of Poincaré series of modules over artinian local rings, cf. [2].

In the following we shall reduce the artinian case to the case where the ring is defined by certain quadratic relations.

DEFINITION. We will say that a local ring is defined by special quadratic relations if it has the form

$$R_0[[X_1,\dots,X_m,Y_1,\dots,Y_m]]/I$$

where R_0 is a field or a complete regular local ring of dimension 1 whose maximal ideal is generated by a prime number, and I is an ideal generated by the quadratic forms

$$\sum_{i,j} \alpha_{ij} X_i Y_j \ , \quad \alpha_{ij} \in R_0$$

where the m×m-matrices (α_{ij}) run through a set of matrices which is closed with respect to transposition.

THEOREM. Let R be an artinian local ring and let M be a finitely generated R-module. Then there exists a local, surjective homomorphism $R^* \to R$ where R^* is a local ring defined by special quadratic relations and

$$P^M_{R^*} = P^R_{R^*} P^M_R$$

PROOF. By Cohen's structure theorem R is an algebra over a ring R_0 with properties described above, and such that the structure map $R_0 \to R$ induces an isomorphism between the residue class fields of R_0 and R. Let $v_1,\ldots,v_m$ be a set of generators for $\underline{m}$, considered as an R_0-module, and consider the following set of $m\times m$-matrices (α_{ij}) with entries in R_0:

$$A = \{(\alpha_{ij}) \mid \sum_{i,j} \alpha_{ij} v_i v_j = 0\}$$

Now put

$$R^* = R_0[[X_1,\ldots,X_m,Y_1,\ldots,Y_m]]/I$$

where I is the ideal generated by all the quadratic forms $\sum \alpha_{ij} X_i Y_j$ where $(\alpha_{ij}) \in A$. Let $\bar{X}_i$ (resp. $\bar{Y}_i$) be the image of X_i (resp. Y_i) in R^* and let $f: R^* \to R$ be the unique homomorphism extending the structure map $R_0 \to R$ and sending $\bar{X}_i$ and $\bar{Y}_i$ to v_i. Let F be a minimal R-free resolution of M. We shall first show that F can be lifted to an R^*-free complex F^* whose differential has coefficients in $\underline{m}^*$, the maximal ideal of R^*.

For each homogeneous component F_q of F we select a basis of F_q as a free R-module. Let D_q be the matrix associated with the differential $F_q \to F_{q-1}$. We now have to lift D_q to matrices D^*_q with entries in $\underline{m}^*$ in such a way that $D^*_{q-1} D^*_q = 0$. We do this in the following way. Observe that since F is a minimal resolution, each entry of D_q is in $\underline{m}$, so for each entry c we can fix elements $\alpha_1,\ldots,\alpha_m$ in R_0 depending on c such that

$$c = \sum_i \alpha_i v_i \quad .$$

To obtain D^*_q we replace each entry c by c^* where $c^* = \sum_i \alpha_i \bar{X}_i$ if q is odd, and $c^* = \sum_i \alpha_i \bar{Y}_i$ if q is even. Then clearly the entries of the product $D^*_{q-1} D^*_q$ are quadratic expressions of the type $\sum \alpha_{ij} \bar{X}_i \bar{Y}_j$ where $\alpha_{ij} \in R_0$. We have

$$f(\sum \alpha_{ij} \bar{X}_i \bar{Y}_j) = \sum \alpha_{ij} v_i v_j$$

On the other hand $f(\sum \alpha_{ij} \bar{X}_i \bar{Y}_j)$ is zero since it is an entry of

$D_{q-1}D_q = 0$. This shows that $(\alpha_{ij}) \in A$, so $\sum \alpha_{ij}\bar{X}_i\bar{Y}_j = 0$, which means that $D^*_{q-1}D^*_q = 0$. The existence of the lifted complex F^* is now established.

Now let Y be a minimal R^*-free resolution of R. Then by a standard spectral sequence argument the total complex $F^*\otimes_{R^*}Y$ is acyclic, so it is a minimal R^*-free resolution of M. Hence

$$\mathrm{Tor}^{R^*}(M,k) \cong F^*\otimes_{R^*} Y\otimes_{R^*} k \cong (F^*\otimes_{R^*} k)\otimes_k (Y\otimes_{R^*} k)$$

$$\cong (F\otimes_R k)\otimes_k (Y\otimes_{R^*} k) \cong \mathrm{Tor}^R(M,k)\otimes_k \mathrm{Tor}^{R^*}(R,k).$$

This proves the formula in the theorem. □

REMARKS.

1. Any R-free complex C which is bounded below, i.e. $C_p = 0$ for all p sufficiently small, can be lifted to an R^*-free complex.
2. There exists a minimal R^*-algebra resolution of R. This follows from the fact that the homomorphism $R^* \to R$ is large in the sense of Levin [3], i.e. the induced map

$$\mathrm{Tor}^{R^*}(k,k) \to \mathrm{Tor}^R(k,k)$$

is surjective. Then by a result of L. Avramov and H. Rahbar-Rochandel [4], the acyclic closure of the angmented algebra $R^* \to R$ is a minimal resolution. See also [3, Theorem 2.5].
3. The integer m in the definition of R^* can be chosen to be the length of $\underline{m}$. In that case the ideal I can be generated by less than or equal to m^2 elements. (This comes from the fact that R_0 is a principal ideal ring and A is an R_0-submodule of the free R_0-module of all $m\times m$-matrices with entries in R_0.) That this estimate is "best possible" is clear from the following example:

$$R = k[t_1,\dots,t_n]/(t_1,\dots,t_n)^2$$

Choose $m = n$. Then we have

$$R^* = k[[X_1,\dots,X_m,Y_1,\dots,Y_m]]/(\dots,X_iY_j,\dots) \quad 1<i<m,\ 1<j<m.$$

REFERENCES

[1] T.H. Gulliksen, Massey Operations and the Poincaré Series of Certain Local Rings. J. Algebra 22 (1972), 223-232.

[2] G. Levin, Poincaré Series of Modules over Local Rings. Proc. Amer. Math. Soc. 72 (1978), 6-10.

[3] G. Levin, Large homomorphisms of local rings. Math. Scand. 46 (1980), 209-215.

[4] H. Rahbar-Rochandel, Thèse, Caen 1979, appendice p.52, prop. 213.

(Note by the editor: For complementary results, cf. the mathematical introduction to these proceedings.)

Department of Mathematics
University of Oslo
Blindern - Oslo 3
(NORWAY)

THE RADICAL OF $\pi_*(\Omega S) \otimes \mathbb{Q}$, II

by

Stephen Halperin

1. INTRODUCTION

Let S be a simply connected topological space whose rational homology is finite dimensional in each degree. The rational homotopy Lie algebra $\pi_*(\Omega S) \otimes \mathbb{Q}$ (equipped with the Samelson product) is then also finite dimensional in each degree.

Quille [Q] showed that if no further restrictions are placed on S then all possible graded connected Lie algebras (finite dimensional in each degree) can occur. By contrast, severe restrictions are imposed on $\pi_*(\Omega S) \otimes \mathbb{Q}$ if we assume that the rational Lusternik-Schnirelmann category of S, $\mathrm{cat}_0(S)$, is finite. (This is defined in general in [F-H] and coincides with the classical $L-S$ category of the localization $S_{\mathbb{Q}}$ when S is a CW complex-[T].)

This article continues the study of the Lie structure of $\pi_*(\Omega S) \otimes \mathbb{Q}$ (with the hypothesis $\mathrm{cat}_0(S)$. finite) begun in [F-H-T] and [F-H-T-T]. The techniques derive from the earlier two papers, and the fact that I have carried out this stage of the investigation alone is due only to the several thousand miles of ocean between Europe and Canada which made impractical a collaborative effort.

In [F-H-T] we showed that if $\mathrm{cat}_0(S) \leq m$ then the sum R, of all the solvable graded ideals in $\pi_*(\Omega S) \otimes \mathbb{Q}$ is itself solvable, and that the length of the derived series of R (solv length) is $\leq 4 + 2\log_2(m)$. The ideal R is the radical of the homotopy Lie algebra. We also established the

Theorem. If $\mathrm{cat}_0(S) \leq m$ then a graded ideal $I \subset \pi_*(\Omega S) \otimes \mathbb{Q}$ is solvable if and only if either $\dim \pi_*(\Omega S) \otimes \mathbb{Q}$ is finite or if for all k

$$\sum_{i=k}^{2k-1} \dim I_{2i} < m .$$

Here it will be shown that the radical is in fact nilpotent (and thus that every solvable ideal is nilpotent) and the length of the lower central series (the nil length) of R is $\leq [2\sqrt{2}m(m+1)]^{8+4\log_2 m}$, if $\mathrm{cat}_0(S) \leq M$. Explicitly we have

Theorem 1. Suppose S simply connected, with finite dimensional rational homology in each degree, and that $\mathrm{cat}_0(S) \leq m$. Let $I \subset \pi_*(\Omega S) \otimes \mathbb{Q}$ be a graded ideal. The following conditions are equivalent:

(i) For all k, $\sum_{i=k}^{2k-1} \dim I_i < [2m(m+1)]^2$.

(ii) Every finitely generated subalgebra of I is finite dimensional.

(iii) I is solvable.

(iv) I is nilpotent.

Under these conditions, moreover

$$\text{nil length } I \le [2\sqrt{2}m(m+1)]^{2d} \le [2\sqrt{2}m(m+1)]^{8+4\log_2 m},$$

where d is the length of the derived series of I.

<u>Corollary</u>. Let I be a graded ideal in $\pi_*(\Omega S) \otimes \mathbb{Q}$. Then either there is a constant $C > 0$ such that

(1) $$\sum_{i=1}^{p} \dim I_i \le C \log_2 p, \quad p \ge 2,$$

or there is a constant $K > 0$ such that

(2) $$\sum_{i=1}^{p} \dim I_i \ge K p, \quad p \ge 1.$$

The ideal I is solvable if and only if it satisfies the first set of inequalities.

<u>Proof</u>. If I is solvable the first set of inequalities follow from Theorem 1(i). If I is not solvable, then Theorem 1(ii) yields a finitely generated subalgebra, I, of infinite dimension, and the second set of inequalities results. □

Suppose now $\phi: S \to T$ is continuous, inducing $\phi_{\#}: \pi_*(\Omega S) \otimes \mathbb{Q} \to \pi_*(\Omega T) \otimes \mathbb{Q}$. Denote by R_S and R_T the radicals. Then we have

<u>Theorem 2</u>. If S and T satisfy the hypotheses of Theorem 1, and if $\operatorname{Im} \phi_{\#}$ is an ideal in $\pi_*(\Omega T) \otimes \mathbb{Q}$ while $\ker \phi_{\#} \subset R_S$ then

$$R_S = (\phi_{\#}^{-1})(R_T).$$

<u>Proof</u>. Since R_T satisfies (1) and since $\ker \phi_{\#} \subset R_S$ also satisfies (1) it follows that $(\phi_{\#}^{-1})(R_T)$ satisfies (1). Hence it is a solvable ideal; i.e. $(\phi_{\#}^{-1})(R_T) \subset R_S$.

Conversely, since $\operatorname{Im} \phi_{\#}$ is an ideal, for every homogeneous $\alpha \in \pi_*(\Omega T) \otimes \mathbb{Q}$: $[\alpha, \phi_{\#}(R_S)] + \phi_{\#}(R_S)$ is an ideal $J(\alpha)$ in $\operatorname{Im} \phi_{\#}$. This ideal satisfies (1). Hence so does $\phi_{\#}^{-1}(J(\alpha))$. Thus $\phi_{\#}^{-1}(J(\alpha)) \subset R_S$ and $J(\alpha) \subset \phi_{\#}(R_S)$. This shows that $\phi_{\#}(R_S)$ is an ideal in $\pi_*(\Omega T) \otimes \mathbb{Q}$. Since $\phi_{\#}(R_S)$ satisfies (1) it is contained in R_T. □

<u>Corollary</u>. Let $F \xrightarrow{i} E \xrightarrow{\pi} B$ be a Serre fibration for which F and E satisfy the hypotheses of Theorem 1. Then

$$(i_{\#})^{-1}(R_E) = R_F.$$

<u>Proof</u>. Clearly, $\operatorname{Im} i_{\#} = \ker \pi_{\#}$ is an ideal, as is $\ker \pi_{\#}$. Moreover ([F-H]) $\dim \ker \pi_{\#}$ is finite, so $\ker \pi_{\#} \subset R_F$. □

The study of rational homotopy theory has, as one of its sources, Chen's work on iterated integrals. Either this way or via the standard methods, it is possible to compute the real homotopy Lie algebra $\pi_*(\Omega S) \otimes \mathbb{R}$ from the algebra of differential forms on a smooth manifold, S. This does not affect the radical:

Proposition 1. Let S satisfy the hypotheses of Theorem 1 and let $R \subset \pi_*(\Omega S) \otimes \mathbb{Q}$ be the radical. If $\mathbb{k} \supset \mathbb{Q}$ is an extension field then $R \otimes \mathbb{k}$ is the sum of the solvable ideals in $\pi_*(\Omega S) \otimes \mathbb{k}$.

Proof. Write $L = \pi_*(\Omega S) \otimes \mathbb{Q}$ and let I be the sum of the solvable ideals in $L \otimes \mathbb{k}$. Fix a $\mathbb{Q}$-basis $\{v_\alpha\}$ of $\mathbb{k}$. For $x \in I$ write $x = \Sigma x_\alpha \otimes v_\alpha$, $x_\alpha \in L$. Now the proof of Theorem 1 still works with $\mathbb{k}$ instead of $\mathbb{Q}$ as coefficients; in particular there is a constant N such that for any k,

$$\# \{i \in [k, 2k-1] \mid I_i \neq 0\} < N .$$

This will then apply to the ideals I_α generated by the x_α in L, and by Theorem 1(ii), $I_\alpha \subset R$. In particular, each $x_\alpha \in R$ and $x \in R \otimes \mathbb{k}$. □

Remark. Part of Theorem 1 may be phrased (in the language of infinite dimensional Lie algebras) by saying that for a graded ideal I of $\pi_*(\Omega S) \otimes \mathbb{Q}$:

I locally finite $\Leftrightarrow$ I locally nilpotent $\Leftrightarrow$ I solvable $\Leftrightarrow$ I nilpotent.

The rest of the paper is devoted to the proof of Theorem 1. At points we shall need explicitly the theory of minimal models and rational category: the reader is referred to [FHT; §3] for the notation and a review of the facts we require. Here I shall simply recall that we use Sullivan's convention: if X is a graded space then

$$\Lambda X = \text{exterior algebra } (X^{odd}) \otimes \text{ symmetric algebra } (X^{even}) .$$

The ground field id $\mathbb{Q}$ throughout.

The main ingredient in the proof of Theorem 1 is

Theorem 3. Let S be as in Theorem 1, and suppose $I \subset \pi_*(\Omega S) \otimes \mathbb{Q}$ is a graded solvable ideal. Then for each k

$$\sum_{i=k}^{2k} \dim I_{2i+1} \leq [2m(m+1)]^2 - 2m^3 .$$

In what follows, Theorem 3 will be first reduced (in §2) to a technical statement about minimal models. Two key lemmata are then proved (they are analogues of the lemmata of [FHT; §5]), and are subsequently used in § 4 to complete the proof. Finally, Theorem 1 is deduced from Theorem 3 in § 5. The space S will be supposed throughout to satisfy the hypotheses of Theorem 1 and 2.

If, moreover, $\pi_*(\Omega S) \otimes \mathbb{Q}$ is finite dimensional, then Theorem 3 is trivial because ([F-H],[H])

$$\dim I_{odd} \leq \dim \pi_{odd}(\Omega S) \otimes \mathbb{Q} \leq \dim \pi_{even}(\Omega S) \otimes \mathbb{Q} \leq m .$$

Thus in our proof of Theorem 3 we suppose that $\pi_*(\Omega S) \otimes \mathbb{Q}$ is infinite dimensional.

2. THE REDUCTION STEP

In order to prove Theorem 3, it is clearly sufficient to establish

$$\sum_{i=\ell}^{\ell'} \dim I_{2i+1} \leq 2m^2(m+1)^2 - m^3 \quad \text{if} \quad 3\ell \geq 2\ell' . \tag{2.1}$$

Fix $\ell \leq \ell'$ with $3\ell \geq 2\ell'$. If $\alpha, \beta \in \Sigma_{i=\ell}^{\ell'} I_{2i+1}$ are homogeneous then the degree of $[\alpha,\beta]$ is even and lies in an interval of the form $[2r, 4r-2]$. It follows from [F-H-T-T] that all such brackets span a space of dimension at most $m-1$. An obvious inductive procedure now gives a sequence $\beta_1, \beta_2, \ldots, \beta_N$ of linearly independent homogeneous elements in $\Sigma_{i=\ell}^{\ell'} I_{2i+1}$ such that

$$[\beta_i, \beta_j] = 0 , \qquad i \neq j , \tag{2.2}$$

and $mN \geq \Sigma_{i=\ell}^{\ell'} \dim I_{2i+1}$. We are thus reduced to proving that

$$N \leq 2m(m+1)^2 - m^2.$$

Moreover, it follows from (2.2) that the sub Lie algebra, E, generated by the β_i's consists of the span of the β_i's (in odd degrees) together with an evenly graded space of dimension at most $m-1$ spanned by elements of the form $[\beta_i, \beta_i]$.

Finally, for any homogeneous $\alpha \in \pi_*(\Omega S) \otimes \mathbb{Q}$, and for any sequence i_γ $(1 \leq i_\gamma \leq N)$ we observe that

$$[\beta_{i_t}, [\beta_{i_{t-1}}, [\ldots[\beta_{i_1}, \alpha]\ldots] = 0 , \qquad t \geq 4m . \tag{2.4}$$

Indeed, by (2.2) we may suppose $\deg \beta_{i_t} \leq \ldots \leq \deg \beta_{i_1}$. The sequence of elements $\gamma_\nu = [\beta_{i_\nu} [\ldots[\beta_{i_1}, \alpha]\ldots]$ $(2m \leq \nu \leq 4m)$ have distinct degrees at least m of which are even, and which lie in an interval of the form $[r, 2r-2]$. By [F-H-T-T] one of the γ_ν is zero, and so $\gamma_\nu = 0$, $\nu \geq 4m$, which proves (2.4).

Now let $(\Lambda T, d)$ be the minimal model of S. Then each pair T^{p+1}, $\pi_p(\Omega S) \otimes \mathbb{Q}$ is equipped with a scalar product. Moreover, if $d_2 x \in \Lambda^2 T$ is the quadratic part of dx, $x \in T$, then

$$\langle d_2 x; \alpha, \beta \rangle = \pm \langle x; [\alpha, \beta] \rangle . \tag{2.5}$$

Suppose the β_i have been numbered so that

$$a = \deg \beta_1 \leq \ldots \leq \deg \beta_N = b ;$$

then $2\ell + 1 \leq a \leq b \leq 2\ell' + 1$. Divide ΛT by $T^{\leq a}$; then the elements of $T^{[a+1, 2a]}$

become cocycles. Since $b \leq 2a$ (because $2\ell' \leq 3\ell$) we can divide further by the elements $x \in T^{[a+1,2a]}$ satisfying $\langle x,\beta_i\rangle = 0$, $1 \leq i \leq N$. Finally, divide by those elements $x \in T^{[2a+1,3a+1]}$ which have now become cocycles.

In this way we achieve a quotient minimal model $(\Lambda X,d)$ of $(\Lambda T,d)$. The dual spaces $L_p = \mathrm{Hom}\,(X^{p+1};\mathbb{Q})$ become via (2.5) a graded sub Lie algebra, L, of $\pi_*(\Omega S) \otimes \mathbb{Q}$. This Lie algebra contains the β_i (and hence E) by construction. Moreover, by construction, we have

(i) $X^p = 0$, $p \leq a$.

(ii) $X^{[a+1,2a]}$ has a basis $y_1,\dots,y_N$ satisfying $dy_i = 0$ and $\langle y_i,\beta_j\rangle = \delta_{ij}$.

(iii) For degree reasons d maps $X^{[2a+1,3a+1]}$ into $\Lambda^2 X^{[a+1,2a]}$; moreover this map is injective.

Since $\deg\,[\beta_N,\beta_N] \leq 4\ell' + 2 \leq 6\ell + 2 \leq 3a - 1$ it follows that the dual space in X to E is contained in $X^{\leq 3a}$. In view of (iii), and (2.5) we may conclude that $X^{[2a+1,3a+1]}$ has as basis elements $w_1,\dots,w_t$ $(t \leq m-1)$ which form a dual basis to a basis $\alpha_1,\dots,\alpha_t$ for E_{even}.

In particular we have $L = E \oplus L_{>3a}$; i.e. E is complemented by an ideal. Now put

$$Y = X^{[a+1,2a]}, \quad W = X^{[2a+1,3a+1]}, \quad U = X^{\geq 3a+2}.$$

Thus U is dual to the ideal $L_{>3a}$, W is dual to E_{even}, Y is dual to E_{odd}, and $\Lambda X = \Lambda Y \otimes \Lambda W \otimes \Lambda U$.

It follows from (i), (ii) and (iii) above, that the differential d, and its quadratic part d_2 satisfy

$$d(Y) = d_2(Y) = 0. \tag{2.6}$$

$$d = d_2 \colon W \to \Lambda^2 Y. \tag{2.7}$$

Let $\bar{d}_2$ be the quadratic part of the differential in the quotient model $(\Lambda U,\bar{d})$, and let $\theta_i, \theta'_j : U \to U$ be dual to $\mathrm{ad}\beta_i$, $\mathrm{ad}\alpha_j$. Then by (2.5) we get for $u \in U$;

$$d_2 u = \sum_i \pm y_i \otimes \theta_i(u) + \sum_j \pm w_j \otimes \theta'_j(u) + 1 \otimes \bar{d}_2 u. \tag{2.8}$$

From $d_2^2 = 0$ we deduce (for the extension of θ_i to a derivation in ΛU)

$$\bar{d}_2\theta_i = \pm\theta_i\bar{d}_2. \tag{2.9}$$

We shall use these formulae to complete the proof of (2.3) - and hence of Theorem 3 - in § 4. First, however we establish our key lemmata.

3. THE KEY LEMMATA

We retain the notation of § 2. For $p \geq 0$ put

$$S_p = \{\sigma = (\sigma_1,\dots,\sigma_N) \mid \sigma_i \in \mathbb{Z},\ \sigma_i \geq 0,\ \Sigma\sigma_i = p\}.$$

If $\sigma \in S_p$ put $|\sigma| = p$, $\sigma! = \prod_i (\sigma_i!)$ and

$$y^\sigma = \prod_i y_i^{\sigma_i}.$$

(Recall y_i is the basis of Y dual to the β_i.) If $1 \leq i \leq N$ then $(\sigma_1,\dots,\sigma_i+1,\dots,\sigma_N)$ is written $i+\sigma$ and σ is written $\partial_i(i+\sigma)$.

Similarly we set $A_o = \{\phi\}$ and for $1 \leq q \leq N$

$$A_q = \{\omega = (\omega_1,\dots,\omega_q) \mid \omega_i \in \mathbb{Z},\ 1 \leq \omega_1 < \dots < \omega_q \leq N\}.$$

If $\omega \in A_q$ then $\partial_i\omega = (\omega_1\dots\hat{\omega}_i\dots\omega_q)$; it is an element of A_{q-1}. Of course $\partial_1(\omega_1) = \phi$. If $\omega \in A_q$ we write $|\omega| = q$.

3.1. <u>Lemma</u>. There is a system of elements $\Psi(\omega;\sigma) \in \Lambda X$ ($\sigma \in S_m$, $\omega \in A_q$, $0 \leq q \leq N$) which satisfy $\Psi(\phi;\sigma) = y^\sigma$, and

(i) $d\Psi(\omega;\sigma) = \sum_{i=1}^{q} (-1)^{i-1} y_{\omega_i} \wedge \Psi(\partial_i\omega;\sigma)$, $|\omega| = q \geq 1$, $|\sigma| = m$ and

(ii) $\sum_{i=1}^{q+1} (-1)^{i-1}\Psi(\partial_i\omega;\omega_i+\tau) = 0$, $|\omega| = q+1 \geq 1$, $|\tau| = m-1$.

<u>Proof</u>. Extend the projection $(\Lambda X,d) \to (\Lambda X/\Lambda^{>m}X,d)$ to a Sullivan model ρ: $(\Lambda X \otimes \Lambda Z,d) \xrightarrow{\simeq} (\Lambda X/\Lambda^{>m}X,d)$, Because $\mathrm{cat}_o(\Lambda X,d) \leq m$ there are elements $v_\tau \in \Lambda X(\tau \in S_{m+1})$ such that $dv_\tau = y^\tau$.

Then $d\rho\, v_\tau = 0$, and so there are cocycles v'_τ in $\Lambda X \otimes \Lambda Z$ such that $\rho v'_\tau = \rho v_\tau$. For $\sigma \in S_m$ define $\Phi(\phi;\sigma) = y^\sigma$ and $\Phi(i;\sigma) \in \Lambda X \otimes \Lambda Z$ by

$$\Phi(i;\sigma) = v_{i+\sigma} - v'_{i+\sigma}.$$

Then $\rho\Phi(i;\sigma) = 0$, $\Phi(j;i+\sigma) - \Phi(i;j+\sigma) = 0$ and $d\Phi(i;\sigma) = y^{i+\sigma} = y_i y^\sigma = y_i\Phi(\phi;\sigma)$.

Thus in particular $y_j\Phi(i;\sigma) - y_i\Phi(j;\sigma)$ is a cocycle in $\ker\rho$. Since $\ker\rho$ is acyclic it is a coboundary. We now define elements $\Phi(\omega;\sigma)$, $|\omega| = 2$, $|\sigma| = m$ in $\ker\rho$ as follows:

(a) If $\sigma_k = 0$, $k > j$ then $\Phi(i,j;\sigma)$ is any element in $\ker\rho$ for which $d\Phi(i,j;\sigma) = y_i\Phi(j;\sigma) - y_j\Phi(i;\sigma)$.

(b) If $k > j$ is the biggest integer for which $\sigma_k \neq 0$ set $\Phi(i,j;\sigma) = \Phi(i,k;j+\partial_k\sigma) - \Phi(j,k;i+\partial_k\sigma)$.

It is then straightforward to check that the $\Phi(i,j;\sigma)$ satisfy equations (i) and (ii).

The same construction (applied inductively over q) gives for $q > 2$ elements $\Phi(\omega;\sigma) \in \ker\rho$, $\omega \in A_q$, $\sigma \in S_m$ satisfying (i) and (ii). Since $\mathrm{cat}_o(\Lambda X,d) \leq m$ there is a retraction $\phi: (\Lambda X \otimes \Lambda Z, d) \to (\Lambda X, d)$. Put $\Psi(\omega;\sigma) = \phi\Phi(\omega;\sigma)$. □

For the next lemma we need a little more notation. Suppose M is a free ΛY-module. Fix an isomorphic copy, Y', of Y with basis y'_i and let sY be the suspension of Y with basis sy_i: $\deg sy_i = \deg y_i - 1$. Thus ΛsY is an exterior algebra and $\Lambda Y'$ is a polynomial algebra.

Bigrade $N = \Lambda sY \otimes \Lambda Y' \otimes M$ by putting

$$N^{q,p} = \Lambda^q sY \otimes \Lambda^p Y' \otimes M .$$

The elements $f \in N^{q,p}$ can be identified with the collections $f(\omega;\sigma) \in M$, $\omega \in A_q$, $\sigma \in S_p$, via

$$f = \sum_{\omega,\sigma} sy_{\omega_1} \wedge \dots \wedge sy_{\omega_q} \otimes \frac{1}{\sigma!} (y')^\sigma \otimes f(\omega;\sigma) ; \qquad (3.2)$$

here $(y')^\sigma = \prod_i (y'_i)^{\sigma_i}$.

Now define operators δ_1 and δ_2 in N, homogeneous of bidegrees $(1,0)$ and $(1,-1)$ as follows. Let ϕ be the derivation in $\Lambda sY \otimes \Lambda Y'$ given by $\phi(sy_i) = 0$, $\phi(y'_i) = sy_i$ and put $\delta_2 = \phi \otimes \mathrm{id}$. Then set

$$\delta_1(\Phi \otimes \Psi \otimes m) = \sum_{i=1}^{N} sy_i \wedge \Phi \otimes \Psi \otimes y_i \cdot m .$$

A short calculation shows that $\delta_1^2 = \delta_2^2 = \delta_1\delta_2 + \delta_2\delta_1 = 0$. Moreover, $(\Lambda sY \otimes \Lambda Y', \phi)$ is the classical contractible model and so

$$H(N,\delta_2) = 1 \otimes 1 \otimes M . \qquad (3.3)$$

On the other hand, because M is a <u>free</u> ΛY-module, we have

$$H(N,\delta_1) = sy_1 \wedge \dots \wedge sy_N \otimes 1 \otimes F , \qquad (3.4)$$

where $M = \Lambda Y \otimes F$. (This is essentially Lemma 5.6 of [FHT]). In particular

$$H^{q,p}(N,\delta_1) = 0 \quad \text{if} \quad q < N . \qquad (3.5)$$

Finally, we interpret δ_1 and δ_2 in terms of the decomposition (3.2). Indeed, a simple calculation gives for $f \in N^{q,p}$

$$(\delta_1 f)(\omega;\sigma) = \sum_{i=1}^{q+1} (-1)^{i-1} y_{\omega_i} \cdot f(\partial_i\omega;\sigma) \qquad \omega \in A_{q+1}, \sigma \in S_p$$

and

$$(\delta_2 f)(\omega;\sigma) = \sum_{i=1}^{q+1} (-1)^{i-1} f(\partial_i\omega;\omega_i + \sigma) \qquad \omega \in A_{q+1}, \sigma \in S_{p-1} .$$

3.6. <u>Lemma</u>. Suppose $f \in N^{q,p}$ $(2 \le q < N)$ satisfies

$$\sum_{i=1}^{q+1} (-1)^{i-1} y_{\omega_i} \cdot f(\partial_i\omega;\sigma) = 0 \quad \text{and} \quad \sum_{i=1}^{q+1} (-1)^{i-1} f(\partial_i\omega;\omega_i + \sigma) = 0$$

for all ω, σ . There is then an element $g \in N^{q-1,p}$ such that

$$\sum_{i=1}^{q} (-1)^{i-1} y_{\omega_i} \cdot g(\partial_i \omega;\sigma) = f(\omega;\sigma) \quad \text{and} \quad \sum_{i=1}^{q} (-1)^{i-1} g(\partial_i \omega;\omega_i + \sigma) = 0 ,$$

for all ω , σ .

Proof. The lemma is stated in "component form" because that it is how it will be applied. Using the formulae above we see that $\delta_1 f = \delta_2 f = 0$ and we are required to find g such that $\delta_1 g = f$ and $\delta_2 g = 0$.

We do this by induction on p . If $p = o$ we have by (3.5) that $f = \delta_1 g$, $g \in N^{q-1,p}$, because $q < N$. Since $p = 0$ it is automatic that $\delta_2 g = 0$.

Suppose by induction that the lemma is proved for $f \in N^{q,p'}$, $p' < p$, and that $f \in N^{q,p}$. As above (3.5) implies that $f = \delta_1 g_1$, $g_1 \in N^{q-1,p}$. Then $\delta_1(\delta_2 g_1) = -\delta_2 f = 0$ and $\delta_2(\delta_2 g_1) = 0$. Since $\delta_2 g_1$ has bidegree $(q,p-1)$ our induction hypothesis yields $g_2 \in N^{q-1,p-1}$ such that

$$\delta_1 g_2 = \delta_2 g_1 \quad \text{and} \quad \delta_2 g_2 = 0 .$$

Since $q-1 \geq 1$ we apply (3.3) to find g_3 with $\delta_2 g_3 = g_2$.

Put $g = g_1 + \delta_1 g_3$. Then $\delta_1 g = \delta_1 g_1 = f$ and $\delta_2 g = \delta_2 g_1 - \delta_1 \delta_2 g_3 = \delta_2 g_1 - \delta_1 g_2 = 0$. □

4. PROOF OF THEOREM 2.

Recall the notation at the end of § 2. In particular, $\theta_i : U \to U$ is the dual of $\mathrm{ad}\beta_i$. Extend the θ_i to derivations in ΛU . Denote by $F_\rho(\Lambda^n U)$ the linear span of the elements of the form $\theta_{i_1} \circ \ldots \circ \theta_{i_\rho} \Phi$, $\Phi \in \Lambda^n U$. Set

$$F_r(\Lambda^n U) = \Lambda^n U , \qquad r \leq 0 , \tag{4.1}$$

and note that, in view of (2.4)

$$F_{4mn}(\Lambda^n U) = 0 . \tag{4.2}$$

Recall that each α_j is a linear combination of vectors of the form $[\beta_i , \beta_i]$. Since θ'_j is dual to $\mathrm{ad}\alpha_j$ its extension to $\Lambda^n U$ satisfies

$$\theta'_j : F_\rho(\Lambda^n U) \to F_{\rho+2}(\Lambda^n U) . \tag{4.3}$$

Further, by definition

$$\theta_i : F_\rho(\Lambda^n U) \to F_{\rho+1}(\Lambda^n U) \tag{4.4}$$

while by (2.9)

$$\bar{d}_2 : F_\rho(\Lambda^n U) \to F_\rho(\Lambda^{n+1} U) . \tag{4.5}$$

Next, define graded spaces $A_{k,r} \subset \Lambda X$ by

$$A_{k,r} = \Lambda^{\geq k+1} X \oplus \sum_{i+j+n=k} \Lambda^i Y \otimes \Lambda^j W \otimes F_{r-i-2n}(\Lambda^n U) .$$

By (4.2) we have

$$\Lambda^{\geq k} X = A_{k,0} \supset A_{k,1} \supset \dots \supset A_{k,(4m+2)k} = \Lambda^{\geq k+1} X . \tag{4.6}$$

From equations (4.3), (4.4), and (4.5) we deduce

$$d: A_{k,r} \to A_{k+1,r+2} . \tag{4.7}$$

On the other hand, if $F'_\rho(\Lambda^n U)$ is a graded complement for $F_\rho(\Lambda^n U)$ and if we put

$$B_{k,r} = \Lambda^{<k} X \oplus \sum_{i+j+n=k} \Lambda^i Y \otimes \Lambda^j W \otimes F'_{r-i-2n}(\Lambda^n U)$$

then

$$\Lambda X = A_{k,r} \oplus B_{k,r} \qquad \text{all } k, r . \tag{4.8}$$

Moreover the linear maps $y_i \wedge : \Lambda X \to \Lambda X$ satisfy

$$y_i \wedge A_{k,r} \subset A_{k+1,r+1} \quad \text{and} \quad y_i \wedge B_{k,r} \subset B_{k+1,r+1} . \tag{4.9}$$

4.10. Lemma. Suppose the elements $\Psi(\omega;\sigma)$ of Lemma 3.1 can be so chosen that for some $q(2 \leq q < N)$, $\Psi(\omega;\sigma) \in A_{k,r}$ if $|\omega| = q+1$. Then for a new choice (if necessary) we can arrange that $\Psi(\omega,\sigma) \in A_{k,r+1}$ if $|\omega| = q$.

Proof of 4.10. Write (for $|\omega| = q$) $\Psi(\omega;\sigma) = \Psi_A(\omega;\sigma) + \Psi_B(\omega;\sigma)$ with $\Psi_A(\omega;\sigma) \in A_{k,r+1}$ and $\Psi_B(\omega;\sigma) \in B_{k,r+1}$ – this is possible by (4.8). It is immediate that the $\Psi_A(\omega;\sigma)$ (respectively the $\Psi_B(\omega;\sigma)$) satisfy equation (ii) of (3.1).

On the other hand if $|\omega| = q+1$ then (4.7) shows that $d\Psi(\omega;\sigma) \in A_{k+1,r+2}$. From equation (i) of (3.1) and (4.9) we thus deduce for $|\omega| = q+1$ that

$$d\Psi(\omega;\sigma) = \sum_{i=1}^{q+1} (-1)^{i-1} y_{\omega_i} \wedge \Psi_A(\partial_i\omega;\sigma)$$

and

$$0 = \sum_{i=1}^{q+1} (-1)^{i-1} y_{\omega_i} \wedge \Psi_B(\partial_i\omega;\sigma) .$$

Since the $\Psi_B(\omega;\sigma)$ also satisfy equation (ii) of (3.1) and since ΛX is a free ΛY-module we can apply Lemma 3.6. This yields elements $\Omega(\omega;\sigma)$ $(|\omega| = q-1)$ such that (for $\omega \in A_q$)

$$\Psi_B(\omega;\sigma) = \sum_{i=1}^{q} (-1)^{i-1} y_{\omega_i} \wedge \Omega(\partial_i\omega;\sigma) , \qquad |\sigma| = m$$

and

$$\sum_{i=1}^{q} (-1)^{i-1} \Omega(\partial_i\omega;\omega_i + \sigma) = 0 \qquad |\sigma| = m-1 .$$

Modify the $\Psi(\omega;\sigma)$ to $\Psi'(\omega;\sigma)$ by putting

$$\Psi'(\omega;\sigma) = \begin{cases} \Psi(\omega;\sigma) & \text{if } |\omega| \neq q \text{ or } q-1 \\ \Psi_A(\omega;\sigma) & \text{if } |\omega| = q \\ \Psi(\omega;\sigma) - d\Omega(\omega;\sigma) & \text{if } |\omega| = q-1 . \end{cases}$$

It is straightforward to check (given the properties derived for Ψ_A, Ψ_B and Ω) that the conclusions of the lemma are satisfied. □

We now revert to, and complete the proof of Theorem 3. We have to establish (2.3) and we do so by contradiction. Indeed, we note that for $|\omega| = N$, $\Psi(\omega;\sigma) \in A_{1,0}$, by definition. An iterated application of the lemma above shows that (cf. (4.6)) we may suppose that

$$\Psi(\omega;\sigma) \in \Lambda^{\geq k} X, \qquad |\omega| = 2,$$

as long as $2 \leq N - k(k-1)(2m+1)$.

In particular, if $2 \leq N - (m+1)m(2m+1)$ then we may suppose $\Psi(i,j;\sigma) \in \Lambda^{\geq m+1} X$. Write $\Psi(i;\sigma) = f(i;\sigma) + g(i;\sigma)$ with $f(i;\sigma) \in \Lambda^{\leq m} X$ and $g(i;\sigma) \in \Lambda^{\geq m+1} X$. The equation $d\Psi(i,j;\sigma) = y_i \wedge \Psi(j;\sigma) - y_j \Psi(i;\sigma)$ now implies that

$$y_i \wedge f(j;\sigma) = y_j \wedge f(i;\sigma), \qquad \sigma \in S_m .$$

Since ΛX is a free module over the polynomial algebra ΛY, it follows that there are elements $h(\sigma)$, $\sigma \in S_m$, in ΛX such that

$$f(i;\sigma) = y_i \wedge h(\sigma) .$$

On the other hand the $\Psi(i;\sigma)$ satisfy

$$\Psi(i;j+\tau) = \Psi(j;i+\tau), \qquad \tau \in S_{m-1},$$

and hence the same equations are satisfied by the $f(i;\sigma)$, For any i,j we then get $y_i \wedge h(j+\tau) = y_j \wedge h(i+\tau)$, $\tau \in S_{m-1}$.

Now fix σ by putting $\sigma_1 = \sigma_2 = \ldots = \sigma_m = 1$ and $\sigma_i = 0$, $i > m$. What we have just shown implies that y_i divides $h(\sigma)$ for each i. Thus $y_{m+1} \wedge y_1 \wedge \ldots \wedge y_m$ divides $f(m+1;\sigma)$ and hence this element is zero, since it is also in $\Lambda^{\leq m} X$. It follows that

$$y_{m+1} \wedge y^{\sigma} = d\Psi(m+1;\sigma) = dg(m+1;\sigma) \in d(\Lambda^{\geq m+1} X) \subset \Lambda^{\geq m+2} X,$$

a patent contradiction.

We obtained this contradiction by supposing $N \geq 2 + (m+1)m(2m+1)$. It follows that $N < 2 + (m+1)m(2m+1)$, and hence (2.3), and Theorem 3, are proved. □

5. PROOF OF THEOREM 1.

Since for $\dim \pi_*(\Omega S) \otimes \mathbb{Q}$ finite we have necessarily ([F-H],[H])

$$\dim \pi_*(\Omega S) \otimes \mathbb{Q} \le 2m$$

and since in this case the Lie algebra is trivially nilpotent of nil length $\le 2m$ we may restrict to the case that $\dim \pi_*(\Omega S) \otimes \mathbb{Q}$ is infinite.

By the theorem of [F-H-T-T] there are then exactly two mutually exclusive possibilities for a graded ideal I: either I is solvable, or for some k, $\Sigma_{i=k}^{2k-1} \dim I_{2i} \ge m$. In the latter case we can apply [F-H-T; Theorem 2.2] to obtain that for certain $\alpha_1,\dots,\alpha_m \in I$ either the α_i generate an infinite dimensional subalgebra, or for some $\beta \in \pi_*(\Omega S) \otimes \mathbb{Q}$ $(\mathrm{ad}\alpha_i)^n \beta \ne 0$ for all n. In particular, α_i and $[\alpha_i,\beta]$ generate an infinite dimensional subalgebra.

This shows that if I is not solvable then also (ii) fails. It also shows that for some constant C, $\Sigma_{i=k}^{2k-1} \dim I_i \ge Ck$, and so (i) fails. Thus either of (i) or (ii) implies (iii) as, clearly, does (iv).

Conversely, suppose I is solvable. Then (i) is an immediate consequence of Theorem 2, and the theorem of [F-H-T-T]. Thus (iii) $\Rightarrow$ (i). Moreover, since finitely generated nilpotent Lie algebras are finite dimensional, (iv) $\Rightarrow$ (ii).

It remains to prove that if I is solvable with derived series of length d, then I is nilpotent of nil length $\le [2\sqrt{2}m(m+1)]^{2d}$. We shall show by induction on d that

$$[\beta_1[\beta_2[\dots[\beta_q,\alpha]\dots] = 0, \quad \beta_i \in I, \quad \alpha \in \pi_*(\Omega S) \otimes \mathbb{Q}, \tag{5.1}$$

if $q \ge [2\sqrt{2}m(m+1)]^{2d}$.

This is clear for $d = 0$. Assume by induction that it holds for I'; set $F^0 = \pi_*(\Omega S) \otimes \mathbb{Q}$ and $F^{i+1} = [I',F^i]$. Then $F^i = 0$ if $i \ge [2\sqrt{2}m(m+1)]^{2d-2}$.

The induced representation, θ, of I in F^i/F^{i+1} satisfies $\theta(\alpha)\theta(\beta) = \pm\theta(\beta)\theta(\alpha)$, α, β homogeneous elements of I. Thus an element $\theta(\beta_1) \circ \dots \circ \theta(\beta_{2p}) \cdot v$ in F^i/F^{i+1} $(\beta_i \in I)$ can be represented in $\pi_*(\Omega S) \otimes \mathbb{Q}$ by an element

$$[\alpha_1[\alpha_2[\dots[\alpha_{2p},\sigma]\dots]$$

where σ represents v, $\alpha_i \in I$ and $\deg \alpha_1 \le \dots \le \deg \alpha_{2p}$.

The string of elements $\gamma_i = [\alpha_i[\dots[\alpha_{2p},\sigma]\dots]$ $(i \le 2p)$ are all in I and satify $2\deg \gamma_p > \deg \gamma_1 > \dots > \deg \gamma_p$. We have seen above that if I is solvable it satisfies (i). Thus $\gamma_1 = 0$ if $p \ge [2m(m+1)]^2$. This holds in each F^i/F^{i+1} and (5.1) follows at once.

REFERENCES

[F-H] Y. Felix and S. Halperin, Rational LS category and its applications. Trans. AMS 273 (1982), 1-38.

[F-H-T] Y. Felix, S. Halperin, and J.-C. Thomas, The homotopy Lie algebra for finite complexes. Publ. Math. I.H.E.S. 56 (1982), 387-410.

[F-H-T-T] Y. Felix, S. Halperin, D. Tanré, and J.-C. Thomas, The radical of $\pi_*(\Omega S) \otimes \mathbb{Q}$, these proceedings.

[H] S. Halperin, Finiteness in the minimal models of Sullivan. Trans. AMS 230 (1977), 173-199.

[Q] D. Quillen, Rational homotopy theory. Ann. of Math. 90 (1969), 205-295.

[T] G. Toomer, Lusternik-Schnirelmann category and the Milnor-Moore spectral sequence. Math. Z. 138 (1974), 123-143.

S. Halperin
Department of Mathematics
University of Toronto
Toronto, Canada M5S 1A1

HIGH SKELETA OF CW COMPLEXES

by

Stephen Halperin and Gerson Levin

1. Introduction. This paper follows the example of Avramov, Felix, Roos and others ([A], [R]) who have begun a fruitful exchange of theorems between rational homotopy theory and the homology theory of commutative rings.

Let R and S be local rings with the same residue field $\mathbb{k}$. A local homomorphism $R\to S$, inducing the identity on $\mathbb{k}$, is called a Golod homomorphism if the "fibre" $\mathrm{Tor}^R(S;\mathbb{k})$ has trivial Massey operations and if the induced map $\mathrm{Tor}^R(\mathbb{k},\mathbb{k})\to\mathrm{Tor}^S(\mathbb{k},\mathbb{k})$ is injective. (If $\dim_{\mathbb{k}}(\mathrm{Tor}_+^R(S;\mathbb{k}))>1$ this latter condition follows from the first -[A2].) The following theorem was proved by the second named author in [L; Theorem 13].

Theorem A: Let R be a local ring with maximal ideal $\underline{m}$ and residue field $\mathbb{k}$. There is then an integer n_o such that for $n\geq n_o$ the map $R\to R/\underline{m}^n$ is a Golod homomorphism.

For R regular this had already been proved by Golod [G] in 1961, with $n_o=2$. In the completely different context of Gelfand-Fuks cohomology, J. Vey showed that the homotopy fibre of the inclusion of the 2n-skeleton of BU(n) was rationally a wedge of spheres. This result was later generalized by Ruchti [Ru] who considered high skeleta of a finite product of $K(\mathbb{Q},n_i)$'s.

Since the lexicon between local algebra and rational homotopy theory translates Golod homomorphisms to continuous maps whose homotopy fibre is rationally a wedge of spheres, and translates $R\to R/\underline{m}^n$ to the inclusion of the n-skeleton of a CW complex, Theorem A appears to suggest a far reaching generalization of Vey-Ruchti: namely that for any 1-connected CW complex of finite type the inclusion of a high skeleton should have as homotopy fibre a wedge of spheres (up to rational homotopy type).

This, however, fails for the product $\prod_{n=2}^{\infty}S^n$. The point is that there has to be some kind of finiteness assumption on our space, corresponding to the fact that local rings are noetherian. The appropriate hypothesis turns out to be semi-finiteness, which we define as follows:

Definition. A simply connected CW complex R of finite type is semi finite if there is a continuous map $E\to R$ with path connected homotopy fibre F, and such that for some M,N;

$$H^{>M}(E;\mathbb{Q}) = 0 \quad \text{and} \quad H^{>N}(F;\mathbb{Q}) = 0.$$

If $H^M(E;\mathbb{Q})\neq 0$, and $H^N(F;\mathbb{Q})\neq 0$ then (M,N) is a <u>dimension pair</u> for R. While the pair (M,N) may vary with E, the difference M-N depends only on R (given $F'\to E'\to R$ consider $E'\times_R E$).

The correct translation of Theorem A reads

<u>Theorem B</u>: Let R be a simply connected CW complex of finite type which is semi-finite with dimension pair (M,N). If $n\geq \max(M+1), 2N+3)$ then the homotopy fibre of the inclusion $R^n\to R$ of the n skeleton is rationally a wedge of spheres.

<u>Corollary</u>: If a 1-connected CW complex, R, has finitely generated rational cohomology ring, then for all $n\geq n_o$ (some n_o) the homotopy fibre of $R^n\to R$ is rationally a wedge of spheres.

<u>Proof</u>: We show R is semi-finite. Let $\alpha_1,\ldots,\alpha_r$ be the even generators of $H^*(R;\mathbb{Q})$. They define $\phi:R\to K=\prod_i K(\mathbb{Q};|\alpha_i|)$. The fibre, E, fibres over R with fibre $\prod_i K(\mathbb{Q};|\alpha_i|-1)$, which has finite dimensional cohomology. The Eilenberg-Moore spectral sequence converges to $H^*(E;\mathbb{Q})$ from $\mathrm{Tor}^{H^*(K;\mathbb{Q})}(H^*(R;\mathbb{Q});\mathbb{Q})$. Since $H^*(K;\mathbb{Q})$ is a polynomial algebra, and $H^*(R;\mathbb{Q})$ is a finitely generated $H^*(K;\mathbb{Q})$ module, this is finite dimensional.

□

This corollary may be regarded as the strict analogue of Theorem A. We are, however, not limited to the spaces in Corollary 1, and indeed we have

<u>Theorem C</u>: Let R be a simply connected CW complex of finite type such that for some m_o, $\pi_i(R)\otimes\mathbb{Q}=0$, $i>m_o$. Then R is semi-finite.

<u>Corollary</u>. Under the hypotheses of Theorem C there is an n_o such that for $n\geq n_o$ the homotopy fibre of $R^n\to R$ is rationally a wedge of spheres.

<u>Remark</u>: The heart of the proof of Theorem B is Theorem 2.1 in the next section, which is its translation into the homotopy theorey of cgda's. As is usual in this kind of exchange, the basic idea of [L] is still the principal element of the proof, but the detailed techniques needed are quite different.

<u>Remark</u>: Suppose $R = \prod_{i=1}^{r} K(\mathbb{Q};n_1)$ (the case considered by Ruchti). Then R satisfies the hypotheses of both corollaries and so the qualitative Vey-Ruchti result follows from either. In this special case we can take $N = \Sigma_i(2n_i-1)$ and M=0 and so our requirement is $n\geq\Sigma(2n_i-1)+3$; in fact in this case Vey-Ruchti get a better bound for n.

The authors wish to thank L. Avramov and W. Singer for several helpful discussions.

2. Differential algebra and the main theorem. The reader is referred to [B-G], [Ha], [Su], [Ta] for the theory of minimal models and its connection with rational homotopy theory. Here we recall that ΛX=exterior algebra (X^{odd}) $\otimes$ symmetric algebra (X^{even}) denotes the free commutative graded algebra over a graded vector space. (Our ground field, $\Bbbk$, here and throughout is assumed of characteristic zero.) All graded spaces are supposed concentrated in degrees ≥ 0, and $\otimes$ denotes tensor product with respect to $\Bbbk$ (as opposed, e.g., to $\otimes_A$).

A commutative graded differential algebra (A,d_A) (cgda) is connected if $A^\circ=\Bbbk$, simply connected if also $A^1=0$. A cgda morphism, ϕ, is a quism if $H(\phi)$ is an isomorphism. The equivalence class of a cgda (under the equivalence relation generated by quisms) is called its homotopy type. A cgda is said to be a wedge of spheres if it has the homotopy type of a connected cgda H with differential zero and satisfying $H^+\cdot H^+=0$.

If A is a cgda, a KS extension of A is a cgda morphism $A\to A\otimes\Lambda X$ where X admits a well ordered basis x_α such that $dx_\alpha \in A\otimes\Lambda(X_{<\alpha})$. Any dgda morphism $A\xrightarrow{\phi}B$ extends to a commutative diagram (if $H^\circ(A)=H^\circ(B)=\Bbbk$)

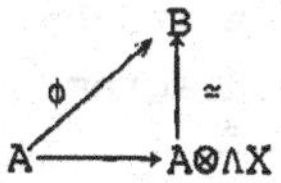

in which the bottom arrow is a KS extension. The quism $A\otimes\Lambda X\xrightarrow{\simeq}B$ is a Sullivan model for ϕ. It can be chosen so that $dx_\alpha \in A^+\otimes\Lambda(X_{<\alpha})+1\otimes(\Lambda^+X)\cdot(\Lambda^+X)$, and in this case is called minimal. A Sullivan model for $\Bbbk\to B$ is called a Sullivan model for B.

If A is augmented then $\Lambda X=\Bbbk\otimes_A(A\otimes\Lambda X)$ becomes a cgda, called the fibre of the KS extension. If the extension is a model for $\phi: A\to B$ then the homotopy type of the fibre is determined by ϕ and is called the Sullivan fibre of ϕ.

The main theorem, from which we get Theorem B as a corollary, reads

2.1 Theorem: Suppose A is a connected cgda admitting a KS extension $A\to A\otimes\Lambda X$ such that

$$H^{>M}(A\otimes\Lambda X) = 0 \quad\text{and}\quad H^{>N}(\Lambda X) = 0$$

(some $M,N\geq 0$). Assume $I\subset A$ is a differential ideal such that $A^{\geq n+1}\subset I\subset A^{\geq n}$ for some $n\geq\max(M+1,2N+3)$.

Then the Sullivan fibre of $A\to A/I$ is a wedge of spheres.

2.2 Remark: Let $A\otimes\Lambda\overline{X}_A\xrightarrow{\simeq}\Bbbk$ be a Sullivan model for the augmentation of A. Then $\otimes_A(A\otimes\Lambda\overline{X}_A)$ is a functor from graded differential A-modules to graded differential vector spaces, and it sends cohomology isomorphisms of modules to cohomology isomorphisms. As a graded space, $M\otimes_A(A\otimes\Lambda\overline{X}_A) = M\otimes\Lambda\overline{X}_A$, and we use this notation for simplicity.

Note that a special case occurs when $\phi: A\to M$ is a cgda morphism, so that M is a differential A-algebra. In this case $M\otimes\Lambda\bar{X}_A$ has the homotopy type of the Sullivan fibre of ϕ. In particular $A/I\otimes\Lambda\bar{X}_A$ has the homotopy type of the Sullivan fibre of $A\to A/I$.

__2.3 Lemma__: Let $r=n-N-1$. Then the inclusion

$$I\otimes_A(A\otimes\Lambda\bar{X}_A) \to A^{\geq r}\otimes_A(A\otimes\Lambda\bar{X}_A)$$

is zero in cohomology.

__Proof__: Let $\bar{d}$ be the differential in ΛX and put

$$F = (\Lambda X)^{<N}\oplus(\ker\bar{d})^N ;\quad B = A\otimes F.$$

Then B is a differential A submodule of $A\otimes\Lambda X$ and a spectral sequence argument (filter by the degree of A) shows that the inclusion $B\hookrightarrow A\otimes\Lambda X$ induces an isomorphism of cohomology. For the same reason

$$H(I\cdot B) = H(I\otimes F)\xrightarrow{\cong}H(I\otimes\Lambda X) = H(I\cdot(A\otimes\Lambda X)) \tag{2.4}$$

and

$$H(A^{\geq r}\cdot B) = H(A^{\geq r}\otimes F)\xrightarrow{\cong}H(A^{\geq r}\otimes\Lambda X) = H(A^{\geq r}\cdot(A\otimes\Lambda X)), \tag{2.5}$$

are isomorphisms.

Now set $W=B^{\geq n}\oplus\hat{B}^{n-1}$, where $\hat{B}^{n-1}$ is a complement for ker d in B^{n-1}. Then $H(W)=H^{\geq n}(B)=H^{\geq n}(A\otimes\Lambda X)=0$, because $n>M$. Moreover, W is a differential A-module, and by 2.2, $H(W\otimes_A(A\otimes\Lambda\bar{X}_A))=0$ as well. Note that $I\cdot B\subset W\subset A^{\geq r}\cdot B$, because $F^j=0$ $(j>N)$ and $r=n-N-1$.

Now consider the commutative diagram

$$\begin{array}{ccc}
(I\cdot B)\otimes_A(A\otimes\Lambda\bar{X}_A) & \xrightarrow{\quad\cong\quad} & (I\otimes\Lambda X)\otimes_A(A\otimes\Lambda\bar{X}_A)\\
\downarrow & & \\
W\otimes_A(A\otimes\Lambda\bar{X}_A) & & \downarrow j\\
\downarrow & & \\
(A^{\geq r}\cdot B)\otimes_A(A\otimes\Lambda\bar{X}_A) & \xrightarrow{\quad\cong\quad} & (A^{\geq r}\otimes\Lambda X)\otimes_A(A\otimes\Lambda\bar{X}_A)
\end{array}$$

in which the horizontal arrows induce cohomology isomorphisms by (2.4), (2.5) and remark 2.2. It follows that the inclusion, j, is zero on cohomology.

On the other hand, because $A\otimes\Lambda\bar{X}_A$ is acyclic, induction on the well ordered basis, x_α, of X gives an isomorphism

$$\begin{aligned}(A\otimes\Lambda X)\otimes_A(A\otimes\Lambda\bar{X}_A) &= A\otimes\Lambda\bar{X}_A\otimes\Lambda X\\ &\cong (A\otimes\Lambda\bar{X}_A,d)\otimes(\Lambda X,\bar{d})\end{aligned}$$

as differential $A\otimes\Lambda\bar{X}_A$ - algebras ([Ha]). Multiplying on both sides by I (respectively by $A^{\geq r}$) identifies j with the map

$$\text{incl.}\otimes\text{id}:(I\otimes\Lambda\bar{X}_A,d)\otimes(\Lambda X,\bar{d})\hookrightarrow(A^{\geq r}\otimes\Lambda\bar{X}_A,d)\otimes(\Lambda X,\bar{d}).$$

It follows that

$$H(\text{incl}):H(I\otimes\Lambda\bar{X}_A) \to H(A^{\geq r}\otimes\Lambda\bar{X}_A) \text{ is zero.}$$

□

__Proof of 2.1__: The commutative diagram

$$\begin{array}{ccccc}
I\otimes\Lambda\bar{X}_A & \longrightarrow & A\otimes\Lambda\bar{X}_A & \longrightarrow & A/I\otimes\Lambda\bar{X}_A\\
\downarrow & & \downarrow = & & \downarrow \text{proj}\\
A^{\geq r}\otimes\Lambda\bar{X}_A & \longrightarrow & A\otimes\Lambda\bar{X}_A & \longrightarrow & A/A^{\geq r}\otimes\Lambda\bar{X}_A
\end{array},$$

and the acyclicity of $A\otimes\Lambda\overline{X}_A$, imply that

$$H(\text{proj}):H^+(A/I\otimes\Lambda\overline{X}_A) \to H^+(A/A^{\geq r}\otimes\Lambda\overline{X}_A)$$

is zero. Thus there are cocycles in $A^{\geq r}/I\otimes\Lambda\overline{X}_A$ which represent a basis of $H^+(A/I\otimes\Lambda\overline{X}_A)$.

Since $n\geq 2N+3$, $2r=2n-2-2N\geq n+1$, and the product of any two of these cocycles is zero. This shows that $H^+(A/I\otimes\Lambda\overline{X}_A)$ has zero multiplication, and also shows that the cocycles we chose define a quism $H(A/I\otimes\Lambda\overline{X}_A)\xrightarrow{\simeq} A/I\otimes\Lambda\overline{X}_A$.

□

3. <u>The second main theorem</u>. Here we establish the result which implies Theorem C. A KS complex is a KS extension of $\Bbbk$.

<u>3.1 Theorem</u>. Let $(\Lambda X,d)$ be a 1-connected KS complex in which $\dim X<\infty$. There is then a KS extension $\Lambda X\to\Lambda X\otimes\Lambda Y$ in which

(i) Y is finite dimensional and concentrated in odd degrees.

(ii) The minimal model of $\Lambda X\otimes\Lambda Y$ is finitely generated, with generators all of odd degree.

In particular, $H(\Lambda Y)$ and $H(\Lambda X\otimes\Lambda Y)$ are finite dimensional.

<u>Proof</u>: Suppose that we have a KS extension $\Lambda X\to\Lambda X\otimes\Lambda Y_n$ with Y_n finite dimensional and concentrated in odd degrees and such that the first even generators of the minimal model of $\Lambda X\otimes\Lambda Y_n$ occur in degree n. Let this minimal model be denoted ΛZ, and put $U=Z^{<n}$.

Extend the projection $\Lambda U\to U\oplus\Bbbk$ (killing decomposables) to a quism $\Lambda U\otimes\Lambda W\xrightarrow{\simeq} U\oplus\Bbbk$. Because U is concentrated in odd degrees, $\Lambda U\otimes\Lambda W$ is the minimal model of a finite wedge of odd spheres. In particular, W is concentrated in odd degrees, and each W^k has finite dimension.

Now restricting the quism to $\Lambda U\otimes\Lambda(W^{<n+2})$ gives a map which is a cohomology isomorphism in degrees $\leq n+1$. The first generator, z, of Z^n is mapped by d to a decomposable $(n+1)$-cocycle in ΛU, which is then killed by the projection. Hence $dz=d\omega$ for some $\omega\in\Lambda U\otimes\Lambda(W^{<n+2})$. Because dz is odd, and $U,W^{<n+2}$ are oddly graded, ω must be decomposable.

In $\Lambda Z\otimes_{\Lambda U}(\Lambda U\otimes\Lambda(W^{<n+2}))=\Lambda Z\otimes\Lambda(W^{<n+2})$, $z-\omega$ is a cocycle. Introduce an odd generator y with $dy=z-\omega$. This gives a KS extension $\Lambda Z\to\Lambda Z\otimes\Lambda(W^{<n+2}\oplus y)$ whose minimal model has for generating space $Z/(z)\oplus W^{<n+2}$. This is finite dimensional, and the number of even generators is reduced by one. Moreover, if $\Lambda Z\xrightarrow{\simeq}\Lambda X\otimes\Lambda Y_n$ is a quism then

$$\Lambda Z\otimes\Lambda(W^{<n+2)}\oplus y) \to (\Lambda X\otimes\Lambda Y_n)\otimes_{\Lambda Z}(\Lambda Z\otimes\Lambda(W^{<n+2}\oplus y))$$

is also a quism, and the cgda on the right has the form $\Lambda X\otimes\Lambda Y_n\otimes\Lambda(W^{<n+2}\oplus y)$.

An obvious induction completes the proof.

□

4. <u>Proof of Theorem B</u>. Denote by A_{PL} Sullivan's functor from spaces to cgda's. If R is as in Theorem B, with n-skeleton R^n let

$$\alpha' : \Lambda Z' \xrightarrow{\simeq} A_{PL}(R)\ ,\quad \beta : \Lambda Y \xrightarrow{\simeq} A_{PL}(R^n)$$

be minimal models. There is then a homotopy commutative cgda diagram

$$\begin{array}{ccc} A_{PL}(R) & \xleftarrow[\alpha']{\simeq} & \Lambda Z' \\ {\scriptstyle A_{PL}(incl)}\downarrow & & \downarrow{\scriptstyle \phi'} \\ A_{PL}(R^n) & \xleftarrow[\beta]{\simeq} & \Lambda Y \end{array} .$$

Because $H^i(R;\mathbb{Q}) \to H^i(R^n;\mathbb{Q})$ is an isomorphism in degrees $<n$ and injective in degree n, the same is true for ϕ'. Thus we can tensor $\Lambda Z'$ with an acyclic KS complex concentrated in degrees $\geq n$ to get a KS complex ΛZ, and we can extend α' and ϕ' to $\alpha : \Lambda Z \to A_{PL}(R)$ and $\phi : \Lambda Z \to \Lambda Y$ so that ϕ is <u>surjective</u>. Note that α is still a quism and $A_{PL}(incl) \circ \alpha$ is homotopic to $\beta\phi$.

On the other hand put $J = (\Lambda Y)^{>n} \oplus J^n$, where J^n complements the cocycles in $(\Lambda Y)^n$. Then $H(\Lambda Y/J) = H^{\leq n}(\Lambda Y) = H^{\leq n}(R^n;\mathbb{Q}) = H^*(R^n;\mathbb{Q}) = H(\Lambda Y)$. Put $I = \phi^{-1}(J)$; since ϕ is injective in degrees $<n$ we have

$$(\Lambda Z)^{\geq n+1} \subset I \subset (\Lambda Z)^{\geq n}.$$

Because ϕ is surjective, $\Lambda Z/I = \Lambda Y/J$ and the commutative diagram

$$\begin{array}{ccc} & \Lambda Z & \\ {\scriptstyle \phi}\swarrow & & \searrow{\scriptstyle proj} \\ \Lambda Y & \xrightarrow{\simeq} & \Lambda Z/I \end{array}$$

identifies the Sullivan fibres of ϕ and the projection. Since $\beta\phi \simeq A_{PL}(incl)\alpha$, the Sullivan fibres of $A_{PL}(incl)$ and of ϕ are identified, while by [Ha;§20] the Sullivan fibre of $A_{PL}(incl)$ is a Sullivan model for the homotopy fibre of $R^n \hookrightarrow R$. To prove that this is rationally a wedge of spheres we have only to prove ([B-G]) that its Sullivan model is a wedge of spheres.

We are thus reduced to showing that the Sullivan fibre of $\Lambda Z \to \Lambda Z/I$ is a wedge of spheres. But the map $E \to R$ guaranteed by the semi-finiteness of R gives a cgda morphism

$$\Lambda Z \xrightarrow{\simeq} A_{PL}(R) \to A_{PL}(E).$$

Form its Sullivan model $\Lambda Z \otimes \Lambda X \xrightarrow{\simeq} A_{PL}(E)$. Then $H^{>M}(\Lambda Z \otimes \Lambda X) = H^{>M}(E;\mathbb{Q}) = 0$. Moreover, again by [Ha;§20], $(\Lambda X, \bar{d})$ is a Sullivan model for the homotopy fibre, F, of $E \to R$. It follows that $H^{>N}(\Lambda X) = H^{>N}(F;\mathbb{Q}) = 0$.

Since $n \geq \max\,(2N+3, M+1)$ and since $(\Lambda Z)^{\geq n+1} \subset I \subset (\Lambda Z)^{\geq n}$ we can apply Theorem 2.1 with $A = \Lambda Z$ to deduce Theorem B.

□

5. <u>Proof of Theorem C</u>. This is an immediate consequence of Theorem 3.1.

Department of Mathematics
University of Toronto,
Toronto, Canada
M5S 1A1

Department of Mathematics,
Brooklyn College,
City University of New York,
New York, N.Y., 10453
U.S.A.

References

[A] L. Avramov, Local algebra and rational homotopy, in Homotopie Algebrique et Algebre Locale, Asterisque 113/114 (1984) p. 15-43.

[A2] L. Avramov, Golod homomorphisms, these proceedings.

[B-G] A.K. Bousfield and V.K.A.M. Gugenheim, On PL de Rham theory and rational homotopy type, Amer. Math. Soc. Memoirs No. 179 (1976).

[G] E.S. Golod, On the homology of some local rings, Dokl. Akad. Nauk SSSR, 144, (1962), p. 479-482. English Transl. Soviet Math. (3) 1962 p. 745-748.

[Ha] S. Halperin, Lectures on minimal models, Mémories de la Soc. Math. de France 9/10, Paris 1984.

[L] G. Levin, Local rings and Golod homomorphisms, J. of Alg., 37, (1975) p. 266-289.

[R] J.E. Roos, Homology of loop spaces and of local rings, Proc. of the 18th Scand. Conf. of Math. (1980); Progress in Math 11, p. 441-468, Birkhauser, 1981.

[Ru] R. Ruchti, On formal spaces and their loop space, Ill. J. of Math., 22 (1978) p. 96-107.

[Su] D. Sullivan, Infinitesimal computations in topology, Publ. Math. I.H.E.S. 47 (1977), p. 269-331.

MATRIC MASSEY PRODUCTS AND FORMAL MODULI I

by

O.A. Laudal

<u>Introduction</u>. It is now folklore that the hull of a deformation functor of an algebraic geometric object, in some way is determined by the appropriate cohomology of the object and its "Massey products", see [M], [May]. The first hints in this direction occurs in Douadys exposé in [Car] (1961).

In 1975 I proved that, in fact, there is a kind of Massey product structure induced by the obstruction calculus characterizing this hull, see [La1].

Independently many authors have published results in this direction, see f.ex. [Pal], [S & S], for references.

This, and a forthcoming paper, are concerned with the problem of actual calculation of these Massey products in two special cases, that of a k-algebra A and of an A-module E.

In §1 we recall the general machinery of [La1] which is common for all the cases we have in mind.

In §2 we prove that the usual matric Massey products, properly adjusted to our needs, for $\mathrm{Ext}_A^{\bullet}(E,E)$ determine the formal moduli of the A-module E, i.e. the hull of the deformation functor of E. As a corollary we obtain the following result

(2.10) Any complete local ring A is uniquely determined by $\mathrm{Ext}_A^i(k,k)$, $i = 1,2$ and the matric Massey products $\overset{n}{\otimes}\,\mathrm{Ext}^1 \dashrightarrow \mathrm{Ext}^2$.

§1 Formal moduli and Massey products

Let X be some algebraic geometric object, say a k-algebra A or an A-module E, and consider the deformation functor

$$\mathrm{Def}_X : \underline{1} \to \underline{\mathrm{Sets}}$$

see [La1].

Let $A^i = A^i(k,X;O_X)$ be the corresponding cohomology. If X is a k-algebra A, then $A^i = H^i(k,A;A)$ is the André cohomology, and if X is an A-module E, $A^i = \mathrm{Ext}^i_A(E,E)$.

By [La1], (4.2.4), we know that the formal moduli of X, i.e. the hull of Def_X, is determined by a morphism of complete local k-algebras

$$o : T^2 = \mathrm{Sym}_k(A^{2*})^\wedge \to T^1 = \mathrm{Sym}_k(A^{1*})^\wedge$$

constructed using only the "obstruction calculus" of $A^\bullet$.

In fact, (4.2.4) of [La1] implies that the formal moduli H has the form

$$H \simeq T^1 \otimes_{T^2} k ,$$

provided A^1 and A^2 has countable dimensions as k-vector spaces.

We shall assume, in what follows, that

$$\dim_k A^i < \infty \quad \text{for } i = 1,2.$$

Pick a basis $\{x_1,\dots,x_d\}$ of A^{1*} and a basis $\{y_1,\dots,y_r\}$ of A^{2*}. Denote by $\{x_1^*,\dots,y_d^*\}$ and $\{y_1^*,\dots,y_r^*\}$ the corresponding dual bases of A^1 resp. A^2.

Put $f_j = o(y_j)$, $j = 1,\dots,r$. Then by (4.2.4) of [La] the ideal $(\underline{f})$ of T^1 generated by the f_j's is contained in $\underline{m}^2_{T^1}$. Moreover $H \simeq T^1/(\underline{f})$. Now, for any surjective homomorphism of local artinian k-algebras $\pi : R \to S$, such that $\underline{m}_R \cdot \ker \pi = 0$, consider the diagram

$$\begin{array}{ccc} \mathrm{Mor}(H,R) & \to & \mathrm{Def}_E(R) \\ \downarrow & & \downarrow \\ \mathrm{Mor}(H,S) & \to & \mathrm{Def}_E(S). \end{array}$$

Suppose given a morphism $\phi: H \to S$ corresponding to the lifting $X_\phi \in \mathrm{Def}_X(S)$, then in the diagram below, we may always lift the map ϕ' to a map $\bar{\phi}$ making the resulting diagram commutative

(1)

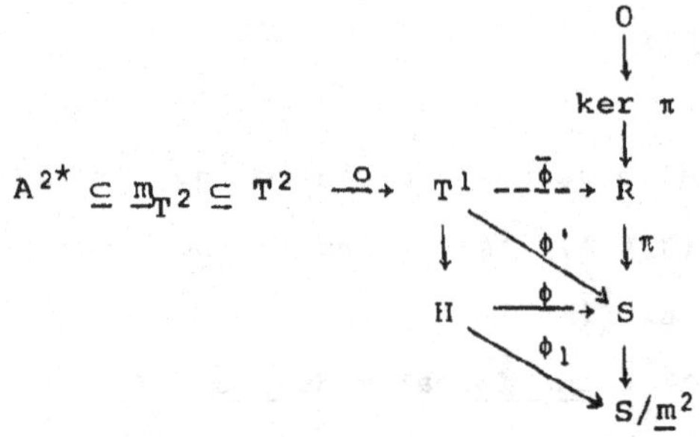

The obstruction for lifting X_ϕ to R is, by construction of o, and functoriality, given by the restriction of $o \circ \bar{\phi}$ to A^{2*}. In fact $o \circ \bar{\phi}$ induces a linear map $A^{2*} \to \ker \pi$, i.e. an element $o(X_\phi, \pi) \in A^2 \otimes \ker \pi = A^2(k, X; O_X \otimes \ker \pi)$, which is the uniquely defined obstruction. Notice that we have the following identity

(2) $$o(X_\phi, \pi) = \sum_k y_j^* \otimes \phi(\bar{f}_j).$$

Notice also that the image X_{ϕ_1} of X_ϕ by the map $\mathrm{Def}_X(S) \to \mathrm{Def}_X(S/\underline{m}^2)$ corresponds to the map ϕ_1. Moreover ϕ_1 is uniquely determined by the induced map on the cotangent level

$$t_\phi : A^{1*} = \underline{m}_H/\underline{m}_H^2 \to \underline{m}/\underline{m}^2$$

thus by an element $t_\phi \in A^1 \otimes \underline{m}/\underline{m}^2$ which under the isomorphism $\mathrm{Def}_X(S/\underline{m}^2) = A^1 \otimes \underline{m}/\underline{m}^2$ corresponds to X_{ϕ_1}. If $t_\phi = \Sigma_{i=1}^d x_i^* \otimes t_i$, $t_i \in \underline{m}/\underline{m}^2$ then $\phi_1(x_i) = t_i$, $i = 1, \ldots, d$.

On the other hand, having fixed a basis $\{\bar{v}_1, \ldots, \bar{v}_p\}$ for $\underline{m}/\underline{m}^2$, we find that $t_\phi = \Sigma_{l=1}^p \alpha_l \otimes \bar{v}_l$, $\alpha_l \in A^1$.

Thus there is a one to one correspondence between maps ϕ_1 and sequences $\alpha_1, \ldots, \alpha_p$ of elements of A^1.

Pick an $\underline{n} = (n_1, \ldots, n_d) \in \underline{N}^d$ with $|\underline{n}| = \Sigma_{i=1}^d n_i = N$ and let $i_1 < i_2 < \cdots < i_p$ be the indices i for which $n_i \neq 0$.

Consider the ideal $J_{\underline{n}} \subseteq k[u_1,\dots,u_d]$ generated by the set of monomials $\{u_1^{t_1}\cdots u_d^{t_d} \mid \exists i,\ t_i > n_i\}$.

Put $R_{\underline{n}} = k[u_1,\dots,u_d]/J_{\underline{n}}$, $S_{\underline{n}} = R_{\underline{n}}/(u_1^{n_1}\cdots u_d^{n_d})$ and let $v_1 = \bar{u}_{i_1}$ be the image of u_{i_1} in $R_{\underline{n}}$ (resp. $S_{\underline{n}}$). Obviously $v_1,\dots,v_p$ generates $R_{\underline{n}}$ (resp. $S_{\underline{n}}$) as k-algebra, and induces a basis $\{\bar{v}_1,\dots,\bar{v}_p\}$ of $\underline{m}_{\underline{n}}/\underline{m}_{\underline{n}}^2$. Fix this basis.

Now let $\alpha_1,\dots,\alpha_p \in A^1$ and consider the corresponding map $\phi_1 : H \to S_{\underline{n}}/\underline{m}_{\underline{n}}^2$.

<u>Definition (1.1)</u>. Any map $\phi_{\underline{n}}$ making the following diagram commutative

$$\begin{array}{ccccc} T^2 & \xrightarrow{o} & T^1 & & R_{\underline{n}} \\ & & \downarrow \rho & & \downarrow \pi_{\underline{n}} \\ & & H & \xrightarrow{\phi_{\underline{n}}} & S_{\underline{n}} \\ & & & \searrow^{\phi_1} & \downarrow \\ & & & & S_{\underline{n}}/\underline{m}_{\underline{n}}^2 \end{array}$$

is called a defining system for the Massey product

$$\langle \alpha_1,\dots,\alpha_p;\underline{n}\rangle = o(X_{\phi_{\underline{n}}},\pi_{\underline{n}}) \in A^2.$$

When $\alpha_l = x_{i_l}^*$ we shall write $\langle \underline{x}^*;\underline{n}\rangle$ for the Massey product $\langle x_{i_1}^*,\dots,x_{i_p}^*;\underline{n}\rangle$.

Suppose now that for some $N > 2$ and every $j = 1,\dots,r$ we have

$$f_j = \sum_{|\underline{n}|=N} a_{j,\underline{n}}\, \underline{x}^{\underline{n}} + \text{higher terms}$$

and consider any map $\bar{\phi}_{\underline{n}} : T^1 \to R_{\underline{n}}$ such that $\bar{\phi}_{\underline{n}} \circ \pi_{\underline{n}} = \rho \circ \phi_{\underline{n}}$. Then

$\bar{\phi}_{\underline{n}}(f_j) = a_{j,\underline{n}}\, \bar{u}_{i_1}^{n_{i_1}} \cdots \bar{u}_{i_p}^{n_{i_p}} \in \ker \pi_{\underline{n}} \simeq k$. Applying the identity (2) we find

(3) $$a_{j,\underline{n}} = y_j(\langle \underline{x}^*;\underline{n}\rangle).$$

It follows that if we let f_j^N be the degree N (leading) form of f_j , then

(4) $$f_j^N(\underline{x}) = \sum_{|\underline{n}|=N} y_j\langle \underline{x}^*;\underline{n}\rangle \cdot \underline{x}^{\underline{n}} .$$

Consider the diagram:

(5) $$\begin{array}{ccccc} T^2 & \vec{o} & T^1 & \xrightarrow{\ \bar{\phi}_{N-1}\ } & k[u_1,\ldots,u_d]/\underline{m}^{N+1} \\ & & \rho\downarrow & & \downarrow\pi \\ & & H & \xrightarrow[\ \phi_{N-1}\ ]{} & k[u_1,\ldots,u_d]/\underline{m}^{N} \end{array}$$

where $\rho\circ\phi_{N-1}(x_i) \equiv u_i (\mathrm{mod}\ \underline{m})$, $\underline{m} = (u_1,\ldots,u_d)$.

Let $X_{\phi_{N-1}} \in \mathrm{Def}_X(k[u_1,\ldots,u_d]/\underline{m}^N)$ correspond to ϕ_{N-1}. Notice that by assumption $X_{\phi_{N-1}}$ is a lifting of the universal lifting of X to $k[u_1,\ldots,u_d]/\underline{m}^2$. Notice also that $\ker\pi = \underline{m}^N/\underline{m}^{N+1} = \bigoplus_{|\underline{n}|=N} k\cdot(\underline{u}^{\underline{n}})$.

An easy argument then shows that the obstruction for lifting $X_{\phi_{N-1}}$ to $k[u_1,\ldots,u_d]/\underline{m}^{N+1}$ is given by:

(6) $$\begin{aligned} o(X_{\phi_{N-1}},\pi) &= \sum_{|\underline{n}|=N} \langle \underline{x}^*;\underline{n}\rangle \otimes \underline{u}^{\underline{n}} = \sum_j y_j^* \otimes \Big(\sum_{|\underline{n}|=N} y_j\langle \underline{x}^*;\underline{n}\rangle \underline{u}^{\underline{n}}\Big) \\ &= \sum_j y_j^* \otimes f_j^N(\underline{u}) . \end{aligned}$$

Now consider the diagram

(7) $$\begin{array}{cccccl} T^2 & \overset{o}{\leftarrow} & T^1 & \xrightarrow{\ \bar{\phi}_N\ } & R_{N+1} & = k[u_1,\ldots,u_d]/(\underline{m}^{N+2}+\underline{m}(f_1^N,\ldots,f_r^N)) \\ & & \downarrow & & \downarrow \pi'_{N+1} & \\ & & H & \xrightarrow{\ \phi_N\ } & S_N & = k[u_1,\ldots,u_d]/(\underline{m}^{N+1}+(f_1^N,\ldots,f_r^N)) \\ & & & \searrow^{\phi_{N-1}} & \downarrow \pi_N & \\ & & & & S_{N-1} & = k[u_1,\ldots,u_d]/\underline{m}^N \end{array}$$

Since S_N is $k[u_1,\ldots,u_d]/\underline{m}^{N+1}$ divided by the ideal generated by the obstruction for lifting $X_{\phi_{N-1}}$, we may lift $X_{\phi_{N-1}}$ to S_N, therefore we may find maps ϕ_N and $\bar{\phi}_N$ making the diagram commutative.

Pick a monomial basis $\{\underline{u}^{\underline{n}}\}_{\underline{n}\in\bar{B}_{N-1}}$ for S_{N-1} (take simply all $\underline{u}^{\underline{n}}$ with $|\underline{n}| \leq N-1$) and pick a monomial basis $\{\underline{u}^{\underline{n}}\}_{\underline{n}\in B_N}$ for $\ker \pi_N = \underline{m}^N/\underline{m}^{N+1}+(f_1,\ldots,f_r^N)$. Put $\bar{B}_N = \bar{B}_{N-1} \cup B_N$. For every $\underline{n}$ with $|\underline{n}| \leq N$ we have a unique relation in S_N

(8) $$\underline{u}^{\underline{n}} = \sum_{\underline{m}\in\bar{B}_N} \beta_{\underline{n},\underline{m}}\, \underline{u}^{\underline{m}}$$

Since by construction $o(X_{\phi_{N-1}}, \pi_N) = 0$, this relation together with (6) implies that for every $\underline{m} \in B_N$ (or $\bar{B}_N$ if one insists),

(9) $$\sum_{|\underline{n}|=N} \beta_{\underline{n},\underline{m}} \langle \underline{x}^*; \underline{n}\rangle = 0$$

Write
$$\ker \pi'_{N+1} = (\underline{m}^{N+1}+(f_1^N,\ldots,f_r^N))/(\underline{m}^{N+2}+\underline{m}(f_1^N,\ldots,f_r^N))$$
$$= (f_1^N,\ldots,f_r^N)/\underline{m}(f_1^N,\ldots,f_r^N) \oplus I_{N+1}$$

Pick a monomial basis for $I_{N+1} = \underline{m}^{N+1}/(\underline{m}^{N+2}+\underline{m}^{N+1} \cap \underline{m}(f_1^N,\ldots,f_r^N))$ of the form $\{\underline{u}^{\underline{n}}\}_{\underline{n}\in B'_{N+1}}$. We may assume that for $\underline{n} \in B'_{N+1}$, $\underline{u}^{\underline{n}}$ is of the form $u_k \cdot \underline{u}^{\underline{m}}$ for some $\underline{m} \in B_N$. Put $\bar{B}'_{N+1} = \bar{B}_N \cup B'_{N+1}$. For every $\underline{n}$ with $|\underline{n}| \leq N+1$ we have a unique relation in R_{N+1}

(10) $$\underline{u}^{\underline{n}} = \sum_{\underline{m}\in\bar{B}'_{N+1}} \beta'_{\underline{n},\underline{m}}\, \underline{u}^{\underline{m}} + \sum_j \beta'_{\underline{n},j} f_j^N.$$

Let

(11) $$f_j^{N+1} = \bar{\phi}_N(f_j) = f_j^N + \sum_{\underline{n}\in\bar{B}'_{N+1}} b_{j,\underline{n}}\, \underline{u}^{\underline{n}}$$

then by definition of o, the obstruction for lifting X_{ϕ_N} to R_{N+1} is

(12) $$o(X_{\phi_N}, \pi'_{N+1}) = \sum_j y_j^* \otimes f_j^{N+1}$$
$$= \sum_j y_j^* \otimes f_j^N + \sum_{\underline{m}\in\bar{B}'_{N+1}} (\sum_j{}^* y \otimes b_{j,\underline{n}}\, \underline{u}^{\underline{n}})$$

<u>Definition (1.2)</u>. The map ϕ_N is called a defining system for the Massey products

$$\langle \underline{x}^*;\underline{n}\rangle = \sum_j b_{j,\underline{n}}\, y_j^* \in A^2, \quad \text{for} \quad \underline{n} \in B'_{N+1}.$$

With these notations we have:

$$(13) \qquad f_j^{N+1} = \sum_{\underline{m}\in B'_N} y_j \langle \underline{x}^*;\underline{m}\rangle \underline{u}^{\underline{m}} + \sum_{\underline{n}\in B'_{N+1}} y_j \langle \underline{x}^*;\underline{n}\rangle \underline{u}^{\underline{n}}$$

where we have put $B'_N = \{\underline{n} \mid |\underline{n}| = N\}$.

Consider the diagram

$$(14) \qquad \begin{array}{ccccl}
T^2 \subseteq T^1 & \xrightarrow{\bar{\phi}_{N+1}} & R_{N+2} & = & k[u_1,\dots,u_d]/(\underline{m}^{N+3} + \underline{m}\cdot(f_1^{N+1},\dots,f_r^{N+1})) \\
\downarrow & & \downarrow \pi'_{N+2} & & \\
H & \xrightarrow{\phi_{N+1}} & S_{N+1} & = & R_{N+1}/(f_1^{N+1},\dots,f_r^{N+1}) \\
& \searrow^{\phi_N} & \downarrow \pi_{N+1} & & \\
& & S_N & &
\end{array}$$

Since S_{N+1} is R_{N+1} divided by the ideal generated by the obstruction for lifting X_{ϕ_N} to R_{N+1}, we may lift X_{ϕ_N} to S_{N+1}, therefore we may find maps ϕ_{N+1} and $\bar{\phi}_{N+1}$ making the diagram above commutative.

Pick a monomial basis $\{\underline{u}^{\underline{n}}\}_{\underline{n}\in B_{N+1}}$ for $\ker \pi_{N+1}$ such that $B_{N+1} \subseteq B'_{N+1}$. Put $\bar{B}_{N+1} = \bar{B}_N \cup B_{N+1}$. Then $\{\underline{u}^{\underline{n}}\}_{\underline{n}\in\bar{B}_{N+1}}$ is a monomial basis for S_{N+1}. For every $\underline{n}$ with $|\underline{n}| < N+1$ we therefore have a unique relation in S_{N+1}

$$(15) \qquad \underline{u}^{\underline{n}} = \sum_{\underline{m}\in\bar{B}_{N+1}} \beta_{\underline{n},\underline{m}}\, \underline{u}^{\underline{m}}$$

Since by construction $o(X_{\phi_N}, \pi_{N+1}) = 0$, this implies for every $\underline{m} \in B_{N+1}$ the following identity:

$$(16) \qquad \sum_{\underline{n}\in B'_{N+1}} \beta_{\underline{n},\underline{m}} \langle \underline{x}^*;\underline{n}\rangle = 0$$

which is analoguous to (9).

Write, again, $\ker \pi'_{N+2} = (\underline{m}^{N+2}+(f_1^{N+1},\dots,f_r^{N+1}))/(\underline{m}^{N+3}+\underline{m}(f_1^{N+1},\dots,f_r^{N+1}))$
$= (f_1^{N+1},\dots,f_r^{N+1})/\underline{m}(f_1^{N+1},\dots,f_r^{N+1}) \oplus I_{N+2}$.

Pick a monomial basis for $I_{N+2} = \underline{m}^{N+2}/(\underline{m}^{N+3}+\underline{m}^{N+2}\cap \underline{m}(f_1^{N+1},\dots,f_r^{N+1}))$ of the form $\{\underline{u}^{\underline{n}}\}_{\underline{n}\in B'_{N+2}}$, where we may assume that for $\underline{n} \in B'_{N+2}$, $\underline{u}^{\underline{n}}$ is of the form $u_k \cdot \underline{u}^{\underline{m}}$ for some $\underline{m} \in B_{N+1}$ and some k. Put $\bar{B}'_{N+2} = \bar{B}'_{N+1} \cup B'_{N+2}$. For every $\underline{n}$ with $|\underline{n}| < N+2$ we have a unique relation in R_{N+2}

$$(17)\qquad u^{\underline{n}} = \sum_{\underline{m}\in\bar{B}'_{N+2}} \beta'_{\underline{n},\underline{m}}\, \underline{u}^{\underline{m}} + \sum_j \beta'_{\underline{n},j}\, f_j^{N+1}$$

of the same form as (10).

Let

$$(18)\qquad f_j^{N+2} = \bar{\phi}_{N+1}(f_j) = f_j^{N+1} + \sum_{\underline{n}\in B'_{N+2}} c_{j,\underline{n}}\, \underline{u}^{\underline{n}} .$$

Again, by definition of o, the obstruction for lifting $X_{\phi_{N+2}}$ to R_{N+2} is

$$(19)\qquad \begin{aligned} o(X_{\phi_{N+1}},\pi'_{N+2}) &= \sum_j y_j^* \otimes f_j^{N+2} \\ &= \sum_j y_j^* \otimes f_j^{N+1} + \sum_{\underline{n}\in B'_{N+2}} (\sum_j y_j^* \otimes c_{j,\underline{n}} \underline{u}^{\underline{n}}) \end{aligned}$$

<u>Definition (1.3)</u>. The map ϕ_{N+1} is called a defining system for the Massey products $\langle \underline{x}^*; \underline{n}\rangle = \Sigma_j c_{j,\underline{n}}\, y_j^* \in A^2$, for $\underline{n} \in B'_{N+2}$.

With these notations, we have the following

$$(20)\qquad f_j^{N+2} = \sum_{\underline{l}\in B'_N} \langle \underline{x}^*;\underline{l}\rangle \underline{u}^{\underline{l}} + \sum_{\underline{m}\in B'_{N+1}} y_j \langle \underline{x}^*;\underline{m}\rangle \underline{u}^{\underline{m}} + \sum_{\underline{n}\in B'_{N+2}} y_j \langle \underline{x}^*;\underline{m}\rangle \underline{u}^{\underline{n}}$$

Clearly this process may be continued indefinitely. For every $k > 0$ we obtain a diagram

$$\begin{array}{ccccc} T^2 & \supseteq & T^1 & \xrightarrow{\bar\phi_{N+k}} & R_{N+k+1} \\ & & \downarrow & & \downarrow \pi'_{N+k+1} \\ & & H & \xrightarrow{\phi_{N+k}} & S_{N+k} \\ & & & & \downarrow \pi_{N+k} \\ & & & & S_{N+k-1} \end{array} \tag{21}$$

a monomial basis $\{\underline{u}^{\underline{n}}\}_{\underline{n}\in\bar{B}_{N+k}}$ for S_{N+k}, such that for every $\underline{n}$ with $|\underline{n}| \leq N+k$ there is a unique relation in S_{N+k}

(22) $$\underline{u}^{\underline{n}} = \sum_{\underline{m}\in\bar{B}_{N+k}} \beta_{\underline{n},\underline{m}}\, \underline{u}^{\underline{m}} ,$$

inducing the identity

(23) $$\sum_{\underline{n}\in B'_{N+k}} \beta_{\underline{n},\underline{m}} \langle \underline{x}^*;\underline{n}\rangle = 0 , \quad \underline{m} \in B_{N+k}.$$

And there is a corresponding basis $\{\underline{u}^{\underline{n}}\}_{\underline{m}\in B'_{N+k+1}}$ for the component I_{N+k+1} of $\ker \pi'_{N+k+1} = (f_1^{N+k},\dots,f_r^{N+k})/\underline{m}(f_1^{N+k},\dots,f_r^{N+k}) \oplus I_{N+k+1}$ such that in R_{N+k+1} we have for every $\underline{n}$ with $|\underline{n}| \leq N+k+1$

(24) $$\underline{u}^{\underline{n}} = \sum_{\underline{m}\in\bar{B}'_{N+k+1}} \beta'_{\underline{n},\underline{m}}\, \underline{u}^{\underline{n}} + \sum_j \beta'_{\underline{n},j}\, f_j^{N+k}$$

where we have put $\bar{B}_{N+k+1} = \bar{B}_{N+k} \cup B'_{N+k+1}$.

Moreover,

(25) $$f_j^{N+k+1} = \bar{\phi}_{N+k}(f_j) = f_j^{N+k} + \sum_{\underline{n}\in B'_{N+k+1}} \omega_{j,\underline{n}}\, \underline{u}^{\underline{n}}.$$

The obstruction for lifting $X_{\phi_{N+k}}$ to R_{N+k+1} is

(26) $$\begin{aligned} o(X_{\phi_{N+k}},\pi'_{N+k+1}) &= \sum_j y_j^* \otimes f_j^{N+k+1} \\ &= \sum_j y_j^* \otimes f_j^{N+k} + \sum_{\underline{n}\in B'_{N+k+1}} \left(\textstyle\sum_j y_j^* \otimes \omega_{j,\underline{n}}\underline{u}^{\underline{n}}\right) \end{aligned}$$

<u>Definition (1.4)</u>. The map ϕ_{N+k} is called a defining system for the Massey products $\langle \underline{x}^*;\underline{n}\rangle = \Sigma_j\, \omega_{j,\underline{n}}\, y_j^* \in A^2$ for $\underline{n} \in B'_{N+k+1}$.

In particular, we find for every k,

(27) $$f_j^{N+k+1} = \sum_{l=o}^{k+1} \sum_{\underline{n}\in B'_{N+1}} y_j\langle \underline{x}^*;\underline{n}\rangle \underline{u}^{\underline{n}}$$

Notice that by (4.2.4) of [La1] we have

(28) $$H \simeq \varprojlim_{k} S_{N+k}$$

therefore

(29) $$H \simeq k[[u_1,\dots,u_d]]/(\bar{f}_1,\dots,\bar{f}_d)$$

where

(30) $$\bar{f}_j = \lim_{k\to\infty} f_j^{N+k}.$$

Formally we may therefore write

(31) $$\bar{f}_j = \sum_{l=o}^{\infty} \sum_{\underline{n}\in B'_{N+1}} y_j\langle \underline{x}^*;\underline{n}_1\rangle \underline{u}^{\underline{n}}.$$

§2 Massey products for $\mathrm{Ext}_A^{\cdot}(E,E)$

In this paragraph we shall let A be any k-algebra and we shall let X, in §1, be some A-module E. We shall thus be concerned with the deformation functor of E as an A-module

$$\mathrm{Def}_E\colon \underline{1} \to \underline{\mathrm{Sets}}$$

defined as follows,

$$\mathrm{Def}_E(S) = \left\{ \begin{matrix} S\otimes_k A & \to & \mathrm{End}(E_S) \\ \downarrow & & \downarrow \\ A & \to & \mathrm{End}(E) \end{matrix} \;\middle|\; \begin{matrix} E_S \text{ is } S\text{-flat} \\ \downarrow \\ E_S\otimes_S k = E \end{matrix} \right\} \Big/ \text{iso.}$$

As is well known, the corresponding cohomology is

$$A^i = \mathrm{Ext}_A^i(E,E)$$

The deformation theory for modules, as hinted at on page 150 of [La1], parallels the corresponding theory for algebras. There is a global theory and a relative theory, and the main theorem (4.2.4) of [La1] holds. There are no surprises, and we shall therefore leave the details to the reader.

Pick any free resolution $L.$ of E as an A-module, and consider the associated single complex $\mathrm{Hom}_A^{\bullet}(L.,L.)$ of the double complex $\mathrm{Hom}_A(L.,L.)$. By definition we have

$$\mathrm{Hom}_A^p(L.,L) = \prod_{m>o} \mathrm{Hom}(L_m,L_{m-p})$$

Let $d_i : L_i \to L_{i-1}$ be the differential of $L.$, then

$$d^p : \mathrm{Hom}_A^p(L.,L.) \to \mathrm{Hom}^{p+1}(L.,L.)$$

is defined by

$$d^p(\{\alpha_i^p\}_{i>o}) = d_i o \alpha_{i-1}^p - (-1)^p \alpha_i^p o d_{i-p}$$

Clearly $\mathrm{Hom}_A^{\bullet}(L.,L.)$ is a graded differential associative A-algebra, multiplication being the composition of $\mathrm{Hom}^{\bullet}(L.,L.)$.

<u>Lemma (2.1)</u>. There is a natural isomorphism

$$\mathrm{Ext}_A^i(E,E) \simeq H^i(\mathrm{Hom}_A^{\bullet}(L.,L.)), \qquad i > 0.$$

Consider any surjective morphism $\pi: R \to S$ in $\underline{1}$, such that $\underline{m}_R \cdot \ker \pi = 0$.

Assume there exists a lifting $\{L.\otimes_k S,\ d_i(S)\}$ of the complex $\{L.,d_i\}$, i.e. of the free resolution $L.$ of E.

This means that there exists a commutative diagram of the form

$$\begin{array}{ccccccccccc}
0 & \leftarrow & H_o(L.\otimes_k S) & \leftarrow & L_o\otimes_k S & \overset{d_1(S)}{\leftarrow} & L_1\otimes_k S & \overset{d_2(S)}{\leftarrow} & L_2\otimes_k S & \overset{d_3(S)}{\leftarrow} & \cdots \\
 & & \downarrow & & \downarrow & & \downarrow & & \downarrow & & \\
0 & \leftarrow & H_o(L.) & \leftarrow & L_o & \overset{d_1}{\leftarrow} & L_1 & \overset{d_2}{\leftarrow} & L_2 & \overset{d_3}{\leftarrow} & \cdots
\end{array}$$

where for every i, the composition

$$d_{i+1}(S) \circ d_i(S) = 0.$$

We shall see that any such lifting is, in fact, an $A \otimes_k S$-free resolution of $H_o(L.\otimes_k S) = E_S$, and that E_S is a lifting of $E = H_o(L.)$ to S.

Both contentions are obviously true for $S = k$, so by induction we may assume they hold for S. If we then are able to prove the corresponding statements for R, we are through.

But first we have an existence problem. Given a lifting E_S of E to S it is easy to see that there is a corresponding lifting $\{L.\otimes_k S, d_i(S)\}$ of $\{L.,d_i\}$ to S. By assumption we have conversally that any such lifting $\{L.\otimes_k S, d_i(S)\}$ of $\{L.,d_i\}$ to S determines a lifting $E_S = H_o(L.\otimes_k S)$ and is, itself, an $A \otimes_k S$-free resolution of E_S.

Pick one such lifting $\{L.\otimes_k S, d_i(S)\}$, and let us compute the obstruction for lifting $\{L.\otimes_k S, d_i(S)\}$ to R. This obstruction is then, clearly, an obstruction for lifting E_S to R.

For every i, pick a lifting $d_i'(R): L_i\otimes_k R \to L_{i-1}\otimes_k R$ of $d_i(S)$: $L_i\otimes_k S \to L_{i-1}\otimes_k S$, to R. This is obviously possible, since all L_i are A-free.

Since $d_i(S)\circ d_{i-1}(S) = 0$ and since $I = \ker \pi$ is killed by the maximal ideal $\underline{m}_R$ of R, the composition $d_i'(R)\circ d_{i-1}'(R)$: $L_i\otimes_k R \to L_{i-2}\otimes_k R$ is induced by a unique map

$$O_i: L_i \to L_{i-2} \otimes_k I$$

The family $\{O_i\}_{i>0}$ defines an element

$$O \in \mathrm{Hom}^2(L.,L.) \otimes_k I$$

One checks that $d^2O = 0$, so that O is a 2-cocycle of $\mathrm{Hom}_A^\bullet(L.,L.)$, defining an element

$$o(E_S,\pi) \in \mathrm{Ext}_A^2(E,E) \otimes_k I.$$

It is easily seen that $o(E_S,\pi)$ is independent of the choice of the $d_i'(R)$'s lifting the $d_i(S)$'s.

Moreover, if $o(E_S,\pi) = 0$, there exists an element $\xi \in \mathrm{Hom}_A^1(L.,L.) \otimes_k I$ such that $d\xi = -O$. Put

$$d_i(R) = d_i'(R)+\xi_i ,$$

then one finds

$$d_i(R) \circ d_{i-1}(R) = 0$$

and $\{L.\otimes_k R, d_i(R)\}$ is a lifting of $\{L.\otimes_k S, d_i(S)\}$ to R. Now let $\{L.\otimes_k R; d_i(R)\}$ be any lifting of $\{L.\otimes_k S, d_i(S)\}$ to R, then there is an exact sequence of complexes

$$0 \to \{L.\otimes_k I, d_i \otimes 1_I\} \to \{L.\otimes_k R, d_i(R)\} \to \{L.\otimes_k S, d_i(S)\} \to 0$$

inducing a long exact sequence

$$\begin{array}{llll} \to H_n(L.\otimes_k I) & \to H_n(L.\otimes_k R) & \to H_n(L.\otimes_k S) & \\ \to H_{n-1}(L.\otimes_k I) & \to \;\cdots\cdots & \to H_1(L.\otimes_k S) & \\ \to H_o(L.\otimes I) & \to H_o(L.\otimes R) & \to H_o(L.\otimes_k S) & \to 0 \end{array}$$

from which it follows that

$$H_n(L.\otimes_k R) = 0 \quad \text{for} \quad n \geq 1, \quad \text{and}$$

$$0 \to E \otimes_k I \to H_o(L.\otimes R) \to E_S \to 0$$

is exact.

Therefore $H_o(L.\otimes R) = E$ is a lifting of E_R to R.

Moreover, given two liftings $\{L.\otimes_k R, d_i(R)_1\}$, $1 = 1,2$ of $\{L.\otimes_k S, d_i(S)\}$, corresponding to two liftings E_R^1 and E_R^2 of E_S , the differences $d_i(R)_1 - d_i(R)_2$ induce maps

$$\eta_i : L_i \to L_{i-1} \otimes_k I.$$

The family $\{\eta_i\}_{i>o}$ is a 1-cocycle of $\mathrm{Hom}^{\bullet}(L.,L.)$ defining an element $\bar{\eta} \in \mathrm{Ext}_A^1(E,E)$.

In this way we obtain a surjective map

$$\{\text{liftings of } E_S \text{ to } R\} \times \mathrm{Ext}_A^1(E.E) \to \{\text{liftings of } E_S \text{ to } R\}$$

making the set of liftings of E_S to R a principal homogenous space (torsor) over $\mathrm{Ext}_A^1(E,E)$.

We have established the following,

<u>Proposition (2.2)</u>. Let $E_S \in \mathrm{Def}_E(S)$ correspond to the lifting $\{L.\otimes_k S,\ d_i(S)\}$ of $L.$ to S. Then there is a uniquely defined obstruction

$$o(E_S,\pi) \in \mathrm{Ext}_A^2(E,E) \otimes_k I$$

given in terms of the 2-cocycle O of $\mathrm{Hom}_A^{\bullet}(L.,L.) \otimes I$ defined above, such that $o(E_S,\pi) = 0$ iff E_S may be lifted to R.

Moreover, if $o(E_S,\pi) = 0$ then the set of liftings of E_S to R is a principal homogeneous space (torsor) over $\mathrm{Ext}_A^1(E,E)$.

Thus we have at hand a nice obstruction calculus for Def_E given entirely in terms of the complex $L.$ and its liftings.
Using this we shall apply the constructions of §1 and compute the Massey products, $\langle \underline{x}^*,\underline{n}\rangle$ for $\underline{n} \in B'_{N+k}$. In fact, the $\langle \underline{x}^*,\underline{n}\rangle$ of §1 will turn out to be some generalized "ordinary" Massey products of the differential graded k-algebra $\mathrm{Hom}_A^{\bullet}(L.,L.)$.
Pick a basis $\{x_1,\dots,x_d\}$ of $\mathrm{Ext}_A^1(E,E)^*$ and a basis $\{y_1,\dots,y_r\}$ of $\mathrm{Ext}_A^2(E,E)^*$. Denote by $\{x_1^*,\dots,x_d^*\}$ and $\{y_1^*,\dots,y_r^*\}$ the corresponding dual bases of Ext^1 and Ext^2.
Let for $i = 1,\dots,d$, $X_i \in \mathrm{Hom}_A^1(L.,L.)$ be a cocycle representing x_i^* and let for $j = 1,\dots,r$, $Y_j \in \mathrm{Hom}_A^2(L.,L.)$ be a cocycle representing y_j^*.
Pick an $\underline{n} = (n_1,\dots,n_d)$ with $|\underline{n}| = \Sigma_{i=1}^d n_i = N$ and consider as in §1 the k-algebras $S_{\underline{n}}$ and $R_{\underline{n}}$. Fix the basis $\{\bar{v}_1 \cdots \bar{v}_p\}$ of $\underline{m}_{\underline{n}}/\underline{m}_{\underline{n}}^2$. Recall that we have in $R_{\underline{n}}$ the following slightly confusing identities

$$(1)\qquad v_l = u_{i_l},\quad l = 1,\dots,p.$$

$$u_i = 0 \quad \text{if} \quad i \notin \{i_1,\dots,i_p\}$$

insisted upon because it makes the notations more streamlined later on.

We shall pick a monomial basis for the k-vectorspace $S_{\underline{n}}$ of the form

$$\{u_1^{m_1}\cdots u_d^{m_d} \mid 0 \le m_i \le n_i,\ \underline{m} \ne \underline{n}\}$$

written as

$$\{\underline{u}^{\underline{m}}\}_{\underline{m}\in\bar{B}_{\underline{n}}} .$$

With this done, let $\alpha_1,\ldots,\alpha_p \in \mathrm{Ext}^1_A(E,E)$ and consider the element $\Sigma_{l=1}^{p}\alpha_l \otimes \bar{v}_l \in \mathrm{Ext}^1_A \otimes \underline{m}_{\underline{n}}/\underline{m}^2_{\underline{n}}$.

Let $\phi_1: H \to S_{\underline{n}}/\underline{m}^2_{\underline{n}}$ be the corresponding map and let $E_{\phi_1} \in \mathrm{Def}_E(S_{\underline{n}}/\underline{m}^2_{\underline{n}})$ be the induced deformation of E.

Assume there is given a defining system $\phi_{\underline{n}}: H \to S_{\underline{n}}$ for the Massey product $\langle\alpha_1,\ldots,\alpha_p;\underline{n}\rangle$ (see (1.1)), corresponding to a lifting $E_{\phi_{\underline{n}}} \in \mathrm{Def}_E(S_{\underline{n}})$ of E_{ϕ_1}.

Then E_{ϕ_1} is represented by a lifting $\{L.\otimes S_{\underline{n}}/\underline{m}^2_{\underline{n}};d_i(S_{\underline{n}}/\underline{m}^2_{\underline{n}})\}$ of $L.$ and $E_{\phi_{\underline{n}}}$ is represented by a lifting $\{L.\otimes S_{\underline{n}};d_i(S_{\underline{n}})\}$ of $\{L.\otimes S_{\underline{n}}/\underline{m}^2_{\underline{n}};d_i(S_{\underline{n}}/\underline{m}^2_{\underline{n}})\}$.

The family of $A \otimes_k S_n$-linear maps

$$d_i(S_{\underline{n}}): L_i \otimes_k S_{\underline{n}} \to L_{i-1} \otimes_k S_{\underline{n}}$$

is uniquely determined by the restriction to $L_i \otimes 1$, thus by the family of A-linear maps

$$\alpha_{i,m}: L_i \to L_{i-1}\ ,\quad \underline{m} \in \bar{B}_{\underline{n}}$$

defined by

$$d_i(S_{\underline{n}})|L_i\otimes 1 = \sum_{\underline{m}\in\bar{B}_{\underline{n}}} \alpha_{i,\underline{m}}\otimes \underline{u}^{\underline{m}}.$$

With this notation, we may assume

$$d_i(S_{\underline{n}}/\underline{m}^2{}_{\underline{n}})|L_i\otimes 1 = \sum_{l=1}^{p} \alpha_{i,\underline{\varepsilon}_l}\otimes \underline{u}^{\underline{\varepsilon}_l}$$

where $\underline{\varepsilon}_l = (\underbrace{0,\ldots,1}_{l},0,\ldots,0) \in \bar{B}_{\underline{n}}$.

According to (1) we may also write

$$d_i(S_{\underline{n}}/\underline{m}_{\underline{n}}^2) \mid L_i \otimes 1 = \Sigma\, \alpha_{i,\underline{\varepsilon}_l} \otimes \bar{v}_l .$$

For every $\underline{m} \in \bar{B}_{\underline{n}}$ the family $\{\alpha_{i,\underline{m}}\}_i$ is a cochain

$$\alpha_{\underline{m}} \in \mathrm{Hom}_A^1(L.,L.)$$

such that $\alpha_{\underline{\varepsilon}_l}$ is a cocycle representing the cohomology class α_l, $l = 1,\dots,p$, and $\alpha_{i,\underline{o}} = d_i$, $i > 0$.

Since $d_i(S_{\underline{n}}) \circ d_{i-1}(S_{\underline{n}}) = 0$ for all $i > 0$ we find that the family $\{\alpha_{\underline{m}}\}_{\underline{m}\in\bar{B}_{\underline{n}}}$ satisfies the following identities

$$\sum_{\substack{\underline{m}_1+\underline{m}_2=\underline{m} \\ \underline{m}_i\in\bar{B}_n}} \alpha_{\underline{m}_1} \circ \alpha_{\underline{m}_2} = 0 \quad \text{for all } \underline{m} \in \bar{B}_{\underline{n}} .$$

Moreover the obstruction for lifting $E_{\phi_{\underline{n}}}$ to $R_{\underline{n}}$, i.e. the obstruction $o(E_{\phi_{\underline{n}}},\pi_{\underline{n}})$ for lifting $\{L_i \otimes S_{\underline{n}}, d_i(S_{\underline{n}})\}$ to $R_{\underline{n}}$, is easily seen to be represented by the, (à priori), cocycle

$$\sum_{\substack{\underline{m}_1+\underline{m}_2=\underline{n} \\ \underline{m}_i\in\bar{B}_{\underline{n}}}} \alpha_{\underline{m}_1} \circ \alpha_{\underline{m}_2} \in \mathrm{Hom}_A^2(L.,L.).$$

<u>Proposition (2.3)</u>. Given a sequence of p cohomology classes $\alpha_l \in \mathrm{Ext}_A^1(E,E)$, then a defining system for the Massey product $\langle\alpha_1,\dots,\alpha_p;\underline{n}\rangle$ corresponds to a family $\{\alpha_{\underline{m}}\}_{\underline{m}\in\bar{B}_{\underline{n}}}$ of 1-cochains of $\mathrm{Hom}_A^1(L.,L.)$, such that for every $\underline{m} \in \bar{B}_{\underline{n}}$

$$\star \qquad \sum_{\substack{\underline{m}_1+\underline{m}_2=\underline{m} \\ \underline{m}_i\in\bar{B}_{\underline{n}}}} \alpha_{\underline{m}_1} \circ \alpha_{\underline{m}_2} = 0$$

and such that $\alpha_{i,\underline{o}} = d_i$ for $i > 0$ and $\alpha_{\underline{\varepsilon}_l}$ represents α_l, $l = 1,\dots,p$.

Conversally, any such family $\{\alpha_{\underline{m}}\}_{\underline{m}\in\bar{B}_{\underline{n}}}$ give rise to a defining system for the Massey product $\langle\alpha_1,\dots,\alpha_p;\underline{n}\rangle$.

Moreover, given such a defining system, the Massey product $\langle\alpha_1,\ldots,\alpha_p;\underline{n}\rangle$ is represented by the 2-cocycle

$$\sum_{\substack{\underline{m}_1+\underline{m}_2=\underline{m} \\ \underline{m}_i\in\bar{B}_{\underline{n}}}} \alpha_{\underline{m}_1} \circ \alpha_{\underline{m}_2} \in \mathrm{Hom}^2_A(L.,L.).$$

<u>Proof</u>. This is just the observation that a lifting $E_{\phi_{\underline{n}}}$ of E_{ϕ_1} corresponds to a lifting $\{L.\otimes_k S_{\underline{n}};d_i(S_{\underline{n}})\}$ of $\{L.\otimes_k S_{\underline{n}}/\underline{m}^2_{\underline{n}};d_i(S_{\underline{n}}/\underline{m}^2_{\underline{n}})\}$, thus to families $\{\alpha_{i,\underline{m}}\}_{\underline{m}\in\bar{B}_{\underline{n}}}$ such that

$$** \qquad d_i(S_n)\mid L_i\otimes 1 = \sum_{\underline{m}\in\bar{B}_{\underline{n}}} \alpha_{i,\underline{m}}\otimes \underline{u}^{\underline{m}} .$$

The relation $d_i(S_n)\circ d_{i-1}(S_n) = 0$ translates into $*$. Conversally $*$ proves that $d_i(S_n)$ defined by $**$ defines a lifting $\{L.\otimes S_{\underline{n}};d_i(S_{\underline{n}}/\underline{m}^2_{\underline{n}})\}$, thus also a lifting $E_{\phi_{\underline{n}}}$ of E_{ϕ_1}. Finally any such $E_{\phi_{\underline{n}}}$ corresponds to a map $\phi_{\underline{n}}: H \to S_{\underline{n}}$, i.e. to a defining system. Q.E.D.

<u>Remark (2.4)</u>. In the light of (2.3) we shall let the notion of a defining system for the Massey product $\langle\alpha_1,\ldots,\alpha_p;\underline{n}\rangle$ refer to either the map $\phi_{\underline{n}}$ or the family $\{\alpha_{\underline{m}}\}$ depending on the situation.

<u>Remark (2.5)</u>. Let $\underline{n} = (n_1,\ldots,n_d)$ be given such that $n_i = 0$ for $i \notin \{i_1,\ldots,i_p\}$, then the Massey product $\langle\alpha_1,\ldots,\alpha_p;\underline{n}\rangle$, if defined, depends only upon $\alpha_1,\ldots,\alpha_p$ and the p-uple $(n_{i_1},n_{i_2},\ldots,n_{i_p})$. Given $\alpha_1,\ldots,\alpha_p \in \mathrm{Ext}^1_A(E,E)$ and any p-uple $\underline{m} = (m_1,\ldots,m_p)$ there is no confusion in writing

$$\langle\alpha_1,\ldots,\alpha_p;\underline{m}\rangle .$$

Suppose $p = 1$ and $\underline{n} = (n)$, $\alpha_1=\alpha \in \mathrm{Ext}^1_A(E,E)$, then a defining system for $\langle\alpha;\underline{n}\rangle$ is a family $\{\alpha_{\underline{m}}\}_{o<\underline{m}<\underline{n}-1}$ of 1-cochains

$$\alpha_{\underline{m}} \in \mathrm{Hom}^1_A(L.,L.)$$

such that, $\alpha_{i,\underline{o}} = d_i$, $i > 0$ and $\alpha_{\underline{1}}$ represents α, with the property that for every $0 < \underline{m} < \underline{n}-1$, $\sum_{\underline{m}_1+\underline{m}_2=\underline{m}} \alpha_{\underline{m}_1} \circ \alpha_{\underline{m}_2} = 0$.

If a defining system exists, then

$$\langle\alpha;\underline{n}\rangle = \mathrm{cl}\Big(\sum_{\substack{\underline{m}_1+\underline{m}_2=\underline{n} \\ o<\underline{m}_i<\underline{m}-1}} \alpha_{\underline{m}_1} \circ \alpha_{\underline{m}_2} \Big)$$

In particular for $\underline{n} = (2)$ the Massey product $\langle\alpha,(2)\rangle$ is always defined and is represented by the 2-cocycle $\alpha\circ\alpha$. These are the "Bocksteins".

If $p = 2$ and $\underline{n} = (1,1)$, $\alpha_1, \alpha_2 \in \mathrm{Ext}^1_A(E,E)$ then the family $\{\alpha_{\underline{m}}\}_{\underline{m}\in\{(0,0),(1,0),(0,1)\}}$ where $\alpha_{(0,0)} = \{d_i\}_{i>0}$, $\alpha_{(1,0)}$ represents α, and $\alpha_{(0,1)}$ represents α_2, is a defining system for $\langle\alpha_1,\alpha_2;(1,1)\rangle$ which is represented by $\alpha_{(1,0)}\circ\alpha_{(0,1)} + \alpha_{(0,1)}\circ\alpha_{(1,0)}$. Thus $\langle\alpha_1,\alpha_2;(1,1)\rangle$ is the symmetrized cup-product.

Now, having a purely cohomological expression for the (defined) Massey products $\langle\alpha_1,\ldots,\alpha_p;\underline{n}\rangle$, we shall procede as in §1, computing step by step a set of generators for the ideal of T^1 defining the formal moduli of E.

Assume, as in §1 that the formal power-series $f_j = o(y_j) \in T^1 = k[[x_1,\ldots,x_d]]$ may be written as

$$f_j = \sum_{|\underline{n}|=N} a_{j,\underline{n}}\, \underline{x}^{\underline{n}} + \text{higher terms} \quad j = 1,\ldots,r$$

for some $N > 2$.

Then by §1 (3), $a_{j,\underline{n}} = y_j\langle x^*;n\rangle$ where, by assumption $\langle\underline{x}^*;n\rangle = \langle x^*_{i_1}, x^*_{i_2},\ldots,x^*_{i_p};\underline{n}\rangle$ is (uniquely) defined.

Put $f^N_j = \Sigma_{|\underline{n}|=N}\, a_{j.\underline{n}}\, \underline{u}^{\underline{n}}$ and consider the diagram §1 (7). The map ϕ_{N-1} induces defining systems for all Massey products $\langle\underline{x}^*;\underline{n}\rangle$ for $|\underline{n}| < N$, and corresponds therefore to a family

(2) $$\{\alpha_{\underline{m}}\}_{\underline{m}\in\bar{B}_{N-1}}$$

of 1-cochains of $\mathrm{Hom}^{\bullet}_{A}(L.,L.)$ such that for every $i > 0$, $\alpha_{i,\underline{o}} = d_i$, and $\alpha_{\underline{e}_i}$ is a cocycle representing $x^{*}_{i,\underline{e}_i} = (\underbrace{0,\ldots,1}_{i},0,\ldots 0) \in \underline{N}$. Moreover, for every $\underline{m} \in \bar{B}_{N-1}$

(3) $$\sum_{\substack{\underline{m}_1+\underline{m}_2=\underline{m}\\ \underline{m}_i\in\bar{B}_{N-1}}} \alpha_{\underline{m}_1} \circ \alpha_{\underline{m}_2} = 0$$

Let $d_i(S_{N-1})\colon L_i \otimes_k S_{N-1} \to L_{i-1} \otimes_k S_{N-1}$ be the $A \otimes S_{N-1}$-linear map defined by

$$d_i(S_{N-1}) \mid L_i \otimes 1 = \sum_{\underline{m}\in\bar{B}_{N-1}} \alpha_{i,\underline{m}} u^{\underline{m}} .$$

Then (3) implies that $\{L.\otimes_k S_{N-1};\ d_i(S_{N-1})\}$ is a lifting of the universal deformation of L. to S_2 defined by the map

$$\phi_1\colon H \to k[u_1,\ldots,u_d]/\underline{m}^2 .$$

Recall that ϕ_1 corresponds to the deformation of L., or of E if one wishes, to S_2 defined by the element

$$\sum_{i=1}^{d} x^{*}_{i} \otimes \bar{u}_i \in \mathrm{Ext}^1_A(E,E) \otimes \underline{m}/\underline{m}^2 = \mathrm{Def}_E(S_2).$$

By construction $\{L.\otimes_k S_{N-1}\ ;\ d_i(S_{N-1})\}$ induces the deformation $E_{\phi_{N-1}} \in \mathrm{Def}_E(S_{N-1})$.

Sticking to the notations of §1, and noticing that for every $\underline{n}\in B'_N = \{\underline{m}\in\underline{N}^d \mid |\underline{m}|=N\}$ the Massey product $\langle \underline{x}^{*};\underline{n}\rangle$ is represented by the 2-cycle

$$Y(\underline{n}) = \sum_{\substack{\underline{m}_1+\underline{m}_2=\underline{n}\\ \underline{m}_i\in\bar{B}_{N-1}}} \alpha_{\underline{m}_1} \circ \alpha_{\underline{m}_2}$$

§1 (8) and (9) translates into the following. For every $\underline{m} \in B_N$, $\Sigma_{\underline{n}\in B'_N}\ \beta_{\underline{n},\underline{m}}\ Y(\underline{n})$ is a coboundary.

Now, pick for every $\underline{m} \in B_N$ a 1-cochain $\alpha_{\underline{m}} \in \mathrm{Hom}_A^1(L.,L.)$ such that

(4) $$d\,\alpha_{\underline{m}} = \sum_{\underline{n}\in B_N'} \beta_{\underline{n},\underline{m}}\, Y(\underline{n})$$

and consider the family

(5) $$\{\alpha_{\underline{m}}\}_{\underline{m}\in \bar{B}_N}\,.$$

Let, for every $i > 0$, $d_i(S_N)\colon L_i \otimes S_N \to L_{i-1}\otimes S_N$ be defined by: $d_i(S_N) \mid L_i \otimes 1 = \Sigma_{\underline{m}\in B_N} \alpha_{i,\underline{m}} \otimes \underline{u}^{\underline{m}}$. Then (4) translates into

$$d_i(S_N)\circ d_{i-1}(S_N) = 0.$$

Consequently $\{L.\otimes_k S_N;\, d_i(S_N)\}$ is a lifting of $\{L.\otimes_k S_{N-1};\, d_i(S_{N-1})\}$ to S_N, and induces therefore a lifting $E_{\phi_N} \in \mathrm{Def}_E(S_N)$ of $E_{\phi_{N-1}}$. E_{ϕ_N}, again, corresponds to a map ϕ_N : $H \to S_N$ which we now fix.

According to (1.2) ϕ_N is a defining system for the Massey products $\langle \underline{x}^*;\underline{n}\rangle$ for $\underline{n} \in B_{N+1}'$. Since ϕ_N is induced by, and induces, a family (5), we shall refer to any such family as a defining system for the Massey products $\langle \underline{x}^*;\underline{n}\rangle$, $\underline{n} \in B_{N+1}'$.

By definition, see (1.2), these Massey products are given in terms of the obstruction, see §1 (11), $o(E_{\phi_N}, \pi_{N+1}')$.

By (2.2) this obstruction is defined by the 2-cocycle $O = \{O_i\}$ where

$$O_i = d_i'(R_{N+1})\circ d_{i-1}'(R_{N+1}),$$

$d_i'(R_{N+1})\colon L_i \otimes R_{N+1} \to L_{i-1}\otimes R_{N+1}$ being any lifting of $d_i(S_N)$. Pick $d_i'(R_{N+1}$ such that

$$d_i'(R_{N+1}) \mid L_i \otimes 1 = \sum_{\underline{m}\in \bar{B}_N} \alpha_{i,\underline{m}} \otimes \underline{u}^{\underline{m}}$$

then streight forward calculation, using §1 (10), shows that

$$o_i = \sum_{\underline{n}\in B'_{N+1}} \Big(\sum_{|\underline{m}|<N+1} \sum_{\underline{m}_1+\underline{m}_2=\underline{m}} \beta'_{\underline{m},\underline{n}} \cdot \alpha_{i,\underline{m}_1} o \alpha_{i-1,\underline{m}_2} \Big) u^{\underline{n}}$$

$$+ \sum_{j=1}^{r} \Big(\sum_{|\underline{m}|<N+1} \sum_{\underline{m}_1+\underline{m}_2=\underline{m}} \beta'_{\underline{m},j} \cdot \alpha_{i,\underline{m}_1} o \alpha_{i-1,\underline{m}_2} \Big) f_j^N .$$

Remember that $d_i(S_N) o d_{i-1}/S_N) = 0$.

Comparing this with (1.2) and §1 (11), we have proved the following

Proposition (2.6). Given a defining system $\{\alpha_{\underline{m}}\}_{\underline{m}\in\bar{B}_{N+1}}$ for the Massey products $\langle \underline{x}^*;\underline{n}\rangle$, $\underline{n} \in B'_{N+1}$, $\langle \underline{x}^*;\underline{n}\rangle$ is represented by the 2-cocycle

$$Y(n) = \sum_{|\underline{m}|<N+1} \sum_{\substack{\underline{m}_1+\underline{m}_2=\underline{m}\\ \underline{m}_i\in\bar{B}_n}} \beta'_{\underline{m},\underline{n}} \alpha_{\underline{m}_1} o \alpha_{\underline{m}_2}$$

By §1 (16) we know that for every $\underline{m} \in B_{N+1}$ the 2-cochain

$$\beta_{\underline{m}} = \sum_{\underline{n}\in B'_{N+1}} \beta_{\underline{n},\underline{m}} Y(\underline{n}) \in \mathrm{Hom}_A^2(L.,L.)$$

is a coboundary. Pick one $\alpha_{\underline{m}} \in \mathrm{Hom}_A^2(L.,L.)$ such that $d\,\alpha_{\underline{m}} = \beta_{\underline{m}}$, and consider the family

(6) $$\{\alpha_{\underline{m}}\}_{\underline{m}\in\bar{B}_{N+1}} .$$

Just as above, (6) is seen to correspond to a defining system, ϕ_{N+1}, for the Massey products $\langle \underline{x}^*;\underline{n}\rangle$, $\underline{n} \in B'_{N+2}$.

There are relations §1, (17), (18), (19), and we may copy the procedure above.

We end up with the following,

Proposition (2.7). Given a defining system $\{\alpha_{\underline{m}}\}_{\underline{m}\in\bar{B}_{N+k-1}}$ for the7 Massey products $\langle \underline{x}^*;\underline{n}\rangle$, $\underline{n} \in B'_{N+k}$, $\langle \underline{x}^*;\underline{n}\rangle$, is represented by the 2-cocycle

$$Y(\underline{n}) = \sum_{|m|<N+k} \sum_{\substack{\underline{m}_1+\underline{m}_2=\underline{m}\\ \underline{m}_i\in\bar{B}_{N+k-1}}} \beta'_{\underline{m},\underline{n}} \alpha_{\underline{m}_1} o \alpha_{\underline{m}_2}$$

Moreover, the polynomials

$$f_j^{N+k} = \sum_{l=o}^{k} \sum_{\underline{n}\in B'_{N+l}} y_j\langle \underline{x}^*;\underline{n}\rangle \underline{u}^{\underline{n}} \qquad j = 1,\ldots,r$$

induces identities §1 (22) and (23), such that if we for every $\underline{m} \in B_{N+k}$ pick a cochain $\alpha_{\underline{m}} \in \mathrm{Hom}^1_A(L.,L.)$ with

$$d\,\alpha_{\underline{m}} = \sum_{\underline{n}\in B'_{N+k}} \beta_{\underline{n},\underline{m}}\, Y(\underline{n})$$

then the family

$$\{\alpha_{\underline{m}}\}_{\underline{m}\in \bar{B}_{N+k}}$$

is a difining system for the Massey products $\langle \underline{x}^*;\underline{n}\rangle$, $\underline{n} \in B'_{N+k+1}$.

We may, refering to §1 (28), (29), (30), sum up the content of this §2 as follows

<u>Theorem (2.8)</u>. Given an A-module E, the formal moduli H of E is determined by the Massey products of $\mathrm{Ext}^{\bullet}_A(E,E)$. In fact

$$H \simeq k[[x_1,\ldots,x_d]]/(f_1,\ldots,f_r)$$

where

$$f_j = \sum_{l=2}^{\infty} \sum_{\underline{n}\in B'_l} y_j\langle \underline{x}^*;\underline{n}\rangle\, \underline{x}^{\underline{n}}.$$

<u>Corollary (2.9)</u>. Any complete local k-algebra A with residue field k is determined by $\mathrm{Ext}^i_A(k,k)$, $i = 1,2$ and its Massey-products.

<u>Proof</u>. Obviously A is the formal moduli of k as an A-module.

Q.E.D.

BIBLIOGRAPHY

[Car] Cartan H. Seminaire 1960-61. Institut Henri Poincaré, Paris, 1962.

[La1] Laudal, O.A., Formal Moduli of Algebraic Structures, Lecture Notes in Mathematics No 754, Springer-Verlag 1979.

[La2] Laudal, O.A., Groups and Monoids and their Algebras. Preprint Series, Inst. of Math., University of Oslo, No 12 (1982).

[M] Massey, W.S., Some Higher Order Cohomology Operations, Symposium International de Topologia Algebraica, p.p. 145-154, La Universidad Nacional Autonoma de Mexico and UNESCO, Mexico City 1958.

[May] May, J.P., Matric Massey Products, Journal of Algebra 12 (1969), p.p. 533-568.

[PAL] Palamodov, V.P., Cohomology of analytic algebras (russian), Trudi Moskovskogo Matematitseskogo obtsjestva. (Vol 44) 1982, pp. 3-61.

[S&S] Schlessinger, M. & Stasheff, J., Deformation theory and rational homotopy type & The Tangent Lie Algebra of a Commutative Algebra, Manuscripts, 1982.

Department of Mathematics
University of Oslo
Blindern, OSLO 3
(NORWAY)

A METHOD FOR CONSTRUCTING BAD NOETHERIAN LOCAL RINGS

Christer Lech

We wish to introduce some new ideas for obtaining results involving so-called bad Noetherian local rings. Taken together, these ideas form an alternative to the well-known method by Rotthaus, first applied in [7]. We shall display them by indicating a proof of Theorem 1 below.

THEOREM 1. A complete Noetherian local ring S is the completion of a Noetherian local domain if and only if the following conditions hold.

(i) The prime ring of S is a domain that acts on S without torsion;

(ii) Unless equal to (0), the maximal ideal of S does not belong to (0) as an associated prime ideal.

Before going further, let us state, without proof, a more comprehensive theorem, which, in essence, comprises not only Theorem 1 but also the main results in the articles [3] by Brodmann and Rotthaus and [5] by Larfeldt and the author. The same basic principles can be used for the proof — not, however, in the same simple guise.

THEOREM 2. Let S be a complete Noetherian local ring and $\mathfrak{a}$ an ideal of S. Then, in order that $\mathfrak{a}$ be generated by a prime ideal in some Noetherian local ring having S as completion, it is necessary and sufficient that the following conditions hold.

(i) For $W = \{x \in S;\ \mathfrak{a}:x = \mathfrak{a}\}$, the ideal $\mathfrak{a}S_W$ of S_W is generated by a prime ideal in a flatly embedded subring of S_W;

(ii) Unless equal to $\mathfrak{a}$, the maximal ideal of S does not belong to $\mathfrak{a}$ as an associated prime ideal.

Theorem 1 rounds off a line of research beginning with Akizuki [1], Sect. 3, and having Brodmann-Rotthaus [2], Prop. (15), as its most recent exponent. The author has been strongly impressed and spurred by the last-mentioned result. He has also had the benefit of stimulating discussions with Christel Rotthaus and Ralf Fröberg.

For the proof of Theorem 1 we shall need certain notions concerning subrings R of a local ring $(S,\mathfrak{m})$. In reality these notions will merely concern the corresponding local subrings $R_{R\cap\mathfrak{m}}$. We shall say that the injection $R \to S$ is <u>flat</u> if S is flat as a R-module, <u>torsionfree</u> if $rs \neq 0$ for $r \in R - \{0\}$, $s \in S - \{0\}$, <u>unramified</u> if $\mathfrak{m} = S(R\cap\mathfrak{m})$, <u>residually rational</u> etc if the field extension $\underline{k}(S)/\underline{k}(R_{R\cap\mathfrak{m}})$ is rational etc. Here $\underline{k}(\ )$ indicates the residue field of a local ring. Similarly $\wedge$ will indicate the completion of a local ring and $^-$ the algebraic closure of a field. Following Weil in [8], we shall say that a field extension K/k is <u>regular</u> if K and $\bar{k}$ are linearly disjoint over k.

In the sequel, let S be the complete Noetherian local ring of the theorem and $\mathfrak{m}$ its maximal ideal. Let further $\emptyset$ be the minimal finite subset of Spec(S) whose union consists of all zerodivisors in S.

The theorem has the form of an equivalence between two statements about S. Let us introduce the following third statement as a connecting link:

$$\left.\begin{array}{l}\text{There exists a ring injection } R \to S \text{ which is}\\ (1)\ \text{flat,}\\ (2)\ \text{unramified,}\\ (3)\ \text{residually regular,}\\ (4)\ \text{torsionfree.}\end{array}\right\} \quad (*)$$

This statement is implied in an obvious way by the statement that S is the completion of a Noetherian local domain. Moreover, it implies, through (4) and (2), the validity of the conditions (i) - (ii) of the theorem. When discussing the reverse implications, we may strengthen condition (ii) into

(iii) $\mathfrak{m}$ does not belong to (0) as an associated prime ideal,

as the possibility $\mathfrak{m} = (0)$ causes no difficulties. Thus, to prove the theorem, it suffices to prove the two implications

(i) - (iii) $\Longrightarrow$ (*);

(*) $\Longrightarrow$ 'S is the completion of a Noetherian local domain'.

The second implication is of minor interest from our present methodological point of view: it can be treated by rather general means, and when $\underline{k}(S)$ is countable, it can be got round entirely by strengthening (*) so that rationality takes the place of regularity. In the proof that follows some of the details will be omitted.

The desired Noetherian local domain is obtained by a ring construction embodied in the concept of straightness defined below.

DEFINITIONS. A morphism $A \longrightarrow B$ of commutative rings with 1-elements is called telescopic if there exists a well-ordered generating set $\{x_i \mid i \in I\}$ for B over A such that, for each $j \in I$, $A[\{x_i \mid i \leq j\}]$ is free as a module over $A[\{x_i \mid i < j\}]$. A morphism $A \longrightarrow B$ of local rings is called straight if it is unramified and can be presented as a composition of a telescopic morphism and a subsequent localization.

Since the given injection $R \longrightarrow S$ can be assumed to be local and since every straight local morphism is obviously flat, we get the desired result by combining the following two propositions.

PROPOSITION. Let $k \to K$ be a separabel field extension, and let $R \to R_1$, $R \to R_2$ be two unramified flat local morphisms, both inducing $k \to K$ as residual field extension. Then there is an isomorphism $\hat{R}_1 \to \hat{R}_2$ respecting the $\hat{R}$-algebra structure and the given identity of the residue fields.

PROPOSITION. For any field extension K/k and any local ring R with residue field k there exists a straight local morphism $R \to R'$ with $k \to K$ as residual morphism. For any such morphism $R \to R'$ the statements indicated below hold true.

$$R \text{ Noetherian} \implies R' \text{ Noetherian;}$$
$$K/k \text{ regular} \implies \mathfrak{p}R' \text{ prime } (\forall \mathfrak{p} \in \operatorname{Spec}(R)).$$

The first proposition can be obtained as a consequence of [6], Thm. 82, which ensures the existence of a $\hat{R}$-algebra homomorphism $\hat{R}_1 \to \hat{R}_2$ respecting K. It is not hard to see that any such homomorphism is indeed an isomorphism. The second proposition is in substance contained in [4] apart from the very last assertion, which can be made evident by the device of extending local domains into valuation rings. (The regularity hypothesis might be unnecessarily strong, but separability, at least, is needed.)

Let us now turn to the more fundamental first implication. Each of the properties (1) - (4) expresses a condition on R. Assuming (i) and (iii) to hold, we shall show how to construct a subring of S that satisfies them all. The first three can be summarized as follows: All S-linear maps $S^n \to S$ and $S^n \to \overline{\underline{k}(S)}$ given by matrices over R and $\overline{\underline{k}(R_{R\cap\mathfrak{m}})}$ resp. must have kernels generated by elements in R^n. Thus R must exhibit a sort of completeness in its relation to S, namely by providing 'ultimate' solutions for all linear equations of certain types. Clearly S, as a subring of itself, meets these demands. However, there is a simple way for obtaining a countable subring with the same property. It consists in forming the union of an increasing sequence of countable subrings so chosen that the linear equations arising from one ring have appropriate solutions in the next. The existence of such sequences is clear from the fact that each submodule of S^n is finitely generated $(n = 1,2,3,\ldots)$. The described procedure for complying with the demands of (1) - (3) forms the basis of our construction.

Concerning the property (4), which means that $R \cap \mathfrak{p} = (0)$ $(\mathfrak{p} \in \mathcal{P})$, let us begin by making a few observations. First, (4) holds for the prime ring of S, by (i). Secondly, simple adjunctions of the form $R \longrightarrow R[s]$ will preserve the validity of (4) (in the natural implicative sense) if the element s either represents a transcendental over R in each $S/\mathfrak{p}$ $(\mathfrak{p} \in \mathcal{P})$ or satisfies a relation of the form $as = b$ with $a,b \in R$, $a \neq 0$; let us refer to these two types of adjunctions as transcendental and fractional resp. Finally, the property of a ring extension to preserve the validity of (4) is 'transitive' with respect to arbitrary well-ordered towers, not only finite ones.

Put together, these facts allow us to conclude that (4) holds for any subring of S that can be obtained from the prime ring by a possibly transfinite succession of transcendental and fractional adjunctions. It suffices to show that our basic procedure can be performed within that framework. Thus it is enough to prove the following assertion: For any countable subring R of S enjoying property (4) and any linear equation over R or $\overline{\underline{k}(R_{R\cap\mathfrak{m}})}$ as considered above, it is possible to incorporate a S-linearly complete system of solutions by means of simple adjunctions of the two permitted types.

What resources in transcendentals do we have? It follows from the assumptions (notably (iii)) that S contains a complete discrete valuation ring which maps injectively into each of the rings $S/\mathfrak{p}$ $(\mathfrak{p} \in \mathcal{P})$.

Hence S contains an uncountable set of elements representing independent indeterminates in each $S/\mathfrak{p}$. It is true that these elements are algebraically independent only in the absolute sense, but, for given countable rings of constants, we can maintain the independence by omitting, if necessary, a countable subset of the original (uncountable) set.

Suppose then that R is a countable subring of S for which (4) holds and that $a_1,\dots,a_n$ are elements of R, say with $a_n \neq 0$. Let E be the submodule of S^n determined by the equation $a_1x_1 + \dots + a_nx_n = 0$. We must show the existence of a generating system for E which can be reached from R by permitted adjunctions. Such a system can be obtained by modifying an arbitrary (minimal) generating system in accordance with Nakayama's lemma. Remembering the abundance of transcendentals, and observing that the n-1 vectors $(a_n,0,\dots,0,-a_1)$, $(0,a_n,0,\dots,-a_2)$, ..., $(0,\dots,0,a_n,-a_{n-1})$ belong to E, we can find a generating system whose all n-1 first components can be adjoined by transcendental adjunctions. In view of the defining equation, the remaining n:th components can then be adjoined by fractional adjunctions.

A similar argument applies for equations over $\overline{\underline{k}(R_{R\cap\mathfrak{m}})}$. But here the possibilities of modification are larger. In fact, any element of $\mathfrak{m}^2S^n$ can be used as a modifying vector. This leads to adjunctions that are exclusively transcendental.

REFERENCES

[1] Akizuki, Y., Einige Bemerkungen über primäre Integritätsbereiche mit Teilerkettensatz. Proc. Phys.-Math. Soc. Japan 17, 327 - 336 (1935).

[2] Brodmann, M., Rotthaus, C., Local domains with bad sets of formal prime divisors. J. of Algebra 75, 386 - 394 (1982).

[3] Brodmann, M., Rotthaus, C., A peculiar unmixed domain. Proc. Amer. Math. Soc. 87, 596 - 600 (1983).

[4] Grothendieck, A., Éléments de géométrie algébrique, Chap. 0, § 10.3. Inst. Hautes Études Sci. Publ. Math. N°11 (1961).

[5] Larfeldt, T., Lech, C., Analytic ramifications and flat couples of local rings. Acta Math. 146, 201 - 208 (1981).

[6] Matsumura, H., Commutative algebra, 2nd ed. Benjamin/Cummings, Reading, Mass. 1980.

[7] Rotthaus, C., Nicht ausgezeichnete, universell japanische Ringe. Math. Z. 152, 107 - 125 (1977).

[8] Weil, A., Foundations of algebraic geometry. Amer. Math. Soc., New York 1946.

Department of Mathematics
University of Stockholm
Box 6701
S-113 85 STOCKHOLM
(SWEDEN)

YET ANOTHER PROOF OF A RESULT BY OGOMA

Crister Lech

As a further illustration of the ideas in [2] we shall give a summary proof of the following theorem of Ogoma ([3]).

<u>THEOREM</u> (Ogoma). There exists a non-catenary, normal, Noetherian, local domain of dimension 3.

The proof of Ogoma has been simplified by Heitmann ([1]). Both Ogoma and Heitmann apply a method of Rotthaus (cf [2]). Though different in this respect, our proof will have certain basic features in common with theirs.

<u>Proof</u>. We shall use the following notation:

k is a countable field, $S = k[[X,Z_1,Z_2,Z_3]]/(Z_1Z_2,Z_1Z_3)$;

x,z_1,z_2,z_3 are the natural images of X,Z_1,Z_2,Z_3 in S;

$\underline{M} = (x,z_1,z_2,z_3)$, $\underline{P} = (z_1,z_2,z_3)$, $\underline{P}_1 = (x,z_1)$, $\underline{P}_2 = (x,z_2,z_3) \in \mathrm{Spec}(S)$.

A local subring R of S will exhibit the truth of the assertion if it satisfies the conditions listed below.

(1) $k[x] \subseteq R$;

(2) $R \to S$ is flat and unramified;

(3) $\underline{P}_i = (R \cap \underline{P}_i)S \quad (i = 1,2)$;

(4) $R \cap \underline{P} = (0)$.

In fact, it follows from (1), (2) and (4) that R is a Noetherian local domain having S as completion. In particular, R has dimension 3. Putting, for $i = 1,2$, $\underline{p}_i = R \cap \underline{P}_i$, we have $\underline{P}_i = \underline{p}_i S$ by (3). We conclude that $ht(\underline{p}_2) = 1$ as the primary decomposition $xR = xS \cap R = \underline{P}_1 \cap \underline{P}_2 \cap R = \underline{p}_1 \cap \underline{p}_2$ must be irredundant, and further that $coht(\underline{p}_2) = coht(\underline{P}_2) = 1$. Hence R is non-catenary.

The only singular prime ideals in S are $\underline{M}$ and $\underline{P}$. In view of (4), it is therefore true for every $\underline{p} \in \mathrm{Spec}(R) - \{R \cap \underline{M}\}$ that $\underline{p}S$ has a regular minimal prime ideal. Thus every prime ideal in R is regular, except the maximal one. As $\mathrm{depth}(R) = \mathrm{depth}(S) = 2$, R is normal by the Serre criterion.

To show the existence of a ring R satisfying the conditions, we need only make slight changes in the proof of the implication '(i) - (iii) $\Longrightarrow$ (*)' in [2]. The construction should start with k[x] rather than the prime ring of S; the rôle of $\emptyset$ should be taken over by $\{\underline{P}\}$; generators of $\underline{P}_1$ and $\underline{P}_2$ should be incorporated in R. (The latter task can be performed by transcendental adjunctions, as each generator can be modified by a multiple of x^2.) This finishes our proof.

The method of Rotthaus has the advantage of providing rings R such that $R/\underline{p}$ has a very simple structure for all $\underline{p} \in \mathrm{Spec}(R) - \{(0)\}$. These factor rings are indeed essentially finitely generated over a field. As a consequence, the examples of Ogoma and Heitmann are pseudo-geometric. To attain this property, we could sharpen condition (3) by demanding that $\underline{Q} = (R \cap \underline{Q})S$ for those $\underline{Q} \in \mathrm{Spec}(S)$ that appear as minimal prime ideals of non-zero ideals in S generated by elements in R. It is not hard to see that this stronger condition could also be satisfied. Then we would have $\underline{p}S$ prime for $\underline{p} \in \mathrm{Spec}(R) - \{(0)\}$. If k is chosen to be of characteristic 0, this gives the result.

REFERENCES

[1] Heitmann, R., A non-catenary, normal, local domain. Rocky Mountain J. of Math. 12, 145 - 148 (1982).

[2] Lech, C., A method for constructing bad Noetherian local rings. These Proceedings.

[3] Ogoma, T., Non-catenary pseudo-geometric normal rings. Japan. J. of Math. 6, 147 - 163 (1980).

Department of Mathematics, University of Stockholm, Box 6701,
S-113 85 STOCKHOLM (SWEDEN)

MODELE MINIMAL RELATIF DES FEUILLETAGES

par

Daniel LEHMANN

1. *Introduction.*

Soit V une variété connexe paracompacte C^∞ (dimension n) munie d'un feuilletage $\mathcal{F}$ (codimension q). On notera

$$\iota_b : \Omega_b(\mathcal{F}) \to \Omega_{DR}(V)$$

l'inclusion naturelle de la sous-algèbre différentielle des formes ω basiques pour $\mathcal{F}$ (c'est-à-dire qui vérifient $i_X\omega = 0$ et $L_X\omega = 0$ pour tout champ de vecteurs X tangent aux feuilles de $\mathcal{F}$) dans l'algèbre de de Rham.

Un point de base x_o une fois choisi dans V permettant de munir $\Omega_b(\mathcal{F})$ et $\Omega_{DR}(V)$ d'une augmentation, ι_b admet un "modèle minimal relatif"

$$(\Omega_b(\mathcal{F}) \otimes \mathcal{M}, D, \phi)$$

rendant commutatif le diagramme suivant de morphismes de $\mathbb{R}$-algèbres différentielles graduées :

$$\begin{array}{ccccc}
\Omega_{DR}(\mathcal{F}) & & \xleftarrow{\quad \phi_{\mathcal{F}} \quad} & & (\mathcal{M}, d) \\
\Big\uparrow {\scriptstyle j_{\mathcal{F}}^*} & & & & \Big\uparrow \\
\Omega_{DR}(V) & & \xleftarrow[\sim]{\quad \phi \quad} & & (\Omega_b(\mathcal{F}) \otimes \mathcal{M}, D) \\
& \nwarrow {\scriptstyle \iota_b} & & \nearrow {\scriptstyle \iota} & \\
& & \Omega_b(\mathcal{F}) & &
\end{array}$$

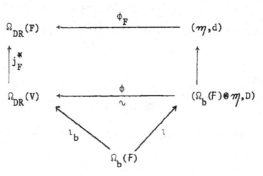

où $(\mathfrak{m},d)$ désigne une $\mathbb{R}$-algèbre minimale au sens de Sullivan [5] , $\Omega_b(F) \otimes \mathfrak{m}$ désigne le produit tensoriel gradué en tant qu'algèbre graduée, mais avec une différentielle D "tordue" $(\forall u \in \mathfrak{m})$, $D(1 \otimes u) - 1 \otimes d_{\mathfrak{m}} u$ appartient à l'idéal de $\Omega_b(F) \otimes \mathfrak{m}$ engendré par l'idéal maximal de $\Omega_b(F)$ relatif au point de base x_o, ϕ est un morphisme de $\Omega_b(F)$-algèbres différentielles graduées induisant un isomorphisme en cohomologie, $j_F : F \to V$ désigne l'inclusion naturelle dans V de la feuille F contenant x_o. Rappelons (cf. S. Halperin [5]) que $(\Omega_b(F) \otimes \mathfrak{m},D)$ est bien définie à $\Omega_b(F)$-isomorphisme près, et ϕ bien définie à $\Omega_b(F)$-homotopie près une fois $(\Omega_b(F) \otimes \mathfrak{m},D)$ fixé dans sa classe d'isomorphisme. D'autre part, V étant connexe, on a le

Lemme 1.

(i) $\Omega_b(F)$ est cohomologiquement connexe,

(ii) ι_b est injectif en cohomologie de dimension 1,

(iii) $\mathfrak{m}$ n'a pas de générateur en degré 0,

(iv) on peut choisir $(\Omega_b(F) \otimes \mathfrak{m},D)$ dans sa classe de $\Omega_b(F)$-isomorphisme de façon que, pour tout élément u de $\mathfrak{m}$, $D(1 \otimes u) - 1 \otimes d_{\mathfrak{m}} u$ appartienne à l'idéal $\Omega_b^+(F) \otimes \mathfrak{m}$ engendré par les formes basiques de degré > 0,

(v) la classe de $\Omega_b(F)$-isomorphie de $(\Omega_b(F) \otimes \mathfrak{m},D)$ ne dépend pas du point de base x_o.

Ce lemme sera démontré au § suivant.

Dans le cas où le feuilletage F est une fibration $f : V \to W$ localement triviale de base une variété W de dimension q, l'inclusion ι_b devient $f^* : \Omega_{DR}(W) \to \Omega_{DR}(V)$, et - au moins si $\pi_1(W)$ opère de façon nilpotente sur $H^*(F,\mathbb{R})$ et si $H^*(F,\mathbb{R})$ ou $H^*(W,\mathbb{R})$ est de dimension finie en chaque degré (si V est compacte par exemple) - $\phi_F : \mathfrak{m} \to \Omega_{DR}(F)$ induit un isomorphisme en cohomologie, de sorte que $(\mathfrak{m},d)$ est le modèle minimal réel de la fibre F : "le modèle de la fibre est égal à la fibre du modèle" (cf. S. Halperin [5]).

Dans le cas général d'un feuilletage, ϕ_F n'a en général plus aucune raison d'induire un isomorphisme en cohomologie, ne serait ce que parce que les feuilles F n'ont plus nécessairement toutes le même type d'homotopie.

On se propose de démontrer le :

Théorème :

Supposons :

(i) V est compacte, orientable, de dimension n,

(ii) la cohomologie $H_b^*(F)$ de l'algèbre $\Omega_b(F)$ vérifie la dualité de Poincaré pour la dimension q égale à la codimension de F, et chaque espace $H_b^i(F)$ est de dimension finie,

(iii) $H^*(\mathcal{M})$ est de dimension cohomologique finie $(H^i(\mathcal{M}) = 0$ pour i suffisamment grand).

Alors :

(i) $H^*(\mathcal{M})$ vérifie la dualité de Poincaré pour la dimension $p = n-q$ des feuilles de F,

(ii) Il existe une application naturelle <u>injective</u>

$$\phi_1 : H^*(\mathcal{M}) \to H^*(V,\underline{\Omega}_b^o)$$

où $H^*(V,\underline{\Omega}_b^o)$ désigne la cohomologie de V à coefficients dans le faisceau $\underline{\Omega}_b^o$ des germes de fonctions basiques.

2. *Rappels sur la suite spectrale d'un feuilletage* (*Vaisman* [10]).

Soit Q un supplémentaire dans $T(V)$ du sous-fibré vectoriel $\tau(F)$ des vecteurs tangents aux feuilles de F, on munit $\Omega_{DR}(V)$ d'une bigraduation en définissant $\Omega_{DR}^{r,s}(V) = \Omega^{r,s}$ comme l'espace des sections C^∞ du fibré $\Lambda^r Q^* \otimes \Lambda^s(\tau F)^*$, et de la filtration par des idéaux différentiels :

$$F_r\Omega_{DR}(V) = \bigoplus_{i \geq r} \Omega^{i,*} .$$

La suite spectrale $(E_i^{r,s},\partial_i)$ associée a son support dans le rectangle $(0 \leq r \leq q,\ 0 \leq s \leq p)$, converge vers $H^*(V,\mathbb{R})$, et $E_0^{r,s} = \Omega^{r,s}$. La restriction à $\Omega^{r,s}$ de la différentielle extérieure d_{DR} admet alors 3 composantes :

$$\Omega^{r,s} \xrightarrow{d_0} \Omega^{r,s+1}, \qquad \Omega^{r,s} \xrightarrow{d_1} \Omega^{r+1,s}, \qquad \Omega^{r,s} \xrightarrow{d_2} \Omega^{r+2,s-1}$$

et la relation $(d_{DR})^2 = 0$ implique en particulier :

$$\begin{cases} (d_o)^2 = 0\,, \\ d_1 d_o + d_o d_1 = 0\,, \\ (d_1)^2 + d_o d_2 + d_2 d_o = 0. \end{cases}$$

La différentielle ∂_o de E_o coïncide avec d_o, de sorte que $E_1^{r,s} = H^s(\Omega^{r,*}, d_o)$.

Notant $\underset{\sim}{\Omega}_b^r$ le faisceau des germes de r-formes basiques et $\underset{\sim}{\Omega}^{r,s}$ le faisceau des germes de section C^∞ de $\Lambda^r Q^* \otimes \Lambda^s(\tau(F))^*$, la suite

$$0 \to \underset{\sim}{\Omega}_b^r \longrightarrow \underset{\sim}{\Omega}^{r,o} \xrightarrow{d_o} \underset{\sim}{\Omega}^{r,1} \xrightarrow{d_o} \cdots$$

constitue une résolution fine de $\underset{\sim}{\Omega}_b^r$ (les faisceaux $\underset{\sim}{\Omega}^{r,s}$ sont fins dès que V est paracompacte, et le lemme de Poincaré dans chaque feuille de F, paramétré par l'espace des feuilles, implique l'exactitude de la suite) : le théorème de de Rham généralisé permet donc d'affirmer que $E_1^{r,s}$ est encore égal à $H^s(V, \underset{\sim}{\Omega}_b^r)$. En particulier, on a :

$$\begin{cases} E_1^{r,o} = \Omega_b^r(F) \\ E_1^{o,s} = H^s(V, \underset{\sim}{\Omega}_b^o)\,. \end{cases}$$

De la relation $(d_1)^2 + d_o d_2 + d_2 d_o = 0$, on déduit que d_1 induit une différentielle $[d_1]$ sur $E_1^{**} = H(\Omega^{**}, d_o)$, qui coïncide avec la différentielle ∂_1 de la suite spectrale, d'où $E_2^{*,*} = H^*(H(\Omega^{**}, d_o), [d_1])$. En particulier, $[d_1] = d_1$ sur $E_1^{r,o} = \Omega_b^r(F)$ de sorte que

$E_2^{*,o} = H_b^*(F)$ est la cohomologie des formes basiques.

3. *Démonstration du lemme 1.*

La suite exactes des termes de bas degré dans la suite spectrale précédente fournit

- un isomorphisme : $H_b^o(F) \cong H^o(V, \mathbb{R})$,

- une suite exacte :

$$0 \to H^1_b(F) \xrightarrow{\iota_b^*} H^1(V,\mathbb{R}) \to E_2^{o,1} \to H^2_b(F) \to H^2(V,\mathbb{R})$$

d'où la partie (i) du lemme, puisque V est connexe et la partie (ii). La partie (iii) est un corollaire de [(i) + (ii)].

Notons $\hat{\Omega}_b(F)$ la $\mathbb{R}$-sous-algèbre différentielle gradué de $\Omega_b(F)$ définie par :

$$\hat{\Omega}^o_b(F) = \mathbb{R}$$

$$\hat{\Omega}^1_b(F) = \text{supplémentaire de } B^1(\Omega_b(F)) \text{ dans } \Omega^1_b(F),$$

$$\hat{\Omega}^i_b(F) = \Omega^i_b(F) \quad \text{pour } i \geq 2.$$

Notons $a : \hat{\Omega}_b(F) \to \Omega_b(F)$ l'inclusion naturelle, qui induit un isomorphisme en cohomologie puisque $\Omega_b(F)$ est cohomologiquement connexe. Dans le modèle minimal relatif de $\iota_b \cdot a$, soit $\hat{\Omega}_b(F) \otimes \mathcal{M}$, il est clair que $D(1\otimes u) - 1 \otimes d_{\mathcal{M}}u$ appartient à $\hat{\Omega}^+_b(F) \otimes \mathcal{M}$. Puisque a induit un isomorphisme en cohomologie, on en déduit la construction d'un modèle minimal relatif de ι_b en faisant la somme amalgamée

$$\hat{\Omega}^+_b(F) \otimes \mathcal{M} \longleftarrow \hat{\Omega}^+_b(F) \xrightarrow{a} \Omega_b(F)$$

d'où la partie (iv) du lemme.

Puisque ι_b est injectif en cohomologie de dimension 1, il existe un supplémentaire $\hat{\Omega}^1_{DR}(V)$ de $B^1\Omega_{DR}(V)$ dans $\Omega^1_{DR}(V)$ tel que $\iota_b(\hat{\Omega}^1_b(F)) \subset \hat{\Omega}^1_{DR}(V)$. Si l'on définit donc

$$\hat{\Omega}^o_{DR}(V) = \mathbb{R} \quad \text{et} \quad \hat{\Omega}^i_{DR}(V) = \Omega^i_{DR}(V) \quad \text{pour } i \geq 2,$$

l'inclusion naturelle $a' : \hat{\Omega}_{DR}(V) \hookrightarrow \Omega_{DR}(V)$ induit un isomorphisme en cohomologie, et ι_b se factorise à travers $\hat{\iota}_b : \hat{\Omega}_b(F) \to \hat{\Omega}_{DR}(V)$ de façon que $a'.\hat{\iota}_b = \iota_b.a$. Le modèle minimal de $\hat{\iota}_b$

$$\begin{array}{ccc} \hat{\Omega}_b(F) \otimes \mathcal{M} & \xrightarrow{\Psi} & \hat{\Omega}_{DR}(V) \\ & \nwarrow \quad \nearrow \hat{\iota}_b & \\ & \hat{\Omega}_b(F) & \end{array}$$

ne faisant intervenir aucun point de base dans V, on en déduit une construction

du modèle minimal $\Omega_b(F) \otimes \mathcal{m} \xrightarrow{\phi} \Omega_{DR}(V)$ de ι_b tel que $a'.\Psi = \phi.(a \otimes 1_{\mathcal{m}})$,

ne dépendant pas du point de base x_o, d'où la partie (v).

4. *Démonstration du théorème.*

Gottlieb a démontré ($[3]$) que si, dans un espace fibré, la base et l'espace total vérifient tous deux la dualité de Poincaré en cohomologie avec dimensions respectives q et $n = p+q$, et si la fibre a une dimension cohomologique finie, alors cette fibre vérifie aussi la dualité de Poincaré en cohomologie pour la dimension p. La conclusion (i) du théorème est une transcription algébrique de ce résultat, qui généralise un théorème de Félix-Halperin $[2]$ (relatif au cas où la pseudo-homotopie de la "base" est de dimension 1) : elle se démontre (Thomas $[9]$) par récurrence sur la dimension de la pseudo-homotopie de la "base" ($\Omega_b(F)$ en l'occurrence), d'où la conclusion (i) du théorème.

La filtration

$$F_r(\Omega_b(F) \otimes \mathcal{m}) = \Omega_b^{\geq r} \otimes \mathcal{m}$$

définit une suite spectrale $\bar{E}_i^{r,s}$ de support inclus dans le rectangle ($0 \leqslant r \leqslant q$, $0 \leqslant s \leqslant p$) convergeant vers $H^*(V,\mathbb{R})$ tandis que ϕ induit un homomorphisme de suites spectrales multiplicatives $\phi_i^{r,s} : \bar{E}_i^{r,s} \to E_i^{r,s}$.

On a alors : $\bar{E}_1^{r,s} = \Omega_b^r(F) \otimes H^s(\mathcal{m})$, (et $\bar{E}_2^{r,s} = H_b^r(F) \otimes H^s(\mathcal{m})$ si $H_b^1(F) = 0$).

Notons $u \in Z^p(\mathcal{m})$ et $v \in \Omega_b^q(F)$ des éléments induisant des classes de cohomologie $[u] \neq 0 \in H^p(\mathcal{m})$ et $[v] \neq 0 \in H_b^q(F)$.

De la commutativité du diagramme

$$
\begin{array}{ccc}
\Omega_b^q(F) \otimes (\Omega_b^o(F) \otimes H^p(\mathcal{m})) & \xrightarrow{\phi_1^{q,o} \otimes \phi_1^{o,p}} & \Omega_b^q(F) \otimes H^p(V,\Omega_b^o(F)) \\
\downarrow & & \downarrow \\
\Omega_b^q(F) \otimes H^p(\mathcal{m}) & \xrightarrow{\phi_1^{q,p}} & H^p(V,\Omega_b^q(F)) \\
\downarrow & & \downarrow \\
\bar{E}_2^{q,p} & \xrightarrow{\phi_2^{q,p}} & E_2^{q,p} \\
\downarrow{\wr} & & \downarrow{\wr} \\
H^n(V,\mathbb{R}) & \xrightarrow[\simeq]{\phi_\infty^{q,p}} & H^n(V,\mathbb{R})
\end{array}
$$

et du fait que $[v].[u] \neq 0$ dans $\bar{E}_2^{q,p}$, on déduit

$$\phi_1^{o,p}(1 \otimes [u]) \neq 0 \in H^p(V,\underset{\sim}{\Omega_b^o}).$$

Puisque $H^*(\mathcal{m})$ vérifie la dualité de Poincaré avec $[u] \neq 0 \in H^p(\mathcal{m})$ et puisque l'application $H^*(\mathcal{m}) \to H^*(V,\underline{\Omega_b^o})$ induite par $\phi_1^{o,*}$ est un homomorphisme d'algèbres, on en déduit la conclusion (ii) du théorème.

5. *Remarques finales.*

1°) Supposons vérifiées simultanément les 2 conditions suivantes (c'est, par exemple, le cas si $\mathcal{F}$ est une fibration de Seifert généralisée, c'est-à-dire si $\mathcal{F}$ a toutes ses feuilles compactes et est localement stable [6]) :

(o) V est compacte orientable,

(i) $\mathcal{F}$ est minimalisable,

(ii) $\dim E_2^{o,p} = 1$.

D'après Kamber et Tondeur [6], $\Omega_b^*(\mathcal{F})$ vérifie automatiquement la dualité de Poincaré. Soit χ une p-forme sur V induisant sur chaque feuille F le volume associé à une métrique riemannienne sur V rendant F minimale. Puisque $d_o\chi = 0$, χ définit $[\chi]_o \in E_1^{o,p}$. D'après [6], $d_1\chi = 0$, χ définit donc $[\chi]_1 \in E_2^{o,p}$, $[\chi]_1 = 0 \in E_2^{o,p}$. Par conséquent, $\phi_1^{o,p}(1 \otimes [u])$ est proportionnel à $[\chi]_1$. L'inclusion $j_F : F \to V$ induit, pour toute feuille F un homomorphisme d'algèbres $j_F^* : H^*(V,\underset{\sim}{\Omega_b^o}) \to H^*(F,\mathbb{R})$ et $j_F^*([\chi]_o) \neq 0$ si F est compacte, puisque $j_F^*(\chi)$ est une forme volume. De la relation $j_F^* \circ \phi_1^{o,p} = \phi_F^*$, on déduit donc la

Proposition 1.-

Si V est compacte orientable, si $\mathcal{F}$ est minimalisable et si $\dim E_2^{o,p} = 1$, $\phi_F^* : H^*(\mathcal{m}) \to H^*(F,\mathbb{R})$ est injectif pour toute feuille F compacte.

2°) De façon plus triviale encore, mais souvent vérifiée en pratique, on a la

Proposition 2.-

Si $\mathcal{m}$ vérifie la dualité de Poincaré en dimension p, s'il est possible de choisir $u \in Z^p(\mathcal{M})$ dans $[u] \neq 0 \in H^p(\mathcal{M})$ et $\phi : \Omega_b(F) \otimes \mathcal{M} \to \Omega_{DR}(V)$ de telle façon que $\phi(1 \otimes u)$ induise une forme <u>volume</u> $\phi_F(u)$ pour une feuille compacte F particulière, alors :

$$\phi_F^* : H^*(\mathcal{M}) \to H^*(F,\mathbb{R})$$

est injectif pour cette feuille compacte.

3°) Lorsque les hypothèses du théorème sont vérifiées, l'application

$$\phi_1^{o,*} : \Omega_o^b \otimes H^*(\mathcal{M}) \to H^*(V,\underset{\sim}{\Omega_b^o})$$

n'est pas nécessairement surjective, comme le prouve l'exemple des droites de pente irrationnelle α sur le tore, chaque fois que α est un nombre de Liouville : Il est aisé de voir que, pour tout nombre α, rationnel ou non, $\mathcal{M} = (\Lambda_1(x), dx=0)$ est le modèle minimal de S^1. Si α est irrationnel, $\Omega_b^o(F)$ ne contient que les constantes tandis que $\Omega_b^1(F)$ ne contient que les formes $\mathbb{R}$-proportionnelles à la forme fermée $dx-\alpha\, dy$ définissant F. Si α est un nombre de Liouville, El Kacimi a alors démontré que $H^*(T,\underset{\sim}{\Omega_b^o})$ a une dimension infinie [1].

4°) On pourrait chercher à interpréter $\mathcal{M}$ comme le modèle de la feuille générique (revêtement commun à toutes les feuilles) lorsque celle-ci existe (cf. Haefliger [4]) : il n'en est rien, puisque pour les droites de pente irrationnelle sur le tore, la feuille générique est $\mathbb{R}$, alors que $\mathcal{M}$ est le modèle de S^1.

5°) Les résultats exposés sont également valables si, au lieu de prendre toutes les formes basiques du feuilletage, on considère seulement celles appartenant à une certaine sous-algèbre différentielle graduée de $\Omega_b^*(F)$, généralement associée à certains types de Γ-structures transverses.

Le modèle $\mathcal{M}$ dépend alors de la structure en question. De plus, à l'étude du modèle minimal relatif, se greffent des questions d'irrationnalité de morphismes entre $\mathbb{R}$-algèbres différentielles graduées admettant des $\mathbb{Q}$-structures données, qui mesurent, en quelque sorte, en quoi ces feuilletages diffèrent de fibrations d'où une théorie de "l'homotopie irrationnelle" (cf. [7]).

R E F E R E N C E S

[1] EL KACIMI-ALAOUI A. - *Cohomologie feuilletée - Exemples de calculs,* Thèse de 3ème cycle, Université de Lille I, 26 juin 1980.

[2] FELIX Y. - HALPERIN S. - L.S.-*category,* (Transactions of American Mathematical Society, 1983).

[3] GOTTLIEB D. - *Poincaré duality and fibrations,* (Proceedings of the American Mathematical Society 76.1.79)

[4] HAEFLIGER A. - *Groupoïdes d'holonomie et classifiants,* (A paraître dans les comptes rendus des journées de Toulouse 1982 sur la géométrie transverse. Astérisque).

[5] HALPERIN S. - *Lectures on minimal models,* Pub. I.R.M.A. Lille I, Vol. 3, fasc. 4, 1977.

[6] KAMBER F. et TONDEUR P. - *Foliations and metrics,* (Differential Geometry) Birkhaüser - 1983).

[7] LEHMANN D. - *Structures de Maurer-Cartan et* Γ_θ*-structures,*
I - feuilletages de Maurer-Cartan (Preprint).
II - espaces classifiants, Astérisque, 116, 1984, 134-148.

[8] RUMMLER D. et SULLIVAN D. - *Currents, flows and diffeomorphisms,* (Topology 14 - 1975).

[9] THOMAS J.C. - Communication privée.

[10] VAISMAN I. - *Variétés riemanniennes feuilletées,* (Czechosl. Math. Jal 21 - 1971).

E.R.A au C.N.R.S. 07 590
UNIVERSITE DES SCIENCES ET TECHNIQUES DE LILLE
U.E.R. DE MATHEMATIQUES PURES ET APPLIQUEES
59655 - VILLENEUVE D'ASCQ CEDEX (France)

LUSTERNIK - SCHNIRELMANN CATEGORY : AN INTRODUCTION

Jean-Michel LEMAIRE (Nice)

The introduction of L.- S. category into rational homotopy theory has lead to important results, due to Yves Félix and Steven Halperin, and also Jean-Claude Thomas, on the structure of the rational homotopy Lie algebra of a finite complex. According to the leading theme of this conference, such results are of interest in local algebra as well. The following notes are meant to serve as an introduction to Félix's and Halperin's fundamental paper [FH] : they therefore contain no original material, but some seasoning of geometry which may make reading [FH] easier.

They also provide an opportunity to thank Jan-Erik Roos for his kind invitation to give this set of lectures in the Nordic Summer School and Research Symposium 1983, which was a most pleasant and profitable one.

§ 1. - BASIC DEFINITIONS AND ELEMENTARY PROPERTIES.

Definition 1.1. Let X be a topological space with base point $* \in X$. A subset $A \subset X$ is categorical if there exists a continuous map $h : X \longrightarrow X$, homotopic to the identity, such that $h(A) = *$.

Definition 1.2. If X admits a finite covering by categorical subsets, one defines cat X, the L. - S. category of X, to be the least number n such that there exists a covering of X by $n+1$ categorical subsets. If no finite categorical covering exists, one sets cat $X = \infty$.

1.3 Examples : cat $X = 0$ iff X is contractible.

cat $S^n = 1$ for all $n \geq 1$.

This definition originates in the work (1934) of the two eponymous authors, who proved that any smooth function on a compact manifold of category n admits at least $n + 1$ critical points. Actually their definition required the categorical sets to be closed, but only contractible in X (i.e. $A \hookrightarrow X$ must be null-homotopic, but the homotopy need not extend to X). Later Fox (1941) modified the definition by requiring the sets to be open, and proved that category so defined is a homotopy invariant, while it is not if one insists that the sets be contractible in themselves.

The definition adopted here is due to George Whitehead (1954) and can be shown to be equivalent to Fox's for cw-complexes.

One can reformulate this definition as follows : let $T_1^n X \subset X^{n+1}$ be the subspace of the $(n + 1)$-fold product which consists of those sequences $(x_0, x_1, \dots, x_n)$ such that $x_i = *$ for some i. Clearly the following definition is equivalent to (1.2).

<u>Definition (1.4)</u>. The category $\operatorname{cat} X$ of a space X is $\leq n$ iff the $(n + 1)$-fold diagonal $\Delta : X \longrightarrow X^{n+1}$ factors through $T_1^n X$ up to homotopy

$$\begin{array}{ccc} X & \xrightarrow{\ \Delta\ } & X^{n+1} \\ & \searrow{\scriptstyle h} \ \cdot & \uparrow \cup \\ & & T_1^n X \end{array}$$

G. Whitehead made the following crucial observation : let SX be the suspension of X, and ΩY be the loop space on Y. Recall that the set of homotopy classes of base-point preserving maps $[SX, Y] = [X, \Omega Y]$ is a group. Then :

<u>Theorem (1.5)</u> ([W]). If $\operatorname{cat} X \leq n$, the group $[SX, Y]$ is nilpotent of class $\leq n$.

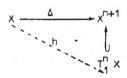

This result shows that category is some kind of "homotopical nilpotency". We shall see that the rational category of a 1-connected space actually is the "homotopical nilpotency" of its Sullivan minimal model. Before we briefly collect some elementary topological facts. From now on we assume that spaces in sight have the homotopy type of cw-complexes of finite type, with a base point. Details can be found in [W].

<u>Lemma 1.6</u> : If X is a homotopy retract of Y, then $\text{cat}\, X \leq \text{cat}\, Y$. In particular, cat is a homotopy invariant. ∎

<u>Lemma 1.7</u> : Let R be a ring, and $u_0, u_1, \ldots, u_n \in \widetilde{H}^*(X\,;\,R)$. If $\text{cat}\, X \leq n$, then $u_0 \cup u_1 \cup \ldots \cup u_n = 0$.

<u>Proof</u> : Since $u_i \in H^*(X, *\,;\,R)$, the cross-product $u_0 \times u_1 \times \ldots \times u_n$ lies in $H^*(X^{n+1}, T_1^n X\,;\,R)$. Let $k : (X^{n+1}, \emptyset) \subset (X^{n+1}, T_1^n)$ be the inclusion (of pairs). Then

$$u_0 \cup u_1 \cup \ldots \cup u_n = \Delta^* k^*(u_0 \times u_1 \times \ldots \times u_n)$$

But if $\text{cat}\, X \leq n$, there exists h with $jh \simeq \Delta$, where $j : T_1^n X \hookrightarrow X$ is the inclusion. Now $\Delta^* k^* = h^* j^* k^*$ and $j^* k^* = 0$ in the long exact sequence of the pair (X^{n+1}, T_1^n). ∎

<u>Lemma 1.8</u>. Let $f : X \longrightarrow Y$ be a map, and $Cf = Y \cup_f CX$ be the mapping cone of f. One has

$$\text{cat}\,(Cf) \leq \text{cat}\, X + 1 .$$

<u>Proof</u> : By (1.6) one may assume that f is a closed cofibration. Then $Cf = Y \cup CX$ and both Y and CX have the homotopy extension property in Cf. Then one easily sees that if $A_0, A_1, \ldots, A_u$ is a categorical covering of Y, then $A_0, A_1, \ldots, A_n, CX$ is a categorical covering of $Y \cup CX$. ∎

<u>Application 1.9</u> : $\text{cat}\, \mathbb{C}P(n) = n$.

<u>Proof</u> : $\mathbb{C}P(n)$ is the mapping cone of the Hopf map $S^{2n-1} \longrightarrow \mathbb{C}P(n-1)$, therefore $\text{cat}\, \mathbb{C}P(n) \leq n$ by (1.8) and induction on n. Now if $c \in H^2(\mathbb{C}P(n)\,;\,\mathbb{Z})$ is the Chern class of the canonical line bundle, one has $c^n \neq 0$ and $\text{cat}\, \mathbb{C}P(n) \geq n$

by (1.7). ∎

One may prove along the same lines that the category of a product of n spheres is n. Incidentally, from (1.5) on we have seen several good reasons to set $\mathrm{cat}\, X \le n$ if X can be covered by $n+1$ categorical subsets, instead of $\mathrm{cat}\, X \le n+1$ which was the original convention.

<u>Lemma (1.10)</u>. Let $F \xrightarrow{i} E \xrightarrow{p} B$ be a fibration sequence such that i is null-homotopic. Then $\mathrm{cat}\; E \le \mathrm{cat}\; B$.

<u>Proof</u> : Plainly it suffices to prove that $p^{-1}(A)$ is categorical in E if A is categorical in B. Let $H : B \times I \longrightarrow B$ be a homotopy with $H_0 = id_B$ and $H_1(A) = *$. By the homotopy lifting property of p, there exists $\widetilde{H} : E \times I \longrightarrow E$ with $\widetilde{H}_0 = id_E$ and $p\widetilde{H} = H(p \times I)$. Hence $\widetilde{H}_1(p^{-1}(A)) \subset F$. W.n.l.g. one may assume that F has the h.e.p. in E, and thus is categorical in E: let $k : E \longrightarrow E$ be such that $k \sim id_E$ and $k(F) = * \in E$. Then $k.\widetilde{H}_1(p^{-1}(A)) = *$ and $k.\widetilde{H}_1 \sim id_E$. ∎

Observe that if i is null-homotopic, $\pi_*(p) : \pi_*(E) \longrightarrow \pi_*(B)$ is injective by exactness of the homotopy exact sequence. The converse is not true in general, but it is true for rational spaces : this is [FH]'s first theorem, the mapping theorem, which we now discuss.

§ 2. - <u>THE MAPPING THEOREM</u>

We begin with recollecting some rational homotopy theory.

Let S be a 1-connected space. The Hurewicz homomorphism

$$h : \pi_*(\Omega S) \longrightarrow H_*(\Omega S\, ;\, \mathbb{Z})$$

is a Lie algebra map, where the Lie structure is given on $\pi_*(\Omega S)$ by the Samelson bracket, and on $H_*(\Omega S)$ by $[a, b] = ab - (-1)^{|a||b|}ba$. A fundamental result of Milnor and Moore asserts that h induces an isomorphism of Lie algebras

$$(2.1) \qquad h_* : \pi_*(\Omega S) \otimes \mathbb{Q} \xrightarrow{\cong} PH_*(\Omega S\, ;\, \mathbb{Q})$$

where P stands for the (Lie subalgebra of) primitive elements.

We can choose a retraction of h_i, for each i :

$$r_i : H_i(\Omega S ; \mathbb{Q}) \longrightarrow \pi_i(\Omega S) \otimes \mathbb{Q}$$

which can be viewed as a class in $H^i(\Omega S ; \pi_i(\Omega S) \otimes \mathbb{Q})$, or as a map

$$r_i : \Omega S \longrightarrow K(\pi_i(\Omega S) \otimes \mathbb{Q}, i)$$

The latter induces an isomorphism on $\pi_i \otimes \mathbb{Q}$. Thus the product map

(2.2) $$r : \Omega S \longrightarrow \prod_{i=1}^{\infty} K(\pi_i(\Omega S \otimes \mathbb{Q}, i)$$

where the product is given the weak topology, induces an isomorphism on rational homotopy groups and therefore is a rational equivalence. Thus any loop space on a 1-connected space has the rational homotopy type of a product of Eilenberg-Mac Lane spaces.

We can now prove the mapping theorem : the following simple proof was found by Yves Félix during the conference ; for a generalization, see [FL].

Theorem 2.3. Let $p : E \longrightarrow B$ be a map between simply connected rational spaces, such that $\pi_*(p)$ is injective. Then $\mathrm{cat}(E) \leq \mathrm{cat}(B)$.

Proof : Recall that a rational space is a space whose homotopy groups are $\mathbb{Q}$-vector spaces. By 1.10, it suffices to prove that the homotopy fibre $i : F \longrightarrow E$ of p is null-homotopic. This will follow from the existence of a section s of the fibre $j : \Omega B \longrightarrow F$ in the fibration sequence for p : indeed $i \sim ijs$ and $ij \sim *$, hence $i \sim *$. To construct s, we observe that $\pi_*(j) : \pi_*(\Omega B) \longrightarrow \pi_*(F)$ is a surjective map of rational vector spaces, by the exactness of the homotopy sequence. Write

$$\pi_*(\Omega B) = U_* \oplus \mathrm{Ker}\, \pi_*(j)$$

so that $\pi_*(j)$ maps U_* isomorphically on to $\pi_*(F)$. Then, by 2.2,

$$\Omega B \sim \prod_{i\geq 1} K(U_i, i) \times \prod_{i\geq 1} K(\mathrm{Ker}\, \pi_i(j), i)$$

and the restriction of j to $\prod_{i\geq 1} K(U_i, i)$ is a homotopy equivalence, whose inverse is the required section s. ■

We shall see in the next section that L. - S. category "localizes" well, that is, if X_0 is the localisation of the homotopy type of X at all primes, then $\mathrm{cat}(X_0) \leq \mathrm{cat}(X)$. Setting $\mathrm{cat}_0(X)$, the rational category of X, to be $\mathrm{cat}(X_0)$, we can reformulate the mapping theorem as follows :

> (1.11') : Let $p : E \longrightarrow B$ be a map such that $\pi_*(p) \otimes \mathbb{Q}$ is injective and the spaces are 1-connected. Then $\mathrm{cat}_0(E) \leq \mathrm{cat}_0(B)$.

We conclude this section with another result of [FH], which we derive from the mapping theorem. The following concept, due to D. Gottlieb, was brought to the attention of rational homotopy theorists by H. Baues :

> Definition (2.4). The Gottlieb group $G_i(X)$ is the subgroup of $\pi_i(X)$ defined as follows : $\alpha : S^i \longrightarrow X$ represents an element in $G_i(X)$ if the map
>
> $$(\mathrm{id}, \alpha) : X \vee S^i \longrightarrow X$$
>
> extends to $X \times S^i$.

We leave to the reader to check that $G_i(X)$ actually is a subgroup of $\pi_i(X)$: indeed $G_i(X)$ is the image of $\pi_i(ev)$, where $ev : \varepsilon_0(X) \longrightarrow X$ is the evaluation at the base point and $\varepsilon_0(X)$ is the group of self-maps of X homotopic to the identity.

> Theorem (2.5) ([FH] Thm III). Let X be a 1-connected space of finite category n. Then
>
> (a) $\forall\, i$, $G_{2i}(X) \otimes \mathbb{Q} = 0$
>
> (b) $\sum_{i=1}^{\infty} \dim_{\mathbb{Q}} G_{2i+1}(X) \otimes \mathbb{Q} \leq n$.

In other words, the groups $G_i(X)$ are torsion except at most n of them which occur in odd dimensions : moreover the sum of their ranks is at most n .

Proof of 2.5. Let us prove (b) first. Let $\alpha_i : S^{2r_i+1} \longrightarrow X$, $i = 1,\dots,s$ represent linearly independant elements in $G_{odd}(X) \otimes \mathbb{Q}$. An easy induction on s shows that

$$\alpha = (\alpha_1,\dots,\alpha_s) : \bigvee_{i=1}^{s} S^{2r_i+1} \longrightarrow X$$

extends to the product $\prod_{i=1}^{s} S^{2r_i+1}$. Indeed, assume that $(\alpha_1,\dots,\alpha_j)$ can be extended to $\check{\alpha}_j : \left(\prod_{i=1}^{j-1} S^{2r_i+1}\right) \vee S^{2r_j+1} \longrightarrow X$. Then a further extension to $\prod^{j} S^{2r_i+1}$ is given by the diagram

$$\begin{array}{ccccc} \left(\prod^{j-1} S^{2r_i+1}\right) \vee S^{2r_j+1} & \xrightarrow{\check{\alpha}_j \vee id} & X \vee S^{2r_j+1} & \xrightarrow{(id,\,\alpha_j)} & X \\ \big\downarrow & & \big\downarrow & \nearrow & \\ \prod^{j} S^{2r_i+1} & \xrightarrow{\check{\alpha}_j \times id} & X \times S^{2r_j+1} & & \end{array}$$

Localizing we obtain a map of rational spaces

$$\prod_{i=1}^{s} S_o^{2r_i+1} \longrightarrow X_o$$

which is injective on homotopy groups. By 1.9, the category of a product of s (rational) spheres is s , therefore $s \leq cat(X_o) = cat_o(X) \leq n$ by the mapping theorem.

A proof of (a) along the same lines is a little more involved : we need James's reduced product construction. Let Z be a connected, pointed cw-complex ; the reduced product $(Z)_\infty$ is the quotient space

$$\coprod_{i=0}^{\infty} Z^i \,/\, ((z_1,\dots,z_{j-1}, *, z_{j+1},\dots,z_i) \sim (z_1,\dots,z_{j-1}, z_j,\dots,z_i))$$

Concatenation gives $(Z)_\infty$ the structure of a topological monoid with unit $* \sim (\ )$. James's theorem asserts that the canonical map $Z \longrightarrow \Omega SZ$ extends to a multiplicative homotopy equivalence $(Z)_\infty \xrightarrow{\simeq} \Omega SZ$. Now if $\alpha : S^r \longrightarrow X$

represents an element in $G_r(X)$, one easily sees that α extends to $\bar{\alpha} : (S^r)_\infty \longrightarrow X$. Localizing we obtain a map

$$\bar{\alpha}_0 : [(S^r)_\infty]_0 \simeq \Omega S_0^{r+1} \longrightarrow X_0$$

If $r = 2s$, $\Omega S^{2s+1} \sim K(\mathbb{Q}, 2s)$ and $\bar{\alpha}_0$ is injective on homotopy groups iff α represents a non-zero element in $G_{2s}(X) \otimes \mathbb{Q} = G_{2s}(X_0)$. But $H^*(K(\mathbb{Q}, 2s) ; \mathbb{Q})$ is a polynomial algebra on one generator of degree $2s$ and therefore $cat(K(\mathbb{Q}, 2s)) = \infty$ by lemma 1.7. Thus the existence of a non-zero element in $G_{2s}(X) \otimes \mathbb{Q}$ would contradict the mapping theorem. ∎

Theorem 2.5 is a key ingredient in the proof of [FH] Thm. IV, which says that the rational homotopy groups of a finite complex are either zero for large enough degrees or grow exponentially.

§ 3. - SPACES OF CATEGORY 1 AND THE HOMOTOPY SUSPENSION DIAGRAM

By definition 1.4 , a space has category ≤ 1 if the diagonal can be factored through the wedge up to homotopy

$$
\begin{array}{ccc}
X & \xrightarrow{\Delta} & X \times X \\
 & \searrow^{h} & \downarrow \\
 & & X \vee X
\end{array}
$$

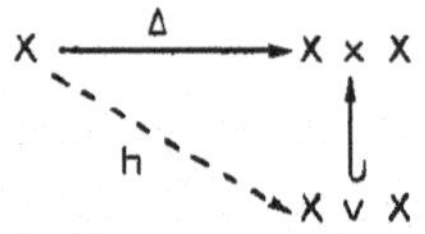

The map h defines a natural composition law $*$ on $[X, Y]$ by $f * g = (f,g) \circ h$

$$X \xrightarrow{h} X \vee X \xrightarrow{(f,g)} Y$$

which admits the trivial map as unit.

One says that h is a co-H.space structure on X . Observe that Whitehead's theorem (1.5) says that $[SX, Y]$ is abelian if X is a co-H.space : incidentally this is why $\pi_n(X)$ is abelian for $n \geq 2$.

Suspensions are canonical examples of co-H.spaces, with the "pinching" map as structure map ; but there are examples of co-H.spaces which do not have the homotopy type of suspensions : the simplest example is $S^3 \cup_\alpha e^7$, where $\alpha \in \pi_6(S^3) = \mathbb{Z}/12\mathbb{Z}$: one may show, using the Hopf invariant, that this space admits a co-H.space structure (which extends the standard one on S^3) iff $\alpha \equiv 0(2)$, but is a suspension only if $\alpha \equiv 0(6)$. This is a torsion phenomenon of course, and over the rationals things are much simpler :

<u>Theorem (3.1) (I. Bernstein)</u>. Every simply connected space of category 1 has the rational homotopy type of a wedge of spheres.

This result will follow as an easy exercice from the characterization of rational category on the Sullivan model that we will discuss in the last section. We now give a proof which avoids models because it leads to interesting side comments. We need the

<u>Bott-Samelson theorem (3.2)</u>. The Pontryagin algebra $H_*(\Omega\, SX\ ;\ k)$, k a field, is isomorphic to the tensor algebra generated by the graded vector space $\widetilde{H}(X\ ;\ k)$.

We can now prove (3.1) for suspensions : it follows from the Milnor-Moore theorem (2.1) and the Bott-Samelson theorem that the Lie algebra $\pi_*(\Omega\, SX) \otimes \mathbb{Q}$ is isomorphic to the free Lie algebra generated by $\widetilde{H}_*(X\ ;\ \mathbb{Q})$; we choose a basis (x_α) of $\widetilde{H}_*(X\ ;\ \mathbb{Q})$, and representatives

$$x_\alpha : S^{|x_\alpha|} \longrightarrow \Omega\, SX .$$

The family of maps (x_α) defines a map

$$x : \bigvee_\alpha S^{|x_\alpha|} \longrightarrow \Omega\, SX$$

whose adjoint $\widetilde{x} : \bigvee_\alpha S^{|x_\alpha|+1} \longrightarrow SX$ induces an isomorphism on rational homotopy groups : since both spaces are simply connected, this is a rational homotopy equivalence. ■

We now introduce the "homology suspension diagram"

$$(3.3)\qquad \begin{array}{ccccc} \Omega X * \Omega X & \xrightarrow{H} & S\Omega X & \xrightarrow{\alpha} & X \\ \| & & \downarrow \nabla & & \downarrow \Delta \\ \Omega X * \Omega X & \xrightarrow{\nabla \circ H} & X \vee X & \xhookrightarrow{j} & X \times X \end{array}$$

in which

α is the evaluation map $\alpha(t, \lambda) = \lambda(t)$

Δ is the diagonal

j is the inclusion

H is the Hopf map $H(\lambda, t, \mu) = (t, \lambda.\mu)$

$\nabla = (\alpha \vee \alpha) \circ \Psi$ where $\Psi : S\Omega X \longrightarrow S\Omega X \vee S\Omega X$ is the pinch map.

<u>Theorem (3.4)</u>. The diagram (3.3) is homotopy commutative, the rows are fibration sequences and the right-hand square is homotopy-cartesian.

By homotopy-cartesian, we mean that if one replaces Δ or j (or both) by a fibration, the pullback square is homotopy equivalent to the given square.

<u>Sketch of proof of 3.4</u>

(a) the right hand square is homotopy commutative

$$\Delta\alpha(t, \omega) = (\omega(t), \omega(t))$$

$$j\nabla(t, \omega) = \begin{cases} (\omega(2t), \omega(0)) & t \leq \frac{1}{2} \\ (\omega(1), \omega(2t-1)) & t \geq \frac{1}{2} \end{cases}$$

The required homotopy is a "simplicial approximation" of the diagonal in $I \times I$

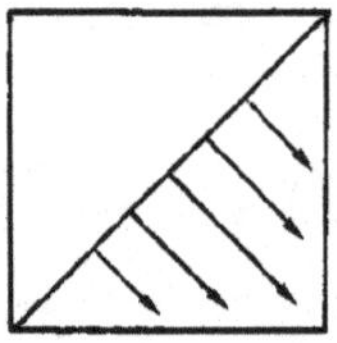

(b) the right-hand square is homotopy catesian : a standard way to replace Δ by a fibration is to consider the evaluation map at the ends

$$X^{[0,1]} \longrightarrow X \times X$$
$$\lambda \longmapsto \lambda(0),\ \lambda(1)\ .$$

Composition with the inclusion of constant paths $\varepsilon : X \longrightarrow X^{[0,1]}$ is Δ, and ε is a homotopy equivalence. Now the pullback is $E^-X \cup_{\Omega X} E^+X$, that is, the set of those paths in X which start or end at the base point. Since E^-X and E^+X are contractible, it is not hard to construct a homotopy equivalence

$$S\Omega X \xrightarrow{\ \sim\ } E^-X \cup_{\Omega X} E^+X$$

(c) it remains to show that the homotopy fibre of j has the homotopy type of the join $\Omega X * \Omega X$: one may consider the fibre square

$$\begin{array}{ccc} E^-X \times \Omega X \cup_{\Omega X \times \Omega X} \Omega X \times E^+X & \longrightarrow & E^-X \times E^+X \\ \downarrow & & \downarrow {\scriptstyle (\lambda,\ \mu)\ \mapsto\ (\lambda(1),\ \mu(0))} \\ X \vee X & \xhookrightarrow{\ j\ } & X \times X \end{array}$$

in which $E^-X \times E^+X$ is contractible, and construct a weak equivalence

$$\Omega X * \Omega X \longrightarrow E^-X \times \Omega X \cup \Omega X \times E^+X\ .$$

We leave the details to the reader (see [Gi]). ■

Now, in the homotopy cartesian square

$$\begin{array}{ccc} S\Omega X & \underset{\sigma}{\overset{\alpha}{\rightleftarrows}} & X \\ \downarrow \nabla & \overset{h}{\swarrow} & \downarrow \Delta \\ X \vee X & \xrightarrow{\ j\ } & X \times X \end{array} \qquad (3.5)$$

the existence of h is equivalent to the existence of a homotopy section σ of α. Thus :

<u>Proposition 3.6</u>. A space has category ≤ 1 iff it is a homotopy retract of a suspension. ■

The proof of (3.1) is achieved if we observe that a retract of a free Lie algebra is free - in fact, any subalgebra of a free Lie algebra is free.

(3.7) Remarks : we call diagram (3.3) the homology suspension diagram because the Serre exact homology sequence for the fibration $\Omega X * \Omega X \longrightarrow S\Omega X \longrightarrow X$ is G. Whitehead's exact sequence for the homology suspension (see [W]).

On the other hand, if we apply the functor $\underline{\pi} = \pi_*(\Omega\ .) \otimes \mathbb{Q}$ to the whole diagram (3.3), setting $L = \underline{\pi}(X)$ we have

$$\underline{\pi}(S\Omega X) = \mathbb{L}(\overline{UL})$$

where $\overline{UL}$ denotes the augmentation ideal of the enveloping algebra of L, and $\mathbb{L}$ is the free Lie algebra functor from vector spaces to Lie algebras, and

$$\underline{\pi}(\Omega X * \Omega X) = \underline{\pi}(S(\Omega X \wedge \Omega X)) = \mathbb{L}(\overline{UL} \otimes \overline{UL})$$

Finally, the map

$$\underline{\pi}(\alpha) : \underline{\pi}(S\Omega X) \longrightarrow \underline{\pi}(X)$$

is surjective, because $\Omega\alpha$ has a section $\beta_{\Omega X} : \Omega X \longrightarrow \Omega S(\Omega X)$. We therefore get a diagram with exact rows

(3.8)
$$\begin{array}{ccccccccc}
0 & \longrightarrow & \mathbb{L}(\overline{UL} \otimes \overline{UL}) & \longrightarrow & \mathbb{L}(\overline{UL}) & \longrightarrow & L & \longrightarrow & 0 \\
 & & \| & & \downarrow & & \downarrow \Delta & & \\
0 & \longrightarrow & \mathbb{L}(\overline{UL} \otimes \overline{UL}) & \longrightarrow & L \amalg L & \xrightarrow{j} & L \times L & \longrightarrow & 0
\end{array}$$

in which the right-hand square is a pullback. Since by Quillen's theorem any graded Lie algebra over $\mathbb{Q}$ occurs as $\underline{\pi}(X)$ for some X, we can conclude

Proposition 3.9. Let L be a graded connected Lie algebra over $\mathbb{Q}$, and

$$j : L \amalg L \longrightarrow L \times L$$

be the canonical map, represented by the unit matrix. Then the Lie subalgebras $\operatorname{Ker} j$ and $j^{-1}(\Delta(L))$ of $L \amalg L$ are free with minimal generating vector spaces isomorphic to $\overline{UL} \otimes \overline{UL}$ and $\overline{UL}$ respectively. ∎

Let us conclude this section with observing that the analogous statement holds for discrete groups – and can be derived from Gruschko's theorem : as a consequence, the fundamental group of any co-H.space (connected) is free.

§ 4. – THE GENERALIZED SUSPENSION DIAGRAM, THE GANEA FILTRATION, AND L.-S. CATEGORY FOR DG ALGEBRAS.

We wish to generalize the homotopy cartesian square (3.5), to get

$$\begin{array}{ccc} X(n) & \xrightarrow{\alpha_n} & X \\ \downarrow{\scriptstyle \nabla_n} & & \downarrow{\scriptstyle \Delta} \\ T_1^n X & \xhookrightarrow{\ j\ } & X^{n+1} \end{array}$$

To achieve this, we may replace either Δ or j by a fibration and take the pull-back $X(n)$, whose homotopy type is then well-defined. We will then have, by general homotopy theoretic nonsense :

(4.1) cat $X \leq n$ iff X is a homotopy retract of $X(n)$.

Of course, we must try to describe $X(n)$ to make (4.1) of any significance. Again we may first replace Δ by the evaluation map at integral points

$$e : X^{[0,n]} \longrightarrow X^{n+1}$$
$$\lambda \longmapsto (\lambda(0), \lambda(1), \ldots, \lambda(n))$$

which is a fibration. Then

$$X(n) = \bigcup_{i=0}^{n} \{\lambda \in X^{[0,n]} \mid \lambda(i) = *\}$$

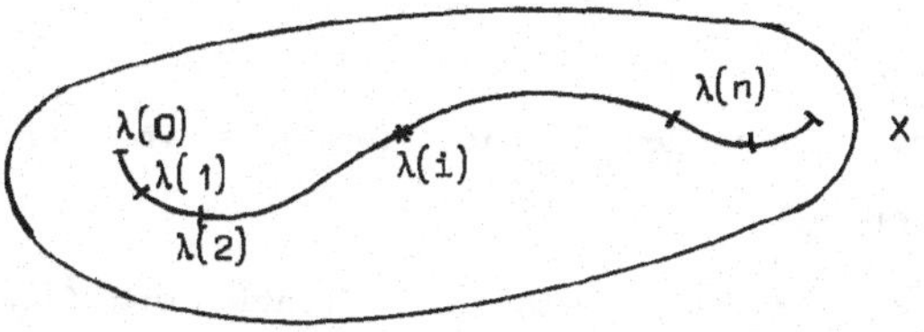

Note that each piece $\{\lambda \mid \lambda(i) = *\}$ is contractible : thus $X(n)$ comes equipped with a standard categorical covering.

<u>Exercise</u> : $\operatorname{cat}(X(\operatorname{cat} X)) = \operatorname{cat} X$.

Moreover the intersection of two pieces has the homotopy type of ΩX . There is a more-than-formal analogy between the structure of $X(n)$ and that of a projective space, say $\mathbb{C}P(n)$, together with its covering by $(n+1)$ affine spaces given by homogenous coordinates : here the intersection of two affine charts admits S^1 as a deformation retract. Indeed $X(n)$ is the n-th projective space of the homotopy associative H.space ΩX , and on the other hand one checks that $\mathbb{C}P(n) \sim K(\mathbb{Z},2)(n)$, with $\Omega K(\mathbb{Z},2) = S^1$.

Another approach to $X(n)$, due to W. Gilbert, is to convert j into a fibration through a construction inspired by the Whitney sum of vector bundles. Let $p_i : E_i \longrightarrow B$, $i = 1,2$ be two fibrations with fibers F_i , and let $\tilde{p}_i : Z_i \longrightarrow B$ be the projection of the mapping cylinder of p_i on B . Let

$$p_1 \circledast p_2 : Z_1 \times E_2 \cup_{E_1 \times E_2} E_1 \times Z_2 = E_1 \circledast E_2 \longrightarrow B \times B$$

be the restriction of $\tilde{p}_1 \times \tilde{p}_2$, and

$$p_1 \# p_2 : E_1 \# E_2 \longrightarrow B$$

be defined by the pullback square

$$\begin{array}{ccc} E_1 \# E_2 & \longrightarrow & E_1 \circledast E_2 \\ \downarrow {\scriptstyle p_1 \# p_2} & & \downarrow {\scriptstyle p_1 \circledast p_2} \\ B & \xrightarrow{\ \Delta\ } & B \times B \end{array}$$

one may check the following

<u>Proposition 4.2</u>

(a) $p_1 \circledast p_2$ and $p_1 \# p_2$ are Hurewicz fibrations with fibre $F_1 * F_2$ (the join of the two fibres)

(b) Let $\pi : E^+X \longrightarrow X$ be the path space fibration with fibre ΩX . Then there is a homotopy equivalence

$$
\begin{array}{ccc}
T_1^n X & \xhookrightarrow{\quad j \quad} & X^{n+1} \\
\big\downarrow \wr & & \| \\
E^+X \divideontimes E^+X \divideontimes \ldots \divideontimes E^+X & \xrightarrow{\pi^{\divideontimes(n+1)}} & X^{n+1}
\end{array}
$$

such that the above triangle commutes up to homotopy

(c) If $p : E \longrightarrow B$ is a fibration with fibre F, there is a homotopy equivalence $E \cup_F CF \xrightarrow{\sim} E \# E^+B$ such that the triangle

$$
\begin{array}{ccc}
E \cup_F CF & \xrightarrow{\sim} & E \# E^+B \\
& \bar{p} \searrow \quad \swarrow p \# \pi & \\
& B &
\end{array}
$$

where $\bar{p}|E = p$, $\bar{p}|CF = *$, commutes up to homotopy. ∎

From this we readily deduce the

<u>Theorem 4.3</u>. In the diagram

$$
\begin{array}{ccccc}
\Omega X^{*(n+1)} & \longrightarrow & X(n) \sim E^+X^{\#(n+1)} & \xrightarrow{\pi^{\#(n+1)}} & X \\
\| & & \big\downarrow & & \big\downarrow \Delta \\
\Omega X^{*(n+1)} & \longrightarrow & T_1^n X \sim E^+X^{\divideontimes(n+1)} & \xrightarrow[j \sim \pi^{\divideontimes(n+1)}]{} & X^{n+1}
\end{array}
$$

the rows are fibration sequences, and the right-hand square is homotopy cartesian. Moreover, the sequence

$$\Omega X^{*(n+1)} \longrightarrow X(n) \longrightarrow X(n+1)$$

is a cofibre sequence for all $n \geq 1$. ∎

This theorem shows that the sequence of spaces $X^{(n)}$ is generated by a "fibre-co-fibre" process, due to T. Ganéa, whose inductive step is the following : assume $\alpha_n : X^{(n)} \longrightarrow X$ has been defined ; convert α_n into a fibration $\bar{\alpha}_n$; then $X(n+1)$ is the mapping cone of the inclusion of the fibre and α_{n+1} is the extension of $\bar{\alpha}_n$ which maps the cone on the fibre to the base-point. Initialize the process with $X(0) = *$ to get the sequence $X(n)$.

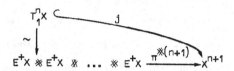

Observe that if X is k-connected, $k \geq 1$, then the fibre $\Omega X^{*(n+1)}$ of $\alpha_n : X(n) \longrightarrow X$ is $(k(n+2)-1)$-connected. Therefore α_n induces an isomorphism on homotopy groups in degrees $\leq k(n+2)-2$. This shows that the homotopy direct limit of the spaces $X(n)$ has the homotopy type of X. The sequence $X(n)$ is a filtration of the homotopy type of X called the Ganéa filtration.

Also the above inductive "fibration-cofibration" process localizes all right at any set of primes P, so that $[X(n)]_P \sim X_P(n)$.

From (4.1) we deduce that

$$\mathrm{cat}(X_P) \leq \mathrm{cat}(X) .$$

In particular we shall define $\mathrm{cat}_0(X)$, the rational L.-S. category of X, to be $\mathrm{cat}(X_0)$: it is the lower bound of $\mathrm{cat}(X')$ when X' runs through the class of spaces with the same rational homotopy type as X.

It turns out that the Ganéa filtration has a nice interpretation in terms of the Sullivan minimal model. Recall that the Sullivan minimal model $\mathcal{M}_X$ of a 1-connected space X is a DG-algebra, freely generated as a graded commutative algebra by a vector space V isomorphic to $\mathrm{Hom}(\pi_*(X), \mathbb{Q})$. Thus $\mathcal{M}_X \cong (\Lambda V, d)$, and the differential d maps V into $\Lambda^{\geq 2}V$: the ideal $\Lambda^{\geq p}V \cong (\bar{\mathcal{M}}_X)^p$ is a differential ideal, and we may consider the DG-algebra $\mathcal{M}_X/\bar{\mathcal{M}}_X^{n+1}$. This algebra represents a rational homotopy type which we denote by $X[n]$, whose minimal model is a model of $\mathcal{M}_X/\bar{\mathcal{M}}_X^{n+1}$. Now :

<u>Theorem 4.4</u> ([FH] VIII and IX)

(a) The Ganéa space $X(n)$ has the rational homotopy type of the wedge of $X[n]$ and a wedge of spheres $\Sigma_n(X)$, such that the composition

$$\Sigma_n(X) \hookrightarrow X[n] \vee \Sigma_n(X) \xrightarrow[\mathbb{Q}]{\sim} X(n) \hookrightarrow X(n+1)$$

is null-homotopic (rationally).

(b) $\mathrm{cat}_0 X \leq n$ iff $\mathcal{M}_X$ is a retract of the minimal model of $\mathcal{M}_X/\bar{\mathcal{M}}_X^{n+1}$. ∎

To prove 4.4, Félix and Halperin first translate the pullback square

$$\begin{array}{ccccc} X(n) & \longrightarrow & X^{[0,n]} & \xleftarrow[\sim]{\varepsilon} & X \\ \downarrow & & \downarrow e & \swarrow \Delta & \\ T_1^n X & \overset{j}{\hookrightarrow} & X^{n+1} & & \end{array}$$

into DG-algebras to obtain a (neither free nor minimal) model of $X(n)$. Next they show that this model is quasi-isomorphic to the direct product of $\mathcal{M}_X/\overline{\mathcal{M}}_X^{n+1}$ and a trivial algebra.

This trivial algebra is huge in general : for instance if $n = 1$, one has $X(1) = S\Omega X$ while $\mathcal{M}_X/\overline{\mathcal{M}}_X^2$ is the trivial algebra $\mathbb{Q} \oplus V$, $d = 0$, $V.V = 0$; thus the reduced homology of $\Sigma_1(X)$ is isomorphic (as vector spaces) to a supplement of $\mathrm{Hom}(V, \mathbb{Q}) = s\underline{\underline{\pi}}(X)$ in $\widetilde{H}_*(S\Omega X) = s\,\overline{U\underline{\underline{\pi}}(X)}$. The only case when $X(1)$ has the rational homotopy type of $X[1]$ is when $\underline{\underline{\pi}}(X) = \overline{U\underline{\underline{\pi}}(X)}$, that is, when $\underline{\underline{\pi}}(X)$ is an abelian algebra on a single generator of odd degree, i.e. $X \underset{\mathbb{Q}}{\sim} K(\mathbb{Q}, 2r)$: in particular, we already mentioned the homotopy equivalence (over the integers !) $CP(\infty)\ (n) \simeq CP(n)$, while clearly $CP(\infty)\ [n] \underset{\mathbb{Q}}{\sim} CP(n)$, for all n .

For a further discussion of the rational homotopy type of $X(n)$, see [L] .

We conclude with pointing out that Theorem (4.4)(b) means that a space has rational category $\leq n$ iff its model is a homotopy retract of a DG-algebra of nilpotency $\leq n$: compare with (1.5) !

This provides an internal definition of the L.-S. category of a DG-algebra. It is an open question whether the model of any space of category n actually is quasi-isomorphic to a DG-algebra of nilpotency n : this is obvious for $n = 1$, and has been checked by Félix and Thomas for $n = 2$.

REFERENCES.

[FH] Yves FELIX and Stephen HALPERIN, Rational L.-S. category and its applications, Trans. A.M.S. 273 (1982) pp. 1-38.

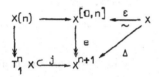

[FL] Yves FELIX and Jean-Michel LEMAIRE, On the mapping theorem for L.-S. category, Topology, 24, 1985, 41-43.

[Ga] Tudor GANEA, L.-S. category and strong category, Ill. J. Math. 11 (1967) pp. 331-348.

[Gi] W. GILBERT, Some examples for weak category and conilpotency, Ill. J. Math. 12 (1968) pp. 421-432.

[J] Ioan JAMES, On category in the sense of L.-S., Topology 17 (1978) pp. 331-348.

[LS] Jean-Michel LEMAIRE and François SIGRIST, Sur les invariants d'homotopie rationnelle liés à la L.-S. catégorie, Comm. Math. Helv. 56 (1981), 103-122.

[W] George WHITEHEAD, Elements of Homotopy Theory, Graduate Texts in Math. Springer-Verlag.

[L] Jean-Michel LEMAIRE, Sur le type d'homotopie rationnelle des espaces de Ganéa, in Homotopie Algébrique et Algèbre Locale, Astérisque nº 113-114 (1984), pp 238-247.

Jean-Michel LEMAIRE
Laboratoire de Mathématiques
U.A. au C.N.R.S. nº 168
Université de Nice
Parc Valrose
F - 06034 NICE-CEDEX

SÉRIES DE BASS DES MODULES DE SYZYGIE

par Jack LESCOT

Tous les anneaux considérés sont des anneaux commutatifs, locaux, nœthériens, de même corps résiduel k.

Soient $(R,\underline{m})$ un anneau local et M un R-module de type fini. On associe à M deux séries formelles :

la série de Poincaré :

$$P_R^M(t) = \sum_{i\geq 0} b_i(M)t^i, \quad \text{où} \quad b_i(M) = \dim_k \mathrm{Tor}_i^R(M,k),$$

la série de Bass :

$$I_R^M(t) = \sum_{i\geq 0} \mu_i(M)t^i, \quad \text{où} \quad \mu_i(M) = \dim_k \mathrm{Ext}_R^i(k,M).$$

Considérons $P_\bullet : \ldots \longrightarrow P_{n+1} \longrightarrow P_n \longrightarrow \ldots \longrightarrow P_o \longrightarrow 0$ une résolution projective minimale de M et soit, pour $n > 0$, $\mathrm{syz}^n(M) = \mathrm{Im}(P_n \longrightarrow P_{n-1})$ le $n^{\text{ième}}$ module de syzygie de M. Il est clair que $b_i(\mathrm{syz}^n(M)) = b_{i+n}(M)$ pour $i \geq 0$. On montre ici comment les $\mu_i(\mathrm{syz}^n(M))$ sont déterminés par M (on note $I_R(t)$ pour $I_R^R(t)$) :

<u>Théorème A</u>.- <u>On a</u> :

$$I_R^{\mathrm{syz}^n(M)}(t) = (b_{n-1}(M)+\ldots+t^{n-1}b_o(M))I_R(t)-t^{n-1}I_R^M(t)+(1+t)t^{n-1}|F_n(M)|(t),$$

<u>où</u> $|F_n(M)|(t)$ <u>désigne la série de Hilbert d'un certain espace vectoriel gradué</u> $F_n(M)$, <u>associé à</u> M.

Soit E une enveloppe injective de k sur R et soit $M^\vee = \mathrm{Hom}_R(M,E)$ le dual de Matlis de M. Il existe un produit homologique associé à M :

$$\mathrm{Tor}_*^R(M,k) \otimes_R \mathrm{Tor}_*^R(M^\vee,k) \longrightarrow \mathrm{Tor}_*^R(R^\vee,k),$$

et $F_n(M)$ est un sous-espace vectoriel de $\mathrm{Tor}_*^R(M^\vee,k)$, défini à l'aide de ce produit. La situation est simple pour le module M si le produit est nul. Dans ce cas $|F_n(M)|(t) = I_R^M(t)$. Ainsi pour $M = k$:

Théorème B.- Soit $(R,\underline{m})$ un anneau local non régulier alors pour tout $n > 0$:

$$I_R^{syz^n(k)}(t) = (b_{n-1}(k) + tb_{n-2}(k) + \ldots + t^{n-1}b_o(k))I_R(t) + t^n P_R^k(t).$$

On utilise ces résultats pour montrer que la dimension syzygétique $\gamma(R)$ introduite par Roos dans [11] est infinie pour la plupart des anneaux qui ne sont pas de Gorenstein.

Afin de mesurer la complexité de l'anneau R vis à vis de la formule du théorème A, on pose la question suivante :

Existe-t-il un entier $\sigma(R)$ tel que pour tout module M de type fini, on ait : $F_{\sigma(R)}(M) = F_p(M) \quad \forall p \geqslant \sigma(R)$?

On montre qu'il en est bien ainsi pour quelques classes d'anneaux.

Voici le plan de cet article :

Dans la première section, on définit le produit homologique associé à M et on étudie $W(M)$ le sous-espace de $Tor_*^R(R^\vee,k)$ engendré par les valeurs du produit. Lorsque $M = R/J$, on a $W(M) = Im\ s_*^\vee$ où $s : R \longrightarrow R/J$ est la projection canonique. Ce résultat est utilisé pour donner une nouvelle démonstration d'une caractérisation des anneaux de Gorenstein, due à Peskine et Szpiro [10].

Dans la deuxième section, on démontre les théorèmes A et B, et dans la troisième section, on étudie l'existence de $\sigma(R)$ pour quelques cas.

D'autres propriétés et applications du produit homologique associé à un module se trouvent dans [9], papier auquel nous ferons référence pour quelques détails.

0. NOTATIONS ET RESULTATS PRELIMINAIRES

Soient $(R,\underline{m})$ un anneau local, k son corps résiduel.

0.1. Soit V un k-espace vectoriel, on note $|V|$ sa dimension. Soit $A = \bigoplus_{p\geqslant 0} A_p$ un k-espace vectoriel gradué. Si pour chaque p, $|A_p| < \infty$ on note $|A|(t) = \sum_{p\geqslant 0} |A_p| t^p$ la série de Hilbert de A.

Soit $f : M \longrightarrow N$ un homomorphisme de R-modules, on note f_* (resp. f_p) l'homomorphisme induit en homologie :

$$f_* : \mathrm{Tor}_*^R(M,k) \longrightarrow \mathrm{Tor}_*^R(N,k) \quad (\text{resp. } f_p : \mathrm{Tor}_p^R(M,k) \longrightarrow \mathrm{Tor}_p^R(N,k)).$$

Soit M un R-module de type fini. Les modules de syzygie de M sont définis à un isomorphisme près à partir d'une résolution projective minimale $P.$ par $\mathrm{syz}^0(M) = M$ et, pour $n > 0$, par $\mathrm{syz}^n(M) = \mathrm{Im}(P_n \longrightarrow P_{n-1})$.

0.2. Duals de Matlis

On choisit pour l'anneau R une enveloppe injective E de son corps résiduel k. Si M est un R-module, soit $M^\vee = \mathrm{Hom}_R(M,E)$ le dual de Matlis de M. Si $f : M \longrightarrow N$ est un R homomorphisme, soit $f^\vee : N^\vee \longrightarrow M^\vee$ l'homomorphisme induit entre les duals de Matlis, ainsi si $a \in N^\vee$ $f^\vee(a) = a \circ f$ (composition des applications). Nous ferons un usage constant de l'isomorphisme canonique $E \simeq R^\vee = \mathrm{Hom}_R(R,E)$ identifiant les éléments de E à des applications de R dans E et vice versa.

La formule de dualité de ([3], chap. VI,5.3) montre :

Pour tout R-module M et pour tout $p \in \mathbb{N}$, il existe des isomorphismes fonctoriels :

$$\mathrm{Tor}_p^R(M^\vee,k) \simeq \mathrm{Ext}_R^p(k,M)^\vee.$$

Comme $k^\vee \simeq k$ il en résulte immédiatement :

Pour tout R-module de type fini $\quad P_R^{M^\vee}(t) = I_R^M(t)$.

En particulier $I_R(t) = P_R^{R^\vee}(t) = P_R^E(t)$. On notera aussi que $P_R^k(t) = I_R^k(t)$.

Le problème du calcul des séries de Bass est ainsi ramené à un calcul de séries de Poincaré (pour des modules qui ne sont pas nécessairement de type fini). Rappelons enfin que $\mathrm{Tor}_*^R(R^\vee,k)$ n'est jamais nul et que la profondeur de l'anneau R peut être définie par [1] :

$$\mathrm{Prof}\, R = \inf \{i \mid \mathrm{Ext}_R^i(k,R) \neq 0\} = \inf \{i \mid \mathrm{Tor}_i^R(R^\vee,k) \neq 0\}.$$

I. LE PRODUIT HOMOLOGIQUE ASSOCIE A UN MODULE

1.1. Soient M un R-module et $\theta : M \otimes_R M^\vee \longrightarrow E \simeq R^\vee$, l'homomorphisme évaluation, $\theta(a \otimes b) = b(a)$. On associe à θ un produit homologique :

$$\mathrm{Tor}_*^R(M,k) \otimes_R \mathrm{Tor}_*^R(M^\vee,k) \longrightarrow \mathrm{Tor}_*^R(R^\vee,k)$$

par composition du produit extérieur :

$$\mathrm{Tor}_*^R(M,k) \otimes_R \mathrm{Tor}_*^R(M^\vee,k) \longrightarrow \mathrm{Tor}_*^R(M \otimes_R M^\vee,k)$$

et de l'homomorphisme induit en homologie par θ :

$$\theta_* : \mathrm{Tor}_*^R(M \otimes_R M^\vee,k) \longrightarrow \mathrm{Tor}_*^R(R^\vee,k),$$

([3], chapitre XI). On note $<,>$ l'application R-bilinéaire correspondante et on l'appelle le produit homologique associé à M.

On désigne par :

$W(M)$ le sous-espace vectoriel gradué de $\mathrm{Tor}_*^R(R^\vee,k)$ image de $\mathrm{Tor}_*^R(M,k) \otimes_R \mathrm{Tor}_*^R(M^\vee,k)$ par le produit associé à M, $W_p(M)$ sa composante de degré p, ($W_p(M) = 0$ si $p<0$),

$W_{p,q}(M)$ le sous-espace vectoriel de $\mathrm{Tor}_{p+q}^R(R^\vee,k)$ image de $\mathrm{Tor}_p^R(M,k) \otimes_R \mathrm{Tor}_q^R(M^\vee,k)$, ($W_{p,q}(M) = 0$ si $p<0$ ou si $q<0$).

On a donc $W_n(M) = \sum_{p+q=n} W_{p,q}(M)$ (la somme n'est pas directe).

On vérifie facilement que $W(R) = \mathrm{Tor}_*^R(R^\vee,k)$ et de même $W(E) = \mathrm{Tor}_*^R(R^\vee,k)$. Notons aussi que $W(M \oplus N) = W(M) + W(N)$.

Les outils essentiels pour travailler avec le produit homologique associé à M sont fournis par la proposition suivante :

1.2. <u>Proposition</u>.- 1) <u>Soit</u> $f : M \longrightarrow N$ <u>un homomorphisme de</u> R-<u>modules</u>. <u>Si</u> $x \in \mathrm{Tor}_*^R(M,k)$ <u>et</u> $y \in \mathrm{Tor}_*^R(N^\vee,k)$ <u>on a</u> $<x,f_*^\vee(y)> = <f_*(x),y>$.

2) <u>Soit</u> $0 \longrightarrow M_1 \xrightarrow{i} M \xrightarrow{s} M_2 \longrightarrow 0$ <u>une suite exacte de R-modules et considérons les homomorphismes de connexion associés à cette suite et à sa duale</u> :

$$\delta : \mathrm{Tor}_{*+1}^R(M_2,k) \longrightarrow \mathrm{Tor}_*^R(M_1,k)$$

$$\delta' : \mathrm{Tor}_{*+1}^R(M_1^\vee,k) \longrightarrow \mathrm{Tor}_*^R(M_2^\vee,k).$$

<u>Si</u> $x \in \mathrm{Tor}_{p+1}^R(M_2,k)$ <u>et</u> $y \in \mathrm{Tor}_{q+1}^R(M_1^\vee,k)$ <u>on a</u> $<x,\delta'(y)> = (-1)^p <\delta(x),y>$.

Preuve : Soit X une R-algèbre différentielle graduée, résolution minimale de k sur R [6]. On choisit $(T_i)_{i\in J}$ une base homogène de X sur R. La structure multiplicative de X permet de définir un produit $(M\otimes_R X)\otimes_R (M^\vee\otimes_R X) \longrightarrow R^\vee\otimes_R X$ de la façon suivante :

soient $a = \sum_{i\in J} a_i T_i$, $a_i \in M$ et $b = \sum_{j\in J} b_j T_j$, $b_j \in M^\vee$, à l'élément $a\otimes b$ on fait correspondre $\sum_{i,j\in J\times J} b_j(a_i)T_iT_j$. On note $<\,,\,>$ l'application R-bilinéaire correspondante. Si a et b sont des cycles, il en est de même de $<a,b>$ et pour les classes d'homologies correspondantes dans $\mathrm{Tor}_*^R(M,k)$, $\mathrm{Tor}_*^R(M^\vee,k)$ et $\mathrm{Tor}_*^R(R^\vee,k)$ on a : $c\ell<a,b> = <c\ell(a),c\ell(b)>$. Dans la démonstration qui suit, pour tout homomorphisme $f : M \longrightarrow N$ on note encore par f l'homomorphisme induit : $M\otimes_R X \longrightarrow N\otimes_R X$.

1) Soient $a = \sum_{i\in J} a_i T_i$ un cycle de $M\otimes_R X$ qui relève x et $b = \sum_{j\in J} b_j T_j$ un cycle de $N^\vee\otimes_R X$ qui relève y. On a $f(a) = \sum_{i\in J} f(a_i)T_i$ et $f^\vee(b) = \sum_{j\in J}(b_j\circ f)T_j$. On obtient $<a,f^\vee(b)> = \sum_{(i,j)\in J\times J}(b_j(f(a_i)))T_iT_j = <f(a),b>$, ce qui démontre le point 1) en prenant les classes d'homologie.

2) Pour tout R-module M désignons par d la différentielle du complexe $M\otimes_R X$. Soit a un cycle de $M_2\otimes_R X$ de degré p+1 qui représente la classe d'homologie x, et soit ℓ un élément de $M\otimes_R X$ de degré p+1 tel que $s(\ell) = a$. Alors $d(\ell)$ est dans $\mathrm{Im}(i)$ et $i^{-1}(d(\ell))$ représente la classe d'homologie $\delta(x)$. De même, soit b un cycle de $M_1^\vee\otimes_R X$ de degré q+1 qui représente la classe d'homologie y, et soit h un élément de $M^\vee\otimes_R X$ de degré q+1 tel que $i^\vee(h) = b$. Alors $d(h)$ est dans $\mathrm{Im}\, s^\vee$ et $(s^\vee)^{-1}(d(h))$ représente la classe d'homologie $\delta'(y)$. On vérifie facilement que $<a,(s^\vee)^{-1}(d(h))> = <\ell,d(h)>$ et que $<i^{-1}(d(\ell)),b> = <d(\ell),h>$. Comme $d(<\ell,h>) = <d(\ell),h> - (-1)^p<\ell,d(h)>$, les cycles $<\ell,d(h)>$ et $(-1)^p<d(h),\ell>$ sont homologues. Ce qui établit le point 2) en prenant les classes d'homologie.

1.3. <u>Corollaire</u>.- <u>Soit</u> M <u>un</u> <u>R-module de type fini</u>.

1) <u>Pour tout</u> $(p,q)\in \mathbb{Z}\times\mathbb{Z}$ <u>on a</u> $W_{p,q}(\mathrm{syz}^r(M)) \subset W_{p+r,q-r}(M)$.

2) <u>Pour tout</u> $j\in\mathbb{Z}$ <u>on a</u> $W_j(\mathrm{syz}^r(M))\subset W_j(M)$ <u>et si</u> $j<r$ $W_j(\mathrm{syz}^r(M)) = 0$.

<u>Preuve</u> : 1) Il suffit d'établir le résultat dans le cas r=1, le résultat général en découlant puisque $\mathrm{syz}^1(\mathrm{syz}^r(M)) = \mathrm{syz}^{r+1}(M)$. Soit $P_\bullet$ une résolu-

tion projective minimale de M. Considérons la suite exacte :
$0 \longrightarrow syz^1(M) \longrightarrow P_o \longrightarrow M \longrightarrow 0$ et soient, comme dans la proposition précédente, les homomorphismes de connexion $\delta = Tor^R_{*+1}(M,k) \longrightarrow Tor^R_*(syz^1(M),k)$, $\delta' : Tor^R_{*+1}((syz^1(M))^\vee,k) \longrightarrow Tor^R_*(M^\vee,k)$. Soient $x \in Tor^R_p(syz^1(M),k)$ et $y \in Tor^R_q((syz^1(M))^\vee,k)$. Comme δ est surjective, on a $x = \delta(x')$ avec $x' \in Tor^R_{p+1}(M,k)$, et alors $\langle x,y\rangle = \langle\delta(x'),y\rangle = (-1)^p \langle x',\delta'(y)\rangle$. Ce qui établit que $\langle x,y\rangle$ est dans $W_{p+1,q-1}(M)$.

2) Comme $W_j(syz^r(M)) = \sum_{p+q=j} W_{p,q}(syz^r(M))$, l'assertion 2) résulte immédiatement de 1).

Pour un module monogène, l'espace $W(M)$ admet une interprétation simple :

1.4. <u>Théorème</u>.- <u>Soient</u> J <u>un idéal de</u> R <u>et</u> $s : R \longrightarrow R/J$ <u>la projection canonique. Alors on a</u> $W(R/J) = \mathrm{Im}\ s^\vee_*$.

<u>Preuve</u> : Soit M un R-module annulé par J.

L'homomorphisme évaluation du début du paragraphe admet la factorisation : $M \otimes_R M^\vee \longrightarrow (R/J)^\vee \xrightarrow{s^\vee} R^\vee$. Par conséquent, le produit associé à M factorise en :

$$Tor^R_*(M,k) \otimes_R Tor^R_*(M^\vee,k) \longrightarrow Tor^R_*((R/J)^\vee,k) \xrightarrow{s^\vee_*} Tor^R_*(R^\vee,k).$$

On en déduit $W(M) \subset \mathrm{Im}\ s^\vee_*$, en particulier $W(R/J) \subset \mathrm{Im}\ s^\vee_*$. Le générateur canonique $\bar{1}$ de R/J définit une classe d'homologie $c\ell(\bar{1})$ dans $Tor^R_o(R/J,k)$. On vérifie facilement que l'application

$$\langle c\ell(\bar{1}),\cdot\rangle : Tor^R_*((R/J)^\vee,k) \longrightarrow Tor^R_*(R^\vee,k) \quad \text{coïncide avec}$$

l'application $s^\vee_*$. On en déduit l'inclusion inverse $\mathrm{Im}\ s^\vee_* \subset W(R/J)$.

Le corollaire suivant est une conséquence immédiate du théorème précédent et de sa démonstration.

1.5. <u>Corollaire</u>.- <u>Soient</u> J <u>un idéal de</u> R, $s : R \longrightarrow R/J$ <u>la surjection canonique. Pour tout</u> R-<u>module</u> M <u>annulé par</u> J <u>on a</u> $W(M) \subset W(R/J) = \mathrm{Im}\ s^\vee_*$.

Donnons maintenant quelques exemples :

1.6.- Soit $s : R \longrightarrow k$ la projection canonique de R sur son corps résiduel. Si R n'est pas un anneau régulier, on a démontré dans [7] (ou [8]) que $\overset{\vee}{s}_*$ était nul. Autrement dit dans ce cas $W(k) = 0$, et par conséquent pour tous les modules de syzygie de k on a aussi $W(syz^r(k)) = 0$ (corollaire 1.3).

1.7.- Soit M un R-module de type fini, de dimension projective finie d. Alors le $d^{\text{ième}}$ module de syzygie de M est un R-module libre et $W(R) = W(syz^d(M)) \subset W(M)$, (corollaire 1.3). Donc nécessairement $W(M) = Tor_*^R(R^\vee, k)$. En particulier sur un anneau local régulier tout R-module M de type fini vérifie $W(M) = Tor_*^R(R^\vee, k)$ et par conséquent $W(M) \neq 0$.

1.8.- Soient N un R-module et $J^\cdot = 0 \longrightarrow J^o \longrightarrow J^1 \longrightarrow \ldots \longrightarrow J^n \longrightarrow \ldots$ une résolution injective minimale de N. Définissons pour $r \geq 0$, les modules de co-syzygie de N par $\text{co-syz}^r(N) = \text{Im}(J^{r-1} \longrightarrow J^r)$. De manière similaire au corollaire 1.3, on démontre que $W(\text{co-syz}^r(N)) \subset W(N)$ (voir [9] pour des détails). Supposons que N soit un R-module de dimension injective finie r(de type fini). Le $r^{\text{ième}}$ module de co-syzygie de N est une somme directe de modules injectifs tous isomorphes à E [1]. Par conséquent $W(E) = W(\text{co-syz}^r(N)) \subset W(N)$ et donc $W(N) = W(E) = Tor_*^R(R^\vee, k)$.

Ce dernier point permet de donner une nouvelle démonstration d'un résultat de Peskine et Szpiro ([10], théorème 5.5).

1.9. Théorème.- Pour qu'un anneau local noethérien R soit de Gorenstein, il faut et il suffit qu'il existe un idéal J de R tel que le R-module monogène R/J soit de dimension injective finie.

Preuve : Rappelons qu'un anneau local noethérien est dit de Gorenstein si il est de dimension injective finie [1]. La condition est clairement nécessaire. Pour la réciproque, notons $s : R \longrightarrow R/J$ la surjection canonique. On sait que $W(R/J) = \text{Im}\ \overset{\vee}{s}_*$ (théorème 1.4). D'autre part, puisque R/J est un R-module de dimension injective finie, $W(R/J) = Tor_*^R(R^\vee, k)$ (1.8). Par conséquent $\overset{\vee}{s}_* : Tor_*^R((R/J)^\vee, k) \longrightarrow Tor_*^R(R^\vee, k)$ est un homomorphisme surjectif. Il en résulte que l'anneau R est de dimension injective finie, donc qu'il est de Gorenstein.

II. UNE FILTRATION SUR $Tor_*^R(M^\vee, k)$

2.1. Définition.- Soit M un R-module, on définit une filtration décroissante $(F_n(M))_{n \in \mathbb{N}}$ sur $Tor_*^R(M^\vee, k)$ de la manière suivante ·

$F_o(M) = \mathrm{Tor}_*^R(M^\vee,k)$ et pour $p > 0$, $F_p(M) = \{y \mid y \in \mathrm{Tor}_*^R(M^\vee,k)$ et $\forall x \in \mathrm{Tor}_j^R(M,k)$, $j < p$, $\langle x,y\rangle = 0\}$. Il est clair que les $F_p(M)$ sont des sous espaces vectoriels gradués de $\mathrm{Tor}_*^R(M^\vee,k)$ et leur intersection $F_\infty(M)$ est caractérisée par : $F_\infty(M) = \{y \mid y \in \mathrm{Tor}_*^R(M^\vee,k)$ et $\forall x \in \mathrm{Tor}_*^R(M,k)$, $\langle x,y\rangle\}$.

2.2. Théorème.- Soient M un R-module de type fini et $(b_p(M))_{p\in\mathbb{N}}$ ses nombres de Betti. Alors la série de Bass de $\mathrm{syz}^n(M)$ est donnée pour $n > 0$ par la formule :

$$I_R^{\mathrm{syz}^n(M)}(t) = (b_{n-1}(M)+tb_{n-2}(M)+\ldots+t^{n-1}b_o(M))I_R(t)-t^{n-1}I_R^M(t)+(1+t)t^{n-1}|F_n(M)|(t).$$

Preuve : Soit $P.$ une résolution projective minimale de M. Considérons pour tout $p \in \mathbb{N}$ les suites exactes : $0 \to \mathrm{syz}^{p+1}(M) \to P_p \to \mathrm{syz}^p(M) \to 0$ et les homomorphismes de connexion associés à ces suites et aux suites duales :

$$\delta \ : \mathrm{Tor}_{*+1}^R(\mathrm{syz}^p(M),k) \to \mathrm{Tor}_*^R(\mathrm{syz}^{p+1}(M),k),$$

$$\delta' \ : \mathrm{Tor}_{*+1}^R((\mathrm{syz}^{p+1}(M))^\vee,k) \to \mathrm{Tor}_*^R((\mathrm{syz}^p(M))^\vee,k).$$

En itérant ces homomorphismes, on obtient pour $n > 0$:

$$\delta^n \ : \mathrm{Tor}_{*+n}^R(M,k) \to \mathrm{Tor}_{*+(n-1)}^R(\mathrm{syz}^1(M),k) \to\ldots\to \mathrm{Tor}_*^R(\mathrm{syz}^n(M),k)$$

$$\delta'^n \ : \mathrm{Tor}_{*+n}^R((\mathrm{syz}^n(M))^\vee,k) \to \mathrm{Tor}_{*+(n-1)}^R((\mathrm{syz}^{n-1}(M))^\vee,k)\to\ldots\to \mathrm{Tor}_*^R(M^\vee,k).$$

Posons en outre δ^o et δ'^o les identités sur $\mathrm{Tor}_*^R(M,k)$ et $\mathrm{Tor}_*^R(M^\vee,k)$. Notons que δ^n est un homomorphisme surjectif de degré $-n$ (isomorphisme sur les composantes de degré supérieur ou égal à n). Soient $x \in \mathrm{Tor}_*^R(M,k)$ et $y \in \mathrm{Tor}_*^R((\mathrm{syz}^n(M))^\vee,k)$ des éléments homogènes, à partir de la proposition 1.2, on obtient : $\langle x,\delta'^n(y)\rangle = \pm \langle\delta^n(x),y\rangle$.

La démonstration du théorème se fait en plusieurs étapes :

Point 1.- On a $F_n(M) = \mathrm{Im}(\delta'^n)$.

Soient z un élément homogène de $\mathrm{Tor}_*^R((\mathrm{syz}^n(M))^\vee,k)$ et $y = \delta'^n(z)$. Si $x \in \mathrm{Tor}_j^R(M,k)$ avec $j < n$, on a $\langle x,y\rangle = \langle x,\delta'^n(z)\rangle = \pm \langle\delta^n(x),z\rangle = 0$. Par suite $\mathrm{Im}(\delta'^n) \subset F_n(M)$.

Pour obtenir l'inclusion inverse, nous avons besoin du

<u>Lemme</u>.- <u>Soit</u> $0 \longrightarrow N_1 \longrightarrow R^n \xrightarrow{s} N \longrightarrow 0$ <u>une présentation minimale de</u> N. <u>Si</u> $y \in Tor_*^R(N^\vee,k)$, <u>on a</u> $s_*^\vee(y) \neq 0$ <u>si et seulement si il existe</u> $x \in Tor_o^R(N,k)$ <u>tel que</u> $\langle x,y\rangle \neq 0$.

Admettons le lemme pour l'instant. Soit $y \in Tor_*^R(M^\vee,k)$ un élément homogène qui n'est pas dans $\mathrm{Im}\,\delta'^n$. On peut trouver un entier r, $0 \leq r < n$, tel que $y \in \mathrm{Im}\,\delta'^r$ et $y \notin \mathrm{Im}\,\delta'^{r+1}$. En particulier, il existe $z \in Tor_*^R((syz^r(M))^\vee,k)$, élément homogène, tel que $y = \delta'^r(z)$, et qui n'est pas dans l'image de $\delta' : Tor_{*+1}^R((syz^{r+1}(M))^\vee,k) \longrightarrow Tor_*^R((syz^r(M))^\vee,k)$, homomorphisme de connexion associé à la suite exacte :
$0 \longrightarrow (syz^r(M))^\vee \xrightarrow{s^\vee} P_r^\vee \longrightarrow (syz^{r+1}(M))^\vee \longrightarrow 0$. Cela signifie aussi que $s_*^\vee(z) \neq 0$. Donc d'après le lemme, il existe $u \in Tor_o^R(syz^r(M),k)$ tel que $\langle u,z\rangle \neq 0$. Soit $x \in Tor_r^R(M,k)$ tel que $\delta^r(x) = u$ alors $\langle x,y\rangle = \langle x,\delta'^r(z)\rangle = \pm \langle u,z\rangle \neq 0$. Par conséquent y n'est pas dans $F_n(M)$.

<u>Preuve du lemme</u> : Soit $y \in Tor_*^R(N^\vee,k)$. Supposons qu'il existe $x \in Tor_o^R(N,k)$ tel que $\langle x,y\rangle \neq 0$. Comme $s_o : Tor_o^R(R^n,k) \longrightarrow Tor_o^R(N,k)$ est surjective, on peut trouver $x' \in Tor_o^R(R^n,k)$ tel que $s_*(x') = x$. On a alors $\langle x',s_*^\vee(y)\rangle = \langle s_*(x'),y\rangle = \langle x,y\rangle \neq 0$, (1.2), et donc $s_*^\vee(y) \neq 0$. Réciproquement supposons $s_*^\vee(y) = 0$. Soient (e_i), $1 \leq i \leq n$, la base canonique de R^n et $j^i : R \longrightarrow Re_i \longrightarrow R^n$. Les éléments e_i définissent des classes d'homologie dans $Tor_o^R(R^n,k)$ et on vérifie que les applications $\langle c\ell(e_i),\cdot\rangle : Tor_*^R((R^n)^\vee,k) \longrightarrow Tor_*^R(R^\vee,k)$ et $j_*^{i,\vee}$ sont identiques. Puisque $s_*^\vee(y) \neq 0$, on peut trouver i tel que $j_*^{i,\vee}(s_*^\vee(y)) \neq 0$. Posons $x = s_*(c\ell(e_i))$ on a $x \in Tor_o^R(N,k)$ et $\langle x,y\rangle = \langle s_*(c\ell(e_i)),y\rangle = \langle c\ell(e_i),s_*^\vee(y)\rangle \neq 0$, (1.2).

<u>Point 2</u>.- <u>Pour chaque</u> r <u>l'homomorphisme</u> δ'^n <u>induit un isomorphisme de degré</u> $-n$

$$\bar{\delta}'^n : F_r(syz^n(M))/F_{r+1}(syz^n(M)) \longrightarrow F_{r+n}(M)\,/\,F_{r+1+n}(M).$$

Comme l'application δ^n est surjective, un élément homogène y de $Tor_*^R((syz^n(M))^\vee,k)$ est dans $F_r(syz^n(M))$ si et seulement si pour tout $x \in Tor_j^R(M,k)$, $j < n+r$, on a : $\langle\delta^n(x),y\rangle = 0$ ou de façon équivalente $\langle x,\delta'^n(y)\rangle = 0$. On a donc l'équivalence : $y \in F_r(syz^n(M)) \Longleftrightarrow \delta'^n(y) \in F_{r+n}(M)$. D'autre part, le point précédent montre que pour tout r : $\delta'^n(F_r(syz^n(M))) = F_{r+n}(M)$. On en déduit le résultat.

Nous sommes maintenant en mesure de démontrer le théorème.

A partir de la longue suite exacte d'homologie associée à la suite exacte $0 \longrightarrow (\mathrm{syz}^{n-1}(M))^{\vee} \longrightarrow P^{\vee}_{n-1} \longrightarrow (\mathrm{syz}^{n}(M))^{\vee} \longrightarrow 0$, et en tenant compte du point 1, on peut écrire :

$$I_R^{\mathrm{syz}^n(M)}(t) = b_{n-1}(M)I_R(t) - I_R^{\mathrm{syz}^{n-1}(M)}(t) + (1+t)|F_1(\mathrm{syz}^{n-1}(M))|(t).$$

Ce qui établit la formule pour $n = 1$. On peut aussi écrire :

$$I_R^{\mathrm{syz}^n(M)}(t) = b_{n-1}(M)I_R(t)+tI_R^{\mathrm{syz}^{n-1}(M)}(t)-(1+t)|F_o(\mathrm{syz}^{n-1}(M))/F_1(\mathrm{syz}^{n-1}(M))|(t).$$

En utilisant le point 2, on obtient :

$$I_R^{\mathrm{syz}^n(M)}(t) = b_{n-1}(M)I_R(t) + tI_R^{\mathrm{syz}^{n-1}(M)}(t)-(1+t)t^{n-1}|F_{n-1}(M)/F_n(M)|(t).$$

Ce qui permet d'obtenir la formule par récurrence.

Remarque 1.- On peut démontrer la formule du théorème 2.2. en utilisant les résultats de Foxby [4] sans toutefois obtenir une description de terme complémentaire $(1+t)t^{n-1}|F_n(M)|(t)$, ce qui est le but recherché ici.

Remarque 2.- On peut définir sur $\mathrm{Tor}_*^R(M,k)$ une filtration similaire à la filtration $F.$. Cette nouvelle filtration intervient dans une formule explicite de la série de Poincaré des modules de co-syzygie d'un module M [9].

2.3. Exemple.- Soit M un R-module de type fini, de dimension injective finie d, $I_R^M(t) = a_o +\ldots+ a_d t^d$. La filtration $F.$ se décrit simplement :
$F_n(M) = \bigoplus_{j < d+1-n} \mathrm{Tor}_j^R(M^{\vee},k)$ (Voir [9] proposition 1.2.9 pour la démonstration).
Ainsi $|F_n(M)|(t) = a_o +\ldots+ a_{d-n}t^{d-n}$ et comme d'autre part
$P_R^M(t) = I_R(t)\cdot I_R^M(t^{-1})$ [4], on obtient des formules explicites pour la série de Bass des modules de syzygie de M.

Il est clair que la filtration $F.$ est triviale ($\forall n$ $F_n(M) = \mathrm{Tor}_*^R(M^{\vee},k)$) si et seulement si le produit homologiques associé à M est nul, autrement dit si $W(M) = 0$. Dans ce cas, la formule du théorème 2.2 se simplifie et on obtient :

2.4. Théorème.- Soit M un R-module de type fini. Si $W(M) = 0$ la série de Bass de $\mathrm{syz}^n(M)$ est donnée pour $n > 0$ par la formule :

$$I_R^{syz^n(M)}(t) = (b_{n-1}(M)+tb_{n-2}(M) +\dots+ t^{n-1}b_o(M))I_R(t) + t^n I_R^M(t).$$

Lorsque R n'est pas un anneau régulier, $W(k) = 0$ (1.6) et on obtient le théorème B de l'introduction puisque $I_R^k(t) = P_R^k(t)$.

Dans ([11], chapitre 7), J.E. Roos introduit la dimension syzygétique d'un anneau local R :

$\gamma(R) = \inf \{t \mid$ tout $t^{\text{ième}}$ module de syzygie de type fini est projectivement équivalent à un $(t+1)^{\text{ième}}$ module de syzygie$\}$, ou $\gamma(R) = \infty$ s'il n'existe pas de tel t. (On dit que M et N sont projectivement équivalents si on peut trouver p et q tels que $M\oplus R^p \simeq N\oplus R^q$). Si l'anneau R est un anneau de Gorenstein, on a $\gamma(R) = \dim R$. Nous allons voir que dans la plupart des autres cas, on a $\gamma(R) = \infty$, ([11], problème 3.8, p. 249).

2.5. Proposition.- Soit R un anneau local de profondeur d. Si $\mu_d(R) \neq 1$ on a $\gamma(R) = \infty$.

Preuve : Si $\gamma(R) < \infty$ et $q > p \geq \gamma(R)$, alors on vérifie facilement que tout $p^{\text{ième}}$ module de syzygie de type fini est projectivement équivalent à un $q^{\text{ième}}$ module syzygie. Soit $M = syz^r(k)$, $r > d$ on va montrer que M n'est pas projectivement équivalent à un $r+2^{\text{ième}}$ module de syzygie. Comme l'anneau R n'est pas régulier, le théorème B permet d'écrire :

$$I_R^{M\oplus R^p}(t) / I_R(t) = (p+b_{r-1}(k)+tb_{r-2}(k) +\dots+ t^{r-1}b_o(k))+t^r P_R^k(t) / I_R(t).$$

C'est une série formelle à cœfficients dans $\mathbb{Q}$. Comme le premier terme non nul de $I_R(t)$ est $\mu_d(R)t^d$, et que le terme constant de $P_R^k(t)$ est 1, seuls les r-d premiers cœfficients de la série ci-dessus sont dans $\mathbb{N}$. Soit $n > d$, en utilisant le théorème 2.2 on vérifie que $I_R^{syz^n(N)\oplus R^q}(t) / I_R(t)$ a au moins ses n-1-d premiers cœfficients dans $\mathbb{N}$. Donc si $M\oplus R^p \simeq syz^n(N)\oplus R^q$ on doit avoir $n \leq r+1$. Par conséquent on a nécessairement $\gamma(R) = \infty$.

III. UNE CONSTANTE ASSOCIEE A L'ANNEAU R

Soit M un R-module de type fini, on a vu (corollaire 1.3) que $W_j(syz^r(M)) = 0$ si $j < r$. On peut espérer que si r est choisi assez grand on ait en fait $W(syz^r(M)) = 0$. On est conduit à définir une constante $\sigma(R)$ de la façon suivante :

3.1. Définition.- Soit $(R,\underline{m})$ un anneau local. Posons

$\sigma(R) = \inf \{r \mid$ pour tout R-module M de type fini, $W(syz^r(M)) = 0\}$.

S'il n'existe pas de tel entier, on pose $\sigma(R) = \infty$.

En utilisant le point 2 de la démonstration du théorème 2.2 (qui permet de comparer les filtrations de M et de $syz^r(M)$) on constate que les assertions, $W(syz^r(M)) = 0$, et, $F_r(M) = F_\infty(M)$, sont équivalentes. Donc $\sigma(R)$ peut aussi être définie par :

$\sigma(R) = \inf \{r \mid$ pour tout R-module M de type fini, $F_r(M) = F_\infty(M)\}$.

Ainsi $\sigma(R)$ mesure la complexité de l'anneau R vis à vis de la formule du théorème 2.2. Comme $W(R) \neq 0$ on a toujours $\sigma(R) \geq 1$.

Question : Pour un anneau local $(R,\underline{m})$, a-t-on toujours $\sigma(R) < \infty$?

Avant de donner des exemples où il en est bien ainsi, notons deux propriétés simples de $\sigma(R)$

a) $\sigma(R) > \text{prof } R = d$.

En effet, il existe des modules M de type fini, de dimension projective d. On a alors $W(syz^d(M)) = W(R) \neq 0$.

b) Soit x un élément de l'idéal $\underline{m}$, x non diviseur de zéro alors

$\sigma(R) \leq \sigma(R/xR) + 1$

On obtient ce résultat en comparant les modules de syzygie sur R à des modules de syzygie sur R/xR (pour les détails voir [9], 1.3.7).

3.2. Proposition.- Soit $(R,\underline{m})$ un anneau local de socle $\underline{a}$; notons $s : R \longrightarrow R/\underline{a}$ la projection canonique. Alors $\sigma(R) = 1$ si et seulement si l'application $\check{s}_*$ est nulle.

Preuve : L'application $\check{s}_*$ est nulle si et seulement si $W(R/a) = 0$ (théorème 1.4). Tout $r^{\text{ième}}$-module de syzygie M avec $r > 0$, est annulé par $\underline{a}$. Donc si $W(R/\underline{a}) = 0$ on a $W(M) = 0$ (corollaire 1.5), et par suite $\sigma(R) = 1$. Réciproquement soit $x_1,\ldots,x_n$ un système générateur de $\underline{m}$; l'application $f : R/\underline{a} \longrightarrow R^n$, $f(y) = (yx_1,\ldots,yx_n)$ est un homomorphisme injectif. Par conséquent $R/\underline{a}$ est un premier module de syzygie. Si

$\sigma(R) = 1$, on a $W(R/\underline{a}) = 0$.

Il existe des anneaux R tels que $\sigma(R) = 1$:

Soit $(R,\underline{m})$ un anneau local tel que $\underline{m}^3 = 0$, de socle $\underline{m}^2$. On suppose en outre que R n'est pas de la forme $B/\underline{b}^3$ où $(B,\underline{b})$ est un anneau régulier. Dans ces conditions $I_R^{R/\underline{m}^2}(t) = (|\underline{m}/\underline{m}^2|-t).P_R^k(t)$ ([2] ou [8] exemple 1.11). Considérant la suite exacte : $0 \longrightarrow (R/\underline{m}^2)^\vee \xrightarrow{s^\vee} R^\vee \longrightarrow (\underline{m}^2)^\vee \longrightarrow 0$, on en déduit que la condition $s_*^\vee = 0$ est satisfaite si et seulement si $I_R(t) = (|\underline{m}^2|-|\underline{m}/\underline{m}^2|t+t^2).P_R^k(t)$. L'existence d'anneaux vérifiant cette dernière condition a été démontrée par Bøgvad [2].

On notera qu'il existe des anneaux artiniens avec $\sigma(R) > 1$: il suffit de prendre le produit fibré de deux anneaux vérifiant la condition précédente et de considérer sa série de Bass qui est calculée dans [7] ou [8].

3.3. <u>Proposition</u>.- <u>Soit</u> $(R,\underline{m})$ <u>un anneau local de Gorenstein de dimension d, alors</u> $\sigma(R) = d+1$.

<u>Preuve</u> : En effet la seule composante non nulle de $\mathrm{Tor}_*^R(R^\vee,k)$ est $\mathrm{Tor}_d^R(R^\vee,k)$. On en déduit que pour tout R-module M, $F_{d+1}(M) = F_\infty(M)$. Comme la profondeur de R est d on a nécessairement $\sigma(R) = d+1$.

3.4. <u>Proposition</u>.- <u>Soit</u> $(R,\underline{m})$ <u>un anneau de Golod</u>, $n = |\underline{m}/\underline{m}^2|$. <u>Alors</u> $\sigma(R) \leqslant n+2$.

(Pour la définition des anneaux de Golod, voir [5] ou [6]).

<u>Preuve</u> : En utilisant l'argument de Ghione et Gulliksen ([5] théorème 1) on montre le

<u>Lemme</u>.- <u>Soient</u> $(R,\underline{m})$ <u>un anneau de Golod et</u> $f : M \longrightarrow N$ <u>un homomorphisme de R-modules. Pour que</u> $f_* = 0$ <u>il faut et il suffit que</u> $f_p = 0$ <u>pour</u> $0 \leqslant p \leqslant 1+|\underline{m}/\underline{m}^2|$. ([9], 1.3.10).

Soient M un R-module de type fini et $P.$ une résolution projective minimale de M. Compte tenu du point 1 de la démonstration du théorème 2.2, pour établir que $F_{n+2}(M) = F_\infty(M)$ il suffit de montrer que pour tout $r \geqslant n+2$, l'homomorphisme de connexion associé à la suite exacte

$0 \longrightarrow (\mathrm{syz}^r(M))^\vee \xrightarrow{s^{r,\vee}} P_r^\vee \longrightarrow (\mathrm{syz}^{r+1}(M))^\vee \longrightarrow 0$ est surjectif, ou ce qui revient au même que $s_*^{r,v} = 0$. Le corollaire 1.3 montre en particulier que $W_{o,q}(syz^r(M)) = 0$ si $q < r$. Par conséquent, pour tout $x \in Tor_o^R(syz^r(M),k)$ et tout $y \in Tor_q^R((syz^r(M))^v,k)$, $q < r$, on a $<x,y> = 0$. En utilisant le lemme de la démonstration du théorème 2, on peut conclure que $s_q^{r,v} = 0$ pour $q < r$ donc finalement que $s_*^{r,v} = 0$ en utilisant le lemme précédent.

Bibliographie

[1] H. BASS.- On the ubiquity of Gorenstein rings, Math. Z., 82 (1963), 8-28.

[2] R. BØGVAD.- Gorenstein rings with transcendental Poincaré series, Math. Scand., 53, 1983, 5-15.

[3] H. CARTAN, S. EILENBERG.- Homological Algebra, Princeton Univ. Press, Princeton, N.J., 1956.

[4] H.B. FOXBY.- Isomorphisms between complexes with applications to the homological theory of modules, Math. Scand., 40 (1977), 5-19.

[5] F. GHIONE, T.H. GULLIKSEN.- Some reduction formulas for the Poincaré series of modules, Atti. Accad. naz. Lincei LVIII Ser., Rend., Cl. Sci. fis. mat. natur., 58 (1975), 82-91.

[6] T.H. GULLIKSEN, G.L. LEVIN.- Homology of local rings, Queen's papers in pure and applied Mathematics, n° 20, Queen's Univ., Kingston, Ontario, (1969).

[7] J. LESCOT.- La série de Bass d'un produit fibré d'anneaux locaux, Comptes-rendus, Acad. Sci., Paris, 293, Série A (1981), 569-571.

[8] J. LESCOT.- La série de Bass d'un produit fibré d'anneaux locaux, Séminaire d'Algèbre P. Dubreil et M.P. Malliavin (1982), Lecture Notes in Mathematics 1029, 218-239, Springer 1983.

[9] J. LESCOT.- Produit homologique associé à un module et applications, Prépublication n° 14 (1983), Dép. de Math. et de Mécanique, Univ. de CAEN.

[10] C. PESKINE, L. SZPIRO.- Dimension projective finie et cohomologie locale, Inst. Hautes Etudes Sci. Publ. Math., Paris, n° 42 (1973), 47-119.

[11] J.E. ROOS.- Finiteness conditions in commutative algebra and solution of a problem of Vasconcelos, Commutative Algebra, Durham 1981, Ed. by R. Sharp, London Math. Soc., Lecture Notes, Vol. 72, 1982, 179-203.

Département de Mathématiques,
Informatique et Mécanique
Université de CAEN

14032 CAEN CEDEX - FRANCE

ON THE SUBALGEBRA GENERATED BY THE ONE-DIMENSIONAL ELEMENTS IN THE YONEDA EXT-ALGEBRA

by

Clas LÖFWALL

INTRODUCTION. Let k be a field and R a ring with a ring epimorphism $R \longrightarrow k$. Then k is a module over R and $E = \mathrm{Ext}_R(k,k)$ is a graded algebra under the Yoneda product. We study the structure of this algebra in two situations. In the first case R is an augmented algebra over k (in general non-commutative) and in the second case $R = (R,m)$ is a local (commutative noetherian) ring with $R/m = k$. We are mainly interested in the second case, but within this theory it is natural to consider certain algebras of the first type. For example the Yoneda algebra E itself is an augmented algebra and the homology theory for E gives information about the structure of E. Another example is the homology algebra of the Koszul complex for R. The Yoneda algebra for local rings has been studied by Levin [9] and Sjödin [15], [18] and Roos [13]. Sjödin determines the structure of E when the local ring is a complete intersection or a Golod ring. In [13] Roos gives an example of a local ring for which E is not finitely generated (this answers negatively a conjecture by Levin [9]). In this example the subalgebra of E generated by the one-dimensional elements plays an important rôle. The main goal for our work will be to "compute" this subalgebra in the two cases described above and examine to what extent it is a good approximation of the whole Yoneda algebra.

Summary

In part 1 of the paper we consider rings of the first type, i.e. augmented algebras over a field k. If R belongs to a certain class of such algebras, we prove that the subalgebra A generated by the one-dimensional elements in the Yoneda algebra $E = \mathrm{Ext}_R(k,k)$ is equal to the free non-commutative algebra on E^1 divided by the two-dimensional relations $\ker(E^1 \otimes E^1 \rightarrow E^2)$. This theorem and its corollaries may essentially be found in Priddy [12]. He studies algebras satisfying $A = E$ and we prove that this condition is equivalent to the "Fröberg formula" (see [5]) being true. When the cube of the augmentation ideal of R is zero, we prove that $E \simeq A \otimes T(V)$ as left A-modules and as right T(V)-modules, where T(V) is the free non-commutative algebra on the graded vector space V. We also give a formula for the Hilbert series H_E of E (which is the same as the Poincaré series of R and is defined by the formal power series $\sum_{n\geq 0}\dim_k(E^n)z^n$) in terms of H_A. As an application we study the homology algebra of the Koszul complex for a local ring with imbedding dimension three. Our methods in part 1 applies to get the rationality of the Poincaré series of a class of local rings, indeed let $R = k[t_1,\ldots,t_n]$/monomials of degree two + $(t_1,\ldots,t_n)^p$ where p is any number ≥ 2 then the Poincaré series of R is rational.

In the second part of the paper we study local commutative rings (R,m). We give an equivalent condition for a local ring-homomorphism $\phi: (R,m) \to (S,n)$ to be a Golod homomorphism in the sense of Levin [8]. Indeed, if ϕ is surjective the condition is as follows:

There is an exact sequence of Hopf algebras

$$k \to T(V) \to \mathrm{Ext}_S(k,k) \to \mathrm{Ext}_R(k,k) \to k$$

where $T(V)$ is the free non-commutative algebra on $V = \coprod_{i\geq 2} V_i$ and $V_{i+1} = \mathrm{Ext}^i_R(S,k)$ for $i\geq 1$.

We prove that for some rings there is a differential R-algebra U, which is a free R-algebra with divided powers, containing the Koszul complex and contained in a minimal resolution such that $\mathrm{Hom}_R(U,k) = A$. This is the case when R is two-homogeneous or $m^4 = 0$ and if both these conditions are fulfilled, i.e. $R = \mathrm{Gr}R$ (with respect to the m-filtration) and $m^4 = 0$, then E is a "semi-tensor-product" (see Smith [19]) of A and a free non-commutative algebra. If furthermore A is nilpotent as Hopf-algebra (see Sjödin [17]) we prove that P_R is rational.

Part 1. Non-commutative algebras

Notations and basic facts

1. k is always a commutative field.
2. A graded vector space $V = \coprod_{i\geq 0} V_i$ is <u>locally finite</u> if $|V_i| = \dim_k V_i < \infty$ for all $i\geq 0$. The <u>Hilbert series</u> of V is the formal power series $H_V(z) = \sum_{i\geq 0} |V_i| z^i$.

3. if V, W are graded vector spaces then $V \otimes W$ is a graded vector space with $(V \otimes W)_n = \coprod_{i+j=n}(V_i \otimes W_j)$ and $\mathrm{Hom}_k(V,W)$ is a graded vector space with $\mathrm{Hom}_k(V,W)_n = \prod_i \mathrm{Hom}_k(V_i, W_{i-n})$, especially $V^* = \mathrm{Hom}_k(V,k)$ is a graded vector space with $V_n^* = \mathrm{Hom}_k(V_n,k)$.

4. We use strict sign convention. This means that when we in a defining formula replace $a \cdot b$ by $b \cdot a$, we must multiply by $(-1)^{\deg(a) \cdot \deg(b)}$. For more details about this principle we refer to Gunnar Sjödin [16] . If a and b are bigraded elements with bigrade (s_1,t_1) and (s_2,t_2) , we use the sign $(-1)^{s_1 s_2 + t_1 t_2}$ when a and b are interchanged in a formula. The graded commutator $[a,b]$ is defined by $ab - (-1)^{s_1 s_2 + t_1 t_2} ba$, for $a \neq b$ and $[a,a] = a^2$ or 0 according to $s_1 + t_1$ odd or even.

5. Let $U \subset X$ be graded vector spaces. Define U^0 as $\{f \in X^* : f(U) = 0\} \simeq$ $\simeq (X/U)^*$. Let ϕ be a linear map $X \to Y$ then,

$$(\ker\phi)^0 = \mathrm{im}\phi^* \quad \text{and} \quad (\mathrm{im}\phi)^0 = \ker\phi^* .$$

If U and V are subspaces of X then,

$$(U + V)^0 = U^0 \cap V^0 \quad \text{and} \quad (U \cap V)^0 = U^0 + V^0 \quad \text{and} \quad U^{00} = U ,$$

where if $V \subset X^*$, V^0 means the set

$\{x \in X;\ f(x) = 0 \text{ for all } f \in V\}$.

6. There is a natural map

$$V^* \otimes V^* \to (V \otimes V)^*$$

defined by

$$f \otimes g(x \otimes y) = (-1)^{st} f(x)g(y) \quad \text{for } f \in (V^*)_s \quad \text{and} \quad g \in (V^*)_t \quad .$$

This map is a <u>monomorphism</u>, indeed suppose $0 \neq \sum_{1\leq i\leq n} f_i \otimes g_i \longrightarrow 0$ and $g_1,\dots,g_n$ are linearly independent, and $n\geq 1$ is minimal with this property.

g_1 linearly dependent of $g_2,\dots,g_n$

$\Longleftrightarrow$

$\operatorname{im} g_1^* \in \operatorname{im} g_2^* + \dots + \operatorname{im} g_n^*$ (where g_i^* is the dual of the map $g_i: V \to k$)

$\Longleftrightarrow$

$\ker g_1 \supset \ker g_2 \cap \dots \cap \ker g_n$

(cf. no.5). So, if we suppose that g_1 is linearly independent of $g_2,\dots,g_n$, then there is an element $y \in V$ such that $g_1(y) \neq 0$ but $g_i(y) = 0$ for $i = 2,\dots,n$. It follows that $f_1(x) = 0$ for all x, i.e. $f_1 = 0$ but this contradicts the minimality of n.

7. A <u>graded algebra</u> R is an associative ring with unit which is also a graded vector space $R = \coprod_{i\geq 0} R_i$ and $1 \in R_0$ and $R_i \cdot R_j \subset R_{i+j}$.

R is <u>augmented</u> if there is a map $\varepsilon: R \longrightarrow k$ of graded algebras such that $\varepsilon(1) = 1$. The kernel of ε is called the <u>augmentation ideal</u> and is written $I(R)$ (or just I). R is <u>locally finite</u> if $|R_i| < \infty$ for all i.

R is <u>connected</u> if $R_0 = k$. If I/I^2 is locally finite, the <u>Hilbert series</u> of R as algebra is $H_R^{alg}(x,y) = \sum_{i\geq 0} H_{I^i/I^{i+1}}(y)x^i$.

8. $V^{\otimes n}$ is short for $V \otimes \dots \otimes V$ (n factors). $T(V)$ is the free non-commutative algebra on V (the tensor algebra). $T(V)$ is a bigraded vector space, $T(V) = \coprod_{n,s\geq 0} (V^{\otimes n})_s$, n is called the "tensor" degree and s is called the "pure" degree and is written $dt(a)$ and $dp(a)$ respectively for $a \in T(V)$. If $x_1,\dots,x_n$ are elements of $T(V)$ we write $(x_1,\dots,x_n)$ instead of $x_1 \otimes \dots \otimes x_n$, and the empty string $(\ )$ is equal to 1.

$T(V)^* = \operatorname{Hom}_k(T(V),k)$ is also a bigraded vector space and it is an algebra with the following definition (cf. no.6):

Let $f,g \in T(V)^{*}$ be homogeneous elements of degree (n,s) and (m,t) respectively. Then

$$f\cdot g(x \otimes y) = (-1)^{n\cdot m + s\cdot t} f(x)\cdot g(y)$$

where $x \in (V^{\otimes n})_s$ and $y \in (V^{\otimes m})_t$.

$T(V)$ is also a Hopf-algebra (see [11]) by requiring the elements of V to be primitive. If V is locally finite, the dual of $T(V)$ is also a Hopf-algebra and we denote it $\hat{T}(V^{*})$. This is even a Hopf-Γ-algebra and it is possible to define $\hat{T}(M)$ for any graded module M over a commutative ring (cf. André [1]): $\hat{T}(M) = T(M)$ as bigraded modules. Let $x = (x_1,\ldots,x_k)$, $y = (y_1,\ldots,y_l)$, $\hat{x} = (x_2,\ldots,x_k)$ and $\hat{y} = (y_2,\ldots,y_l)$. Then

$$x\cdot y = (x_1,\hat{x}\cdot y) + (-1)^{dp(y_1)\cdot dp(x)}(y_1,x\cdot\hat{y})$$

$$\Delta x = \sum_{0\le i\le k}(x_1,\ldots,x_i) \otimes (x_{i+1},\ldots,x_k)$$

$$x^{(n)} = (x_1,\hat{x}\cdot x^{(n-1)}) \quad , \quad dp(x) \text{ even.}$$

This defines the structure of $\hat{T}(M)$ inductively (using the formula $(x+y)^{(n)} = \sum_{i+j=n} x^{(i)}\cdot y^{(j)}$) .

9. If R is any ring, we define the Yoneda product (A,B,C are left R-modules)

$$\mathrm{Ext}_R(B,C) \otimes \mathrm{Ext}_R(A,B) \xrightarrow{\circ} \mathrm{Ext}_R(A,C)$$

in the following way(see [15]): Choose projective resolutions PA of A and PB of B . Let $f \in Z\mathrm{Hom}(PA,B)$ and $g \in Z\mathrm{Hom}(PB,C)$. (If (X,d_X) and (Y,d_Y) are complexes, then $\mathrm{Hom}(X,Y)$ has a differential d defined by:

$$df(x) = d_Y(f(x)) - (-1)^{\deg(f)} f(d_X(x)) \text{).}$$

There is a map $\tilde{f} \in Z\mathrm{Hom}(PA,PB)$, i.e. $d_B \circ \tilde{f} = (-1)^{\deg(f)} \tilde{f} \circ d_A$, such that $\varepsilon \circ \tilde{f} = f$ where $\varepsilon\colon PB \to B$. We now define $[g] \circ [f]$ as $[g \circ \tilde{f}]$.
This product is independent of all choices and gives,for any left R-module A , $\mathrm{Ext}_R(A,A)$ a structure of a graded associative ring.

If R is a graded augmented algebra, $\mathrm{Ext}_R(k,k)$ is a bigraded algebra. The definition of the product is unchanged. If (P,d) is a graded free resolution of k, then d has bidegree $(-1,0)$ so the sign $(-1)^{\deg(f)}$ in the definition of $\tilde{f}$ follows the principle given in no.4.

The subalgebra of $\mathrm{Ext}_R(k,k)$ generated by $\mathrm{Ext}^1_R(k,k)$ is written $[\mathrm{Ext}^1_R(k,k)]$.

10. Let R be a graded augmented algebra with augmentation ideal I. The <u>Bar-resolution</u> (B,d) is a graded free resolution of k over R defined as follows: $B = R \otimes_k T(I) = R\cdot T(I)$, d is R-linear and for $(x_1,\dots,x_n) \in T(I)$

$$d(x_1,\dots,x_n) = x_1\cdot(x_2,\dots,x_n) + \sum_{1\leq i<n}(-1)^i(x_1,\dots,x_i\cdot x_{i+1},\dots,x_n) .$$

B is a bigraded R-module and d has bidegree $(-1,0)$. (B,d) is acyclic, since there is a homotopy operator s defined by

$$s(r\cdot(x_1,\dots,x_n)) = (r,x_1,\dots,x_n) \text{ if } r\in I \text{ and } s(1\cdot(x_1,\dots,x_n)) = 0 \text{ and}$$

s k-linear. We have $\mathrm{Ext}_R(k,k) = H\mathrm{Hom}_R(B,k) = H(T(I)^*,d^*)$ as bigraded vector spaces. We will now examine the Yoneda product. Suppose $f \in T(I)^*_{nj}$ where $dt(f) = n$ and $dp(f) = j$.

$$d^*f(x) = (-1)^{n+1}f(d(x)) = (-1)^{n+1}\sum_{1\leq i\leq n}(-1)^i f(x_1,\dots,x_i\cdot x_{i+1},\dots,x_{n+1})$$

Suppose $d^*f = 0$, we define a "lifting" $\tilde{f}$ by:

$$\tilde{f}(x_1,\dots,x_{n+k}) = (-1)^{n\cdot k + j\cdot dp(x_1,\dots,x_k)} f(x_{k+1},\dots,x_{n+k})\cdot(x_1,\dots,x_k)$$

and $\tilde{f}$ R-linear. $\tilde{f}$ is homogeneous of degree $(-n,-j)$ and a simple computation shows that $d\tilde{f} = (-1)^n\tilde{f}d$. Let $f\in Z^{ns}\mathrm{Hom}_R(B,k)$ and $g\in Z^{kt}\mathrm{Hom}_R(B,k)$ representing elements $[f]\in \mathrm{Ext}^{ns}_R(k,k)$ and $[g]\in \mathrm{Ext}^{kt}_R(k,k)$. Then,

$$[g]\cdot[f] = [g\circ\tilde{f}] \in \mathrm{Ext}^{n+k,s+t}_R(k,k)$$

and

$$g\circ\tilde{f}(x_1,\dots,x_{n+k}) = (-1)^{nk+st}g(x_1,\dots,x_k)f(x_{k+1},\dots,x_{n+k}).$$

One easily proves that

$$d^*(f\cdot g) = d^*f\cdot g + (-1)^{dt(f)}f\cdot d^*g$$

which shows that $H(T(I)^*,d^*)$ is an algebra, and it follows from no.8 and no.9 that

$$\mathrm{Ext}_R(k,k) = H(T(I)^*,d^*)$$

as algebras (see also Cartan [4]).

11. Suppose R is a graded locally finite algebra and X_n , $n\geq 0$, are graded locally finite vector spaces and

$$Y_n = R \otimes X_n \quad \text{and} \quad Y = \coprod_{n\geq 0} Y_n \, .$$

Suppose further there is a homogeneous differential d making (Y,d) to a graded complex such that,

$$\cdots \longrightarrow Y_{n+1,j} \longrightarrow Y_{nj} \longrightarrow Y_{n-1,j} \longrightarrow \cdots$$

is finite for each j (e.g. if $X_{nj} = 0$ for almost all n for each j or if $d: R_s \otimes X_{nt} \longrightarrow R_{s+1} \otimes X_{n-1,t-1}$). Then for each j ,

$$\sum_n (-1)^n |Y_{nj}| = \sum_n (-1)^n |H_{nj}(Y)|$$

which gives the formula,

$$H_R(z) \cdot H_X(-1,z) = H_{H(Y)}(-1,z) \, .$$

This "lemma" is due to Jörgen Backelin. It has several applications, e.g. the following proved in Lemaire [7] :

Let R be a connected augmented locally finite algebra. The construction of a minimal resolution yields a resolution of the type $R \otimes \mathrm{Tor}_n^R(k,k)$ where $\mathrm{Tor}_{nj}^R(k,k) = 0$ for $j<n$. Hence the lemma applies and we get

$$H_R(z) \cdot P_R(-1,z) = 1 \, . \qquad (\, P_R \text{ is the same as the Hilbert series of } \mathrm{Tor}^R(k,k) \,)$$

Suppose now, $X_{nj} = 0$ for $j \neq n$ and $X_{nn} = K_n$ and

$$d: \; R_s \otimes K_n \longrightarrow R_{s+1} \otimes K_{n-1} \, .$$

We have $Y_{nj} = R_{j-n} \otimes K_n$. Put $Y_{nj} = Y'_{n,j-n}$. Then,

$$H_{H(Y)}(-1,z) = H_{H(Y')}(-z,z)$$

and hence we get the formula

$$H_R(z) \cdot H_K(-z) = H_{H(Y')}(-z,z)$$

We are going to use this formula several times in the paper. For instance if (R,m) is a local ring with imbedding dimension r and $R = GrR$ with respect to the m-filtration, we may apply the formula above to the Koszul complex of R. In this case $H_K(-z) = (1-z)^r$ and we get the formula mentioned in [5].

12. Let R be a graded augmented algebra with augmentation ideal $I = \coprod_{i\geq 0} I_i$ Suppose I_0 = Jacobson radical of R_0 (i.e. R_0 is local). If $M = \coprod_{n\geq 0} M_n$ is an R-module with each M_n a finitely generated R_0-module (or $I_0 = 0$) and $IM = M$, then $M = 0$.

<u>Proof</u>: Suppose $M_i = 0$ for $i<n$. $I_0 M_n = M_n$ and M_n finitely generated R_0-module (or $I_0 = 0$) implies $M_n = 0$.

Two-homogeneous algebras

Let R be an augmented graded algebra over k, and let I be the augmentation ideal with degree ≥ 0.

<u>Definition</u>. R is two-homogeneous if there is a decomposition as graded vector spaces $I = V \coprod I^2$ such that $V^2 \cap I^3 = 0$ (I^2 is short for $I \cdot I$).

If R is two-homogeneous then as a k-space $R = k \coprod V \coprod V^2 \coprod I^3$ (and conversely).

<u>Proposition 1.1</u> If R is two-homogeneous then the subalgebra generated by V $(R = k \coprod V \coprod V^2 \coprod I^3)$ is isomorphic to $T(V)/\mathfrak{a}$ where, as a k-space, $\mathfrak{a} = \mathfrak{a}_2 \coprod \mathfrak{a}_3$ with $\mathfrak{a}_2 \subset V \otimes V$ and $\mathfrak{a}_3 \subset \coprod_{i\geq 3} V^{\otimes i}$. Conversely $T(V)/\mathfrak{a}$ with $\mathfrak{a}$ as above is two-homogeneous.

Proof. Consider the morphism $T(V) \rightarrow R$. If $\sum_{i\geq 2} x_i$, $x_i \in V^{\otimes i}$ is in the kernel, then x_2 must be in the kernel since $V^2 \cap I^3 = \{0\}$. Hence the kernel has the stated property. Conversely $T(V)/\mathfrak{a}$ is certainly two-homogeneous if $\mathfrak{a} = \mathfrak{a}_2 \amalg \mathfrak{a}_3$ with $\mathfrak{a}_2 \subset V \otimes V$ and $\mathfrak{a}_3 \subset \coprod_{i\geq 3} V^{\otimes i}$, indeed let I be the augmentation ideal of $T(V)$ and $\bar{I}$, $\bar{V}$ the images of I, V in $T(V)/\mathfrak{a}$. If $\bar{x} \in \bar{V}^2 \cap \bar{I}^3$, then there is $y \in V^2$ and $z \in I^3$ such that $\bar{y} = \bar{z} = \bar{x}$. Hence $y - z \in \mathfrak{a}$, which implies that $y \in \mathfrak{a}$ and $\bar{x} = 0$.

Theorem 1.1 (cf. [12, Theorem 2.5]) Let R be two-homogeneous with augmentation ideal I, $R = k \amalg V \amalg V^2 \amalg I^3$ and let ϕ be the multiplication map $V \otimes V \rightarrow V^2$ then,

$$[\mathrm{Ext}_R^1(k,k)] = T(V^*)/\mathfrak{a}$$

where $\mathfrak{a}$ is the two-sided ideal generated by $\mathrm{im}(\phi^*) \cap V^* \otimes V^*$. ($V^* \otimes V^*$ is embedded in $(V \otimes V)^*$ by the map given in Notations no.6).

Proof. From no.10 in Notations we have that $\mathrm{Ext}_R(k,k)$ as a bigraded algebra is equal to $H(T(I)^*, d^*)$. Let m be the multiplication map $I \otimes I \rightarrow I$. We have $\mathrm{Ext}_R^1(k,k) = \ker(m^*) = V^*$ (we identify in the sequel V^* with $(I^2)^0$ as a subspace of I^*). Hence,

$$(\mathrm{Ext}_R^1(k,k))^2 = V^* \otimes V^*/\mathrm{im}(m^*) \cap V^* \otimes V^* = V^* \otimes V^*/\mathrm{im}(\phi^*) \cap V^* \otimes V^*$$

and generally

$$(\mathrm{Ext}_R^1(k,k))^n = (V^*)^{\otimes n}/\mathrm{im}(d^*) \cap (V^*)^{\otimes n} .$$

Since $d^*(V^*) = 0$ we have $\mathfrak{a} \subset \mathrm{im}(d^*) \cap (V^*)^{\otimes n}$. We will now prove the other inclusion,

$$\mathrm{im}(d^*) \cap (V^*)^{\otimes n} = \{f \in (V^*)^{\otimes n} : \exists\, g \in \coprod_{0\leq i\leq n-2} (V^{\otimes i} \otimes V^2 \otimes V^{\otimes(n-i-2)})^* \text{ and } f = g \circ d\}$$

this set is contained in,

$$\sum_{0\leq i\leq n-2} \{f \in (V^*)^{\otimes n} : \exists\, g \in (V^{\otimes i} \otimes V^2 \otimes V^{\otimes(n-i-2)})^* \text{ and } f = g \circ (1^{\otimes i} \otimes \phi \otimes 1^{\otimes(n-i-2)})\}$$

Call this last set $\sum_{0\leq i\leq n-2} M_i$. Let $f_1 \otimes \ldots \otimes f_n = f \in M_i$, then there is a map g such that,

$$g(v_1\otimes\ldots\otimes v_i\otimes v_{i+1}v_{i+2}\otimes v_{i+3}\otimes\ldots\otimes v_n) = f(v_1\otimes\ldots\otimes v_n) = \pm f_1(v_1)\cdot\ldots\cdot f_n(v_n) .$$

Let x be a fixed element in V^2. Then $x = v\cdot v'$ for some $v,v' \in V$ and

$$g(v_1\otimes\ldots\otimes v_i\otimes x\otimes v_{i+3}\otimes\ldots\otimes v_n) = c(x)f_1(v_1)f_2(v_2)\cdot\ldots\cdot f_i(v_i)f_{i+3}(v_{i+3})\cdot\ldots\cdot f_n(v_n) .$$

$c(x)$ depends only on x . It follows that c is a linear form, and hence $g \in (V^*)^{\otimes i} \otimes (V^2)^* \otimes (V^*)^{\otimes(n-i-2)}$. Hence,

$$\mathrm{im}(d^*) \cap (V^*)^{\otimes n} \subset \sum_{0\le i\le n-2} (V^*)^{\otimes i} \otimes (\mathrm{im}(\phi^*) \cap V^* \otimes V^*) \otimes (V^*)^{\otimes(n-i-2)} \subset \mathfrak{a} .$$

Corollary 1.1 (cf. [12, p. 42]) Let $V = \coprod_{i\ge 0} V_i$ be a locally finite graded vector space and $R = T(V)/\mathfrak{a}$, where $\mathfrak{a} = \coprod_{i\ge 2} \mathfrak{a}_i$ and $\mathfrak{a}_i \subset V^{\otimes i}$. Then R is bigraded and $\mathrm{Ext}_R(k,k)$ is trigraded, where we let the first degree be the homological degree, the second arising from the tensor degree of R and the third arising from the pure degree of R . We have

$$[\mathrm{Ext}^1_R(k,k)] = \coprod_{n,q} \mathrm{Ext}_R^{n,n,q}(k,k) = T(V^*)/(\mathrm{im}\phi^*)$$

where ϕ is the multiplication map $V \otimes V \to V^2$.

Proof. Since R is connected in the first degree, it follows from the Bar-resolution that

$$\mathrm{Ext}_R^{n,n}(k,k) = (V^*)^{\otimes n}/\mathrm{im}(d^*) \cap (V^*)^{\otimes n}$$

and since R is two-homogeneous and V is locally finite

$$\mathrm{im}(d^*) \cap (V^*)^{\otimes n} = (\mathrm{im}(\phi^*)) .$$

Corollary 1.2 (cf. [12, Proposition 9.1]) Let R be a graded augmented k-algebra with augmentation ideal I and suppose I/I^2 is locally finite. Let $A = [\mathrm{Ext}^1_R(k,k)]$. Then,

$$[\mathrm{Ext}^1_A(k,k)] = T(I/I^2)/(\ker(\phi))$$

where $\phi\colon I/I^2 \otimes I/I^2 \to I^2/I^3$ is induced by multiplication.

Proof. A is a graded augmented algebra. Put $V = \mathrm{Ext}_R^1(k,k) = (I/I^2)^*$.

$A = k \coprod \coprod_{i\geq 1} V^i$, so A is two-homogeneous. From the proof of Theorem 1.1 we get that A is a quotient of $T((I/I^2)^*)/(\mathrm{im}\ \phi^*)$ (since I/I^2 is locally finite). Let $\psi: V \otimes V \to V^2$ be the map induced by multiplication in A . Then $\mathrm{im}\ \phi^* = \ker \psi$. Hence,

$$\mathrm{im}\ \psi^* = (\mathrm{im}\ \phi^*)^0 = (\ker \phi)^{00} = \ker \phi$$

(cf. Notations no. 5).

Remark. The corollary holds also for a local commutative noetherian ring R with maximal ideal I , indeed Jan-Erik Roos has proved in [13] that in this case, $\ker(\mathrm{Ext}_R^1(k,k) \otimes \mathrm{Ext}_R^1(k,k) \to \mathrm{Ext}_R^2(k,k)) = \mathrm{im}((I^2/I^3)^* \to (I/I^2)^* \otimes (I/I^2)^*)$.

Corollary 1.3 Let $k[x_1,\ldots x_n]$ denote the free graded strict commutative algebra on the variables $x_1,\ldots,x_n$ with certain degrees $\deg x_i \geq 0$, $i = 1,\ldots,n$. Let

$$f_i = \sum_{j\leq k} b_{ijk} x_j x_k \quad , \; b_{ijk} \in k \quad , \; i = 1,\ldots,r$$

be homogeneous elements. Put

$$R = k[x_1,\ldots,x_n]/(f_1,\ldots,f_r) .$$

Then

$$A = [\mathrm{Ext}_R^1(k,k)] = k\langle T_1,\ldots,T_n\rangle/(\phi_1,\ldots,\phi_s)$$

as bigraded algebras, T_i has bidegree $(1, \deg x_i)$, and

$$\phi_i = \sum_{j\leq k} c_{ijk} [T_j, T_k] \quad , \; c_{ijk} \in k \; , \; i = 1,\ldots,s$$

($[T_j,T_k]$ is the graded commutator defined in Notations no.4) and $(c_{ijk})_{jk}$ $i = 1,\ldots,s$ is a basis of the solutions of the linear equation system

$$\sum_{j\leq k} b_{ijk} X_{jk} = 0 \; , \; i = 1,\ldots,r$$

(hence $s = n(n+1)/2 - r$).

Also $[\mathrm{Ext}_A^1(k,k)] \simeq R$ where R is bigraded giving x_i bigrade $(1, \deg x_i)$.

<u>Proof.</u> Let V be a graded k-space with basis $\{X_1,\dots,X_n,\ \deg X_i = \deg x_i\}$. The map $V \to R$ defined by $X_i \to x_i$ extends to a map $T(V) \to R$. Let $\phi: V \otimes V \to R$ be the restriction of this map to $V \otimes V$. Then $R \simeq T(V)/(\ker \phi)$ and from Theorem 1.1 $A \simeq T(V^*)/(\mathrm{im}\ \phi^*)$. $\ker \phi$ is generated by

$$F_1,\dots,F_r \quad \text{and} \quad [X_i, X_j] = X_i \otimes X_j - (-1)^{\deg x_i \cdot \deg x_j} X_j \otimes X_i \qquad i,j = 1,\dots,n$$

where $F_i = \sum_{j \le k} b_{ijk} X_j X_k$. It follows that $\mathrm{im}\ \phi^* = (\ker \phi)^0$ is generated by elements satisfying the conditions in the corollary. The last statement follows from Corollary 1.2 . Q.E.D.

<u>Homogeneous Koszul algebras</u>

Following Priddy [12] we call an algebra of the form $T(V)/(\mathfrak{a}_2)$ with $\mathfrak{a}_2 \subset V \otimes V$ a "<u>homogeneous pre-Koszul algebra</u>". A homogeneous pre-Koszul algebra R is a "<u>homogeneous Koszul algebra</u>" if $[\mathrm{Ext}^1_R(k,k)] = \mathrm{Ext}_R(k,k)$. In this section we will prove that R is a homogeneous Koszul algebra if and only if the "Fröberg formula" $P_R(x,y) \cdot H_R(-x,y) = 1$ holds. To do this we first construct a complex (U,d), which is the foundation of the material of this paper. The complex coincides with the "Koszul resolution" defined in Priddy [12] when it is acyclic.

From now on R is a two-homogeneous algebra with augmentation ideal I and I/I^2 <u>locally finite</u>. $I = V \coprod I^2$ and $V^2 \cap I^3 = \{0\}$. Let $\phi: V \otimes V \to V^2$ as usual be restriction of multiplication. Define graded locally finite vector spaces K_n as follows: $K_n = 0$ for $n<0$, $K_0 = k$ and $K_1 = V$. Suppose K_r is defined for $r \le n$ and there is an inclusion map $i_n: K_n \to V \otimes K_{n-1}$. Then K_{n+1} is defined by

$$K_{n+1} = \ker(V \otimes K_n \xrightarrow{1 \otimes i_n} V \otimes V \otimes K_{n-1} \xrightarrow{\phi \otimes 1} V^2 \otimes K_{n-1}) .$$

Hence for $n \ge 2$,

$$K_n = \bigcap_{i=0}^{n-2} V^{\otimes i} \otimes \ker \phi \otimes V^{\otimes (n-i-2)} .$$

$U_n = R \otimes K_n$ is a graded free R-module. There is a homogeneous map $d\colon U_n \to U_{n-1}$ with $d^2 = 0$, indeed $d(R \otimes K_0) = 0$ and for $n \geq 1$

$$d\colon R \otimes K_n \xrightarrow{1 \otimes i_n} R \otimes V \otimes K_{n-1} \xrightarrow{m \otimes 1} R \otimes K_{n-1}$$

where $m\colon R \otimes V \to R$ is induced by multiplication in R .

It is easy to see that $d^2 = 0$ and it is formally proved by the following commutative diagram

$$\begin{array}{ccccc}
R \otimes K_n & \xrightarrow{1 \otimes i_n} & R \otimes V \otimes K_{n-1} & \xrightarrow{m \otimes 1} & R \otimes K_{n-1} \\
 & & \downarrow {\scriptstyle 1 \otimes 1 \otimes i_{n-1}} & & \downarrow {\scriptstyle 1 \otimes i_{n-1}} \\
 & & R \otimes V \otimes V \otimes K_{n-2} & \xrightarrow{m \otimes 1 \otimes 1} & R \otimes V \otimes K_{n-2} \\
 & & \downarrow {\scriptstyle 1 \otimes \phi \otimes 1} & & \downarrow {\scriptstyle m \otimes 1} \\
 & & R \otimes V^2 \otimes K_{n-2} & \xrightarrow{m \otimes 1} & R \otimes K_{n-2}
\end{array}$$

$d^2 = (m \otimes 1)(1 \otimes i_{n-1})(m \otimes 1)(1 \otimes i_n)$ and from the definition of K_n,

$$(1 \otimes \phi \otimes 1)(1 \otimes 1 \otimes i_{n-1})(1 \otimes i_n) = 0 .$$

Put $U = \coprod_{n \geq 0} U_n$. U is a bigraded R-module and $d\colon U \to U$ is of degree $(-1,0)$. The complex (U,d) has the following two properties

P1. $\bar{d}\colon U_n/IU_n \to IU_{n-1}/I^2U_{n-1}$ is mono for $n \geq 1$

P2. $Z_nU \subset I^2U_n + BU_n$ for $n \geq 1$.

<u>Proposition 1.2</u> $K = \coprod_{n \geq 0} K_n$ and $[\mathrm{Ext}_R^1(k,k)]$ are isomorphic as bigraded vector spaces.

<u>Proof.</u> K_n is a subspace of $V^{\otimes n}$ and $K_n^* \simeq (V^*)^{\otimes n}/(K_n)^0$. It follows from no.5 in Notations that

$(K_n)^0 = \sum (V^*)^{\otimes i} \otimes (\ker \phi)^0 \otimes (V^*)^{\otimes(n-i-2)} = \sum (V^*)^{\otimes i} \otimes \operatorname{im} \phi^* \otimes (V^*)^{\otimes(n-i-2)}$.

Hence $K^* \simeq T(V^*)/(\operatorname{im} \phi^*)$ as bigraded vector spaces. The statement now follows from Theorem 1.1 .

Proposition 1.3. Suppose R is two-homogeneous,

$$R = k \amalg V \amalg V^2 \amalg I^3 .$$

Consider the following properties for a complex (U,d) of R-modules.

(a) $U = R \otimes_k X$ where X is a graded vector space and $X_0 = k$,

(b) $dX \subset V \otimes X$,

(c) $ZU_n \subset I^2U_n + BU_n$ for $n \geq 1$.

Suppose (U,d) and (U',d') satisfy these conditions. Then there is a unique homomorphism $f\colon (U,d) \to (U',d')$ such that $f_0\colon R \to R$ is the identity map, and $fX \subset X'$. This homomorphism is an isomorphism. Moreover if X and X' are graded algebras such that $(R \otimes X,d)$ and $(R \otimes X',d')$ are differential algebras, i.e.

$$d(u\cdot v) = du\cdot v + (-1)^{\deg(u)}\, u\cdot dv$$

(and the same formula for d'), then the homomorphism $f\colon U \to U'$ is an algebra homomorphism.

Proof. Suppose $f_i\colon X_i \to X_i'$ is defined for $i<n$. From (a) and (b) we get a commutative diagram

$$\begin{array}{ccccccc}
0 & \to & X_n & \xrightarrow{d} & V \otimes X_{n-1} & \xrightarrow{d} & V^2 \otimes X_{n-2} \\
 & & & & \downarrow 1\otimes f_{n-1} & & \downarrow 1\otimes f_{n-2} \\
0 & \to & X_n' & \xrightarrow{d'} & V \otimes X_{n-1}' & \xrightarrow{d'} & V \otimes X_{n-2}'
\end{array} .$$

Because of (c) the rows are exact, and hence $f_n\colon X_n \to X_n'$ is uniquely defined by the diagram. By linearity we extend the maps $\{f_i\}$ to a map $f\colon U \to U'$. The uniqueness of f (or the "five-lemma") shows that f is an isomorphism. If f_{n-1} is an algebra homomorphism, then it is easy to see (using the fact that $d'\colon X_n' \to V \otimes X_{n-1}'$ is injective) that also f_n is an algebra homomorphism. Q.E.D.

Theorem 1.2 Suppose R is two-homogeneous, $R = k \coprod V \coprod V^2 \coprod I^3$, V locally finite and R generated by V as an algebra, then the following are equivalent:

(i) R is a homogeneous Koszul algebra,

(ii) $\operatorname{Ext}^1_R(k,k)$ generates $\operatorname{Ext}_R(k,k)$,

(iii) R is a homogeneous pre-Koszul algebra and $\operatorname{Ext}_R^{n,p,q}(k,k) = 0$ for $n \neq p$,

(iv) R is a homogeneous pre-Koszul algebra and all matric Massey products of $\operatorname{Ext}_R(k,k)$ are zero,

(v) R is a homogeneous pre-Koszul algebra and satisfies the "Fröberg formula" $P_R(x,y)\cdot H_R(-x,y) = 1$,

(vi) the complex (U,d) defined above is acyclic,

(vii) R is a homogeneous pre-Koszul algebra, $R = T(V)/(\mathfrak{a}_2)$ and R has distibutive associated lattices in the sense of Backelin [3] § 2.

Proof. We first prove that (ii) implies that R is a homogeneous pre-Koszul algebra. We know from Proposition 1.1 that $R = T(V)/\mathfrak{a}$ where $\mathfrak{a} = \mathfrak{a}_2 \coprod \mathfrak{a}_3$ with $\mathfrak{a}_2 \subset V \otimes V$ and $\mathfrak{a}_3 \subset \coprod_{i\geq 3} V^{\otimes i}$. Let Z be the kernel of $U_1 \to U_0$, i.e. $Z = \ker(R \otimes V \to R)$. From Proposition 1.2 and (ii) we get $\operatorname{Ext}^2_R(k,k) \simeq K_2 = \mathfrak{a}_2$. Hence $Z/VZ \simeq \mathfrak{a}_2$. The natural map $T(V) \to R \otimes V$ induces an exact sequence

$$0 \to \mathfrak{a}V \to \mathfrak{a} \to Z \to 0$$

Hence $\mathfrak{a}/V\mathfrak{a} + \mathfrak{a}V \simeq Z/VZ \simeq \mathfrak{a}_2$. It follows that $\mathfrak{a}$ is generated by $\mathfrak{a}_2$. Now (i) and (ii) are equivalent by definition and from Corollary 1. it follows that (ii) and (iii) are equivalent. For the proof of (i) <=> (iv) we refer to Priddy [12] and for the proof of (vi) <=> (vii) we refer to Backelin [3] .

(ii) $\Rightarrow$ (vi): Suppose $[\mathrm{Ext}_R^1(k,k)]^i = \mathrm{Ext}_R^i(k,k)$ for $i \leq p$. We claim that $\tilde{H}_i(U) = 0$ for $i \leq p-1$. Suppose $n<p$ and $\tilde{H}_i(U) = 0$ for $i \leq n-1$. We want to prove that $\tilde{H}_n(U) = 0$. Put $Z_n = \ker(U_n \to U_{n-1})$. We have $d: U_{n+1} \to Z_n$. We must show that this map is epi. But,

$$\bar{d}: K_{n+1} = U_{n+1}/IU_{n+1} \to Z_n/IZ_n$$

is mono because of P1 page 13 and the fact that $Z_n \subset IU_n$ (which also follows from P1). From Proposition 1.2 and since $n+1 \leq p$ we get

$$(Z_n/IZ_n)^* \simeq \mathrm{Ext}_R^{n+1}(k,k) \simeq [\mathrm{Ext}_R^1(k,k)]^{n+1} \simeq K_{n+1}^* .$$

Hence $\bar{d}$ is an isomorphism. Now Nakayama's lemma (see Notations no. 12) may be applied. Indeed, since $R = T(V)/(\mathfrak{a}_2)$ with $\mathfrak{a}_2 \subset V \otimes V$ (see above), the "tensor" grade gives R a structure of a graded connected algebra and $d: U_n \to Z_n$ is a homomorphism of graded modules.

(vi) $\Rightarrow$ (v): The complex $U = R \otimes K$ satisfies the conditions given in Notations at the end of no. 11. Since K has one additional grading, we get the following formula

$$H_K(x,y)H_R(-x,y) = H_{HU}(x,-x,y) \tag{1}$$

Since (U,d) is acyclic (v) will follow from (vi) if we can prove that $H_K = P_R$. But $dU \subset IU$ implies that $\mathrm{Hom}(U,k)$ has differential zero and hence $\mathrm{Ext}_R^n(k,k) \simeq K_n^*$.

(v) $\Rightarrow$ (ii): Suppose that $n \geq 1$ and $(\mathrm{Ext}_R^1(k,k))^i = \mathrm{Ext}_R^i(k,k)$ for $i \leq n$. We will prove that $(\mathrm{Ext}_R^1(k,k))^{n+1} = \mathrm{Ext}_R^{n+1}(k,k)$. From the proof of (ii) $\Rightarrow$ (vi) we have that $\tilde{H}_i(U) = 0$ for $i \leq n-1$ and from P2 page 13 it follows that $H_{ij}(U) = 0$ for $j=0,1$ if $i>0$. Hence $H_{ij}(U) = 0$ for $i+j=n+1$, so formula (1) shows that the x^{n+1} - term of $H_K(x,y)H_R(-x,y)$ is zero. The x^{n+1} - term of $P_R(x,y)H_R(-x,y)$ is also zero since we assume that (v) holds. Moreover $K_i \simeq \mathrm{Ext}_R^i(k,k)$ for $i \leq n$ by assumption and Proposition 1.2. Hence $H_K(x,y)$ and $P_R(x,y)$ are equal up to dimension n of the variabel x. It follows that $K_{n+1} \simeq \mathrm{Ext}_R^{n+1}(k,k)$ and hence $(\mathrm{Ext}_R^1(k,k))^{n+1} = \mathrm{Ext}_R^{n+1}(k,k)$.

<u>Remark</u> The proof originates essentially from Christer Lech. He proved (v) <=> (vi) in the ungraded commutative case.

<u>The case $(I(R))^3 = 0$.</u>

Suppose R is a graded augmented algebra with augmentation ideal I and $I^3 = 0$. Then trivially R is two-homogeneous, so the results of the previous sections are valid. In fact $R = T(V)/(\mathfrak{a}_2 + V^{\otimes 3})$ where $\mathfrak{a}_2 \subset V \otimes V$ and $V = I/I^2$, so R is bigraded. Suppose furthermore that V is <u>locally finite</u>. We may construct (U,d) as in the preceding section. We will now define a bigraded vector space C such that $U \otimes T(C)$ has a differential, that yields a minimal resolution of k over R.

We give K_n the tensor degree n, then $U_n = R \otimes K_n$ is bigraded and d is homogeneous of degree zero in both degrees. Put $C_{n+1} = \tilde{H}_n(U)$ ($C_n = 0$ for $n \leq 1$). Since

$$0 \to K_{n+1} \xrightarrow{d} V \otimes K_n \xrightarrow{d} V^2 \otimes K_{n-1}$$

is exact for all $n \geq 0$ and

$$U_n = K_n \amalg V \otimes K_n \amalg V^2 \otimes K_n$$

it follows that C_{n+1} is the cokernel of $d\colon V \otimes K_{n+1} \to V^2 \otimes K_n$. Hence C_n has tensor degree $n+1$. There is a map p such that

$$0 \to K_{n+1} \xrightarrow{d} V \otimes K_n \xrightarrow{d} V^2 \otimes K_{n-1} \xrightarrow{p} C_n \to 0$$

is exact for $n \geq 0$. Choose a k-linear homogeneous map $\alpha\colon C_n \to V^2 \otimes K_{n-1}$ such that $p\alpha = \mathrm{id}$. Observe that α is homogeneous also in the tensor degree. Put $C = \amalg C_n$ and $Y = U \otimes T(C)$. Then $Y_n = (U \otimes T(C))_n =$ $= \amalg_{p \geq 0, \Sigma n_i = n} U_{n_0} \otimes C_{n_1} \otimes \ldots \otimes C_{n_p}$ is a bigraded free R-module.

Define $d'\colon Y_n \to Y_{n-1}$ as follows: d' is R-linear and

$$d'(x \otimes v_1 \otimes \ldots \otimes v_k) = dx \otimes v_1 \otimes \ldots \otimes v_k \quad \text{for } x \in U_n,\ n>0$$

$$d'(1 \otimes v_1 \otimes \ldots \otimes v_k) = \alpha(v_1) \otimes v_2 \otimes \ldots \otimes v_k$$

where $v_i \in C$ for $i=1,\ldots,k$ are homogeneous elements. d' is clearly homogeneous of degree $(-1,0,0)$ and $(d')^2 = 0$, since $\alpha(v) \in V^2Y$ for $v \in C$.

<u>Proposition 1.4</u> (Y,d') is acyclic.

<u>Proof.</u> We have an exact sequence of complexes,

$$0 \to U \to Y \to Y \otimes C \to 0$$

where the differential on $Y \otimes C$ is $d' \otimes 1$. Since $H_0(U) = k$ also $H_0(Y) = k$. Suppose $n \geq 1$ and $H_i(Y) = 0$ for $1 \leq i \leq n-1$. We claim that $H_n(Y) = 0$. The following sequence is exact

$$H_{n+1}(Y \otimes C) \to H_n(U) \to H_n(Y) \to H_n(Y \otimes C)$$

Since C has degree (the "first" one) ≥ 2, $H_n(Y \otimes C) = (H(Y) \otimes C)_n = C_n$. But $Z_n(Y) \cap C_n = \{0\}$. Hence $H_n(Y) \to H_n(Y \otimes C)$ is the zero map. C_{n+1} is a part of $H_{n+1}(Y \otimes C)$ and the restriction of the map $H_{n+1}(Y \otimes C) \to H_n(U)$ to C_{n+1} is α which is epi. Hence $H_n(Y) \to H_n(Y \otimes C)$ is mono and therefore $H_n(Y)$ must be zero. Q.E.D. (The technique of the proof is due to Gerson Levin.)

<u>Corollary 1.4</u> The graded vectorspace $E^{n,r} = \mathrm{Ext}_R^{n,r}(k,k)$ is given by

$$E^{n,n+p} = \coprod_{\Sigma n_i = n} K_{n_0} \otimes C_{n_1} \otimes \ldots \otimes C_{n_p} \quad \text{for } p \geq 0 .$$

Also $E^{n,n+p} = 0$ for $p > n/2$. C is obtained from K by the following formula,

$$xH_C(x,y) + 1 = H_K(x,y)(1 - H_V(y)x + H_{V^2}(y)x^2) .$$

<u>Proof.</u> The definition of the tensor degree of K and C gives the first formula. To prove the second formula we use (1) on page 16. We have $C_{n+1} = \tilde{H}_{n,2}(U)$ where the sum of the indices of $\tilde{H}(U)$ is the tensor degree. Hence,

$$H_K(x,y)H_R(-x,y) = \sum_{n \geq 1} x^{n+2} H_{C_{n+1}}(y) + 1 = xH_C(x,y) + 1 . \text{ Q.E.D.}$$

We could also give a formula for the Poincaré series of R , but first we will define one additional degree of $\mathrm{Ext}_R(k,k)$ and thereby obtain a Poincaré series of R in <u>four</u> variables.

Let $A = K^* = [\mathrm{Ext}_R^1(k,k)] = [E^1]$ and $E = \mathrm{Ext}_R(k,k)$. The trigraded vector space E is filtered by $E \supset E^1E \supset (E^1)^2E \supset \ldots$. Let for $r \geq 0$ Q_r be the trigraded vectorspace $(E^1)^rE/(E^1)^{r+1}E$.

<u>Definition</u>. $P_R(x,y,z,u) = \sum_{r\geq 0} H_{Q_r}(x,y,z)u^r$, where x,y,z are the variables for the homological degree, the tensor degree and the pure degree respectively.

The coefficient before $x^ny^pz^qu^r$ in $P_R(x,y,z,u)$ is the dimension of

$$\coprod_{s+t=q} E^{r,r,s}E^{n-r,p-r,t}/\coprod_{s+t=q} E^{r+1,r+1,s}E^{n-r-1,p-r-1,t} .$$

We are going to compute this series in terms of H_A and for this reason we must consider the left A-module structure on E .

(Y,d') is a minimal resolution of k , so $E = \mathrm{Hom}_R(Y,k)$. Let $f \in \mathrm{Hom}_R(Y_1,k) = \mathrm{Hom}_k(V,k)$ and $g \in \mathrm{Hom}_R(Y_n,k)$. Lift g to a map $\tilde{g}_0\colon Y_n \to Y_0$. We must find a map $\tilde{g}_1 \in \mathrm{Hom}_R(Y_{n+1},Y_1)$ such that $d'\tilde{g}_1 = (-1)^n\tilde{g}_0d'$. Suppose $g \in \mathrm{Hom}_k(K_{n_1} \otimes C_{n_2} \otimes \ldots \otimes C_{n_k}, k)$. It is enough to define $\tilde{g}_1$ on $Y_{n+1} \otimes k$ and extend it linearly to Y_{n+1} . There are two factors of $Y_{n+1} \otimes k$ that are mapped under d' into $R \otimes K_{n_1} \otimes C_{n_2} \otimes \ldots \otimes C_{n_k}$, namely

$K_{n_1+1} \otimes C_{n_2} \otimes \ldots \otimes C_{n_k}$ and $C_{n_1+1} \otimes C_{n_2} \otimes \ldots \otimes C_{n_k}$.

The differential of the last one is in V^2Y_n , hence

$$d\tilde{g}_1(C_{n_1+1} \otimes C_{n_2} \otimes \ldots \otimes C_{n_k}) \in V^2$$

and hence

$$\tilde{g}_1(C_{n_1+1} \otimes C_{n_2} \otimes \ldots \otimes C_{n_k}) \in VY_1$$

and therefore

$$f \circ \tilde{g}_1(C_{n_1+1} \otimes C_{n_2} \otimes \ldots \otimes C_{n_k}) = 0 .$$

An element in $K_{n_1+1} \otimes C_{n_2} \otimes \ldots \otimes C_{n_k}$ may be written as $x \otimes y$ where $x \in K_1$ and $y \in K_{n_1} \otimes C_{n_2} \otimes \ldots \otimes C_{n_k}$. Define $\tilde{g}_1(x \otimes y)$ as

$$(-1)^{n + \deg(x)\deg(y)} g(y) \cdot x \in K_1 ,$$

then $d'\tilde{g}_1 = (-1)^n \tilde{g}_0 d'$ and

$$f \circ \tilde{g}_1(x \otimes y) = (-1)^{n + \deg(x)\deg(y)} f(x)g(y) .$$

Thus we have proved that $E = A \otimes T(C^*)$ as left A-module. We can also prove that $T(C^*)$ is a subalgebra of E and $E = A \otimes T(C^*)$ as right $T(C^*)$-modules (however nothing is said about the product $T(C^*) \cdot A$): Let $g \in \mathrm{Hom}_k((T(C))_n, k)$. $\tilde{g}$ is defined by the following: Let $x \otimes y \in K \otimes T(C)$ such that $y \in (T(C))_n$ and $x \in (K \otimes T(C))_m$ then

$$\tilde{g}(x \otimes y) = (-1)^{n \cdot m + \deg(x)\deg(y)} g(y) \cdot x .$$

<u>Theorem 1.3.</u> Let R be a graded augmented algebra with augmentation ideal I and $I^3 = 0$. Suppose $V = I/I^2$ is locally finite. Let $\phi: V \otimes V \to V^2$ be the map induced by multiplication in R. Then,

$$P_R(x,y,z,u) = xH_A(xyu,z)/(1 + x - H_A(xy,z)(1 - H_V(z)xy + H_{V^2}(z)x^2y^2))$$

where

$$A = [\mathrm{Ext}^1_R(k,k)] = T(V^*)/(\mathrm{im}\,\phi^*)$$

and where $P_R(x,y,z,u)$ is defined on page 19.

<u>Proof.</u> It follows from above that

$$Q_r^{n,n+p} \cong \coprod_{\Sigma n_i = n-r} K_r \otimes C_{n_1} \otimes \ldots \otimes C_{n_p} .$$

Hence,

$$H_{Q_r}(x,y,z) = \sum_{p \geq 0} (xy)^r H_{K_r}(z)(H_C(xy,z))^p y^p = (xy)^r H_{K_r}(z)/(1 - yH_C(xy,z)).$$

According to Corollary 1.4,

$$xyH_C(xy,z) + 1 = H_K(xy,z)(1 - H_V(z)xy + H_{V^2}(z)x^2y^2)$$

and if we use the fact that $H_K = H_A$ we obtain the formula for $P_R(x,y,z,u)$.

Remark. If we put $z = 0$ and $y = 1$ and $u = 1$ in (2) we get the Poincaré series for R_0 obtained in [10] .

Application. (This application is due to Gerson Levin.)
Let (R,m) be a commutative local ring with $|m/m^2| = 3$, which is not an artinian complete intersection. Wiebe [20] has proved that in this case $(H_1(K))^3 = 0$, where K is the Koszul complex. Hence $H(K)$ is a graded augmented algebra with $(I(H(K)))^3 = 0$, and the results above may be applied. From Avramov [2] we get

$$P_R(x) = (1+x)^3 P_{H(K)}(x,1,x,1) . \tag{3}$$

Hence the problem of rationality for P_R is reduced to that of $H_A(x,x)$ where $A = [\mathrm{Ext}^1_{H(K)}(k,k)]$.

Theorem 1.4. Let (R,m) be a local ring with imbedding dimension 3 , which is not an artinian complete intersection. Let K be the Koszul complex. Then $H(K) = T(V)/(\mathfrak{a}_2 + V^{\otimes 3})$ where $\mathfrak{a}_2 \subset V \otimes V$ and $V = I(H(K))/(I(H(K)))^2$. Suppose $S = T(V)/(\mathfrak{a}_2)$ is a homogeneous Koszul algebra then,

$$P_R(x) = x(1+x)^3 / \big((1+x)H_S(-x,x) - H_{H(K)}(-x,x)\big) . \tag{4}$$

Proof. By Theorem 1.1 $A = [\mathrm{Ext}^1_{H(K)}(k,k)] = [\mathrm{Ext}^1_S(k,k)]$, and by Theorem 1.2 $H_A(x,y)H_S(-x,y) = 1$. Hence by Theorem 1.3

$$P_{H(K)}(x,1,x,1) = x/\big((1+x)H_S(-x,x) - H_{H(K)}(-x,x)\big)$$

and (4) follows from (3) .

Remark. Levin [9] has proved the rationality of P_R in imbedding dimension three if $(H_1(K))^2 = 0$. His formula is the same as (4) (in this case "$S = H(K)$"). Avramov has proved the rationality of P_R for "three generators and three relations", and again the formula for P_R is given by (4) .

A Poincaré series.

We end Part 1 with a theorem, that is a generalization of the method above to compute the Poincaré series for $H(K)$ to algebras that not necessarily satisfy the condition $I^3 = 0$.

Theorem 1.5. Suppose $S = T(V)/(\mathfrak{a}_2)$ is a homogeneous Koszul algebra ($\mathfrak{a}_2 \subset V \otimes V$ and V is locally finite). Let $p \geq 3$ and

$$R = T(V)/(\mathfrak{a}_2 + V^{\otimes p})$$

then

$$P_R(-x,y,z,u) = x^{p-2} \cdot (H_S(xyu,z))^{-1}/(x^{p-2} - 1 + (H_S(xy,z))^{-1} H_R(xy,z))$$

and particularly

$$P_R(-x,1,z,1) = x^{p-2}/((x^{p-2} - 1)H_S(x,z) + H_R(x,z)) .$$

Proof. Since R is two-homogeneous we may construct the complex (U,d) as above (see pp. 12-13). Put $C_{n+1} = \tilde{H}_n(U)$ ($C_n = 0$ for $n \leq 1$). By Theorem 1.2 there is an exact sequence

$$0 \to K_i \xrightarrow{d} V \otimes K_{i-1} \xrightarrow{d} \dots \xrightarrow{d} V^{p-1} \otimes K_{n-1} \to C_n \to 0$$

for $n \geq 0$ and $i = n+p-2$. Hence C_n has tensor degree $n+p-2$. Exactly as for $p = 3$ we get a minimal resolution of the form $U \otimes T(C)$ and $\mathrm{Ext}_R(k,k) = A \otimes T(C^*)$ as left A-modules where $A = [\mathrm{Ext}_R^1(k,k)] = \mathrm{Ext}_S(k,k)$.
As in the proof of Theorem 1.3 we get

$$P_R(x,y,z,u) = H_A(xyu,z)/(1 - y^{p-2} H_C(xy,z))$$

We have $C_{n+1} = \tilde{H}_{n,p-1}(U)$ where $n+p-1$ is the tensor degree, hence by (1) we get

$$H_A(x,z)H_R(-x,z) = \textstyle\sum_{n \geq 1} x^{n+p-1}(-1)^{p-1} H_{C_{n+1}}(z) + 1 .$$

Hence

$$x^{p-2} H_C(-x,z) = 1 - H_A(-x,z)H_R(x,z) .$$

We now put this into the formula above for P_R and use the fact that $H_A(-x,z) = (H_S(x,z))^{-1}$, which follows from Theorem 1.2 .

Corollary 1.5. Let for $p\geq 3$

$$R = k[x_1,\dots,x_n]/\mathfrak{a} + (x_1,\dots,x_n)^p$$

where $\deg(x_i) = 0$ and $\mathfrak{a}$ is generated by elements of the form x_ix_j for $1\leq i,j\leq n$ then,

$$P_R(-x) = x^{p-2}/((x^{p-2} - 1)H_S(x) + H_R(x))$$

where $S = k[x_1,\dots,x_n]/\mathfrak{a}$.

Proof. S is a homogeneous Koszul algebra according to Fröberg [5] .

Part 2. Commutative local rings.

Notations.

(R,m) is always a local noetherian commutative ring with maximal ideal m . k = R/m is the residue-field.

An R-algebra U is a graded (skew) commutative differential algebra over R , $U = \coprod_{i\geq 0} U_i$ and each U_i is a free finitely generated R-module and $U_0 = R$ (for more details and other related definitions such as reduced cycles $\tilde{Z}U$, reduced homology $\tilde{H}U$, divided polynomial algebra, the process of adjoining variables to kill cycles, free extension, etc. we refer to [6]).

All R-algebras are considered as augmented with the natural augmentation $R \rightarrow R/m$.

We use the following sign notation : $\bar{x} = (-1)^{\deg(x)+1}\cdot x$, instead of $(-1)^{\deg(x)}$ we write $(-1)^x$.

If M and N are submodules of L , we write M/N instead of $M/M\cap N$.

Serre R-algebras.

Definition. An R-algebra U is called an S-R-algebra (Serre R-algebra) if the following property holds

$$du \in m^2 U \Rightarrow u \in mU \quad \text{or} \quad u \in U_0$$

(cf. condition (c), appendix to chapter IV in Serre [14]).

Proposition 2.1 Let X be an S-R-algebra and $s \in \tilde{Z}X$, $s \notin BX + m^2X$. Then $X' = X\langle S: dS = s\rangle$ is an S-R-algebra.

Proof. Suppose $n \geq 1$ and $d(x_0 + x_1S + \ldots + x_nS^{(n)}) \in m^2X'$, where $x' = x_0 + x_1S + \ldots + x_nS^{(n)}$ is a homogeneous element in X' ($S^{(i)} = 0$ for $i>1$ if $\deg(S)$ is odd). We get $dx_{n-1} - \bar{x}_n s \in m^2X$, hence x_n is not a unit since $s \notin BX + m^2X$. Hence from $dx_n \in m^2X$ we get $x_n \in mX$. As a consequence of the definition of an S-R-algebra we have $\tilde{Z}X \subset mX$, hence $s \in mX$ and we get $dx_{n-1} \in m^2X$. It follows that $x_{n-1} \in mX$, since $\deg(x_{n-1}) > \deg(x_n) \geq 0$ Hence $dx_{n-2} \in m^2X$ etc. , finally we get $x' \in mX'$. Q.E.D.

Corollary 2.1. (Proposition 3 in appendix to chapter IV in Serre [14]). The Koszul complex is an S-R-algebra.

Proof. R is trivially an S-R-algebra. The Koszul complex is obtained by killing cycles that represents a basis for m/m^2 .

Corollary 2.2. There is a free extension U of the Koszul complex K (i.e. $U = K\langle \ldots S_i \ldots : dS_i = s_i\rangle$) such that U is an S-R-algebra and $\tilde{Z}U \subset m^2U + BU$.

Proof. Put $U^1 = K$ and suppose U^i is constructed for $i \geq 1$ such that U^i is an S-R-algebra and $\tilde{Z}_jU^i \subset m^2U^i + BU^i$ for $j<i$. Let $s_1, \ldots, s_n$ represent a basis for

$$Z_iU^i/(m^2U^i + BU^i) .$$

Define U^{i+1} as $U^i\langle S_1,\dots,S_n : dS_j = s_j\rangle$. It is clear that U^{i+1} is an S-R-algebra (Proposition 2.1) and

$$\tilde{Z}_j U^{i+1} \subset m^2 U^{i+1} + BU^{i+1} \quad \text{for } j \leq i$$

Finally, define U as the union of all U^i.

<u>Proposition 2.2</u>. Let U be an S-R-algebra. It is possible to imbed U in a minimal resolution Y of k, which is a free differential U-module (i.e. $Y = U \otimes_R F$, F is a graded free R-module, $F_0 = R$, and $d(u \otimes f) = du \otimes f - \bar{u} \cdot df$).

<u>Proof</u>. Put $Y^0 = U$ and suppose $i \geq 0$ and Y^i is constructed such that $\tilde{H}_j(Y^i) = 0$ for $j<i$ and $Y^i = U \otimes_R F^i$, where F^i is graded free, $F_0^i = R$ and $F_j^i = 0$ for $j>i$ and

$$d(u \cdot y) = (du) \cdot y - \bar{u} \cdot dy \quad \text{for } u \in U \text{ and } y \in Y^i$$

(Y^i is considered as a U-module in a natural way). Choose a free R-module F_{i+1} such that

$$F_{i+1} \otimes k \simeq \tilde{Z}_i Y^i / (m\tilde{Z}_i Y^i + B_i Y^i) .$$

Choose a map $\alpha : F_{i+1} \to \tilde{Z}_i Y^i$ such that the following diagram commutes

$$\begin{array}{ccc} F_{i+1} & \xrightarrow{\alpha} & \tilde{Z}_i Y^i \\ \downarrow & & \downarrow \\ F_{i+1}/mF_{i+1} & \longrightarrow & \tilde{Z}_i Y^i / (m\tilde{Z}_i Y^i + B_i Y^i) . \end{array} \qquad (5)$$

Put $Y^{i+1} = Y^i \coprod U \otimes F_{i+1} = U \otimes F^{i+1}$, where $F^{i+1} = F^i \coprod F_{i+1}$.

Define d on $U \otimes F_{i+1}$ by the "derivation" formula,

$$d(u \otimes f) = du \otimes f - \bar{u} \cdot \alpha(f) .$$

Then $d^2 = 0$ and for $u \in U$ and $y = u' \otimes f \in Y^{i+1}$ we get

$d(u \cdot y) = d(uu' \otimes f) = d(uu') \otimes f - \overline{uu'} \cdot \alpha(f) = (du) \cdot y - \bar{u} \cdot (du' \otimes f) + \bar{u} \cdot \bar{u}' \cdot \alpha(f) =$
$= (du) \cdot y - \bar{u} \cdot dy$.

We have $B_iY^{i+1} = \mathrm{im}(\alpha) + B_iY^i$. From (5) we get

$$\mathrm{im}(\alpha) + B_iY^i + m\tilde{Z}_iY^i = \tilde{Z}_iY^i .$$

Hence by Nakayamas lemma

$$\tilde{Z}_iY^i = \mathrm{im}(\alpha) + B_iY^i = B_iY^{i+1} .$$

Hence $\tilde{H}_i(Y^{i+1}) = 0$ and also $\tilde{H}_j(Y^{i+1}) = \tilde{H}_j(Y^i) = 0$ for $j<i$.

Hence Y^{i+1} satisfies the induction hypothesis. Finally put $Y = \bigcup_{i\geq 0} Y^i$. It follows that Y is <u>acyclic</u> (so far we have <u>not</u> used the fact that U is an S-R-algebra).

We will now prove that $\tilde{Z}Y \subset mY$. We have $\tilde{Z}U \subset mU$ and suppose $\tilde{Z}Y^j \subset mY^j$ for $j<i$. We will prove that $\tilde{Z}Y^i \subset mY^i$. Suppose $i\geq 1$ and $r\geq i$ and $x \in Z_r(Y^i)$,

$$x \in U_r \otimes F_0 \amalg U_{r-1} \otimes F_1 \amalg \dots \amalg U_{r-i} \otimes F_i .$$

We write $x = \sum_{0\leq k\leq i} a_k \otimes b_k$ where $b_k = (b_{k1},\dots,b_{kn_k})$ is a basis for F_k and $a_k = (a_{k1},\dots,a_{kn_k})$ is an n_k-tuple of elements in U_{r-k} and

$$a_k \otimes b_k = \sum_{1\leq l\leq n_k} a_{kl} \otimes b_{kl} .$$

We have $d(a_i) \otimes b_i = 0$, hence $d(a_i) = 0$ and hence $a_i \in mU_{r-i}$ if $r>i$. If $r=i$ we have $a_i \in R$ and $\alpha(a_ib_i) \in B(Y^{i-1})$, hence by (5) $a_i \in m$. Suppose now $a_k \in mU_{r-k}$ for $k\geq p+1$ $(p+1\leq i)$. We have $db_k \in \tilde{Z}Y^{i-1} \subset mY^{i-1}$ for all k, so $d(\sum_{0\leq k\leq p} a_k \otimes b_k) \in m^2Y^i$. Hence $d(a_p) \otimes b_p \in m^2Y^i$, hence $d(a_p) \in m^2U_{r-p}$ and hence $a_p \in mU_{r-p}$ since $p<r$. Hence we have proved that $a_k \in mU_{r-k}$ for all k such that $0\leq k\leq i$, and we get $x \in mY^i$. Q.E.D.

It is probably well-known that the Ext-algebra is generated by the one-dimensional elements if and only if any minimal resolution of k is an S-R-algebra, but anyway it follows from a lemma by Roos [13], which is a more general result.

Lemma 2.1. Let (R,m) be a local ring. Then $[\mathrm{Ext}_R^1(k,k)]$ is the kernel of the natural map:

$$\mathrm{Ext}_R(k,k) \rightarrow \mathrm{Ext}_R(R/m^2,k)$$

Proof. See [13, Cor 1, page 291].

Let Y be a minimal resolution of k. The dual of the map above is the natural map

$$H(Y/m^2Y) \rightarrow H(Y/mY) = Y/mY .$$

By the lemma $[\mathrm{Ext}_R^1(k,k)] = \mathrm{Ext}_R(k,k)$ if and only if this map is zero in degrees greater than zero, which is the same as saying that Y is an S-R-algebra.

Corollary 2.3. $[\mathrm{Ext}_R^1(k,k)] = \mathrm{Ext}_R(k,k)$ if and only if any S-R-algebra U satisfying $\tilde{Z}U \subset m^2U + BU$ is a minimal resolution.

Proof. The "if-part" clearly follows from above and Corollary 2.2. Suppose U is an S-R-algebra satisfying $\tilde{Z}U \subset m^2U + BU$. Proposition 2.2 gives a minimal resolution Y of the form $Y = U \otimes F$. Suppose $F \neq 0$ and i is the lowest index such that $F_i \neq 0$. We may choose $\alpha\colon F_i \rightarrow \tilde{Z}U$ such that $\alpha(F_i) \subset m^2U$. Take now $y \in F_i$, $y \notin mF_i$. We have $dy \in m^2Y$, hence Y is not an S-R-algebra and therefore we conclude from above that $\mathrm{Ext}_R(k,k)$ is not generated by its one-dimensional elements.

Proposition 2.3. Let U be any R-algebra such that there is a minimal resolution Y, which is a free differential U-module (cf. Proposition 2.2), $Y = U \otimes F$, $F_0 = R$, then,

(a) $U \otimes k$ is a subalgebra of $Y \otimes k$, $F \otimes k$ is a quotient Hopf algebra of $Y \otimes k$ and $F \otimes k$ is the algebra cokernel of $U \otimes k \rightarrow Y \otimes k$.

(b) If $HHom(1,\varepsilon)\colon HHom_R(U,U) \to Hom_R(U,k)$ is epi ($\varepsilon\colon U \to k$ is the augmentation), we have an exact sequence of Hopf algebras:

$$k \to U \otimes k \to Y \otimes k \to F \otimes k \to k$$

(c) If $dF \subset m^2 Y$ and U is an S-R-algebra then, $F \otimes k$ is a subalgebra of $Y \otimes k$, $U \otimes k$ is a quotient Hopf algebra of $Y \otimes k$ and $U \otimes k$ is the algebra cokernel $F \otimes k \to Y \otimes k$.

(d) If both the conditions in (b) and (c) are valid, then we have a split exact sequence of Hopf algebras:

$$k \to Hom_R(F,k) \to Hom_R(Y,k) \to Hom_R(U,k) \to k$$

i.e. $Hom_R(Y,k) = Hom_R(F,k) \odot Hom_R(U,k)$ (semi-tensor product, see Smith [19])

<u>Remark.</u> (a) , (b) and (c) may also be formulated in terms of the dual Hopf algebra $Hom_R(Y,k)$.

<u>Proof.</u> (a). The algebra structure on $Y \otimes k$ is determined by a map of complexes $\phi\colon Y \otimes Y \to Y$ lifting the identity on k. We define a map of this kind by induction. Suppose ϕ is defined up to degree n, such that

$$\phi(u_1y_1 \otimes u_2y_2) = (-1)^{y_1u_2}\, u_1u_2\phi(y_1 \otimes y_2)$$

where $u_i \in U$ and $y_i \in Y$ for $i=1,2$. Define ϕ on $(Y \otimes Y)_{n+1}$, first on $(F \otimes F)_{n+1}$ and then in general by

$$\phi(u_1f_1 \otimes u_2f_2) = (-1)^{f_1u_2}\, u_1u_2\phi(f_1 \otimes f_2)$$

where $u_i \in U$ and $f_i \in F$ for $i=1,2$. It is straight-forward to check that this defines a lifting $\phi\colon Y \otimes Y \to Y$. It is clear from this that $U \otimes k$ is a subalgebra of $Y \otimes k$ and that $\coprod_{i\geq 1} U_i \otimes F \otimes k$ is an ideal in $Y \otimes k$, hence $F \otimes k$ is the algebra cokernel of the map $U \otimes k \to Y \otimes k$. Proving $F \otimes k$ is a coalgebra quotient of $Y \otimes k$ is the same as proving $Hom(F,k)$ is a subalgebra of $Hom(Y,k)$.

Let $f \in \mathrm{Hom}(F,k)$. Define $\tilde{f} \in \mathrm{ZHom}(Y,Y)$ inductively. Suppose $\tilde{f}$ is defined up to dimension n , such that $\tilde{f}$ is zero on U and $\tilde{f}(uy) = (-1)^{uf}.u\tilde{f}(y)$ for $u \in U$ and $y \in Y$. Define $\tilde{f}$ on Y_{n+1} first on F_{n+1} in some way, as zero on U , and generally by

$$\tilde{f}(u \otimes v) = (-1^{uf}.u\tilde{f}(v) \quad \text{for } u \in U \text{ and } v \in F .$$

Again, it is easy to see that this defines a lifting $\tilde{f} \in \mathrm{ZHom}(Y,Y)$ of f . It follows that $\tilde{f}$ maps the ideal $\coprod_{i \geq 1} U_i \otimes F \otimes k$ in itself. Hence $f' \circ \tilde{\tilde{f}}$ is zero on this ideal if f and f' is, and thus we have proved that $\mathrm{Hom}(F,k)$ is a subalgebra of $\mathrm{Hom}(Y,k)$.

(b). We must prove that $\mathrm{Hom}(Y,k) \rightarrow \mathrm{Hom}(U,k)$ is an algebra map, i.e. $\mathrm{Hom}(\coprod_{j \geq 1} U \otimes F_j, k)$ is a two-sided ideal. It is a left ideal since a lifting of a map which is zero on U may be chosen to be zero on U , and it is a right ideal since, by assumption, to every $f \in \mathrm{Hom}(U,k)$ there is a lifting $\tilde{f} \in \mathrm{ZHom}(Y,Y)$ such that $\tilde{f}$ maps U into U .

(c). We first prove that it is possible to define the map $\phi: Y \otimes Y \rightarrow Y$, such that $\phi(F \otimes F) \subset F + mY$. Since $d(F \otimes F) \subset m^2(Y \otimes Y)$ it is enough to prove that $dy \in m^2Y \Rightarrow y \in F + mY$. But $dF \subset m^2Y$ so it is enough to show that if $y \in \coprod_{i \geq 1} U_i \otimes F$ and $dy \in m^2Y$ then $y \in mY$. Suppose

$$y = \textstyle\sum_{1 \leq i \leq r} y_i \quad \text{where } y_i \in U_i \otimes F_{r-i} \quad \text{for } 1 \leq i \leq r$$

and $dy \in m^2Y$. Then $y_1 \in mU_1 \otimes F_{r-1}$, since U is an S-R-algebra. Suppose $y_i \in mU_i \otimes F_{r-i}$ for $i \leq s-1$ then,

$$d(\textstyle\sum_{s \leq i \leq r} y_i) \in m^2Y$$

and as before we get $y_s \in mU_s \otimes F_{r-s}$. Hence $y \in mY$.

From $\phi(F \otimes F) \subset F + mY$ it follows that $F \otimes k$ is a subalgebra of $Y \otimes k$ and also that $U \otimes k$ is the algebra cokernel of the map $F \otimes k \rightarrow Y \otimes k$, since $\coprod_{j \geq 1} U \otimes F_j \otimes k$ is an ideal. The fact that $U \otimes k$ is a quotient coalgebra of $Y \otimes k$ will follow from Theorem 2.1 below, which says that

$Hom(U,k)$ is equal to the subalgebra generated by $Hom(U_1,k)$ in $Hom(Y,k)$.

(d). This follows from (b) and (c). Q.E.D.

On the structure of Ext(k,k).

We want to apply the proposition to the case when all Massey products of $\tilde{H}U$ are zero. To do this we start with the following proposition, whose proof has been obtained in collaboration with Gunnar Sjödin.

Proposition 2.4. Let U be an R-algebra. Suppose all Massey products of $\tilde{H}U$ are zero. Then there is a differential on $U \otimes_R \hat{T}(F)$, where F is a locally finite free R-module, such that $U \otimes_R \hat{T}(F)$ is a differential acyclic R-algebra. Here $\hat{T}(F)$ is the "shuffle" algebra defined in Notations no. 8 , and $\hat{T}(F)$ is graded by means of its pure degree.

Proof. Since products of reduced cycles in U are boundaries, $\tilde{H}U$ is annihilated by $m = \tilde{Z}_0U$. Let for every $i \geq 1$ F_i be a free R-module such that $F_{i+1} \otimes k \simeq \tilde{H}_iU$. Choose a map $\lambda_i: F_{i+1} \to \tilde{Z}_iU$, that lifts the isomorphism $F_{i+1} \otimes k \to \tilde{H}_iU$. We are going to extend this map to a map $\lambda: \hat{T}(F) \to U$ and by means of this define a differential on $U \otimes_R \hat{T}(F)$, such that $U \otimes_R \hat{T}(F)$ becomes a differential R-algebra (it is well-known that it is possible to define λ such that $U \otimes_R \hat{T}(F)$ becomes a complex). We will define λ and d on $\hat{T}(F)$ inductively with respect to the tensor degree of $\hat{T}(F)$ such that,

(i) λ is R-linear, zero on k and zero on the decomposable elements of $\hat{T}(F)$ as Γ-algebra.

(ii) $$d \circ \lambda = (\lambda \otimes \lambda) \circ \Delta$$

i.e. $d\lambda(x_1,\dots,x_k) = \sum_{1 \leq i \leq k-1} \overline{\lambda(x_1,\dots,x_i)}\lambda(x_{i+1},\dots,x_k)$.

(iii) $$d = d \otimes 1 + \lambda$$

i.e. $d(x_1,\dots,x_k) = \sum_{1 \leq i \leq k} \lambda(x_1,\dots,x_i) \otimes x_{i+1} \otimes \dots \otimes x_k$.

(iv) $$d(u \otimes f) = du \otimes f - \bar{u} \cdot df \quad \text{for } u \in U \text{ and } f \in \hat{T}(F).$$

It is easy to see that (ii) - (iv) implies that $d^2 = 0$ and the acyclicity of $U \otimes \hat{T}(F)$ is proved in the same way as Proposition 1.3. Suppose now λ is defined on $\hat{T}(F)$ for elements of tensor degree $\leq n-1$ such that (i) - (iv) holds. Then $\lambda \otimes \lambda \circ \Delta$ is zero on the decomposable elements of tensor degree n, indeed let x,y be elements in $\hat{T}(F)$ of positive degree then,

$$(\lambda \otimes \lambda) \circ \Delta(x \cdot y) = (\lambda \otimes \lambda)(\Delta x \cdot \Delta y) = 0 + (\lambda \otimes \lambda)(x \otimes y + (-1)^{xy} y \otimes x) = 0$$

and for $x = (x_1,\dots,x_k)$ of even positive (pure) degree,

$$(\lambda \otimes \lambda) \circ \Delta(x^{(n)}) = (\lambda \otimes \lambda)(\Delta x)^{(n)} \quad \text{where} \quad (\Delta x)^{(n)} =$$

$$= \sum \pm\left[(x_1,\dots,x_{i_1})^{(n_1)} \cdot \ldots \cdot (x_1,\dots x_{i_p})^{(n_p)}\right] \otimes \left[(x_{i_1+1},\dots,x_k)^{(n_1)} \cdot \ldots \cdot (x_{i_{p+1}},\dots,x_k)^{(n_p)}\right]$$

the only term, that might not be zero is $(\lambda \otimes \lambda)(x \otimes x)$, which appears for $n = 2$, but x is of even degree implies that $\lambda(x)$ is of odd degree and hence $\lambda(x) \cdot \lambda(x) = 0$.

We have $(\lambda \otimes \lambda) \circ \Delta(x) \in \langle \lambda(x_1),\dots,\lambda(x_k) \rangle$ = the set of k-fold Massey products defined by $\lambda(x_1),\dots,\lambda(x_k)$, indeed a defining system is $\{A_{ij}\} = \{\lambda(x_i,\dots,x_j)\}$. Hence, for every $x \in (\hat{T}(F))_n$ = the elements of tensor degree n, there is a $\lambda(x)$ such that (ii) holds. By Lemma 2.3 below we may choose a basis of the free R-module $(\hat{T}(F))_n$ that, as a subset, contains a basis for the decomposable elements of tensor degree n. Since $(\lambda \otimes \lambda) \circ \Delta$ is R-linear, it is enough to define λ on a basis. It follows that it is possible to define λ on $(\hat{T}(F))_n$ such that (i) and (ii) holds.

<u>Lemma 2.2</u>. Suppose λ is zero on the decomposable elements of $\hat{T}(F)$ as an <u>algebra</u> then, $(U \otimes \hat{T}(F), d)$ is a differential algebra where d is given by (iii) and (iv) above.

<u>Proof</u>. We must prove that

$$d(x \cdot y) = (dx) \cdot y + (-1)^x x \cdot (dy) \quad \text{for} \quad x,y \in U \otimes \hat{T}(F) .$$

It is easy to see that it is enough to prove the formula for $x,y \in \hat{T}(F)$. We prove it by induction on $dt(x \cdot y)$.

Let $x = (x_1,\dots,x_k)$, $y = (y_1,\dots,y_n)$, $\hat{x} = (x_1,\dots,x_{k-1})$ and $\hat{y} = (y_1,\dots,y_{n-1})$. We get

$$d(x\cdot y) = d(x\cdot\hat{y} \otimes y_n + (-1)^{x_k y}\, \hat{x}\cdot y \otimes x_k) = d(x\cdot\hat{y}) \otimes y_n +$$

$$+ (-1)^{x_k y} d(\hat{x}\cdot y) \otimes x_k + \lambda(x\cdot y) = (dx\cdot\hat{y}) \otimes y_n + (-1)^{x_k y}(d\hat{x}\cdot y) \otimes x_k +$$

$$+ (-1)^{x}\left[(x\cdot d\hat{y}) \otimes y_n + (-1)^{x_k y + x_k}(\hat{x}\cdot dy) \otimes x_k\right] = (dx)\cdot y + (-1)^{x} x\cdot(dy) .$$

The first equality follows from the definition of the product in $\hat{T}(F)$, for the second equality we use (iii) and for the third we use (i) and the induction hypothesis. For the fourth equality we use (iii) twice indeed,

$$(dx\cdot\hat{y}) \otimes y_n + (-1)^{x_k y}(d\hat{x}\cdot y) \otimes x_k =$$

$$= ((d\hat{x} \otimes x_k)\cdot\hat{y}) \otimes y_n + \lambda(x)\cdot y + (-1)^{x_k y}(d\hat{x}\cdot y) \otimes x_k =$$

$$= (d\hat{x} \otimes x_k)\cdot y + \lambda(x)\cdot y = (dx)\cdot y ,$$

the expression $(-1)^{x}[\dots]$ is treated in the same way.

Lemma 2.3. $\hat{T}(F)$ is a free Γ-algebra.

Proof. $\hat{T}(F) \otimes k = \hat{T}(F \otimes k)$ is a free Γ-algebra (see e.g. André [1]). Let $\{S_i\}$ be a basis for $\hat{T}(F) \otimes k$ as Γ-algebra. Choose $\{x_i\}$, such that $\bar{x}_i = S_i$. Let B be the set of all monomials $x_{i_1}^{(n_1)}\cdot\ldots\cdot x_{i_p}^{(n_p)}$. Since $\overline{x\cdot y} = \bar{x}\cdot\bar{y}$ and $\overline{x^{(n)}} = \bar{x}^{(n)}$, $\bar{B}$ is a basis for $\hat{T}(F) \otimes k$ as vector space. By Nakayama's lemma, we get that B is a basis for $\hat{T}(F)$ as R-module.

This ends the proof of Proposition 2.4.

Corollary 2.4. Let ϕ: $(R,m) \to (S,n)$ be a local ringhomomorphism, then ϕ is a Golod homomorphism in the sense of Levin [8] if and only if there is an exact sequence of Hopf algebras

$$k \to T(V) \to \mathrm{Ext}_S(k,k) \to \mathrm{Ext}_R(k,k) \to k$$

where the right map is induced by ϕ and V is a graded vector space such that

$$V_1 = (n/mS)^* \quad \text{and} \quad V_{i+1} = (\mathrm{Tor}_i^R(k,S))^* \quad \text{for} \quad i \geq 1$$

(for the "if-part" it is supposed that n/mS and $\mathrm{Tor}_i^R(k,S)$ are annihilated by n).

Remark. The corollary may be applied to trivial extensions. In this case there is also a homomorphism $\psi: S \to R$ such that $\psi \circ \phi = \mathrm{id}$, hence the sequence splits to the right (i.e. it is a semi-tensor product).

Proof. Suppose $\phi: R \to S$ is Golod. Let X be a minimal resolution of k over R , $X^* = X \otimes_R S$ and F a free S-module such that $F_{i+1} \otimes_S k \simeq \simeq \tilde{H}_i(X^*)$ for $i \geq 0$. We know from Levin [8] that all Massey products of $\tilde{H}_i(mX^*)$ are zero, which enables us to apply Proposition 2.4 and also to choose λ (and d) such that $(X^* \otimes \hat{T}(F),d)$ becomes a minimal algebra resolution. Since the lifting property holds for X^* , we may apply Proposition 2.3 (b) and get the desired exact sequence, where the coalgebra structure on $T(V)$ is the natural one. That also the algebra structure of $T(V)$ is the natural one follows from an explicit formula for the lifting of an element in $\mathrm{Hom}_S(\hat{T}(F),k)$ (this is formally the same as the one used in Part 1 page 20).

Conversely given the exact sequence we get a formula for the Poincaré series of S which shows that ϕ is Golod (see Levin [8]).

Theorem 2.1. Suppose R is a local ring with a minimal resolution of k of the form $Y = U \otimes F$, where U is an S-R-algebra and $dF \subset m^2Y$. Then $\mathrm{Hom}_R(U,k)$ is equal to the subalgebra of $\mathrm{Hom}_R(Y,k)$ generated by $\mathrm{Hom}_R(Y_1,k)$, i.e. $\mathrm{Hom}_R(U,k) = [\mathrm{Ext}_R^1(k,k)]$.

Proof. It is possible to give a direct proof by proving that the image of the multiplication map

$$\mathrm{Hom}_R(U_1,k) \otimes \mathrm{Hom}_R(U_{n-1},k) \to \mathrm{Hom}_R(Y_n,k)$$

is $\mathrm{Hom}_R(U_n,k)$. Firstly $dF \subset m^2Y$ implies that the image is contained in $\mathrm{Hom}_R(U_n,k)$, and secondly one proves that the dual of the map

$$\mathrm{Hom}_R(U_1,k) \otimes \mathrm{Hom}_R(U_{n-1},k) \longrightarrow \mathrm{Hom}_R(U_n,k)$$

is a monomorphism, which follows from the fact that U is an S-R-algebra. It is however also possible to use Lemma 2.1. We have to prove that the algebra cokernel of the natural map $\phi\colon H(Y/m^2Y) \longrightarrow Y/mY$ is exactly $U \otimes k$. From the proof of (c) in Proposition 2.3 we have

$$dy \in m^2Y \Rightarrow y \in F + mY .$$

Hence $\mathrm{im}(\phi)$ is equal to $F \otimes k$ and Proposition 2.3 (c) gives that $U \otimes k$ is the algebra cokernel of ϕ .

<u>Theorem 2.2</u>. Suppose (R,m) is a local ring and $m^4 = 0$, then there is a graded vector space $V = \coprod_{i\geq 1} V_i$, such that $T(V)$ is a sub-Hopf-algebra of $\mathrm{Ext}_R(k,k)$ and

$$\mathrm{Ext}_R(k,k) \simeq [\mathrm{Ext}_R^1(k,k)] \otimes T(V)$$

as coalgebras and also as left $[\mathrm{Ext}_R^1(k,k)]$ -modules and as right $T(V)$-modules.

<u>Proof</u>. Apply Corollary 2.2 to get an S-R-algebra U with the property $\tilde{Z}U \subset m^2U + BU$. Since $m^4 = 0$ all Massey products of $\tilde{H}U$ are trivially zero. Proposition 2.4 gives an algebra resolution of the form $U \otimes_R \hat{T}(F)$. Since U is an S-R-algebra, $BU \subset mU$ and since $\tilde{Z}U \subset m^2U + BU$ the map λ , defining the differential d of $U \otimes \hat{T}(F)$, may be chosen such that $\mathrm{im}(\lambda) \subset m^2U$. Hence $(U \otimes \hat{T}(F),d)$ is a <u>minimal</u> algebra resolution. Thus

$$\mathrm{Tor}^R(k,k) \simeq (U \otimes k) \otimes \hat{T}(F \otimes k)$$

as algebras, and if we put $V = (F \otimes k)^*$ we get an isomorphism as coalgebras

$$\mathrm{Ext}_R(k,k) \simeq \mathrm{Hom}(U,k) \otimes T(V) \simeq [\mathrm{Ext}_R^1(k,k)] \otimes T(V) .$$

The last isomorphism follows from Theorem 2.1. The assertion about the algebra structure of $\mathrm{Ext}_R(k,k)$ follows from explicit lifting formulas as on the pages 19 and 20. It is however also possible to apply the theorems 4.4 and 4.7 in Milnor-Moore [11] :

At first we consider the Hopf algebra map

$$A = [\mathrm{Ext}^1_R(k,k)] \longrightarrow \mathrm{Ext}_R(k,k) = B .$$

It follows from the proof of Proposition 2.3 that $T(V) = k \otimes_A B$. Theorem 4.4 in [11] now gives that $B \simeq A \otimes T(V)$ as left A-modules. Secondly we consider the coalgebra map

$$B = \mathrm{Ext}_R(k,k) \longrightarrow [\mathrm{Ext}^1_R(k,k)] = A$$

(arising from the algebra map $U \longrightarrow Y$). It follows from Proposition 2.3 (a) that $T(V) = k \,\square_A\, B$. Theorem 4.7 in [11] gives that $B \simeq A \otimes T(V)$ as right $T(V)$-modules.

<u>Remark</u>. Compare this theorem with the result obtained for non-commutative rings with $m^3 = 0$ in Part 1 page 20.

<u>Definition</u>. Let (R,m) be a local ring, then GrR is the graded associated ring with respect to the filtration $F_pR = m^pR$.

<u>Theorem 2.3</u>. Suppose (R,m) is a local ring with $m^3 = 0$. Then

$$A = [\mathrm{Ext}^1_R(k,k)] = [\mathrm{Ext}^1_{GrR}(k,k)]$$

and there is a graded vector space V , such that $T(V)$ is a sub-Hopf-algebra of $\mathrm{Ext}_R(k,k)$ and of $\mathrm{Ext}_{GrR}(k,k)$ and

$$\mathrm{Ext}_R(k,k) \simeq \mathrm{Ext}_{GrR}(k,k) \simeq A \otimes T(V)$$

as coalgebras and as left A-modules and as right $T(V)$-modules. Moreover

$$P_R(x) = P_{GrR}(x) = xH_A(x)/(1 + x - H_A(x)H_R(-x)) .$$

<u>Remark</u>. The fact that $P_R(x) = P_{GrR}(x)$ was first proved by Levin.

<u>Proof</u>. We construct (U,d) as in Corollary 2.2. We have (GrU, Grd_1) is a GrR-algebra, where $d_1 \colon U \to s^{-1}U$ is a map of filtered modules and $F_p(s^{-1}U) = F_{p+1}(U) = m^{p+1}U$. It is easily seen to be an S-GrR-algebra. Since $\tilde{Z}U \subset m^2U + BU$ we also have $\tilde{Z}GrU \subset m^2GrU + BGrU$, and hence $\tilde{H}U = \tilde{H}GrU = m^2U/BU$.

It trivially holds that $U \otimes_R k = GrU \otimes_{GrR} k$ and

$$\mathrm{Hom}_R(U,k) = \mathrm{Hom}_{GrR}(GrU,k) .$$

Hence everything follows from Theorem 2.1 and 2.2. The formula for P_R follows from Theorem 1.3. Q.E.D.

Next we are going to look for sufficient conditions for a ring to satisfy the premises of Theorem 2.1.

Definition. A local ring (R,m) is said to have "property E" if there is a minimal resolution (Y,d) of k of the form $Y = U \otimes F$, where U is an S-R-algebra and F is a free graded R-module with $dF \subset m^2Y$.

Lemma 2.4. A local ring (R,m) has property E if there is an S-R-algebra U satisfying

$$du \in m^4U \Rightarrow u \in m^2U + BU \quad \text{or} \quad u \in U_0 , \tag{6}$$

if furthermore

$$du \in m^3U \Rightarrow u \in m^2U + BU \quad \text{or} \quad u \in U_0 \tag{7}$$

then

$$[\mathrm{Ext}^1_R(k,k)] = [\mathrm{Ext}^1_{GrR}(k,k)] .$$

Proof. The condition (6) implies that $\tilde{Z}U \subset m^2U + BU$, in particular we have $B_0U = m$. We use the same notations as in the proof of Proposition 2.2. Suppose $\tilde{Z}Y^i \subset m^2Y^i + BY^i$ and $dF^i \subset m^2Y^{i-1}$ for $i<j$, then $d: F_j \to Y^{j-1}$ may be chosen such that $dF_j \subset m^2Y^{j-1}$. We must prove that also

$$\tilde{Z}Y^j \subset m^2Y^j + BY^j .$$

Suppose $x \in \tilde{Z}_rY^j$, $x = \sum_{0\le k\le j} a_k \otimes b_k$ where $a_k \in U_{r-k}$ and $b_k \in F_k$ for $0\le k\le j$. We know from Proposition 2.2 that $x \in mY^j$. If $r=j$ we have $a_j \in m = B_0U$, say $a_j = du$. Then $x - d(u \otimes b_j) \in \tilde{Z}Y^{j-1} \subset m^2Y^j + BY^j$ by the induction hypothesis. Suppose $r>j$ and $a_k \in m^2U$ for $k\ge p+1$. Since $db_k \in m^2Y^j$ we get

$da_p \otimes b_p \in m^4U \otimes F_p$. Hence $a_p \in m^2U + BU$ and it follows that x is homologous to $x' = \sum_{0\leq k\leq j} a'_k \otimes b'_k$ with $a'_k \in m^2U$ for $k \geq p$. At last we get

$$x \in m^2Y^j + BY^j \ .$$

Suppose (7) is true. Put $R' = GrR$ and $(U',d') = (GrU, Grd_1)$. U' is an S-R-algebra satisfying

$$\tilde{Z}U' \in \bar{m}^2U' + BU'$$

and since d' is homogeneous we also have

$$d'u' \in \bar{m}^3U' \Rightarrow u \in \bar{m}^2U' + ZU' \ .$$

Combining these facts we get that (6) is true also for U' . Hence

$$[\mathrm{Ext}^1_{GrR}(k,k)] = \mathrm{Hom}_{R'}(U',k) = \mathrm{Hom}_R(U,k) = [\mathrm{Ext}^1_R(k,k)] \ . \text{ Q.E.D.}$$

Example. It is not always true that $[\mathrm{Ext}^1_R(k,k)] = [\mathrm{Ext}^1_{GrR}(k,k)]$, as the following example shows:

$$R = k[x,y]/(x^2 + y^3, xy)$$

R is a complete intersection and

$$\mathrm{Ext}_R(k,k) = [\mathrm{Ext}^1_R(k,k)] = k\langle T_1,T_2\rangle/(T_2^2,[T_1^2,T_2])$$

(cf. Sjödin [15]). On the other hand

$$GrR = k[x,y]/(x^2,xy,y^4)$$

and by Theorem 1.1

$$[\mathrm{Ext}^1_{GrR}(k,k)] = k\langle T_1,T_2\rangle/(T_2^2) \ .$$

Theorem 2.4. Let (R,m) be a local ring. R has property E if any of the following conditions is satisfied:

(i) $R = k[[t_1,\ldots,t_n]]/(f_1,\ldots,f_r) + \mathfrak{a}$ where f_i for $i=1,\ldots,r$ are linear forms in $\{t_jt_k\}$ and $\mathfrak{a} \subset (t_1,\ldots,t_n)^3$.

(ii) $m^4 = 0$.

(iii) R is a complete intersection.

(iv) R is "trivially" Golod, i.e. it is possible to choose cycles in the Koszul complex K representing a basis for $H(K)$, such that all products of these cycles are zero.

(v) $|m/m^2| \leq 2$ or $R = \tilde{R}/\mathfrak{a}$ where $(\tilde{R},\tilde{m})$ is regular, $\mathfrak{a} \subset \tilde{m}^2$ and $\mathfrak{a}$ is generated by two elements.

Proof. (i). R is two-homogeneous with V equal to the k-space generated by $(t_1,\ldots,t_n)$ (V is concentrated in degree zero),

$$R = k \coprod V \coprod V^2 \coprod I^3 .$$

We construct an S-R-algebra as in Corollary 2.2 and we use the same notation. Suppose $U^i = X^i \coprod V \otimes X^i \coprod V^2 \otimes X^i \coprod I^3 \otimes X^i$ for some X^i and that the differential is homogeneous of degree +1 with respect to this "grading", i.e. $dX^i \subset V \otimes X^i$. Then $s_1,\ldots,s_n$ may be chosen in $V \otimes X^i$ and it follows that U^{i+1} has the same properties as U^i and hence also $U = R \otimes X$ has these properties where X is the union of all X^i. Suppose now $u \in U$ and $du \in m^3U = I^3 \otimes X$, then $u \in m^2U + ZU$, but $\tilde{Z}U \subset m^2U + BU$. Hence (7) of Lemma 2.4 is true and we are done.

(ii). This case follows trivially from Lemma 2.4 and we have also proved it in the proof of Theorem 2.2.

(iii). Let K be the Koszul complex and $s_1,\ldots,s_n$ represent a basis for $H_1(K)$, where $s_1,\ldots,s_r$ represent a basis for

$$Z_1(K)/(m^2K + BK)$$

and $s_i \in m^2K$ for $i=r+1,\ldots,n$, then a minimal resolution of k is given by

$$K<S_1,\ldots,S_r : dS_i = s_i> \otimes_R R<S_{r+1},\ldots,S_n>$$

where $dS_i = s_i$ for $i=r+1,\ldots,n$. It follows from Proposition 2.1 that

$$K<S_1,\ldots,S_r : dS_i = s_i>$$

is an S-R-algebra.

(iv). Let X be any R-algebra that has "trivially" all Massey products equal to zero, i.e. there is a set $\{s_i\}$ of cycles representing a basis for $\tilde{H}X$, such that $s_i \cdot s_j = 0$ for all i and j. Let $s \in \{s_i\}$. We <u>claim</u>:

$$Y = X\langle S: dS = s\rangle$$

has also trivially all Massey products zero. Indeed, suppose

$$y = x_0 + x_1 S + x_2 S^{(2)} + \ldots + x_n S^{(n)}$$

is a homogeneous cycle in Y. We get $dx_n = 0$, hence x_n is homologous to $x_n' = \sum r_i s_i$ where $r_i \in R$, so y is homologous to

$$x_0 + x_1 S + \ldots + x''_{n-1} S^{(n-1)} + x_n' S^{(n)} .$$

Since $d(x_n' S^{(n)}) = 0$, the same argument may be repeated and we get eventually that y is homologous to

$$x_0' + x_1' S + \ldots + x_n' S^{(n)}$$

where $x_j' \in \sum R s_i$ for all j. If we choose cycles representing a basis for $\tilde{H}Y$ in this manner, it follows that the product of any two of these is zero. Thus the claim above is proved. If R is trivially Golod, we may construct a complex U as in Corollary 2.2 such that U has trivially all Massey products zero. Since $\tilde{Z}U \subset m^2 U + BU$ it follows as in the proof of Theorem 2.2 that R has property E.

(v). This follows from (iii) and (iv) (for a proof, see [6]). Q.E.D.

<u>Remark</u>. The condition (i) of the theorem is equivalent to R complete and two-homogeneous. Indeed if R satisfies (i) then clearly R is complete and two-homogeneous. Conversely suppose R is complete and two-homogeneous, then R is a quotient of $k[[t_1,\ldots,t_n]]$, since R is an algebra over k. By definition of a two-homogeneous algebra we may choose the variables $t_1,\ldots,t_n$ in such a way that the relations between them are of the given form.

It is possible to prove (i) of the theorem by using the complex $(R \otimes K, d)$ defined on page 12-13 for two-homogeneous algebras. In the commutative case this complex has additional structure. K is defined there as a subspace of $T(V)$ and, using the definition of $\hat{T}(V)$ on page 5, it is fairly easy to see that K is a sub-Hopf-Γ-algebra of $\hat{T}(V)$ (where the elements of V have degree one). This also follows from [16, Theorem 2, page 17], since the dual of the map $K \to \hat{T}(V)$ is the natural map $T(V^*) \to T(V^*)/(\mathrm{im}\phi^*)$ and this map is equal to $W(f)$ for a map of graded Lie-algebras where $W(\cdot)$ is the universal enveloping algebra functor. Moreover it is an almost direct consequence of the definitions on page 5, that the differential on $R \otimes K$ is compatible with the algebra structure and the structure of divided powers. Proposition 1.3 page 14 tells us that this differential algebra structure is unique and also that the differential algebra defined in the proof of (i) is isomorphic to $(R \otimes K, d)$.

<u>Corollary 2.5</u>. Suppose $R = k[[t_1,\dots,t_n]]/\mathfrak{a}+\mathfrak{b}$ where $\mathfrak{a}$ is generated by linear forms in $\{t_i \cdot t_j\}$ and $\mathfrak{b} \subset (t_1,\dots,t_n)^3$ and one of the following two conditions is fulfilled:

(a) $(t_1,\dots,t_n)^4 \subset \mathfrak{b}$

(b) $k[t_1,\dots,t_n]/\mathfrak{a}$ is a homogeneous Koszul algebra and there is a k such that $(t_1,\dots,t_n)^{2k} \subset \mathfrak{b} \subset (t_1,\dots,t_n)^{k+1}$.

Then there is a graded vector space $V = \coprod_{i \geq 1} V_i$ and a split exact sequence of Hopf algebras

$$k \to T(V) \to \mathrm{Ext}_R(k,k) \underset{\longleftarrow}{\longrightarrow} [\mathrm{Ext}^1_R(k,k)] \to k$$

i.e. $\mathrm{Ext}_R(k,k)$ is a semi-tensor product of $T(V)$ and $[\mathrm{Ext}^1_R(k,k)]$.

Proof. Suppose (a) is fulfilled. We want to apply Proposition 2.3 (d) so we must prove that the complex defined above has the lifting property. But this is easy: Let $f \in \mathrm{Hom}_R(U_n,k) = \mathrm{Hom}_k(K_n,k)$. Put

$$\tilde{f}(x_1 \otimes \ldots \otimes x_{n+r}) = (-1)^{nr} f(x_{r+1} \otimes \ldots \otimes x_{n+r})(x_1 \otimes \ldots \otimes x_r)$$

and extend f to $R \otimes K_{n+r}$ by linearity. We get the split exact sequence and the argument for the fact that the Hopf algebra structure on $T(V)$ is the natural one is the same as in the proof of Theorem 2.2.

Suppose (b) is fulfilled. Let R' be the ring $k[[t_1,\ldots,t_n]]/\mathfrak{a}$. Let U' be the minimal resolution of k over R' given by Theorem 1.2. Then $\tilde{H}U = \tilde{H}(R \otimes_{R'} U')$ has all Massey products zero(trivially), since we may choose cycles in $(t_1,\ldots,t_n)^k U$ representing a basis for $\tilde{H}U$. Hence $R' \to R$ is a Golod homomorphism (see [8]) and we may apply Corollary 2.4 and use the fact that $[\mathrm{Ext}_R^1(k,k)] = [\mathrm{Ext}_{R'}(k,k)] = \mathrm{Hom}_R(U,k)$, which follows from Theorem 1.1 and 1.2. Q.E.D.

$[\mathrm{Ext}_R^1(k,k)]$ as enveloping algebra of a graded Lie-algebra.

It has been proved by André [1] and Sjödin [16] , that $\mathrm{Ext}_R(k,k) = WL =$ = universal enveloping algebra of a graded Lie-algebra $L = \coprod_{i\geq 1} L_i$ where $L_1 = \mathrm{Ext}_R^1(k,k)$. Let $\mathfrak{g}$ be the graded Lie-algebra generated by L_1. We claim that $W\mathfrak{g} = [\mathrm{Ext}_R^1(k,k)]$.

The following lemma is due to Gunnar Sjödin.

Lemma 2.4. Suppose $\mathfrak{g} \subset L$ are graded Lie-algebras, then the subalgebra of WL generated by $\mathfrak{g}$ is equal to $W\mathfrak{g}$.

Proof. $\mathfrak{g}$ is imbedded in WL by the theorem of Poincaré-Birkhoff-Witt (PBW) (see e.g. Sjödin [16]). Let A be the subalgebra of WL generated by $\mathfrak{g}$. There is an epimorphism of algebras $W\mathfrak{g} \to A$ which makes the following diagram commutative

$$\begin{array}{ccc} \mathfrak{g} & \to & WL \\ \downarrow & & \uparrow \\ W\mathfrak{g} & \to & A \end{array}$$

The map $W\mathfrak{g} \to A \to WL$ is a monomorphism, because of PBW. Hence also $W\mathfrak{g} \to A$ is a monomorphism. Q.E.D.

We know, by PBW, that the Hilbert series of

$$W\mathfrak{g} = [\mathrm{Ext}_R^1(k,k)]$$

has the form

$$(1+x)^{\delta_0}(1+x^3)^{\delta_2}\cdot\ldots/(1-x^2)^{\delta_1}(1-x^4)^{\delta_3}\cdot\ldots$$

where $\delta_i = \dim_k \mathfrak{g}_{i+1}$.

<u>Definition.</u> The natural numbers $\{\delta_i\}_{i\geq 0}$ are called the subdeviations of R .

<u>Definition.</u> (cf. Sjödin [17]) $[\mathrm{Ext}_R^1(k,k)] = W\mathfrak{g}$ is <u>nilpotent</u> if $\mathfrak{g}$ is nilpotent, i.e. $\delta_i = 0$ for sufficiently large i .

Since $\mathfrak{g}$ (defined above) is generated by $\mathfrak{g}_1$, $W\mathfrak{g}$ is nilpotent if $\delta_i = 0$ for <u>some</u> i .

<u>Theorem 2.5</u>. Let (R,m) be a local ring and K its Koszul complex. Then

$$\delta_0 = \dim_k(m/m^2)\ ,\ \delta_1 = \dim_k(Z_1K/(m^2K + BK))$$

and

$$\delta_2 = \dim_k(Z_2K/(m^2K + BK))\ .$$

<u>Proof.</u> We will use Lemma 2.1. Consider the minimal algebra resolution Y obtained by killing cycles (see [6]). Let ϕ be the natural map

$$H(Y/m^2Y) \to Y/mY\ .$$

By Lemma 2.1, the algebra cokernel of ϕ is equal to the dual of $[\mathrm{Ext}_R^1(k,k)]$. We are going to compute the dimension of the ideal generated by $\mathrm{im}(\phi)$ in degrees ≤ 3 . Choose a basis $\{\bar{s}_1,\ldots,\bar{s}_{\varepsilon_1}\}$ for Z_1K/BK such that $\{\bar{s}_1,\ldots,\bar{s}_{n_1}\}$ is a basis for $Z_1K/(m^2K + BK)$ and $s_{n_1+1},\ldots,s_{\varepsilon_1} \in m^2K$. Put

$$E = K\langle S_i : dS_i = s_i \ , \ i = 1,\dots,\varepsilon_1\rangle \ .$$

Choose a basis $\{\bar{u}_1,\dots,\bar{u}_{\varepsilon_2}\}$ for Z_2E/BE such that $\{\bar{u}_1,\dots,\bar{u}_{n_2}\}$ is a basis for $Z_2E/(m^2E + BE)$ and $u_{n_2+1},\dots,u_{\varepsilon_2} \in m^2E$. Put

$$F = E\langle U_i : dU_i = u_i \ , \ i = 1,\dots,\varepsilon_2\rangle \ .$$

We have $F_j = Y_j$ for $j \leq 3$. Since K is an S-R-algebra, there are no elements of K/mK in $\mathrm{im}(\phi)$. Suppose

$$y = \sum r_i S_i + x \quad \text{with } r_i \in R \text{ for } i=1,\dots,\varepsilon_1 \text{ and } x \in K$$

is an element of degree two in Y such that $dy \in m^2Y$. Then

$$\sum r_i s_i \in m^2K + BK$$

which implies that

$$\sum_{1 \leq i \leq n_1} r_i s_i \in m^2K + BK$$

and hence $r_i \in m$ for $i=1,\dots,n_1$, it follows that $dx \in m^2K$ hence $x \in mK$.
We have proved that $\mathrm{im}(\phi)$ is generated by $S_{n_1+1},\dots,S_{\varepsilon_1}$ in degree two. Suppose

$$z = \sum r_i U_i + \sum x_i S_i + x_0 \quad \text{with } r_i \in R \text{ for } i=1,\dots,\varepsilon_2 \text{ and } x_i \in K \text{ for } i=0,\dots,\varepsilon_1 .$$

We get

$$\sum r_i u_i \in m^2E + BE$$

and as above this implies that $r_i \in m$ for $i=1,\dots n_2$. It follows that

$$d(\sum x_i S_i + x_0) \in m^2E$$

hence $dx_i \in m^2K$, hence $x_i \in mK$ and finally also $x_0 \in mK$. We have proved that $\mathrm{im}(\phi)$ is generated by $U_{n_2+1},\dots,U_{\varepsilon_2}$ in degree three. Thus we have proved that a basis for the algebra cokernel of ϕ in degree ≤ 3 , is given by the polynomials of degree ≤ 3 in the variables $T_1,\dots,T_{\varepsilon_0}$ (the Koszul variables) , $S_1,\dots,S_{n_1}$ and $U_1,\dots,U_{n_2}$. Hence $\delta_0 = \varepsilon_0$, $\delta_1 = n_1$ and $\delta_2 = n_2$.

It remains to prove that

$$Z_2E/(m^2E + BE) \simeq Z_2K/(m^2K + BK) \ .$$

Suppose

$$y = \sum r_i S_i + x \quad \text{with} \quad x \in K$$

is an element in Z_2E . Then $r_i \in m$ for all i , so y is homologous to an element x' in Z_2K . Suppose

$$x' \in m^2E + BE \ , \ x' + d(\sum x_i S_i + x_0) \in m^2E \ .$$

Then

$$dx_i \in m^2K \text{ for all } i \implies x_i \in mK \text{ for all } i \implies x' \in m^2E + BK \implies x' \in m^2K + BK$$

Remark 1. Let Y be the minimal algebra resolution obtained by killing cycles. Let Y^n be the subalgebra of Y generated by variables of degree $\leq n$. Then

$$\delta'_n = \dim_k \tilde{Z}_n Y^n/(m^2Y^n + BY^n)$$

are invariants of R , since Y^n are invariant complexes (see [6]). It is easy to see, using Lemma 2.1, that $\delta_n \leq \delta'_n$. We have proved equality for $n \leq 2$ and it is possible to use the same technique as in the proof above to prove equality for $n \leq 5$ but the equality is probably not true in general.

Remark 2. If the complex U obtained from Corollary 2.2 satisfies

$$\mathrm{Hom}_R(U,k) = [\mathrm{Ext}^1_R(k,k)]$$

we get the subdeviations as

$$\delta_i = \dim_k \ \tilde{Z}_i U^i/(m^2U^i + BU^i)$$

with notations as in the corollary. This is the case if, for instance, some of (i) - (v) of Theorem 2.4 is fulfilled.

Christer Lech has derived an equivalent condition to $\delta_2 = 0$ when

$$R = k [[t_1,\dots,t_n]] /(f_1,\dots,f_r)$$

where f_i $i=1,\dots,r$ are linear forms in $\{t_i t_j\}$. We are able to prove his result by means of Theorem 2.5.

Theorem 2.6. Let

$$R = k[[t_1,\dots,t_n]]/(f_1,\dots,f_r)$$

where f_i $i=1,\dots,r$ are linear forms in $\{t_i t_j\}$ and $f_1,\dots,f_r$ are linearly independent. Then

$$\delta_2 = 0 \iff \{t_i f_j\} \text{ linearly independent.}$$

Proof. We may apply Notations no. 11 to the bigraded Koszul complex for GrR and get

$$(1+x)^n H_R(-x) = H_{H(K)}(x,-x) = 1 - h_{11}x^2 + (-h_{21} + h_{12})x^3 + \dots$$

where $h_{ij} = \dim_k H_{ij}(K)$, here i is the Koszul degree and j is the ring degree. We have $h_{11} = r$, $h_{12} = 0$ and $h_{21} = \delta_2$, the last equality by Theorem 2.5. Let c be the dimension of the linear space spanned by $\{t_i f_j\}$. From above we get

$$-\delta_2 = -\left(\binom{n+2}{3} - c\right) + n\left(\binom{n+1}{2} - r\right) - \binom{n}{2}n + \binom{n}{3} = c - nr .$$

Thus

$$\delta_2 = 0 \iff c = nr \iff \{t_i f_j\} \text{ linearly independent.}$$

Corollary 2.6. Let

$$R = k[[t_1,\dots,t_n]]/(f_1,\dots,f_r) + \mathfrak{a}$$

where $\mathfrak{a} \subset (t_1,\dots,t_n)^3$. Then

$$nr - \delta_2 = \text{the dimension of the k-space spanned by } \{t_i f_j\} .$$

Proof. Let

$$R' = k[[t_1,\dots,t_n]]/(f_1,\dots,f_r) .$$

By Theorem 1.1 $[\mathrm{Ext}^1_R(k,k)] = [\mathrm{Ext}^1_{R'}(k,k)]$.

Theorem 2.7. Let (R,m) be a local ring with $m^4 = 0$ such that $[\mathrm{Ext}^1_R(k,k)]$ is nilpotent and one of the following conditions is valid

(i) R has characteristic $p>0$,

(ii) $R \simeq \mathrm{Gr} R$.

Then P_R is rational.

Proof. Let U be given by Corollary 2.2. Since $m^4 = 0$, $\tilde{H}U$ has all Massey products zero. Hence

$$P_R(x) = H_U(x)/(1 - H_{s\tilde{H}U}(x))$$

where $(s\tilde{H}U)_{i+1} = \tilde{H}_i U$. From Theorem 2.1 and 2.4 we have

$$\mathrm{Hom}_R(U,k) = [\mathrm{Ext}_R^1(k,k)] \ .$$

The condition $[\mathrm{Ext}_R^1(k,k)]$ is nilpotent implies $H_U(x)$ is rational and that U is a finite extension of K . Now we apply a theorem by Gulliksen [6'] which says that $H_{HU}(x)$ is rational if U is a finite free extension of K and one of the following conditions is valid:

(a) R has characteristic $p>0$,

(b) it is possible to extend "derivations" to U .

Suppose $R \simeq \mathrm{Gr} R$. We will prove that (b) is satisfied. We use the same notations as in the proof of (i) of Theorem 2.4. Suppose j is a derivation on U^i such that j is homogeneous in the ring degree, i.e.

$$j(X^i) \subset X^{i-\deg(j)} \ .$$

Let S be a variable of degree $i+1$. In order to extend j to $U^i\langle S:\, dS = s\rangle$, it is enough to show that $j(s)$ is a boundary (see [6, Lemma 1.3.2 page 16]). But we may choose s in $V \otimes X^i$, hence $j(s)$ is a cycle in $V \otimes X^{i-\deg(j)}$ and this is a boundary since $Z_j U^i \subset m^2 U^i + BU^i$ for $j<i$, in fact $j(s) \in dX$ and hence we may choose the extension of j such that it is homogeneous in the ring degree (since this construction is unique, j must satisfy the lifting formula on page 41). Q.E.D.

Corollary 2.7. Let (R,m) be a local ring with $m^4 = 0$ and $R \simeq GrR$, i.e.

$$R = k[[t_1,\dots,t_n]] /(f_1,\dots,f_r,g_1,\dots,g_s) + (t_1, \dots, t_n)^4$$

where f_i are homogeneous elements of degree two (in the variables t_i of degree 1) and g_i are homogeneous elements of degree three. Suppose $\{t_i f_j\}$ is linearly independent. Then P_R is rational.

Proof. By Corollary 2.6 we have $\delta_2 = 0$. Hence $[\mathrm{Ext}_R^1(k,k)]$ is nilpotent, and from Theorem 2.7 we get that P_R is rational.

References

[1] M. ANDRÉ, Hopf algebras with divided powers, J. Algebra 18 (1971), 19-50.

[2] L.L. AVRAMOV, On the Hopf algebra of a local ring, Izvestija Akad. Nauk SSSR ser. mat. 38 (1974), 253-277.

[3] J. BACKELIN, A distributiveness property of augmented algebras, and some related homological results, part of thesis, Stockholm 1982 (defended January 21, 1983).

[4] H. CARTAN, Homologie et cohomologie d'une algèbre graduée, Séminaire Henri Cartan, 11e année 58/59, exposé 15.

[5] R. FRÖBERG, Determination of a class of Poincaré series, Math.Scand. 37 (1975), 29-39.

[6] T.H. GULLIKSEN, G. LEVIN, Homology of local rings, Queen's Paper No. 20, Queen's University, Kingston, Ontario, (1969).

[6'] T.H. GULLIKSEN, On the Hilbert series of the homology of differential graded algebras, Math. Scand., 46 (1980), 15-22.

[7] G. LEMAIRE, Algèbres connexes et Homologie des Espaces de Lacets, Lecture Notes in Mathematics, No. 422, Springer Verlag, Berlin, 1974.

[8] G. LEVIN, Local rings and Golod homomorphisms, J. Algebra 37 (1975), 266-289.

[9] G. LEVIN, Two conjectures in the homology of local rings, J Algebra 30 (1974), 56-74.

[10] C. LÖFWALL, The Poincaré series for a class of local rings, Preprint series, University of Stockholm, No. 8, 1975.

[11] J.W. MILNOR, J.C. MOORE, On the structure of Hopf algebras, Ann. of Math. 81 (1965), 211-264.

[12] S. PRIDDY, Koszul resolutions, Trans. A.M.S. 152 (1970), 39-60.

[13] J-E. ROOS, Relations between the Poincaré-Betti series of loop spaces and of local rings, Lecture Notes in Math., 740, Springer-Verlag, Berlin, 1979.

[14] J-P. SERRE, Algèbre locale multiplicités, Lecture notes in Mathematics, No. 11, 3 ed., Springer Verlag, Berlin, 1975.

[15] G. SJÖDIN, A set of generators for $Ext_R(k,k)$, Math. Scand., 38 (1976), 1-12.

[16] G. SJÖDIN, Hopf algebras and derivations, J. Algebra, 64 (1980), 218-229.

[17] G. SJÖDIN, A characterization of local complete intersections in terms of the Ext-algebra, J. Algebra, 64 (1980), 214-217.

[18] G. SJÖDIN, The Ext-algebra of a Golod ring, Journal of Pure and Applied Algebra, 38, 1985, 337-351.

[19] L. SMITH, Split extensions of Hopf algebras and semi-tensor products, Math. Scand. 26 (1970), 17-41.

[20] H. WIEBE, Über homologische invarianten lokaler ringe, Math. Ann. 179 (1969), 257-274.

Department of Mathematics
University of Stockholm
Box 6701
S-113 85 STOCKHOLM
(SWEDEN)

THE GENERAL EXTENSION OF A LOCAL RING AND MIXED MULTIPLICITIES.

D.REES.

The present paper is an extended version of a lecture given at Stockholm during a Research Symposium associated with the Nordic Summer School in August 1983. During the preparation of the manuscript, the author was able to extend considerably the main result of that lecture, and this extension appears as theorem 4.6 below. It will also be described in this introduction.

The proof of this result requires the definition and proof of the existence of elements of ideals which are "sufficiently general". The earlier part of the paper is therefore devoted to a desecription of a set-up in which such "sufficiently general" elements can be defined.

We start with a local ring (Q,m,k,d), this indicating that Q has maximal ideal m, residue field k $(=Q/m)$ and Krull dimension d. With Q we associate another local ring Q_g termed its general extension. This is constructed by adjoining to Q a countable set of indeterminates $X_1, X_2, \ldots$ (also denoted by $X(1), X(2), \ldots$ for typographical reasons). Q_g is then the localisation of the ring $Q[X_1, X_2 \ldots]$ at the prime ideal $m[X_1, X_2 \ldots]$. Q_g can also be described as the direct limit of rings Q_N similarly defined in terms of a finite set of indeterminates $X_1, \ldots, X_N$. A general result of A.Grothendieck implies that Q_g is noetherian, and the elementary results proved in the first section mostly stem from this fact.

In the first section of the paper, we introduce the notion of an independent set of general elements of a set of ideals $\underline{I} = (I_1, \ldots, I_s)$ of Q. This is done in two stages. First choose an arbitrary basis $u(j,k)$ $(k = 1, \ldots, n_j)$ for each of the ideals $I_1, \ldots, I_s$. Now write $N_j = n_1 + \ldots + n_{j-1}$. We now term the set of elements

$$y_j = \sum_{k=1}^{n_j} X(N_j + k)u(j,k) \qquad j = 1, \ldots, s$$

a standard set of general elements of $\underline{I}$. An independent set of general elements of $\underline{I}$ is a set of elements of the form

$$x_j = T(y_j) \qquad j = 1, \ldots, s$$

where T is an automorphism of Q_g over Q. This definition is independent of the choice of the standard set $y_1, \ldots, y_s$ chosen, and also of the order of the ideals $I_1, \ldots, I_s$.

One consequence of this definition is the following. Let $X(\underline{I})$ denote the ideal $(x_1, \ldots, x_s)$ of Q_g, and let P range over the set of prime ideals of Q_g minimal over $X(\underline{I})$. Then the set of ideals $p = P \cap Q$ of Q is independent of the choice of $x_1,\dots,x_s$ and sc depends only on $\underline{I}$. We will be concerned with a sub-set of this set of ideals in a particular case.First we place a restriction on the set $\underline{I}$.We say that $\underline{I}$ is a general set of ideals if an independent set of general elements $(x_1,\dots,x_s)$ of $\underline{I}$ is a sub-set of a set of parameters of Q_g(this condition is independent of the choice of $x_1,\dots,x_s$).Next,we place a restriction on P.We require that P be good,this meaning that

$$\mathrm{ht}P + \dim(Q_g/P) = \dim(Q_g).$$

We now consider the set of prime ideals $p = Q \cap P$,where P ranges over the good prime ideals of Q_g minimal over $(x_1,\dots,x_s)$.Suppose p is such a prime ideal.Then p is a good prime ideal of Q.Further,if $\mathrm{ht}p = t$,then p contains t of the ideals I_j and is minimal over their union.Finally there is only one of the prime ideals P meeting Q in p,and the local ring $(Q_g)_P/p(Q_g)_P$ is regular of dimension s-t.

In the third section of the paper,we consider another ingredient of the main theorem.This is the concept of an m-valuation on Q.The definition given here differs from that given in the lecture referred to above.First we consider a valuation v,with value group the additive group of integers, on the field of fractions F_p of Q/p,where p is a prime ideal of Q.We subject v to the further restrictions that it is ≥ 0 on Q/p and >0 on m/p.We now lift v to Q,the value of v on p being ∞.We term p the limit ideal of v and denote it by L(v).If $p \neq m$,we term the lifting of v an m-valuation.We will apply the same term to the function v(J) on the set of ideals J of Q definied by $v(J)= \operatorname{Min}_{x \in J}(v(x))$.If J is restricted to be m-primary,v(J) is finite.The set of functions v(J),where J is restricted to be m-primary,are then linearly independent.We now consider two classes of m-valuations.We say that an m-valuation v is good if it satisfies the condition that

$$t(v) + \mathrm{ht}L(v) = d-1$$

where t(v) is the transcendence degree of the residue field K_v of v over k.Note that the left-hand side of the above equation is at most equal to the right-hand side in all cases.If v is good,then L(v) is a good prime ideal of Q.Finally,an m-valuation is proper if htL(v) =0.

Now we come to the final section,which is devoted to the statement and proof of the main theorem.First we need to define a multiplicity function.Let $\underline{I} = (I_1,\dots,I_d)$ be an independent set of ideals with s=d, and let M be a finitely generated Q-module. Since $\underline{I}$ is independent,it follows that if $x_1,\dots,x_d$ is an independent set of general elements of $\underline{I}$,it is a set of parameters of Q_g.Then we define

$$e(\underline{I}\ ;\ M) = e(x_1,\dots,x_d\ ;M_g)$$

where $M_g = M \otimes_Q Q_g$. Then $e(\underline{I},M)$ is independent of the choice of $(x_1,\ldots,x_d)$. It is an additive function on the category of finitely generated Q-modules. It is also a "multi-linear" function on the variables $I_1,\ldots,I_d$ in the sense that, if $\underline{I} = (I_1,\ldots,I_{j-1},I_j,\ldots,I_d)$ and $\underline{I}' = (I_1,\ldots I_{j-1},I_j',\ldots,I_d)$ are both independent sets of ideals and $\underline{I}'' = (I_1,\ldots,I_{j-1},I_jI_j',\ldots,I_d)$, then $\underline{I}''$ is also a general set of ideals and

$$e(\underline{I}'' ;M) = e(\underline{I} ; M) + e(\underline{I}' ;M).$$

This multiplicity function is a generalisation of Teissier's mixed multiplicity, reducing to it if $I_1,\ldots,I_d$ are all m-primary..

Now suppose that $\underline{I}$ is an independent set $(I_1,\ldots,I_{d-1})$ containing d-1 ideals, and that J is any ideal of Q such that $I_1,\ldots,I_{d-1},J$ is independent. Our objective is to express $e(\underline{I},J ;M)$ in terms of m-valuations. First we must introduce some notation. Let v be a good m-valuation, and let $\underline{I}$ be as above. Then we define $\underline{I}'(v)$ to be the set of ideals I_j contained in $p = L(v)$. Next, we define $\underline{I}''(v)$ to be the set of ideals I_j+p/p of Q/p, where I_j ranges over $\underline{I}-\underline{I}'(v)$. Finally, write Q_v for Q_p and M_v for M_p. Then our expression is

$$e(\underline{I},J ;M) = \sum_v d(\underline{I}''(v),v)e(\underline{I}'(v)Q_v;M_v)v(J)$$

the sum being over all good m-valuations on Q. In this formula, $d(\underline{I}'',v)$ is a function taking non-negative integer values and defined for all independent sets of ideals $\underline{I}'' = (I_1'',\ldots,I_{d(v)-1}'')$ of $Q/L(v)$, where $d(v) = \dim(Q/L(v))$. The existence of such a formula implies its uniqueness, and so determines the coefficients $d(\underline{I}'',v)$. The function $d(\underline{I}'',v)$ is also multilinear in the sense defined above. If $I_1,\ldots,I_{d-1}$ are all m-primary, then the valuations occurring in the above sum are proper, and it was this case which was dealt with in the lecture.

1. The General Extension of a local ring.

In what follows Q will be a noetheriag local ring with maximal ideal m, residue field k and Krull dimension d, this being summarised by the notation (Q,m,k,d).

Let $X_1,X_2,\ldots$ be a countable set of indeteminates over Q(as stated in the introduction, we will occasionally write X(j) in place of X_j). Then we will denote by Q_N the localisation of $Q[X_1,\ldots,X_N]$ at the prime ideal $m[X_1,\ldots,X_N]$. It is clear that $Q \subsetneq Q_1 \subsetneq Q_2 \subsetneq \ldots$. We will denote by Q_g the union of the rings Q_N, or, what is the same thing, the localisation of $Q[X_1,X_2,\ldots]$ at $m[X_1,X_2,\ldots]$. The maximal ideal of $Q_g(Q_N)$ is $mQ_g(mQ_N)$ and we will denote them by $m_g(m_N)$. The residue fields of Q_g,Q_N are respectively isomorphic to $k(X_1,X_2,\ldots)$ and $k(X_1,X_2,\ldots,X_N)$ and will be denoted by

k_g and k_N respectively. Finally we note that Q_g, Q_N both have Krull dimension d.

Q_N is clearly noetherian and, in fact, the same is true of Q_g. This is a consequence of a more general result of A.Grothendieck([2], Prop 10.3.1. p20). However we give an ad hoc proof.

LEMMA 1.1 <u>If</u> I <u>is a finitely generated ideal of</u> Q_g, <u>then</u>

$$\bigcap_1^{\infty}(I + m_g{}^n) = I$$

Let $u_1,\ldots,u_r$ be a basis of I and let x be an element of $\bigcap_1^{\infty}(I+m_g{}^n)$. Then we can find N such that $u_1,\ldots,u_r,x$ all belong to Q_N. Hence, as $m_g = mQ_g$, it follows that all the ideals $I+m_g{}^n$ have a basis in Q_N. Therefore, if $I' = I \cap Q_N$,

$$x \in \bigcap_1^{\infty}(I+m_g{}^n) \cap Q_N = \bigcap_1^{\infty}(I'+m^n)Q_N = I' \subseteq I$$

since Q_N is noetherian,

THEOREM 1.2. Q_g <u>is noetherian</u>.

Since the maximal ideal of Q_g is finitely generated, the completion $\bar{Q}_g$ of Q_g is noetherian. Hence, if I is any ideal of Q_g, $I\bar{Q}_g$ has a finite basis contained in I. Let I' be the ideal of Q_g generated by this basis. Then

$$I \subseteq I\bar{Q}_g \cap Q_g = I'\bar{Q}_g \cap Q_g \subseteq \bigcap_1^{\infty}(I'+m_g{}^n) = I' \subseteq I$$

and hence I = I' is finitely generated.

We continue with a number of elementary results concerning Q_g. In many cases the proof is left to the reader.

First suppose that h: Q--->Q' is a homomorphism of a local ring Q into a local ring Q'. Then we have a natural extension h_g of h mapping Q_g into Q'_g obtained by mapping the indeterminates $X_1, X_2, \ldots$ into themselves Further, if I is the kernel of h, then $I_g = IQ_g$ is the kernel of h_g. In particular, taking Q' = Q/I, we obtain an isomorphism

$$(Q/I)_g = Q_g/I_g.$$

As a consequence, if dimI is defined to be dim(Q/I), $\dim I_g = \dim I$.

Next, suppose that p is a prime ideal of Q. Then there is a natural isomorphism between $(Q_p)_g$ and the localisation of Q_g at the prime ideal p_g, both being considered as Q-algebras. This implies that $\mathrm{ht}p_g = \mathrm{ht}p$.

Now we make use of the fact that Q_g is noetherian. Let I be an ideal of Q_g. Then I has a basis consisting of elements of Q_N if N is large enough. Denote by I_N the ideal $I \cap Q_N$ of Q_N (so that $I \cap Q = I_o$). Then, for N large enough, $I = I_N Q_g = (I_N)_g$ (here we have identified $(Q_N)_g$ with Q_g in the natural way.). The least integer N for which this is true will be termed the index of definition of I and will be denoted by i(I).

Note that in addition to the natural isomorphism between Q_g and $(Q_N)_g$ used in the last paragraph, we also have a natural isomorphism between Q_g and $(Q_g)_N$. Both isomorphisms are isomorphisms as Q-algebras.

Next, let M be a finitely generated Q-module. Then we define M_N and

M_g as, respectively, $M\otimes_Q Q_N$ and $M\otimes_Q Q_g$. We can consider $M \dashrightarrow M_N$ and $M \dashrightarrow M_g$ as functors from the category of finitely generated Q-modules to, respectively, the categories of finitely generated Q_N-modules and finitely generated Q_g-modules. These functors are both faithful and exact. This simply reflects the fact that Q_N and Q_g are faithfully flat extensions of Q.

Nowsuppose that M is of finite length. Then, since $L((Q/m)_g)$ and $L((Q/m)_N)$ both have value 1, the above implies that $L(M)=L(M_N)=L(M_g)$. Hence Hilbert functions and, consequently, multiplicities are preserved by the passage from M to M_N or M_g. To be precise, if I is an m-primary ideal of Q, then

$$e(I_N;M_N) = e(I;M) = e(I_g;M_g)$$

As a consequence, it follows that if Q is Cohen-Macaulay, so are Q_N and Q_g. We simply use the fact that Q is Cohen-Macaulay if, for some m-primary ideal I generated by d elements $L(Q/I) = e(I;Q)$.

We give a proof of the last result of this section.

THEOREM 1.3. *If* M *is a finitely generated* Q-*module, then the associated prime ideals of* M_g *are the prime ideals* P_g, *where* P *ranges over the associated prime ideals of* M.

We recall that P is an associated prime of M if M contains an element u whose annihilator is P. By taking u to be a suitable element of M we see that P_g is an associated prime ideal of M_g if P is an associated prime ideal of M. Conversely, suppose that P' is an associated prime ideal of M_g, and that u is an element of M_g whose annihilator is P'. We can take u to be of the form u(X), where u(X) denotes a polynomial in $X_1, X_2, \ldots$ with coefficients in M. Further, P' has a set of generators which are polynomials in $X_1, X_2, \ldots$ over Q. Let f(X) be one of these generators. We define the content c(u) of u to be the sub-module of M generated by the coefficients of u(X) and the content c(f) of f to be the ideal of Q generated by the coefficients of f(X). Then a classical result (see Northcott[4]) states that, for some r

$$c(f)^{r+1}c(u) = c(f)^r c(fu) = 0.$$

Bur the annihilator of c(u) is contained in the annihilator P' of u. Hence P' contains $c(f)^{r+1}$ and therefore c(f). This implies that P' is generated by elements of Q and therefore $P' = P_g$, where $P = P' \cap Q$. Now P is the intersection of the annihilators of the individual coefficients of u(X) and hence, being prime, is the annihilator of one of them. Hence P is a prime ideal associated with M.

Corollary. *If* I *is an ideal of* Q_g *such that* $i(I) \leq N$, *then* $i(P) \leq N$ *for all associated primes* P *of* I.

By the prime ideals associated with I we mean the prime ideals associated with the module Q_g/I. Now, if $I_N = I \cap Q_N$, $I \doteq I_N Q_g$. Hence Q_g/I

is isomorphic to $(Q_N/I_N)_g$ and the result follows from the theorem.

2. General elements of ideals.

DEFINITIONS. Let $\underline{I} = (I_1,\dots,I_s)$ be a set of ideals of Q. Then, by a standard set of general elements of $\underline{I}$ we mean a set of elements $x_1,\dots,x_s$ of Q_g constructed as follows. For each j choose a set of generators $u(j,k)$ $(k = 1,\dots,n_j)$ of I_j. Let $N_j = n_1+n_2+\dots+n_{j-1}$. Then we take

$$x_j = \sum_{k=1}^{n_j} X(N_j+k)u(j,k) \qquad (j= 1,\dots,s)$$

We term a set of elements $x_1,\dots,x_s$ of Q_g an independent set of general elements of I if there exists an automorphism T of Q_g over Q such that the elements $z_j = T(x_j)$ $(j=1,\dots,s)$ form a standard set of general elements of I.

THEOREM 2.1. If $x_1,\dots,x_s;x'_1,\dots,x'_s$ are two independent sets of general elements of the set $\underline{I} = (I_1,\dots,I_s)$ of ideals of Q, then there exists an automorphism T of Q_g over Q such that

$$T(x_j) = x'_j$$

Without loss of generality, we can assume that both sets of elements are standard sets of general elements of $\underline{I}$, that is, we can take

$$x_j = \sum_{k=1}^{n_j} X(N_j+k)u(j,k) \qquad (j=1,\dots,s)$$

$$x'_j = \sum_{k=1}^{m_j} X(M_j+k)w(j,k) \qquad (j=1,\dots,s)$$

where $(u(j,1),\dots,u(j,n_j)),(w(j,1),\dots,w(j,m_j))$ are two sets of generators of I_j for $j = 1,\dots,s$ and $N_j=n_1+\dots+n_{j-1}, M_j=m_1+\dots+m_{j-1}$. Further, we can assume that $(u(j,1),\dots,u(j,n_j))$ is a minimal basis of I_j for each j, so that $n_j \leq m_j$ for each j.

We now construct T as a product of two automorphisms T_2T_1. T_1 is determined by a permutation of $X(1),X(2),\dots$, which fixes $X(r)$ if $r>M_{s+1}$, sends $X(N_j+k)$ to $X(M_j+k)$ for $j = 1,\dots,s$ and $k = 1,\dots,n_j$ and otherwise is completed arbitrarily. Now T_2 has to satisfy

$$T_2\Big(\sum_{k=1}^{m_j} X(M_j+k)u(j,k)\Big) = \sum_{i=1}^{m_j} X(M_j+i)w(j,i)$$

We will, in fact, describe T_2^{-1}. We can write

$$u(j,k) = \sum_{i=1}^{m_j} a_j(k,i)w(j,i)$$

for $j = 1,\dots,s$ and $k = 1,\dots,n_j$. Since the elements $u(j,k)(k=1,\dots,n_j)$ form a minimal basis of I_j, it follows that, for each j, the $n_j x m_j$ matrix $(a_j(k,i))$ must have an $n_j x n_j$ sub-matrix with determinant a unit of Q. Hence, by adding suitable rows of an $m_j x m_j$ identity matrix, we can complete it to an $m_j x m_j$ matrix $A_j = (a_j(k,i))$ which has determinant a unit of Q, for each j. With this extension of the definition of $a_j(k,i)$ we now define T_2^{-1}, by first defining

$$T_2^{-1}(X(M_j+i)) = \sum_{k=1}^{m_j} a_j(k,i)X(M_j+k)$$

for $j = 1,\dots,s, i = 1,\dots,m_j$ and $T_2^{-1}(X(r))=X(r)$ if $r>M_{s+1}$. This can now be extended to an automorphism which satisfies

$$T_2^{-1}\left(\sum X(M_j+i)w(j,i)\right) = \sum a_j(k,i)X(M_j+k)w(i,j) = \sum X(M_j+k)u(j,k)$$

as required.

In the following corollaries, we will denote the ideal $(x_1,\dots,x_s)$ by $X(\underline{I})$.

Corollary i) <u>To within isomorphism as a Q-algebra</u>, $Q_g/X(\underline{I})$ <u>depends only on the set</u> $\underline{I}$ <u>and not on the choice of</u> $x_1,\dots,x_s$.

Corollary ii) <u>The ideal</u> $Q\cap X(\underline{I})$ <u>depends only on</u> I <u>and not on the choice of</u> $x_1,\dots,x_s$

Corollary iii) <u>If</u> P <u>ranges over either the set of all prime ideals associated with</u> X(I) <u>or the set of minimal prime ideals over</u> $X(\underline{I})$, <u>then the set of ideals</u> $P\cap Q$ <u>depends only on</u> <u>I.</u>

In what follows we will be concerned with a particular situation which we will now describe. We will say that a set of ideals $\underline{I}=(I_1,\dots,I_s)$ is an independent set of ideals if an independent set of general elements $x_1,\dots,x_s$ of $\underline{I}$ is a sub-set of a set of parameters of Q_g. This property is clearly independent of the choice of $x_1,\dots,x_s$. It follows that the prime ideals minimal over X(I) all have height at most s and dimension at most d-s, with equality in at least one case. We further term a prime ideal P of a local ring a good prime ideal if

$$\text{ht}P + \dim P = \dim Q.$$

LEMMA 2.2. <u>Let</u> $\underline{I} = (I_1,\dots,I_s)$ <u>be an independent set of ideals of</u> Q, <u>and let</u> $(x_1,\dots,x_s) = X(\underline{I})$ <u>be an independent set of general elements of</u> $\underline{I}$. <u>Let</u> P <u>be a good prime ideal minimal over</u> $X(\underline{I})$, <u>and suppose that</u> $p=P\cap Q$ is <u>is a minimal prime ideal of</u> Q. <u>Then</u>

i) p <u>is the only minimal prime ideal of</u> Q <u>contained in</u> P,

ii) $\dim Q/p = d$,

iii) P <u>is the only prime ideal of</u> Q_g <u>minimal over</u> X(I) <u>whose intersection with</u> Q <u>is</u> p,

iv) <u>If</u> R <u>denotes the ring</u> $(Q_g)_P/p(Q_g)_P$, <u>then</u> R <u>is a regular local ring of dimension</u> s, <u>whose maximal ideal is generated by the images of</u> $x_1,\dots,x_s$ <u>in</u> R.

i) Clearly, any ideal of Q contained in P is contained in p.

ii) It is enough to show that $\dim(Q_g/p_g) = d$. It is at most d since $\dim(Q_g) = d$. Since P is a good prime ideal of height s, there exists a chain of prime ideals $P_0 \subset P_1 \subset \dots \subset P_d = m_g$ with $P_s = P$. Then P_0 is a minimal prime ideal of Q_g contained in P and so meets Q in p. Hene it must be p_g. Hence $\dim(Q_g/p_g)$ is at least d.

We will prove iii) and iv) together. By applying a suitable automorphism of Q_g over Q, we may assume that $X(\underline{I})$ is standard, and hence that x_i is a linear form in the indeterminates $X_1, X_2, \ldots$ whose coefficients generate I_i, the indeterminates entering in different x_i being distinct.

Choose N so that $x_1, \ldots, x_s$ belong to Q_N. Then, denoting $P \cap Q_N$ by P_N, $P = P_N Q_g$. Hence $\mathrm{ht}P_N = \mathrm{ht}P = s$ and $\dim P_N = \dim P = d-s$, i.e. P_N is good. Further, we can replace Q by Q/p and assume that Q is a domain. Finally we will write R_N for the localisation of Q_N at P_N. Then $R = (R_N)_g$. Hence we can replace Q_g by Q_N to obtain iii), iv), with the assumption that Q is a domain.

We can construct R_N in two stages. For the first stage we take the ring of fractions of Q_N with respect to the set of non-zero elements of Q. Then, if F denotes the field of fractions of Q, R_N is the localisation of $F[X_1, \ldots, X_N]$ at a prime ideal minimal over the ideal $(x_1, \ldots, x_s)$. But the latter is generated by the s linearly independent linear forms $x_1, \ldots, x_s$ in $X_1, \ldots, X_N$ over F, and so is prime of height s. Since any prime ideal of Q_g minimal over $(x_1, \ldots, x_s)$ is generated by elements of Q_N, it follows that there is only one prime ideal of Q_g minimal over $X(\underline{I})$ meeting Q in zero, and iii) is proved. Further R_N is a localisation of $F[X_1, \ldots, X_N]$ at a prime ideal of height s and so is regular of height s. Further its maximal ideal is clearly generated by the images of $x_1, \ldots, x_s$. This proves iv).

THEOREM 2.3. *Let* $\underline{I} = (I_1, \ldots, I_s)$ *be an independent set of ideals of* Q *and let* $X(I) = (x_1, \ldots, x_s)$ *be an independent set of general elements of* $\underline{I}$ *(or the ideal they generate). Let* P *be a good prime ideal of* Q_g *minimal over* $X(\underline{I})$, *and let* $p = P \cap Q$. *Suppose that* $I_1, \ldots, I_s$ *are so numbered that* $I_1, \ldots, I_t$ *are contained in* p *but* $I_{t+1}, \ldots, I_s$ *are not. Then*

i) p *is a good prime ideal of* Q *of height* t, *and* p_g *is minimal over* $(x_1, \ldots, x_t)$,

ii) P *is the only prime ideal minimal over* $X(\underline{I})$ *meeting* Q *in* p,

iii) p_g *is the only prime ideal minimal over* $(x_1, \ldots, x_t)$ *contained in* P,

iv) *the local ring* $R = (Q_g)_P / p(Q_g)_P$ *is a regular local ring of dimension* $s-t$, *whose maximal ideal is generated by the images of* $(x_{t+1}, \ldots, x_s)$.

As in the proof of the lemma we will take $x_1, \ldots, x_s$ in standard form, which we will make explicit, and write

$$x_j = \sum_{k=1}^{n_j} u(j,k) X(N_j + k) \qquad j = 1, \ldots, s.$$

where, as earlier, $N_j = n_1 + \ldots + n_{j-1}$. Let $N_{s+1} = N$ and $N_{t+1} = N'$.

Write $Q' = Q/p$, and P' for the prime ideal P/p_g of Q'_g. Let Q_N, P_N have the same meanings as in the lemma, N being as already defined. We define Q'_N, P'_N similarly.

Now Q'_N/P'_N is a localisation of a finitely generated extension of Q' and P'_N meets Q' in the zero ideal. Let F', E' be the fields of fractions of Q' and Q'_N/P'_N respectively. Then E' is the field of fractions of $F'[X_1,\dots,X_N]/P''$, where P'' is a minimal prime ideal over $(y_1,\dots,y_s)$, where y_i is the image of x_i in $F'[X_1,\dots,X_s]$. But, as $I_1,\dots,I_t$ are all contained in P, and hence in $p, y_1,\dots,y_t$ are all zero, and since $I_{t+1},\dots,I_s$ are not contained in $p, y_{t+1},\dots,y_s$ are non-zero, and, as in the lemma, are linearly independent over F. It follows that P'' is generated by $y_{t+1},\dots,y_s$ and so has height s-t. Hence, the transcendence degree $t(E'/F')$ of E' over F' is N-s+t. Further the residue field of Q'_N/P'_N is isomorphic to $k(X_1,\dots,X_N)$, and so has transcendence degree N over k, the residue field of Q'. We now make two applications, the first to p, P_N and the second to the local rings Q' and Q'_N/P'_N, of the dimension inequality. We obtain

$$\mathrm{ht}p + N \geq \mathrm{ht}P_N + t(E'/F') = s+N-s+t \quad \text{i.e., } \mathrm{ht}p \geq t$$

$$\dim Q' + t(E'/F') = \dim Q' + N-s+t \geq \dim(Q'_N/P'_N)+N = d-s+N$$

whence $\dim Q' = \dim Q/p \geq d-t$. These two inequalities imply equality and hence that p is a good prime ideal. Further p_g contains $x_1,\dots,x_t$ which is part of a set of parameters, and hence all minimal prime ideals over it have dimension $\leq$ d-t.. Hence p_g must be minimal over $(x_1,\dots,x_t)$ and we have proved i).

Now we introduce the ring $Q^*_g = Q_g/(x_1,\dots,x_t)$ and write Q^*_N for $Q_N,/(x_1,\dots,x_t)$. We denote images of ideals of $Q_g, Q_N,$ modulo $(x_1,\dots,x_t)$ by adjoining "*". Then $p^*_N,$ is a minimal prime ideal of $Q^*_N,$ and has dimension d-t. P^* is clearly a minimal prime ideal of $x^*_{t+1},\dots,x^*_s$, which, in turn are general elements of the images $I^*_{t+1},\dots,I^*_s$ of $I_{t+1}Q_N,\dots, I_sQ_N$ in Q^*_N. Further, $x^*_{t+1},\dots,x^*_s$ is a sub-set of a set of parameters in Q^*_g, so that the set of ideals $\underline{I}^* = (I^*_{t+1},\dots,I^*_s)$ is an independent set of ideals. Finally, $\dim P^* = \dim P = d-s$, and $\mathrm{ht}P^* = s-t$, so that P^* is a good prime ideal of Q^*_g.. Hence we can apply the lemma. Statement iv) is immediate, and iii) follows if we observe that the minimal prime ideals of $(x_1,\dots,x_t)$ all have sets of generators in $Q_N,$. Finally we consider ii). The lemma implies that P is the only prime ideal minimal over $(x_1,\dots,x_s)$ which meets $Q_N,$ in $p_N,$. Let P_1 be another prime ideal of Q_g, minimal over $(x_1,\dots,x_s)$. If the intersection of P_1 with Q contains p, its intersection with $Q_N,$ contains $p_N,$ and so is either $p_N,$ in which case P_1 must be P, or properly contains it, and so has dimension <d-t. But this implies that it must contain at least one of the ideals $I_{t+1},\dots I_s$ and so the intersection of P_1 with Q properly contains p.

To conclude this section, we consider the relationship of independent sets of general elements to joint reductions. Let $\underline{I}$ be a set of d ideals $(I_1,\dots,I_d)$, not necessarily distinct. We now introduce some notation. Let $R = (r_1,\dots,r_d)$ be a set of d integers. Then, by $\underline{I}^R$ we mean

$$I_1^{r_1} I_2^{r_2} \dots I_d^{r_d}$$

where, if $r \leq 0$, $I_j^r = Q$. If R is as above, with the restriction that $r_j > 0$, then R(j) will denote the set of integers $(r_1,\dots,r_j - 1,\dots r_d)$. Then we term a set of elements $y_1,\dots,y_d$ of Q a joint reduction of $\underline{I}$ if $y_j \in I_j$ for each j and there exists a set of positive integers $R=(r_1,\dots,r_d)$ such that

$$\underline{I}^R = y_1\underline{I}^{R(1)} + y_2\underline{I}^{R(2)} + \dots + y_d\underline{I}^{R(d)}$$

Our object is to prove that an independent set of general elements $x_1,\dots,x_d$ of $\underline{I}$ is a joint reduction of $\underline{I}_g$.

We need some further definitions which now follow.

Let $T = (t_1,\dots,t_d)$ be a set of independent indeterminates over Q. Then, if $R=(r_1,\dots,r_d)$ is a set of integers, we will write T^R for the product

$$t_1^{r_1} \dots t_d^{r_d},$$

negative exponents being allowed. We will denote t_j^{-1} by u_j and $(u_1,\dots,u_d)$ by U. We now define the multigraded ring $R(\underline{I})$ to be the sub-ring of Q[T,U] consisting of all finite sums $\sum a(R)T^R$, with a(R) in $\underline{I}^R$, $r_1,\dots,r_d$ being allowed to be positive, zero, or negative. This ring is graded by sets of d integers $(r_1,\dots,r_d)$, r_j being termed the jth degree of R(I).

LEMMA 2.4. <u>Let x_j be a general element of I_j. Then there exists an integer r_j^o such that, if</u> $R = (r_1,\dots,r_d)$ <u>with</u> $r_j > r_j^o$, r_i <u>for</u> $i \neq j$ <u>being unrestricted, such that</u>

$$\underline{I}^R Q_g \cap x_j Q_g = x_j \underline{I}^{R(j)} Q_g$$

Since $u_1 u_2 \dots u_d = u$ is a non-zero-divisor of $R(\underline{I})$, the set of prime ideals associated with $u^n R(\underline{I})$ is independent of n. We divide this set into two sub-sets, S_1 consisting of those which contain $I_j t_j R(\underline{I})$ and S_2 consisting of those which do not contain $I_j t_j R(\underline{I})$.

Now we replace Q by Q_g and $\underline{I}$ by $\underline{I}_g = (I_1 Q_g,\dots,I_d Q_g)$. We will still divide the set of prime ideals associated with $uR(I_g)$ into two classes, but note that, if S_i consists of $p_1,\dots,p_s$, when we consider $R(\underline{I})$, then the passage to $R(I_g)$ replaces these by $(p_1)_g,\dots,(p_s)_g$ and this set will now be S_i. Hence the element $x_j t_j$ of I_g is not contained in any ideal in S_2. This implies that, if we denote by M_n the module

$$(u^n R(\underline{I}_g) : x_j t_j) / u^n R(\underline{I}_g)$$

then M_n is annihilated by $(I_j t_j)^N$ for some integer N depending on n. But

M_n is a finitely generated $R(I_g)$-module and hence any element of M_m of degree R is zero if the jth degree r_j is sufficiently large.

Now let B denote the ideal of $R(\underline{I}_g)$ consisting of all finite sums $\sum a(R)T^R$ with a(R) contained in $x_j Q_g \cap \underline{I}_g^R$. Then B has a finite basis consisting of elements of the form $x_j b_j T^R$ with b_j in Q_g. It follows that we can find an integer q such that the elements $b_j u^q$ all belong to $R(\underline{I}_g)$ and hence $B = x_j t_j R(I_g):u^q$. Now suppose that zT^R belongs to B. Then

$$u^q.zT^R = x_j t_j W$$

where W is a homogeneous element of $R(\underline{I}_g)$ whose j^{th} degree is r_j-q-1. Then, by the first part of the proof, W will belong to $uR(I_g)$ providing that r_j-q-1 is sufficiently large. Hence we can find r_j^o such that, if $r_j > r_j^o$, $W \in u^q R(\underline{I}_g)$ and hence zT^R belongs to $x_j t_j R(\underline{I}_g)$. Therefore, if $r_j > r_j^o$.

$$x_j Q_g \cap (\underline{I}_g)^R = x_j (\underline{I}_g)^{R(j)}.$$

THEOREM 2.5. <u>If</u> $x_1,\ldots,x_d$ <u>is an independent set of general elements of</u> $\underline{I} = (I_1,\ldots,I_d)$, <u>then</u> $x_1,\ldots,x_d$ <u>is a joint reduction of</u> $\underline{I}_g$.

We will prove this result by induction on d. We commence with the observation that, if $I_1\ldots I_d$ is nilpotent, the result is trivial. This disposes of the case d=0 and we can in fact impose the restriction that $(0:I_1\ldots I_d) = 0$, which implies that each of the ideals $I_1,\ldots,I_d$ contains a non-zero-divisor, and hence that $x_1,\ldots,x_d$ are all non-zero-divisors. For suppose that the result has been proved for all local rings Q satisfying this condition and sets of ideals $I_1,\ldots,I_{d'}$, where $d' = \dim Q$ is at most d. Then we will show that it then holds without this restriction for all local rings Q of dimension $d' \leq d$. Let $J = \bigcup_{q=1}^{\infty} (0:(I_1\ldots I_{d'})^q)$. Then we can choose q so that $J = (0:(I_1\ldots I_{d'})^q)$ Consider the ring Q'/J, and the set of ideals $\underline{I}' = (I'_1,\ldots,I'_{d'})$, where $I'_j = I_j+J/J$. Then $\dim Q' \leq \dim Q$ and the ideals $I'_1,\ldots,I'_{d'}$ either each contain a non-zero-divisor, or their product is zero, which implies that $I_1\ldots I_{d'}$ is nilpotent, when the result is true. Hence we may assume that the result is true for Q'. This implies that, if $r_1,\ldots,r_d$ are all sufficiently large, then

$$\underline{I}_g^R \subseteq x_1 \underline{I}_g^{R(1)} + \ldots x_{d'} \underline{I}_g^{R(d')} + J$$

and we obtain the result for $Q, I_1,\ldots I_{d'}$, if we increase $r_1,\ldots,r_d$ each by q.

We now come to the inductive part of the proof. Suppose that d>0. By taking $x_1,\ldots,x_d$ in standard form, we can find N such that x_1 belongs to Q_N. Now take $Q' = Q_N/x_1Q_N$, $I'_j = I_jQ_N + x_1Q_N/x_1Q_N$ for j= 2,...,d, and let x'_j be the image of x_j in Q'_g, so that $x'_2,\ldots,x'_d$ form an independent set of general elements of the set $\underline{I}' = (I'_2,\ldots,I'_d)$ of Q'. Then we can apply our inductive assumption, and hence obtain, for $r_1,\ldots,r_d$ sufficiently large

$$\underline{I}_g^R \subseteq x_1 Q_g + x_2 \underline{I}_g^{R(2)} + \ldots + x_d \underline{I}_g^{R(d)}$$

and this can be written

$$\underline{I}_g^R = (x_1 Q_g \cap \underline{I}_g^R) + x_2 \underline{I}_g^{R(2)} + \ldots + x_d \underline{I}_g^{R(d)}$$

and, hence, by lemma 2.4, at the expense of increasing $r_1, \ldots, r_d$ further,

$$\underline{I}_g^R = x_1 \underline{I}_g^{R(1)} + \ldots + x_d \underline{I}_g^{R(d)}.$$

which completes the proof.

3. m-valuations.

We commence this section by defining m-valuations on a local ring (Q,m,k,d). We begin with a valuation v on the field of fractions F_p of Q/p, where p is a prime ideal of Q. We place the following restrictions on v:

i) its value group is the additive group of integers,

ii) it takes values ≥ 0 on $Q|p$, and >0 on $m|p$.

Now suppose that I is an ideal of Q. Then, if I is not contained in p, we define $v(I)$ to be the minimum value of v on the ideal $I+p/p$. If I is m-primary, and $p \neq m$, this is always finite (the case $p=m$ will be implicitly excluded). We will apply the term m-valuation to v, or to the function $v(I)$ on the set of ideals of Q. Note that $v(I)$ has the obvious properties

a) $v(I_1 I_2) = v(I_1) + v(I_2)$

b) if $I_1 \supseteq I_2$, then

$v(I_1) \leq v(I_2)$

c) if $I \subsetneq p$, then $v(I) = v(I+m^r)$ for r large.

The ideal p consisting of all elements in Q for which $v(x) = \infty$, will be termed the limit ideal of v and written $L(v)$.

LEMMA 3.1. *Let $v_1, \ldots, v_N$ be a set of N distinct m-valuations on Q. Then, if*

$$\sum_1^N a_i v_i(I) = 0$$

for all m-primary ideals I, $a_1, a_2, \ldots, a_N$ are all zero.

Let r be the number of distinct prime ideals in the set $L(v_1), \ldots, L(v_N)$. Then, we proceed by induction on r. First suppose that $r=1$. Then, by replacing Q by Q/p, we may assume that $v_1, \ldots, v_N$ are derived from valuations on the field of fractions F of Q.

Now let c be any non-zero element of Q. Then, by c) above, if r is sufficiently large, $v_i(cQ+m^r) = v_i(c)$ for $i = 1, \ldots, N$, and hence

$$\sum_1^N a_i v_i(c) = 0 \qquad \text{for all } c \neq 0 \text{ in } Q.$$

from which it follows immediately that

$$\sum_1^N a_i v_i(x) = 0 \qquad \text{for all } x \neq 0 \text{ in } F.$$

Then, by the theorem of independence of valuations, $a_1 = \ldots = a_N = 0$. (Note that, since the value group of each valuation group is Z, any two valuations among $v_1,\ldots,v_N$ are linearly independent.)

Now suppose that $r > 1$. Choose p minimal among the ideals $L(v_i)$, and suppose that $v_1,\ldots,v_N$ are so numbered that $L(v_i)=p$ if $1\le i\le N'<N$, and that $L(v_j)\neq p$ if $j>N'$. Now let c be any element of Q not contained in p but contained in $L(v_j)$ for all $j>N'$. Then we can find n such that

$$v_i(c) = v_i(cQ+m^n) \qquad 1\le i \le N'$$

$$v_j(cQ+m^n) = nv_j(m) \qquad j > N'$$

and these equations remain true if n is replaced by $n+1$. Hence

$$\sum_{i=1}^{N} a_i v_i(c) = (n+1)\sum_{i=1}^{N} a_i v_i(cQ+m^m) - n.\sum_{i=1}^{N} a_i v_i(cQ+m^{n+1}) = 0,$$

for all c choosen in the way described. Clearly we can replace c by ca, cb where a,b are any elements of Q not in p and hence we have

$$\sum_{i=1}^{N'} a_i v_i(x) = 0$$

for any non-zero x in the field of fractions of Q/p, implying as before, that $a_1,\ldots,a_{N'}$ are all zero. We have thus reduced r to $r-1$, and the statement of the lemma follows by our inductive hypothesis.

We are now going to consider the relationship between m_g-valuations v on Q_g and their restrictions v_o, v_N to Q, Q_N respectively. For this purpose we require the lemma which follows, due in a more general form to Abhyankar. The proof given is due to Zariski and Samuel[12], vol.II, appendix 2. First we introduce some notation. Let (Q,m,k,d) be a local domain with field of fractions F and let E be a finitely generated extension of F. Let v be a valuation on E, ≥ 0 on Q and >0 on m. Then the residue field K_v of v is an extension of k. We write $t(v/Q)$ for the transcendence degree $t(K_v/k)$.

LEMMA 3.2 If $d = \dim Q$, then

$$d+t(E/F)-1 \ge t(v/Q)$$

Choose elements $z_1,\ldots,z_s$ in O_v such that their images in K_v are algebraically independent over k. Let p be the centre of v on $Q[z_1,\ldots,z_s]$, and let Q' be the local ring $Q[z_1,\ldots,z_s]_p$. Let E' be the field of fractions of Q', so that $t(E'/F)\le t(E/F)$. We now apply the dimension inequality to the pair of rings Q,Q' and obtain

$$d+t(E/F) \ge d+t(E'/F)\ge \dim Q'+s$$

Since $\dim Q'>0$, it follows that s is bounded above by $d+t(E/F)-1$, proving that $t(v/Q)$ is finite, and we can now take $s = t(v/Q)$.

We now turn to the rings Q, Q_g. Suppose that Q is a local domain and that v is an m-valuation on Q with $L(v)= 0$. Then we define the general extension v_g of v to Q_g as follows. First, let f be an element of Q_g of the form $\sum a_\mu.\mu$, where μ runs over monomials in $X_1, X_2,\ldots$, the sum being finite, and the coefficients a_μ belonging to Q. Then we define $v_g(f)$ to be $\operatorname{Min} v(a_\mu)$, We now have a unique extension to Q_g itself. If

$L(v) \neq 0$, then we use the same definition with $Q/L(v)$ replacing Q.

LEMMA 3.3. Let Q be a local domain and let v be an m-valuation on Q such that $L(v) = 0$ and $t(v/Q) = 0$. Let w be an extension of v to Q_g such that $L(w) = 0$. Then $w = v_g$.

Suppose this is false. Let N be the least integer such that Q_{N+1} contains an element $f = \sum a_\mu \cdot \mu$ with $w(f)$ not equal to, and hence, greater than $\operatorname{Min} w(a_\mu)$. Then w_N is the general extension of v to Q_N and w_{N+1} is not the general extension of w_N to Q_{N+1}. Replacing Q by Q_N, we can assume that $N = 0$ and hence that there exists a polynomial in X_1 with coefficients in Q such that $w_1(f) > v_g(f)$. Multiplying f by a suitable power of the generator of the maximal ideal of O_v, we can find a polynomial $f(X_1)$ with coefficients in O_v such that $v_g(f(X_1))=0$. But this implies that X_1 is algebraic over the residue field of O_v, and hence over k, modulo the maximal ideal of O_w, and hence that w is not an m_g-valuation.

THEOREM 3.4. Let w be an m_g-valuation on Q_g. Then there exists an integer N such that w is the general extension of the restriction w_N of w to Q_N.

We find N in two stages. Let $P=L(w)$. Then we can find N' such that P is generated by elements of $Q_{N'}$. Then, by replacing Q by $Q_{N'}/P_{N'}$, where $P_{N'} = P \cap Q_{N'}$, we may assume that $L(w)=0$. Next, we can find $N>N'$ such that the field of fractions of Q_N contains elements contained in O_w whose images form a transcendence basis of K_w over k_g. Let these elements be $z_1,\ldots,z_s$ and let Q' be the localisation of $Q_N[z_1,\ldots,z_s]$ at the centre of w_N. Then $t(w_N/Q') = 0$ and w is a valuation on Q'_g. Hence w is the general extension of w_N on the field of fractions of Q'_g and this is the same as that of Q_g.

4. Mixed Multiplicities.

DEFINITION. Let Q be a local ring, and let $\underline{I} = (I_1,\ldots,I_d)$ be a set of d ideals of Q subject to the condition that, if $x_1,\ldots,x_d$ is an independent set of general elements of $\underline{I}$, then $X(\underline{I}) = (x_1,\ldots,x_d)$ is m_g-primary. (This condition states that $\underline{I}$ is an independent set of ideals). Let M be a finitely generated Q-module. Then we define the multiplicity $e(\underline{I};M)$ to be

$$e(x_1,\ldots,x_d;M_g)$$

This is clearly independent of the choice of $x_1,\ldots,x_d$. The definition of $e(x_1,\ldots,x_d;M)$ that we shall use is that due to Wright [11], i.e., if $d=0$, then $e(\emptyset;M_g) = L(M)$, while if $d>0$ we define it inductively by

$$e(x_1,\ldots,x_d;M_g)=e(\bar{x}_2,\ldots,\bar{x}_d;M_g/x_1M_g)-e(\bar{x}_2,\ldots,\bar{x}_d;(0:x_1)_{M_g})$$

where $\bar{x}_2,\dots,\bar{x}_d$ are the images of $x_2,\dots,x_d$ in $\bar{Q}=Q_g/x_1Q_g$ and M_g/x_1M_g and $(0:x_1)_{M_g}$ are both considered as $\bar{Q}$-modules. It is clear that $e(\underline{I};M)$ s an additive function on the category of finitely generated Q-modules, and is non-negative (see for example, Nothcott[5], p308 theorem 6). We also observe that we can choose N such that x_1 belongs to Q_N and $(\bar{x}_2,\dots,\bar{x}_d)$ is an independent set of general elements of the set of ideals $(\bar{I}_2,\dots,\bar{I}_d)$, where $\bar{I}_j = I_jQ_N+x_1Q_N/x_1Q_N$ of $\bar{Q}_N= Q_N/x_1Q_N$. This is clearest if we take $x_1,\dots,x_d$ in standard form. Then we can write the inductive definition in the form

$$e(\underline{I};M) = e(\bar{\underline{I}};M')-e(\bar{\underline{I}};M'')$$

where $\bar{\underline{I}} = (\bar{I}_2,\dots,\bar{I}_d)$, $M' = M_N/x_1M_N$, $M'' = (0:x_1)_{M_N}$, the two terms on the right-hand side referring to multiplicities over the ring $\bar{Q}_N$. This enables us to reduce the proof of certain results to the case d=1, where the definition in terms of Hilbert functions is often more convenient. This is because, if x is a general element of an ideal I of a 1-dimensional local ring, then xQ_g is a reduction of I_g and hence we can define $e(I;M) = L(I^nM/I^{n+1}M)$ for large n. We recall without proof two results which hold in that case, and which we will use in the next lemma

a) $e(I_1I_2;M) = e(I_1;M)+e(I_2;M)$

b) if I_1,I_2 have the same integral closure, then

$$e(I_1;M) = e(I_2;M)$$

LEMMA 4.1. a) Let $\underline{I} = (I_1,\dots,I_d)$ and $\underline{I}' = (I_1,\dots,I_{d-1},I_d')$ be independent sets of d ideals, and let $\underline{I}'' = (I_1,\dots,I_{d-1},I_dI_d')$. Then $\underline{I}''$ is an independent set, and

$$e(\underline{I}'';M) = e(\underline{I};M)+e(\underline{I}';M)$$

b) Let $\underline{I} = (I_1,\dots,I_d)$ be an independent set of ideals, and let $\underline{I}^* = (I_1^*,\dots,I_d^*)$, where I_j^* is the integral closure of I_j. Then

$$e(\underline{I}^*;M) = e(\underline{I};M)$$

In both cases we can reduce the proof to the case d=1, In the first case, we must have $ht(x_1Q_g+\dots+x_{d-1}Q_g) = d-1$ and $\dim(Q_g/x_1Q_g+\dots+x_{d-1}Q_g)$ equal to 1. and in the case d=1 a single ideal is an independent set if and only if it is m-primary. Since the product of two m-primary ideals is m-primary, the first statement of a) follows, while the second is simply a) above.

For b), we replace the ideals $I_1,\dots,I_d$ by $I_1^*,\dots,I_d^*$ one at a time, using the symmetry of $e(\underline{I};M)$ in $I_1,\dots,I_d$, Note that, if we write $X_{d-1}(\underline{I})$ for $x_1Q_g+\dots+x_{d-1}Q_g$, then $I_d^*Q_g+X_{d-1}(\underline{I})/X_{d-1}(\underline{I})$ is not necessarily the integral closure of $I_dQ_g+X_{d-1}(\underline{I})/X_{d-1}(\underline{I})$, but has the same integral closure, so that we can apply b) above..

We now come to the main theorem of the paper,which will require some preliminary explanation.We fix an independent set $\underline{I} = (I_1,\dots,I_{d-1})$ consisting of d-1 ideals of Q,and we let J range over all ideals such that $(I_1,\dots,I_{d-1},J)$ is also m-primary(for example,J could be any m-primary ideal).Let $x_1,\dots,x_{d-1}$ be an independent set of general elements of $\underline{I}$,and let y be a general element of J such that $x_1,\dots,x_{d-1},y$ is an independent set of general elements of $\underline{I}$,J.Our concern is with the multiplicity function

$$e(\underline{I},J;M) = e(x_1,\dots,x_{d-1},y;M)$$

considered as an additive function on the category of finitely generated Q-modules and as a finction of J.As earlier,we will denote by $X(\underline{I})$ the ideal $(x_1,\dots,x_{d-1})$ of Q_g.

In general terms,the theorem states that we have an equation

$$e(\underline{I},J;M) = \sum a_v(\underline{I};M)v(J)$$

where $a_v(\underline{I};M)$ is a non-negative integer depending on $\underline{I}$,M,defined for all m-valuations v of Q,which is zero for given $\underline{I}$ for all save a finite set of v.Lemma 3.1 implies that the above equation uniquely determines the coefficients $a_v(\underline{I};M)$,and can be used to define the function $a_v(\underline{I};M)$,and to derive many of its properties.

We now introduce some further notation.Let P range over the set of all good minimal prime ideals of $X(\underline{I})$.We also write p = Q P,and note that,as a consequence of theorem 2.3,p is also a good prime ideal. We will,for convenience of notation,denote by Q_P and M_P the ring $(Q_g)_P$ and the module $(M_g)_P$.We will also denote by Q(P),M(P) and J(P) the ring Q_g/P,the Q(P)-module M_g/PM_g and the ideal JQ_g+P/P of Q(P).These notations will play only a temporary role.The lemma following proves the existence of the formula for $e(\underline{I},J;M)$.

LEMMA 4.2. <u>There exists an expression for</u> $e(\underline{I},J;M)$ <u>of the form</u>

$$\sum_v a_v(\underline{I};M)v(J)$$

<u>where the sum is over all</u> m-<u>valuations on</u> Q,<u>and</u> $a_v(\underline{I};M)$ <u>is a non-negative integer equal to zero for all save a finite number of</u> v.

We start with the associativity formula for multiplicities([5], p342,Theorem 18),which we apply to the partition $x_1,\dots,x_{d-1}|y$ of the set of parameters $x_1,\dots,x_{d-1},y$.We can write it in the form

$$e(\underline{I},J;M) = \sum_P e(J(P);Q(P))e(IQ_P;M_P)$$

the some being over all good prime ideals P minimal over X(I).Note that the first term does not involve M.We now consider the first term. The ring Q(P) is a 1-dimensional local domain.Further

$$e(J(P);Q(P)) = e(yQ(P);Q(P)).$$

In the 1-dimensional case we have a formula for e(yQ(P);Q(P)) due to Northcott([3],Theorem 6(with the observation that lengths and multiplicities are equal in 1-dimensional domains). In our notation, this runs as follows. Let w range over the valuations associated with the integral closure of Q(P) in its field of fractions (recall that this ring has finitely many maximal ideals, and its localisation at each of these is a discrete valuation ring). Let K_w denote the residue field of w. Then K_w is a finite extension of k_g, and Northcott's formula is

$$e(yQ(P);Q(P)) = \sum_w r(Q(P),w)[K_w:k_g]w(y).$$

Each of the valuations w determines an m-valuation on Q_g whose limit ideal is P. The factor r(Q(P),w) is a ramification factor about which all we need note is that it is a positive integer. Now we turn to w(y). We can choose an integer N such that w is the general extension of w_N and we can assume that y is choosen to be a general element of JQ_N. It then follows that w(y) = v(J), where v is the restriction of w to Q. Since different w may determine the same v, we now have an expression

$$e(J(P);Q(P)) = \sum_v d(\underline{I},v)v(J)$$

where $d(\underline{I},v)$ is a non-negative integer equal to zero for all save a finite set of v with limit ideal p. Again referring to theorem 2.3, we recall that different P meet Q in different p. Hence, we have derived a formula

$$e(\underline{I},J;M) = \sum_v a_v(\underline{I};M)v(J)$$

where $a_v(\underline{I};M) = d(\underline{I},v)e(\underline{I}Q_p;M_p)$.

The expression we have derived for $a_v(\underline{I};M)$ is, for the moment, not particularly convenient, and our objective is to simplify the two factors. We will first consider the factor $e(IQ_p;M_p)$. We now associate with each v a partition of the set $\underline{I}$ into two sub-sets $\underline{I}'(v)$ and $\underline{I}''(v)$. We define I'(v) to be the set of ideals I_j contained in the limit ideal of v (that is, p). We will also write Q_v, M_v for Q_p, M_p to indicate their dependence on v. $\underline{I}''(v) = \underline{I}-\underline{I}'(v)$. We now have the following lemma.

LEMMA 4.3. $$e(IQ_p;M_p) = e(\underline{I}'(v);M(v))$$

Suppose we number the ideals $I_1,\ldots,I_{d-1}$ so that $\underline{I}'(v)$ consists of $I_1,\ldots,I_t$ and I"(v) consists of $I_{t+1},\ldots,I_{d-1}$. Then if we refer back to theorem 2.3, we see that $x_1Q_p+\ldots+x_tQ_p$ has only one minimal prime ideal, namely pQ_p and further, the ring Q_p/pQ_p is a regular local ring whose maximal ideal is generated bythe images of $x_{t+1},\ldots,x_{d-1}$. We now apply the associativity formula to $e(\underline{I}Q_p;M_p)$ with the partition $x_1,\ldots,x_t|x_{t+1},\ldots,x_{d-1}$ of $x_1,\ldots,x_{d-1}$. Then we obtain

$$\begin{aligned} e(\underline{I}Q_p;M_p) &= e(x_1,\ldots,x_{d-1};M_p)=e(x_{t+1},\ldots,x_{d-1};Q_p/pQ_p)e(x_1,\ldots,x_t;(M_p)_{pQ_p}) \\ &= e(I'(v)Q_g;(M_p)_g) = e(I'(v);M_v) \end{aligned}$$

Now we turn to the other factor $d(\underline{I},v)$. Our object is to showthat this factor depends only on the set I"(v) and not on the whole set $\underline{I}$.

To this end we will consider the ring Q/p and the set of ideals I_j+p/p. Note that, if I_j belongs to $\underline{I}'(v)$, this ideal will be zero. Hence we have essentially reduced the set $\underline{I}$ to the set $\underline{I}''(v)$. We now change notation and use $\underline{I}''(v)$ to denote the set of ideals I_j+p/p, where I_j belongs to the original set $\underline{I}''(v)$. We now calculate $e(\underline{I}''(v),J+p/p;M/pM)$. Again, renumber $I_1,\ldots,I_{d-1}$ so that $\underline{I}''(v)$ consists of the ideals I_j+p/p for $j= t+1,\ldots,d-1$. of $Q/p = Q/L(v)$. Consider the set of good prime ideals of $(Q/p)_g$ minimal over $(x_{t+1},\ldots,x_{d-1})+p_g/p_g$. These are the prime ideals of the form P'/p_g, where P' is a prime ideal of Q_g which contains $(x_{t+1},\ldots,x_{d-1})+p_g$ of height $d-1$. Hence P' ranges over the good minimal prime ideals of $(x_1,\ldots,x_{d-1})$ which meet Q in a prime ideal p' containing p. We can now use lemmas 4.2 and 4.3 to calculate $e(\underline{I}''(v),J+p/p;M/pM)$ as

$$\sum d(\underline{I}''(v'),v')e(\bar{I}(v');M_{v'}/pM_{v'})v'(J)$$

where v' ranges over the m-valuations of Q such that $L(v') \supseteq p$, and $\bar{I}(v')$ is the set of ideals I_j+p/p such that $I_j \subseteq L(v')$. In particular, suppose that we take $v' = v$. First we note that $\bar{I}(v)$ is empty, and hence that the factor $e(\bar{I}(v);M_v/pM_v)$ is simply the dimension of the vector space $M_p/pM_p = M_v/pM_v$ over the field of fractions of Q/p. Next consider the term $d(\underline{I}''(v),v)$. This is defined, by the above, as the sum $\sum r(Q(P),w)[K_w:k_g]$, taken over those valuations w on $Q(P) = Q_g/P$ whose restriction to Q/p is v. This is precisely the same as $d(\underline{I},v)$ and hence we have proved that $d(I,v)$ depends only on the set $\underline{I}''(v)$ as we have redefined it.

This suggests that we approach $d(I,v)$ differently. First suppose that Q is a domain and that $L(v)= 0$. Then we define $d(\underline{I},v)$ as above. Now suppose that v is an m-valuation such that $L(v)=p \neq 0$. Now let $I(v)$ be an independent set of ideals of Q/p containing $\dim(Q/p)-1$ ideals. Then, by replacing Q by Q/p, we can apply the definition of $d(\underline{I},v)$ in the case $L(v)=0$ to this case. If we follow this procedure then we see that, with $\underline{I}$ an independent set of $d-1$ ideals $(I_1,\ldots,I_{d-1})$ of Q, $d(\underline{I},v) = d(I''(v),v)$. We summarise this in the statement of the following lemma.

LEMMA 4.4. $$e(\underline{I},J;M)= \sum d(I''(v),v)e(I'(v),M_v)v(J)$$

where $I''(v)$ is the set of ideals I_j+p/p of Q/p which are non-zero. $I''(v)$ contains $\dim p-1$ ideals, and is an independent set of ideals of Q/p, and v is a valuation on the field of fractions of Q/p.

To conclude this preliminary discussion, we require one further definition and a lemma, which indicates that the valuations v occurring in the above formula belong to a restricted class of m-valuations.

DEFINITION. Let $t(v)$ denote the transcendence degree of the residue field K_v of v over k. Then we will say that v is a good m-valuation if

$$t(v) + ht(L(v)) = d-1.$$

We have already seen that $t(v) \leq \dim(L(v))$. Hence this definition is equivalent to the two statements that $L(v)$ is good and $t(v) = \dim(L(v))-1$.

We now prove that the valuations occurring in lemma 4.4 can be restricted to be good.

LEMMA 4.5. *If in lemma* 4.4, $d(I''(v),v)e(I'(v),M_v)$ *is non-zero*, v *is good.*

We recall that the valuations v in lemma 4.4 are obtained as the restrictions of m_g-valuations w on Q_g such that $P = L(w)$ satisfies the condition that Q_g/P has dimension 1, and further the residue field K_w of w is a finite algebraic extension of k_g. We now choose N so that P is generated by elements of Q_N and we consider the restriction w_N of w to Q_N. It follows from theorem 3.4 that w is the general extension of w_N and hence that the residue field K_N of w_N is algebraic over k_N. Now let $p = Q \cap P = Q \cap P_N$, and write Q' for Q/p. Further let $E = Q_N/P_N$ and let F', E' be the fields of fractions of Q', R. Then we have already seen that R is a localisation of a finitely generated extension of Q', and that $t(E'/F') = N+t-d+1$, where $d-t$ is the dimension of Q'. Further R is 1-dimensional. Let R' be the localisation of $O_v[R]$ at the centre of w. Then R' is a localisation of O_v at a finitely generated extension of O_v and, since it contains a 1-dimensional local domain, it is also 1-dimensional. Since O_v is a regular local ring, we can apply the dimension equality

$$\dim O_v + t(E'/F') = \dim R' + t(k_{R'}/K_v)$$

i.e.

$$1 + n-t-d+1 = 1 + t(k_{R'}/K_v)$$

where $k_{R'}$ is the residue field of R'. But $R \subseteq R' \subseteq O_{w_N}$ and hence $k_{R'}$ has transcendence degree N over k. Hence K_v has transcendence degree $d-t-1$ over k and t is the height of p, which proves the result.

We now collect together the lemmas above and parts of theorem 2.3 to state our main theorem.

THEOREM 4.6. *Let* $\underline{I} = (I_1,\dots,I_{d-1})$ *be an independent set of ideals of* Q. *Let* v *be a good* m-*valuation such that* $ht(L(v))=t$. *Let* $\underline{I}'(v)$ *be the set of ideals of* $\underline{I}$ *contained in* $L(v)$, *and let* $\underline{I}''(v)$ *be the set of ideals of* $Q/L(v)$ *of the form* $I_j+L(v)/L(v)$ *where* I_j *does not belong to* $\underline{I}'(v)$. *Then, if* J *is any ideal of* Q *such that* $\underline{I},J$ *is an independent set of ideals, and* M *is any finitely generated* Q-*module, then we can write* $e(\underline{I},J;M)$ *in the form*

$$e(\underline{I},J;M) = \sum d(\underline{I}''(v),v)e(\underline{I}'(v)Q_v;M_v)v(J)$$

where v *ranges over all good* m-*valuations on* Q, *and* $e(I'(v)Q_v;M_v)$ *is to to be taken as zero if either the number of ideals in* $\underline{I}'(v)$ *is not* equal to ht(L(v)) or if $\underline{I}'(v)Q_v$ is not an independent set of ideals of ideals of Q_v. Further, $d(\underline{I}'',v)$ is a function, taking non-negative integer values, defined for all independent sets I" of ideals of Q/L(v) which contain d_v-1 ideals($d_v = \dim(Q/L(v))$). The function $d(\underline{I}'',v)$ has the following additional property. If $\underline{I}$ is an independent set of ideals of Q/L(v) containing d_v-2 ideals, and K_1,K_2 are two ideals of Q/L(v) such that $\underline{I}$ K_i (i =1,2) is an independent set of ideals of Q/L(v), so that I,K_1K_2 is also independent, then

$$d(\underline{I},K_1,K_2,v) = d(\underline{I},K_1,v)+d(\underline{I},K_2,v)$$

The greater part of the proof of this theorem is contained in the proofs of lemmas 4.2 to 4.5. The statement as to when $e(\underline{I}'(v),M_v)$ is to be taken to be zero has the effect of reducing the number of terms in the sum to those occurring in the expression occurring in lemma 4.2. The only statement requiring further proof is the last one. To prove this ,we first fix v, and write v_1 in place of v. Secondly, we can, without loss of generality, assume that $L(v_1) = 0$ and that Q is a domain. Now let J be any m-primary ideal, so that $\underline{I},K_i,J$ is an independent set of ideals of Q containing d ideals, and the same is true of $\underline{I},K_1K_2,J$. We now make use of the symmetry of the function e to obtain

$$\begin{aligned} e(\underline{I},K_1K_2,J;M) &= e(\underline{I},J,K_1K_2;M) \\ &= e(\underline{I},J,K_1;M)+e(\underline{I},J,K_2;M) \\ &= e(\underline{I},K_1,J;M)+e(\underline{I},K_2,J;M) \end{aligned}$$

Now suppose we expand the terms $e(\underline{I},K_1K_2,J;M)$ and $e(\underline{I},K_i,J;M)$ (i = 1,2) in the form given in the theorem aboveand further take M = Q. The fact that the functions v(J) on the set of m-primary ideals are linearly independent according to lemma 3.1 implies that we can replace e above by the coefficient of $v_1(J)$. Since we are assuming that $L(v_1)=0$, it follows that Q_{v_1} is the field of fractions of Q and that $e((\underline{I},K_i)'(v_1);Q) = e((\underline{I},K_1K_2)'(v_1);Q) = 1$. Hence we are left with the equation

$$d(\underline{I},K_1K_2,v_1) = d(\underline{I},K_1,v_1)+d(\underline{I},K_2,v_1).$$

Note that d(I,v) is also symmetric in the set of ideals $\underline{I}$.

We conclude this paper by applying the formula given in the last theorem to two special situations and re-derive some results already in the literature.

The first situation that we consider is when $I_1,\ldots,I_{d-1}$ are all m-primary, and hence necessarily form an independent set of ideals of Q. First we consider the multiplicity function $e(\underline{I},J;M)$ where J is also assumed to be m-primary. If, $x_1,\ldots,x_{d-1},y$ is an independent set of general elements of $\underline{I}$,J, then, by theorem 2.5, they form a joint reduction of $\underline{I}$,J.But it was proved in [9] that this implies that the multiplicity of the ideal $(x_1,\ldots,x_{d-1},y)$ is the same as the mixed multiplicity of the set of ideals $\underline{I}$,J in the sense used by Teissier [10].Hence the multiplicity function of this paper is a generalisation of Teissier's mixed multiplicity function.Next we consider the formula given in theorem 4.6.Let v be an m-valuation on Q,and consider L(v).If ht(L(v))=t, and v(J) occurs in the formula with non-zero coefficient,then v is good and contains t of the ideals $I_1,\ldots,I_{d-1}$.Since these are all m-primary, this is only possible if t=0,i.e.L(v) must be a minimal prime ideal of Q and dim(Q/L(v))=d. We will refer to an m-valuation such that this is true as a proper m-valuation.Now consider the factor $e(\underline{I}'(v)Q_v,M_v)$.The set I'(v) is empty ,and hence this factor reduces to $L(M_v)$ where M_v is considered as a Q_v-module,and Q_v is a primary artinian ring.Hence the formula reduces to

$$e(\underline{I},J;M) = \sum_v d(\underline{I},v)L(M_v)v(J)$$

the sum being over all proper m-valuations.We note that we do not require J to be m-primary.Hence,if x is an element of Q,we can take J = xQ providing that dimQ/xQ = d-1(since X_1x is a general element of xQ) and we can take y=x.Now consider the ring Q/xQ and write $\underline{I}_x$ for the set of ideals I_j+xQ/xQ.Then the images of $x_1,\ldots,x_{d-1}$ in $(Q/xQ)_g$ is an independent set of general elements of $\underline{I}_x$. Hence we have

$$e(\underline{I},xQ;M) = e(\underline{I}_x;M/xM) - e(I_x,(0:x)_M)$$

which we denote by d($\underline{I}$,x,M).We now obtain the formula

$$d(\underline{I},x,M) = \sum_v d(\underline{I},v)L(M_v)v(x).$$

This formula,and in fact,most of the results of this paper are treated in greater detail in [8].If the ideals $I_1,\ldots,I_{d-1}$ are all equal,and M=Q,we obtain the main result of [6].

Now we consider the case where Q has dimension 2.In this case $\underline{I}$ consists of just one ideal which we will denote by I.If I is m-primary we obtain a special case of the above.Hence we will suppose that I is not m-primary and consider the set of good prime ideals minimal over I.Let p be such a prime ideal Then dim(Q/p) = 1 and there are only a finite number of m-valuations of Q with L(v)=p.Further,the good minimal prime ideal P of xQ_g meeting Q in p,x being a general element of I,is also 1-dimensional,and contains p_g and so is equal to it. Hence ,by the proof of theorem 4.2,the contribution of the terms with L(v)=p to the sum in the formula is $e(J+p/p)e(IQ_p,M_p)$. Note that,according to the definition adopted here,$e(IQ_p,M_p)$ has to be interpreted as $e(x(Q_p)g;M_{pg})$ where x is a general element of I,but,since Q_p is 1-dimensional,$x(Q_p)_g$ is a reduction of I_g and so this definition reduces to the usual one. Now we consider the other possible valuations v occurring in the formula.These must be proper valuations .Hence the formula reduces to

$$e(\underline{I},J;M) = \sum_{p} e(IQ_p;M_p)e(J+p/p) + \sum d(I,v)L(M_v)v(J)$$

the first sum being over the ht 1 prime ideals minimal over I, and the second sum being over the proper m-valuations on Q (note that the first sum could be reduced further, but seems simplest in the above form).

In conclusion, the author would like to thank the Matematisk Institut, Aarhus, for its hospitality during April and May 1979, when the ideas behind this paper were first developed and the first draft was written, and the organisers of the Nordic Summer School and Research Symposium in Stockholm during August 1983 for an invitation to give a lecture in the Research Symposium which formed the basis of the version, still not final, presented here.

References

1. S.Abhyankar. On the valuations centered in a local domain. Amer.J.Math.78(1956) pp321-348

2. A .Grothendieck. Elements de Géomètrie Algebrique. Publ.Math.IHES No 11(1961)

3. D.G.Northcott. A general theory of one-dimensional local rings. Proc.Glasgow Math.Association 2(1956) pp159-169.

4 " " A generalisation of a theorem on the content of polynomials. Proc.Cam.Phil.Soc.55(1959) pp282-288

5 " " Lectures on Rings, modules and multiplicities. Cambridge University Press, Cambridge(1968)

6 D.Rees Degree Functions in Local rings. Proc.Cam.Phil.Soc.57(1961) pp1-7

7 " " Multiplicities, Hilbert Functions and degree functions (in "Commutative Algebra; Durham 1981") London Math.Soc.lecture note series no72(1982) pp 70-78.

8 " " Asymptotic Properties of Ideals (Lectures given in Nagoya 1983) To be published in the L.M.S Lecture Note series

9. " " Generalisations of Reductions and Mixed Multiplicities. Jour.London.Math.Soc.29(1984)397-414.

10. B.Teissier. Cycles evanescents, sections planes et Conditions de Whitney. In "Singularites a Cargese 1972 Asterisque 7-8(1973)

11. D.J.Wright General Multiplicity Theory Proc.London.Math.Soc.(3) 15(1965) pp269-288

12 O.Zariski and P.Samuel Commutative Algebra. volume II D.von Nostrand (Princeton)1960.

COHOMOLOGIE DE HARRISON ET TYPE D'HOMOTOPIE RATIONNELLE

Daniel TANRÉ

ERA C.N.R.S. 07 590
Université des Sciences et Techniques de LILLE
U.E.R. de Mathématiques Pures et Appliquées
59655 - VILLENEUVE D'ASCQ CEDEX (France)

La théorie de la déformation permet l'étude des types d'homotopie rationnelle à algèbre de cohomologie (ou algèbre de Lie d'homotopie) rationnelle fixée [S-S]. Halperin et Stasheff [H-S] ont obtenu les premiers résultats dans ce domaine ; rappelons d'abord la terminologie utilisée :
un espace dont le type d'homotopie rationnelle est entièrement déterminé par la donnée de son algèbre de cohomologie (resp. algèbre de Lie d'homotopie) rationnelle est appelé formel (resp. coformel). Un espace est intrinsèquement formel si son algèbre de cohomologie est réalisée par un seul type d'homotopie rationnelle.

Nous montrons ici que les obstructions d'Halperin-Stasheff à la formalité s'interprètent comme classe de cohomologie de Harrison. Cette dernière semble être le cadre le mieux adapté à cette situation ; à partir du résultat ci-dessus, elle a permis à D. Merle [Me] d'unifier les théories d'obstructions d'Halperin-Stasheff [H-S] et Félix [Fe] et celle introduite par Lemaire et Sigrist [L-S] dans le cadre des modèles de Quillen.

La construction du modèle bigradué [H-S] est illustrée par un exemple à cohomologie non bornée : $\mathbb{C}P(\infty)/\mathbb{C}P(2)$. Nous rendons triviale la première déformation possible de ce modèle. La cohomologie étant non bornée, il existe une infinité de déformations possibles. Seule l'utilisation de la cohomologie de Harrison permet d'obtenir leur trivialité. La démonstration complète passe par une détermination explicite de tout le modèle bigradué et par l'interprétation de $\mathbb{C}P(\infty)/\mathbb{C}P(2)$ comme espace total d'une fibration ; elle fera l'objet d'une publication ultérieure. Plus généralement, le résultat obtenu concerne les espaces projectifs tronqués $\mathbb{C}P(\infty)/\mathbb{C}P(n)$; il s'énonce [Ta 1] :

Théorème : Rationnellement, il existe deux espaces $\mathbb{C}P(\infty)/\mathbb{C}P(n)$ et E_n ayant même algèbre de Lie d'homotopie rationnelle que $\mathbb{C}P(\infty)/\mathbb{C}P(n)$. $\mathbb{C}P(\infty)/\mathbb{C}P(n)$ n'est pas coformel ; E_n est l'espace coformel associé. Ils sont tous deux intrinsèquement formels.

Ce texte reprend une partie de ma thèse d'Etat soutenue à Lille, le 26 janvier 1982.

Notations : Nous emploierons les conventions usuelles de signe pour les objets gradués : si un élément de degré p est permuté avec un élément de degré q, le signe $(-1)^{pq}$ apparaît. En particulier, le signe correspondant à toute permutation σ d'objets gradués est noté $\varepsilon(\sigma)$ et appelé signe de Koszul de σ.

L'expression adgc signifie algèbre différentielle graduée commutative. Hormis le paragraphe I, les espaces vectoriels considérés sont sur le corps $\mathbb{Q}$ des rationnels. Si V est un espace vectoriel gradué, de base $(x_1,\ldots,x_n)$, on désigne par $\Lambda V = \Lambda(x_1,\ldots,x_n)$ l'algèbre graduée commutative libre, $\mathbb{L}(V) = \mathbb{L}(x_1,\ldots,x_n)$ l'algèbre de Lie libre, $T(V) = T(x_1,\ldots,x_n)$ l'algèbre tensorielle, engendrées par V. L'espace vectoriel dual est noté $\# V$, la suspension sV : $(sV)^n = V^{n+1}$, le degré d'un élément x par $|x|$.

D'une manière générale, les notations utilisées sont celles de [Ta].

I - COHOMOLOGIES DE HOCHSCHILD ET DE HARRISON.

Soient $\mathbb{k}$ un corps et V un $\mathbb{k}$-espace vectoriel gradué.

Définition.- Un (p,q)-mixage σ est une permutation de l'ensemble des entiers $\{1,\ldots,p+q\}$ telle que :

$$\sigma(i) < \sigma(j) \quad \text{si} \quad 1 \leq i < j \leq p \quad \text{ou} \quad p+1 \leq i < j \leq p+q.$$

Définition.- L'espace gradué $T(V)$ est une algèbre graduée commutative pour le produit mixé défini par :

$$(v_1 \otimes \ldots \otimes v_p) * (v_{p+1} \otimes \ldots \otimes v_n) = \sum_\sigma \varepsilon(\sigma) v_{\sigma^{-1}(1)} \otimes \ldots \otimes v_{\sigma^{-1}(n)}$$

où σ parcourt les $(p,n-p)$ mixages, $\varepsilon(\sigma)$ est le signe de Koszul de σ, les éléments v_i sont homogènes.

Soit A une $\mathbb{k}$-algèbre graduée commutative, de type fini, ($\dim A^p$ finie pour tout p), connexe, ($A^0 = \mathbb{k}$), et soit M un A-module gradué.

La <u>cohomologie de Hochschild</u>, Hoch(A;M), de A à coefficients dans M provient du complexe suivant :

. si $a_j \in A$, $a_1 \otimes \dots \otimes a_n$ est muni du degré $1 + \sum_{j=1}^{n} (|a_j| - 1)$;

. $\mathrm{Hom}^p(\overset{n}{\otimes} A,M)$ est l'ensemble des applications $\Bbbk$-linéaires de degré p telles que $f(a_1 \otimes \dots \otimes a_n) = 0$ si $a_i = 1$;

. $(\delta f)(a_1 \otimes \dots \otimes a_{n+1}) = a_1 f(a_2 \otimes \dots \otimes a_{n+1}) + (-1)^{\nu(n)} f(a_1 \otimes \dots \otimes a_n).a_{n+1}$
$+ \sum_{j=1}^{n} (-1)^{\nu(j)} f(a_1 \otimes \dots \otimes a_j.a_{j+1} \otimes \dots \otimes a_{n+1})$,
avec $\nu(j) = \sum_{i=1}^{j} (|a_i| - 1)$;

. $Z^{n,p}(A;M)$ est formé des δ-cocycles de $\mathrm{Hom}^p(\overset{n}{\otimes} A,M)$

$$\boxed{\mathrm{Hoch}^{n,p}(A\ ;\ M) = Z^{n,p}(A\ ;\ M)/\delta\,\mathrm{Hom}^{p-1}(\overset{n-1}{\otimes} A,M)} \quad .$$

La <u>cohomologie de Harrison</u>, Harr(A;M), de A à coefficients dans M s'obtient à partir d'un sous-complexe du complexe de Hochschild, défini comme suit :

. $\mathrm{Hom}^p_s(\overset{n}{\otimes} A,M)$ est formé des éléments de $\mathrm{Hom}^p(\overset{n}{\otimes} A,M)$ s'annulant sur les décomposables du produit mixé ; il est stable pour la différentielle δ ;

. $Z^{n,p}_s(A;M) = Z^{n,p}(A;M) \cap \mathrm{Hom}^p_s(\overset{n}{\otimes} A,M)$

$$\boxed{\mathrm{Harr}^{n,p}(A\ ;\ M) = Z^{n,p}_s(A\ ;\ M)/\delta\,\mathrm{Hom}^{p-1}_s(\overset{n-1}{\otimes} A,M)} \quad .$$

Elle est reliée à la cohomologie de Hochschild par :

Théorème (M. Barr ; [Ba]).- Si $\Bbbk$ est un corps de caractéristique 0, l'application naturelle :

$$\mathrm{Harr}(A;M) \to \mathrm{Hoch}(A;M)$$

est injective.

Pour terminer ces rappels, notons que ce résultat est faux en caractéristique p et que la cohomologie de Harrison ne provient pas d'un foncteur dérivé (à droite) du foncteur $\mathrm{Hom}_A(A;-)$. Avec un saut d'un degré, Harr(A;A) est le complexe cotangent $D(A/k;A)$ [Qu].

II - THEORIE DE L'OBSTRUCTION D'HALPERIN-STASHEFF.

Soit $\rho : (\Lambda V,d) \to (A,0)$ le modèle minimal de Sullivan [Su] d'une algèbre graduée commutative connexe. Halperin et Stasheff [H-S] définissent, pour un bon choix de V, une graduation supplémentaire sur V, $V = \bigoplus_{p\geq 0} V_p$; celle-ci s'étend en graduation d'algèbre à ΛV et vérifie : $dV_p \subset (\Lambda V)_{p-1}$; ρ est bihomogène de degré 0 ;

$H(\rho) : H_0(\Lambda V,d) \to A$ est un isomorphisme ; $H_+(\Lambda V,d) = 0$

$(\Lambda V,d)$ est le modèle bigradué de A.

L'appendice illustre la construction des premiers générateurs du modèle bigradué de $H(\mathbb{CP}(\infty)/\mathbb{CP}(2);\mathbb{Q})$.

Si (A,d_A) est une adgc, cohomologiquement connexe, Halperin et Stasheff [H-S] construisent un modèle de Sullivan (non minimal) $\pi : (\Lambda V,D) \to (A,d_A)$, en partant du modèle bigradué $\rho : (\Lambda V,d) \to (H(A,d_A),0)$. Ce modèle vérifie : $(D-d)(V_n) \subset \bigoplus_{m\leq n-2} (\Lambda V)_m$; si $v \in V_o$, la classe de cohomologie de $\pi(v)$ est égale à $\rho(v)$.

$(\Lambda V,D)$ est le modèle filtré de (A,d_A).

Chacun de ces deux modèles vérifie un théorème d'unicité [H-S]. Appelons TJ-graduation et TJ-filtration (pour Tate-Jozefiak) les graduation et filtration supplémentaires.

Définitions : Une adgc (A,d_A) est formelle si elle a même modèle minimal que $(H(A,d_A),0)$. L'algèbre H est intrinsèquement formelle si toutes les adgc d'algèbre de cohomologie isomorphe à H sont formelles ; autrement dit, H est réalisée par un seul type d'homotopie.

Résumons la théorie d'obstructions à la formalité d'Halperin-Stasheff.

Soit $\pi : (\Lambda V,D) \to (A,d_A)$ un modèle filtré construit à partir du bigradué $\rho : (\Lambda V,d) \to (H(A,d_A),0) = (H,0)$; notons $\eta : H \to \Lambda V_o$ une section de ρ. Définissons maintenant une application

$\lambda : \mathrm{Hom}^o(V_{p-1},H) \to \mathrm{Hom}^1(V_p,H)$ par $\boxed{\lambda(\Psi)(v) = \rho\, \theta_\Psi dv}$, où θ_Ψ est l'unique dérivation de $(\Lambda V)_{\leq p-1}$, de degré 0, définie par : $\theta_\Psi = 0$ sur $V_{<p-1}$, $\theta_\Psi = \Pi\Psi$ sur V_{p-1}.

Soit p le premier indice tel que D-d soit non nulle sur V_p. (D-d) définit un élément de $\mathrm{Hom}^1(V_p,H)$, on note $O_p(D)$ sa classe dans $\mathrm{Hom}^1(V_p,H)/\mathrm{Im}\,\lambda$. Si $O_p(D) = 0$, il existe un automorphisme $\mu = e^{\theta_\Psi}$ de ΛV tel que $D\mu = \mu d$ sur $(\Lambda V)_{\leq p}$. On remplace alors D par $\mu^{-1}D\mu$ pour construire $O_{p+1}(D)$.

$O_p(D)$ <u>représente l'obstruction à l'existence d'un automorphisme</u> μ de ΛV tel que $D\mu = \mu d$ sur $(\Lambda V)_{\leq p}$.

Théorème [H-S].- Supposons $H(\Lambda V,d)$ de type fini. Si $O_p(D) = 0$ pour tout p, $(\Lambda V,D)$ est formelle. Si $O_p(D) = 0$ pour tout p et tout modèle filtré $(\Lambda V,D)$ construit sur le modèle bigradué $(\Lambda V,d)$, $H(\Lambda V,d)$ est intrinsèquement formelle.

Exemple : Illustrons cette théorie à l'aide de l'appendice. Pour des raisons de degré, la première déformation possible de d baisse la TJ-graduation de 3 unités. Elle est donnée par :

$$D_3v_1 = \alpha x_3^4 x_4 \quad ; \quad D_3v_2 = \beta\, x_3^4 x_5 \quad ; \quad D_3v_3 = \gamma\, x_3^6 ,$$

où α, β, γ sont des rationnels.

Pour que D_3 se prolonge en Z_4, il faut et il suffit que $\alpha = \beta = \gamma$. En posant : $\Psi(z_1) = \alpha x_3^4$; $\Psi(z_2) = \alpha x_3^3 x_4$, on obtient :

$$\lambda(\Psi)(v_1) = \rho\theta_\Psi(y_1y_2 + x_4z_1 + x_3z_2) = 2\alpha x_3^4 x_4 = 2D_3v_1 \quad ;$$

$$\lambda(\Psi)(v_2) = 2D_3v_2 \quad ; \quad \lambda(\Psi)(v_3) = 2D_3v_3 .$$

La déformation D_3 peut donc être rendue triviale par un automorphisme.

<u>*Liaison entre les obstructions d'Halperin-Stasheff et de Félix*</u> :

La théorie développée par Félix [Fe] est également menée à partir du modèle bigradué. La différence entre les deux approches tient au fait qu'Halperin et Stasheff considèrent toutes les applications de degré 1 de V_p dans H, Félix ne prenant en compte que celles se prolongeant à V_{p+1}. Au niveau des obstructions, les résultats sont identiques, comme le montre le corollaire page 26 de [Fe].

III - COHOMOLOGIE DE HARRISON ET FORMALITE INTRINSEQUE.

Pour la fin de ce paragraphe, A est une algèbre graduée commutative, connexe, de type fini.

La liaison entre les obstructions précédentes et la cohomologie de Harrison passe par l'utilisation d'un modèle particulier : l'algèbre des cochaînes sur le modèle de Quillen ou modèle FHS ([Ch], [Fe 1], [Ta] page 67).
Décrivons-le pour $(A,0)$:

$\rho : (\Lambda Z,d) \to (A,0)$ est un morphisme d'adgc induisant un isomorphisme en cohomologie avec :
$(\Lambda Z,d) = (\Lambda s^{-1} \# \mathbb{L}(W), d_1 + d_2)$, $W \oplus \mathbb{Q} = s^{-1} \# A$, d_1 est linéaire en Z et d_2 quadratique.

L'injection canonique de l'algèbre de Lie libre $\mathbb{L}(W)$ dans l'algèbre tensorielle $T(W)$ fournit par dualité et désuspension :
$j : s^{-1}T(\# W) \cong s^{-1} \# T(W) \to s^{-1} \# \mathbb{L}(W)$.

Si $(y_i)_{i\in I}$ est une base homogène de $\# W$, $j(s^{-1}(y_{i_o} \otimes \ldots \otimes y_{i_p}))$, noté $y_{i_o \ldots i_p}$ fournit un système de générateurs de $s^{-1} \# \mathbb{L}(W)$; l'application induite $j : A \to \Lambda Z_o$ est une section de ρ $(j\rho = id)$.

Soit $y_i y_j = \sum_{k\in|i,j|} c_{ij}(k) y_k$, $c_{ij}(k) \in \mathbb{Q}$, la loi d'algèbre de A.

En détaillant la définition de l'algèbre des cochaînes, on obtient ([Ta] page 71) :

$$d_2 y_{i_o \ldots i_m} = \sum_{p=0}^{m-1} \bar{y}_{i_o \ldots i_p} y_{i_{p+1} \ldots i_m} \quad ;$$

$$d_1 y_{i_o \ldots i_m} = \sum_{j=0}^{m-2} \sum_{k\varepsilon|i_j,i_{j+1}|} c_{i_j i_{j+1}}(k)(-1)^{\nu(j)} y_{i_o \ldots i_{j-1} k\, i_{j+2} \ldots i_m} .$$

$(\Lambda Z, d_2)$ est le modèle bigradué du bouquet de sphères d'homologie $\# A$ ([Ta] page 66). La TJ-graduation correspond à la longueur des crochets par orthogonalité :

$$Z_p = s^{-1}(\mathbb{L}^{p+2}(W))^{\perp}.$$

Le lien avec la cohomologie de Harrison apparaît dans la :

Proposition.- Soit $\gamma : Z_m \to \Lambda Z_o$ une application linéaire de degré 1, on désigne par θ_γ la dérivation d'algèbre associée et par $\tilde{\gamma} \in \mathrm{Hom}_s^1(\overset{m+1}{\otimes} A, A)$ le composé $\tilde{\gamma} = \rho\, \gamma j$. Si δ est la différentielle du complexe de Harrison, alors :

1) $\delta\tilde{\gamma}(s^{-1}(y_{i_o}\otimes\ldots\otimes y_{i_{m+1}})) = \rho\theta_\gamma(d_2+d_1)y_{i_o\ldots i_{m+1}}$;

2) γ peut être étendue à Z_{m+1} ssi $\delta\tilde{\gamma} = 0$;

3) γ peut être rendue triviale par un automorphisme de ΛZ ssi il existe $\tilde{\gamma}'$ telle que $\tilde{\gamma} = \delta\tilde{\gamma}'$.

Démonstration :

1) Par définition, on a :

$$\delta\tilde{\gamma}(s^{-1}(y_{i_o}\otimes\ldots\otimes y_{i_{m+1}})) = y_{i_o}\tilde{\gamma}(s^{-1}(y_{i_1}\otimes\ldots\otimes y_{i_{m+1}})) +$$

$$(-1)^{|y_{i_o\ldots i_m}|}\tilde{\gamma}(s^{-1}(y_{i_o}\otimes\ldots\otimes y_{i_m}))y_{i_{m+1}} + \sum_{j=0}^{m-1}(-1)^{\nu(j)}\tilde{\gamma}(s^{-1}(y_{i_o}\otimes\ldots\otimes y_{i_j}y_{i_{j+1}}\otimes\ldots\otimes y_{i_{m+1}}))$$

$$= \rho\theta_\gamma\Big[(-1)^{|y_{i_o}|}y_{i_o}y_{i_1\ldots i_{m+1}} + (-1)^{|y_{i_o\ldots i_m}|}y_{i_o\ldots i_m}y_{i_{m+1}} +$$

$$\sum_{j=0}^{m-1}(-1)^{\nu(j)}\sum_{k\in|i_j,i_{j+1}|}c_{i_j i_{j+1}}(k)y_{i_o\ldots i_{j-1}\,k\,i_{j+2}\ldots i_{m+1}}\Big]$$

$$= \rho\theta_\gamma(d_2+d_1)y_{i_o\ldots i_{m+1}}.$$

2) γ peut être étendue à Z_{m+1} ssi $\theta_\gamma(d_2+d_1)y_{i_o\ldots i_{m+1}}$ est un (d_2+d_1)-cocycle, i.e. $\rho\theta_\gamma(d_2+d_1)y_{i_o\ldots i_{m+1}} = 0$.

3) La dernière propriété se déduit directement de la comparaison de $\lambda(\gamma')$ et $\delta\tilde{\gamma}'$; la construction de λ et le théorème d'Halperin-Stasheff se transcrivent tels quels au modèle FHS.

De la proposition ci-dessus et du théorème d'Halperin-Stasheff, on déduit directement :

Théorème.- Si $Harr^{m,1}(A;A) = 0$ pour tout $m > 2$, alors A est intrinsèquement formelle.

Remarque : La réciproque du théorème est fausse en général. En effet, l'hypothèse $Harr^{*,1}(A;A) = 0$ signifie que toute application $\gamma \in Hom_s^1(\overset{m+1}{\otimes} A,A)$

prolongeable en colonne Z_{m+1} peut être rendue triviale par un automorphisme. Or, $d_2+d_1+\gamma$ ne donne pas nécessairement une différentielle, sauf dans un cas particulier :

Proposition.- Si $Harr^{m,2}(A;A) = 0$ pour $m > 4$, les propriétés suivantes sont équivalentes :

(i) l'algèbre A est intrinsèquement formelle,
(ii) $Harr^{m,1}(A;A) = 0$ pour $m > 2$.

Nous laissons la démonstration au lecteur ; celle faite par Félix ([Fe], page 21) s'adapte ici sans problème. De même, l'exemple de l'annexe 1 de [Fe] illus tre la remarque ci-dessus.

Appendice : Un simple calcul donne la description par générateurs et relations de l'algèbre de cohomologie rationnelle de $\mathbb{C}P(\infty)/\mathbb{C}P(2)$:

$$\Lambda(x_3, x_4, x_5)/R \quad ,$$

où $|x_i| = 2i$, R est l'idéal engendré par $x_4^2 - x_3x_5$, $x_4x_5 - x_3^3$, $x_5^2 - x_3^2x_4$.

Les premiers générateurs du modèle bigradué s'écrivent :

Z_o	6	$dx_3 = 0$
	8	$dx_4 = 0$
	10	$dx_5 = 0$
Z_1	15	$dy_1 = x_4^2 - \{x_3\ x_5\}$
	17	$dy_2 = x_4\ x_5 - \{x_3^3\}$
	19	$dy_3 = x_5^2 - \{x_3^2\ x_4\}$.
Z_2	24	$dz_1 = y_1\ x_5 - y_2\ x_4 + \{y_3\ x_3\}$
	26	$dz_2 = y_2\ x_5 - y_3\ x_4 - \{y_1\ x_3^2\}$
Z_3	31	$dv_1 = y_1\ y_2 + x_4\ z_1 + \{x_3\ z_2\}$
	33	$dv_2 = z_1\ x_5 + y_1\ y_3 + x_4\ z_2$
	35	$dv_3 = z_2\ x_5 + y_2\ y_3 + \{x_3^2\ z_1\}$
Z_4	38	$dw_1 = y_1\ z_1 - x_4\ v_1 + \{x_3\ v_2\}$
	40	$dw_2 = y_1\ z_2 - x_4\ v_2 + v_1\ x_5$
	40	$dw_5 = z_1\ y_2 + y_1\ z_2 - x_4\ v_2 + \{x_3\ v_3\}$

42 $dw_3 = v_2\, x_5 - z_1\, y_3 - x_4\, v_3$

42 $dw_6 = y_2\, z_2 + z_1\, y_3 - v_2\, x_5 + \{x_3^2\, v_1\}$

44 $dw_4 = v_3\, x_5 - z_2\, y_3 - \{x_3^2\, v_2\}.$

Z_5 45 $du_1 = w_1\, x_4 - v_1\, y_1 + \{w_2\, x_3\}$

47 $du_2 = w_2\, x_4 + w_1\, x_5 - v_2\, y_1$

47 $du_7 = x_4(w_5 - w_2) - v_1\, y_2 + \{x_3(w_3 + w_6)\}$

47 $du_{10} = 1/2\, z_1^2 - v_2\, y_1 + w_5\, x_4 + \{x_3\, w_3\}$

49 $du_3 = x_4\, w_3 - v_3\, y_1 + v_1\, y_3 + w_2\, x_5 + \{x_3^2\, w_1 + x_3\, w_4\}$

49 $du_6 = x_4\, w_6 + z_1\, z_2 - w_5\, x_5 + \{x_3^2\, w_1 + x_3\, w_4\}$

49 $du_8 = v_2\, y_2 - y_3\, v_1 + x_5\, w_2 - x_5\, w_5 - x_4\, w_6 + \{x_3\, w_4\}$

51 $du_4 = x_4\, w_4 + v_2\, y_3 + w_3\, x_5$

51 $du_9 = w_6\, x_5 - v_2\, y_3 - 1/2\, z_2^2 - \{x_3^2\, w_2\}$

51 $du_{11} = y_2\, v_3 + x_5\, w_3 + x_5\, w_6 - \{x_3^2(w_5 - w_2)\}$

53 $du_5 = v_3\, y_3 + w_4\, x_5 + \{x_3^2\, w_3\}.$

Comme annoncé dans l'introduction, la cohomologie de Harrison alliée aux techniques des modèles minimaux (KS-modèles,...) permet d'établir la formalité intrinsèque de cette algèbre.

BIBLIOGRAPHIE

[Ba] Michael BARR - Harrison homology, Hochschild homology and Triples, Journal of Algebra 8, (1968), 314-323.

[Ch] Kuo Tsai CHEN - Extension of C^{∞} function Algebra by Integrals and Malcev completion of π_1, Advances in Math. 23, (1977), 181-210.

[Fe] Yves FELIX - Dénombrement des types de K-homotopie. Théorie de la déformation, Mémoires SMF, nouvelle série n° 3, (1980).

[Fe 1] Yves FELIX - Modèles bifiltrés. Can. J. Math. 33, n° 26, (1981), 1448-1458.

[Fe 2] Yves FELIX - Espaces formels et π-formels. Conférence Marseille-Luminy (à paraître SMF).

[H-S] Steve HALPERIN, James STASHEFF - Obstructions to homotopy equivalences, Advances in Math. 32, (1979), 233-279.

[Ha] D.K. HARRISON - Commutative algebras and cohomology T.A.M.S. 104, (1962), 191-204.

[Ho] G. HOCHSCHILD - On the cohomology groups of an associative algebra. Ann. of Math. 46, (1945), 58-67.

[L-S] Jean-Michel LEMAIRE, François SIGRIST - Dénombrement des types d'homotopie rationnelle. C.R.A.S. t. 287 A, (1978), Paris.

[Me] Pierre MERLE - Formalité des espaces et des applications continues. Thèse de 3ème cycle, Nice, (1983).

[Qu] Daniel QUILLEN - On the (co)-homology of commutative rings, Proc. Symp. Pure Math. 17, A.M.S. Providence (1970), 65-87.

[S-S] Michael SCHLESSINGER, James STASHEFF - Deformation theory and rational homotopy type (à paraître).

[Su] Denis SULLIVAN - Infinitesimal computations in Topology, Publ. I.H.E.S. 47, (1977), 269-331.

[Ta] Daniel TANRÉ - Homotopie rationnelle : Modèles de Chen, Quillen, Sullivan. Lecture notes in Math. 1025, (1983), Springer Verlag.

[Ta 1] Daniel TANRÉ - Thèse, Lille (1982).

COHOMOLOGIE DE L'ESPACE DES SECTIONS D'UN FIBRE ET COHOMOLOGIE DE GELFAND-FUCHS D'UNE VARIETE

par

Micheline VIGUÉ-POIRRIER (*)

Résumé : Soit $F \hookrightarrow E \xrightarrow{\pi} X$ un fibré nilpotent, où les espaces sont connexes par arcs, nilpotents et ont le type d'homotopie de C.W. complexes de type fini. On suppose que $H^+(X,\mathbb{Q}) \neq 0$, qu'il existe $n \geqslant 1$ tel que X a le type d'homotopie d'un complexe simplicial de dimension n, et F a le type d'homotopie de $\bigvee_{i=1}^{r} S^{k_i+1}$ où $r \geqslant 2$ et $\inf(k_i) \geqslant n$. Soit Γ l'espace des sections continues du fibré. On démontre qu'il existe $N \in \mathbb{N}$ et une constante réelle $C > 1$ tels que si $p \geqslant N$, on a $\sum_{i=0}^{p} \dim H^i(\Gamma,\mathbb{Q}) \geqslant C^p$ dans les deux cas suivants : ou bien, le fibré est trivial (i.e. $\Gamma = F^X$), ou bien X a le type d'homotopie d'un bouquet $S^d \vee Y$ (où Y est un complexe simplicial de dimension $\leqslant n$ et $d \geqslant 1$). On en déduit que la suite des dimensions des groupes de la cohomologie de Gelfand-Fuchs d'une variété M, C^∞, compacte, connexe, nilpotente de dimension $\geqslant 2$ telle que $H^+(M,\mathbb{R}) \neq 0$, et dont toutes les classes de Pontryagin sont nulles est à croissance exponentielle.

CLASSIFICATION AMS : 55 P 62, 55 R 05, 57 R 32

MOTS CLES : Modèle minimal de Sullivan, fibré nilpotent, cohomologie de Gelfand-Fuchs.

(*) ERA au CNRS 07 590

0. *Introduction.*

Dans [16], Thom étudie le type d'homotopie de l'espace des applications continues d'un espace X dans un espace F, homotopes à une application donnée.

Dans [14], Sullivan décrit une algèbre différentielle graduée commutative, modèle de l'espace des sections d'un fibré algébrique donné.

Dans [5], Haefliger détermine le type d'homotopie rationnelle de l'espace Γ des sections homotopes à une section donnée pour un fibré nilpotent $\Pi : E \to X$. Si on a un tel fibré $\Pi : E \to X$ tels que les espaces soient connexes par arcs, tel que $H^*(X,\mathbb{Q})$ soit de dimension finie, et X a un modèle (A,d_A) tel que $\dim A^n < \infty$ pour tout n, il démontre qu'une certaine algèbre différentielle graduée commutative $(\Lambda SZ,D)$ est un modèle de Sullivan de l'espace Γ.

Tous les espaces considérés dans ce papier sont connexes par arcs, nilpotents et ont le type d'homotopie de C.W. complexes de type fini ; ce qui nous permettra d'utiliser, de manière biunivoque, le dictionnaire établi par Sullivan entre la topologie et l'algèbre, voir §.1.

Nous nous intéresserons à des fibrés nilpotents $F \hookrightarrow E \xrightarrow{\Pi} X$ où la base X a le type d'homotopie d'un complexe simplicial de dimension $n \geq 1$ et la fibre F est n-connexe.

Utilisant les résultats de [5] et [18], nous montrerons

Théorème 3.3. Soit X un espace nilpotent ayant le type d'homotopie rationnelle d'un complexe simplicial de dimension $n \geq 1$ et tel que $H^+(X,\mathbb{Q}) \neq 0$. Soit F un espace ayant le type d'homotopie rationnelle d'un bouquet de sphères $\bigvee_{i=1}^{r} S^{k_i+1}$ où $r \geq 2$ et $\inf(k_i) \geq n$. Alors, si F^X est l'espace des applications continues de X dans F muni de la topologie compacte ouverte, il existe $N \in \mathbb{N}$ et une constante réelle $C > 1$ tels que $p \geq N$, on a $\sum_0^p \dim H^i(F^X,\mathbb{Q}) \geq C^p$.

Théorème 3.4. Soit un fibré nilpotent $F \hookrightarrow E \xrightarrow{\Pi} X$ ayant les propriétés suivantes, il existe $n \geq 1$ tel que :

1) X a le type d'homotopie rationnelle d'un bouquet $S^d \vee Y$ où $1 \leq d \leq n$ et Y est un complexe simplicial de dimension $\leq n$.

2) F a le type d'homotopie rationnelle d'un bouquet de sphères $\bigvee_{i=1}^{r} S^{k_i+1}$ où $r \geq 2$, $\inf(k_i) \geq n$.

Alors, si Γ est l'espace des sections continues du fibré il existe $N \in \mathbb{N}$ et une constante réelle $C > 1$ tels que si $p \geq N$, on a $\sum_{i=0}^{p} \dim H^i(\Gamma,\mathbb{Q}) \geq C^p$.

En corollaire, on obtient un résultat prolongeant ceux de [11].

Théorème 3.7. Soit M une variété C^∞ compacte, connexe, nilpotente de dimension ≥ 2 et telle que $H^+(M,\mathbb{R}) \neq 0$. On suppose que toutes les classes de Pontryagin de M sont nulles. Alors, si $C^*_{\mathbb{R}}(L_M)$ est l'algèbre des formes multilinéaires continues sur l'algèbre de Lie des champs de vecteurs sur M, il existe $N \in \mathbb{N}$ et une constante réelle $A > 1$ tels que si $p \geq N$, on a $\sum_{i=0}^{p} \dim H^i(C^*(L_M)) \geq A^p$.

1. *Théorie du modèle minimal de Sullivan.*

Nous rappelons brièvement les résultats de la théorie de Sullivan qui seront nécessaires dans la suite. Les détails se trouvent dans [14], [9], [6], [7], [17].

Les algèbres considérées sont des k-algèbres graduées $A = \bigoplus_{n \geq 0} A^n$ où $k = \mathbb{Q}$ ou $\mathbb{R}$, commutatives dans le sens suivant : si $a \in A^p$, $b \in B^q$, alors $b.a = (-1)^{pq} a.b$. On notera $|a| = p$ le degré de $a \in A^p$. Si A et B sont deux algèbres commutatives graduées, la multiplication dans $A \otimes B$ est définie par : $(a \otimes b)(a' \otimes b') = (-1)^{|b|.|a'|} aa' \otimes bb'$. Une algèbre est dite connexe si $A^0 = k$. Une algèbre différentielle graduée commutative (A,d) (en abrégé A.D.G.C.) est une algèbre graduée commutative munie d'une différentielle d

de degré +1 vérifiant $d(a.b) = (da).b+(-1)^{|a|}a.(db)$. On dit qu'une A.D.G.C. (M,d) est un modèle de (A,d_A) s'il existe un homomorphisme d'A.D.G.C. $\varphi : (M,d) \to (A,d_A)$ induisant un isomorphisme en cohomologie. Une A.D.G.C. (A,d_A) est dite libre s'il existe un espace vectoriel gradué $V = \bigoplus_{n\geq 1} V^n$ tel que $A = \Lambda V$ est le produit tensoriel de l'algèbre extérieure construite sur $\bigoplus_n V^{2n+1}$ et de l'algèbre symétrique construite sur $\bigoplus_n V^{2n}$. On démontre, [6], que toute (A,d_A) telle que $H^0(A,d_A) = k$ possède un modèle minimal unique à isomorphisme près. Dans le cas où $H^1(A) = 0$, c'est une A.D.G.C. libre $(\Lambda V,d)$ caractérisée par le fait que $d(V) \subset \Lambda^{\geq 2}V$.

Dans [14], Sullivan définit un foncteur contrevariant, noté A( ), de la catégorie des ensembles simpliciaux dans celle des A.D.G.C. sur $\mathbb{Q}$. Si X est un espace topologique, on considère Sing X, l'ensemble simplicial des simplexes singuliers de X et on note encore $A(X)$ l'algèbre $A(\text{Sing } X)$. De plus, l'intégration des formes différentielles définit un isomorphisme d'algèbres graduées de $H^*(A(X))$ sur la cohomologie rationnelle singulière $H^*(X,\mathbb{Q})$. Une A.D.G.C. (A,d_A) est appelée modèle de l'espace X si (A,d_A) est un modèle de $A(X)$.

Si on se restreint à des espaces topologiques nilpotents ayant le type d'homotopie d'un C.W. complexe de type fini, on peut associer à un tel espace X, un $\mathbb{Q}$-espace $X_{\mathbb{Q}}$ ayant même homotopie rationnelle et même cohomologie rationnelle que X. Le foncteur de Sullivan A induit une équivalence de catégories entre la catégorie homotopique rationnelle (dont les objets sont les Q-espaces), et la catégorie des $\mathbb{Q}$-A.D.G.C. libres $(\Lambda Z,d)$ telles que $\dim Z^n < \infty$ pour tout n, et il existe un ensemble bien ordonné I tel que $Z = \bigoplus_{\alpha\in I} Z_\alpha$; pour tout α, il existe $n_\alpha \in \mathbb{N}$ tel que $Z_\alpha \subset Z^{n_\alpha}$; n_α est une fonction croissante de α ; $d(Z_\alpha) \subset \Lambda(\bigoplus_{\beta<\alpha} Z_\beta)$. Le modèle minimal d'un espace nilpotent X correspond à la décomposition de Postnikov de X. En particulier, on a : $Z^n = \mathrm{Hom}(\Pi_n(X),\mathbb{Q})$.

Soit maintenant $F \overset{j}{\hookrightarrow} E \xrightarrow{\Pi} X$ un fibré dont tous les espaces sont connexes par arcs. On suppose que $H^*(X,\mathbb{Q})$ ou bien $H^*(F,\mathbb{Q})$ sont de dimension finie en chaque degré, et que $\Pi_1(X)$ agit de manière nilpotente sur $H^*(F)$. Soit $(B,d_B) \xrightarrow{\alpha} A(X)$ un modèle de X, alors il existe un espace vectoriel gradué Z, une différentielle d sur $B \otimes \Lambda Z$, et un morphisme d'A.D.G.C. $\psi : (B \otimes \Lambda Z,d) \to A(E)$ tels que le carré suivant commute :

$$\begin{array}{ccccc} A(X) & \xrightarrow{A(p)} & A(E) & \xrightarrow{A(j)} & A(F) \\ \uparrow{\scriptstyle \alpha} & & \uparrow{\scriptstyle \psi} & & \\ (B,d_B) & \overset{i}{\hookrightarrow} & (B \otimes \Lambda Z,d) & \xrightarrow{q} & (\Lambda Z,\bar{d}) \end{array}$$

L'inclusion i et la projection q sont des morphismes d'A.D.G.C. et ψ induit un isomorphisme en cohomologie. Soit $\bar{\psi} : (\Lambda Z,\bar{d}) \to A(F)$ l'application induite par ψ. On montre, dans [6], que $(\Lambda Z,\bar{d}) \xrightarrow{\bar{\psi}} A(F)$ est le modèle minimal de F. On dit que $(B,d_B) \hookrightarrow (B \otimes \Lambda Z,d)$ est le modèle minimal de base B du fibré $E \xrightarrow{\Pi} X$.

De plus, il existe un ensemble bien ordonné I tel que $Z = \bigoplus_{\alpha \in I} Z_\alpha$, $d(Z_\alpha) \subset B \otimes \Lambda(\bigoplus_{\beta<\alpha} Z_\beta)$, les éléments de Z_α sont de même degré n_α et n_α est une fonction croissante de α (Propriété "N").

Un fibré nilpotent possède un tel modèle minimal [2], (remarque 6.7).

2. *Modèle de Sullivan de l'espace des sections d'un fibré d'après Haefliger* ([3], [4], [5], [14]).

Soit $F \hookrightarrow E \xrightarrow{\Pi} X$ un fibré nilpotent ayant une section s dont tous les espaces sont connexes par arcs, nilpotents, et ont le type d'homotopie d'un C.W. complexe de type fini.

On va expliciter les techniques de [5] donnant un modèle de l'espace des sections du fibré homotopes à s. Soit $\gamma : (C,d_C) \to A(X)$ un modèle de X tel que $\dim C < \infty$.

D'après les résultats du §.1, le fibré possède un modèle minimal $(C,d_C) \hookrightarrow (C \otimes \Lambda Z,d) \to (\Lambda Z,d_o)$; de plus, le morphisme, qui est l'identité sur C et 0 sur Z, est un modèle de la section s.

Les calculs algébriques de [5] (décrits aussi dans [13]) peuvent se résumer ainsi : plus généralement, on se donne une A.D.G.C. (C,d_C) de dimension totale finie, et un morphisme d'A.D.G.C. de (C,d_C) dans une autre A.D.G.C. D'après les résultats de [6], chapitre 6, ce morphisme a un modèle minimal $(C,d_C) \hookrightarrow (C \otimes \Lambda Z,d) \to (\Lambda Z,d_o)$ ayant la propriété "N" décrite à la fin du §.1. Dans [4] et [5], Haefliger associe à $(C \otimes \Lambda Z,d)$ une A.D.G.C. libre notée $\Sigma^*(C \otimes Z)$, construite sur l'espace vectoriel gradué $W = \bigoplus_{k\in\mathbb{Z}} W^k$ où $W^k = \bigoplus_{j-i=k} (\mathrm{Hom}(C^i,\mathbb{Q}) \otimes Z^j)$. Si $\dim Z^p < \infty$ pour tout p, on a $\dim W^k < \infty$, pour tout k. En général, W contient des éléments de degrés ≤ 0, sauf s'il existe $n > 0$ tel que $Z^p = 0$, $p \leq n$ et $C^p = 0$ $p > n$.

L'application $\varepsilon' : C \otimes \Lambda Z \to C \otimes \Sigma^*(C \otimes Z) \simeq \mathrm{Hom}(C^*,\Sigma^*(C \otimes Z)$ définie en posant $\varepsilon'(c) = c$ si $c \in C$, $\varepsilon'(z)(c^*) = c^* \otimes z$ pour tout $c^* \in C^*$, $z \in Z$, s'étend en un C-morphisme d'algèbres graduées. Si (c_i) est une base de C et (c_i') la base duale, on a :

$$\varepsilon'(z) = \sum_i c_i \otimes (c_i' \otimes z).$$

Il est clair que $C^* = \mathrm{Hom}(C,\mathbb{Q})$ a une structure de coalgèbre différentielle graduée si on la munit de la différentielle d^* de degré -1 définie par :

$$< d^*c^*,c > = (-1)^{|c^*|} <c^*,dc> \quad \text{si } c \in C, \quad c^* \in C^* .$$

Il existe une unique différentielle D' sur $\Sigma^*(C \otimes Z)$ faisant de $\varepsilon' : (C \otimes \Lambda Z, d) \to (C, d_C) \otimes (\Sigma^*(C \otimes Z), D')$ un morphisme d'A.D.G.C. On a , si $c^* \in C^*$, $z \in Z$

$$D'(c^* \otimes z) = \varepsilon'(dz)(c^*) + (-1)^{|z|}\, d^*c^* \otimes z$$

Considérons dans $(\Sigma^*(C \otimes Z), D')$ l'idéal engendré par les éléments des W^k de degrés ≤ 0, et la partie linéaire $Q(D')(\underset{k\leq 0}{\oplus} W^k)$ de leurs différentielles. L'A.D.G.C. quotient notée $(\Sigma_+^*(C \otimes Z, D)$ est connexe.

Théorème 2.1. Haefliger [5] : soit $F \hookrightarrow E \to X$ un fibré nilpotent ayant une section s. Soit $(C, d_C) \to A(X)$ un modèle de X de dimension totale finie, soit $(C, d_C) \to (C \otimes \Lambda Z, d)$ le modèle minimal du fibré, de base (C, d_C). Alors l'A.D.G.C. $(\Sigma_+^*(C \otimes Z), D)$ décrite ci-dessus est un modèle de l'espace des sections du fibré homotopes à s.

2.2. *Changement du modèle de la base.*

Soit $(B, d_B) \to (B \otimes \Lambda Z, d)$ le modèle minimal d'un morphisme d'A.D.G.C. de source une A.D.G.C. B de dimension totale finie. On suppose que Z^p est de dimension finie pour tout p, et que $Z^1 = 0$.

On se donne une A.D.G.C. (A, d_A) de dimension totale finie et un morphisme $\alpha : (B, d_B) \to (A, d_A)$. On en déduit un diagramme commutatif :

$$\begin{array}{ccccc} (B, d_B) & \longrightarrow & (B \otimes \Lambda Z, d) & \longrightarrow & (\Lambda Z, d_o) \\ \big\downarrow \alpha & & \big\downarrow \alpha\otimes Id & & \big\| \\ (A, d_A) & \longrightarrow & (A \otimes \Lambda Z, \delta) & \longrightarrow & (\Lambda Z, d_o) \end{array}$$

où $\delta z = (\alpha \otimes Id)(dz)$, pour tout z.

Par dualité, on définit une application linéaire :

$$(\alpha^*)^i \otimes \mathrm{Id} : (A^i)^* \otimes Z^j \to (B^i)^* \otimes Z^j$$

pour tout i, tout j, qui se prolonge en un morphisme $\Sigma^*(\alpha)'$:

$$(\Sigma^*(A \otimes Z),\Delta') \to (\Sigma^*(B \otimes Z),D')$$

où les différentielles Δ' et D' sont celles définies dans la construction de Haefliger décrite ci-dessus. On vérifie que $\Sigma^*(\alpha)'$ commute aux différentielles et que

$$\Sigma^*(\alpha)'(Q(\Delta')(\bigoplus_i ((A^i)^* \otimes Z^i)) \subset Q(D')(\bigoplus_i (B^i)^* \otimes Z^i).$$

Donc, par passage au quotient, le morphisme $\Sigma^*(\alpha)'$ définit un morphisme $\Sigma^*(\alpha) = (\Sigma^*_+(A \otimes Z),\Delta) \to (\Sigma^*_+(B \otimes Z),D)$.

Il est clair que si on a des morphismes d'A.D.G.C. :

$$(C,d_C) \xrightarrow{\beta} (B,d_B) \xrightarrow{\alpha} (A,d_A) \quad \text{alors} \quad \Sigma^*(\alpha \circ \beta) = \Sigma^*(\beta) \circ \Sigma^*(\alpha).$$

Lemme 2.2. Si α induit un isomorphisme en cohomologie, alors $\Sigma^*(\alpha)$ induit un isomorphisme en cohomologie.

Démonstration : Elle généralise celle de la proposition 3 du §.5.5 de [4]. On remarque que s'il existe $q > 1$ tel que $Z = Z^q$ et $dZ \in B$, alors le lemme est vrai, car les différentielles D et Δ sont égales respectivement aux transposées de $d_B \otimes 1$ et $d_A \otimes 1$.

Dans le cas général, rappelons qu'il existe un ensemble bien ordonné I tel que $Z = \bigoplus_{\alpha\in I} Z_\alpha$; pour tout α, il existe n_α tel que $Z_\alpha \subset Z^{n_\alpha}$ et la fonction $\alpha \to n_\alpha$ est croissante ; $d(Z_\alpha) \subset B \otimes \Lambda(\bigoplus_{\beta<\alpha} Z_\beta)$.

Soit $\alpha_o \in I$ fixé, la construction Σ^*_+ appliquée à $(B,d_B) \hookrightarrow (B \otimes \Lambda(\bigoplus_{\alpha\leq\alpha_o} Z_\alpha),d)$ est une A.D.G.C. de la forme

$$\left[\Sigma^*_+(B \otimes (\bigoplus_{\alpha<\alpha_o} Z_\alpha)) \otimes \Sigma^*_+(B \otimes Z_{\alpha_o}),D\right]$$

où $\Sigma_+^*(B \otimes (\bigoplus_{\alpha<\alpha_o} Z_\alpha))$ est stable par D et apparaît comme la construction de Haefliger Σ_+^* appliquée à $(B,d_B) \hookrightarrow (B \otimes \Lambda(\bigoplus_{\alpha<\alpha_o} Z_\alpha),d)$.

Si on quotiente $(\Sigma_+^*(B \otimes (\bigoplus_{\alpha\leqslant\alpha_o} Z_\alpha)),D)$ par l'idéal engendré par les éléments de degrés > 0 de $\Sigma_+^*(B \otimes (\bigoplus_{\alpha<\alpha_o} Z_\alpha))$, on obtient une A.D.G.C. quotient $(\Sigma_+^*(B \otimes Z_{\alpha_o}),D_o)$ qui est la construction de Haefliger relativement à l'inclusion : $(B,d_B) \hookrightarrow (B \otimes \Lambda Z_{\alpha_o}, d_B \otimes 0)$.

Les A.D.G.C. considérées sont toutes connexes ; donc, d'après les théorèmes d'isomorphismes de [6], si le lemme est vrai pour $\Sigma^*(\alpha)_{|\Sigma_+^*(A \otimes (\bigoplus_{\alpha<\alpha_o} Z_\alpha))}$, il est vrai pour $\Sigma^*(\alpha)_{|\Sigma_+^*(A \otimes (\bigoplus_{\alpha\leqslant\alpha} Z_\alpha))}$. On conclut par récurrence.

On se restreint maintenant à des fibrés nilpotents $F \hookrightarrow E \xrightarrow{\Pi} X$, tels que :

1) X a le type d'homotopie d'un complexe simplicial de dimension $n \geqslant 1$ et $H^+(X,\mathbb{Q}) \neq 0$.

2) F est n-connexe.

Sous ces hypothèses, l'espace Γ des sections du fibré est non vide, connexe et nilpotent.

Lemme 2.3. Soit X un espace connexe nilpotent ayant le type d'homotopie d'un complexe simplicial de dimension n, $H^+(X,\mathbb{Q}) \neq 0$.

Soit $(\Lambda U,\delta)$ le modèle minimal de X, alors il existe une A.D.G.C. (A,d_A) et un morphisme $\alpha : (\Lambda U,\delta) \longrightarrow (A,d_A)$ induisant un isomorphisme en cohomologie tels que:

i) $\dim A < \infty$, $A^p = 0$ si $p > n$.

ii) il existe $a \in A^+$, $|a| = \inf \{ p > 0 | A^p \neq 0 \}$, $d_A a = 0$

Démonstration: Ce lemme est bien connu, il suffit de prendre pour (A,d_A), l'A.D.G.C. quotient de $(\Lambda U,\delta)$ par l'idéal I défini par $I^p = (\Lambda U)^p$ $p > n$, I^n est un supplémentaire de Ker δ dans $(\Lambda U)^n$ et $I^p = 0$ si $p < n$.

Compte-tenu des lemmes 2.2. et 2.3., et de la remarque finale de [5], on peut supposer que le fibré nilpotent vérifiant 1) et 2) possède un modèle minimal : $(A,d_A) \to (A \otimes \Lambda Z,d) \to (\Lambda Z,d_o)$ où :

i) $\dim A < \infty$, $A^p = 0$ si $p > n$, il existe $a \in A^+$, $da = 0$,

$$|a| = \inf\{i > 0 \mid A^i \neq 0\}$$

ii) $Z^p = 0$ $p \leq n$.

Explicitons les calculs du théorème 2.1.

Soient $a_o = 1$, $a_1,\dots,a_m$ une base homogène de l'espace vectoriel gradué A, numérotée par ordre croissant des degrés, telle que $d_A a_1 = 0$. On définit des constantes $\beta_{ij} \in \mathbb{Q}$ par $d_A a_i = \sum_{j>i} \beta_{ij}\, a_j$. (On a $\beta_{1j} = 0$, pour tout j).

Notons $\alpha_i = |a_i|$ le degré de a_i, on a $0 = \alpha_o \leq \alpha_1 \leq \dots \leq \alpha_m = n$ Soit (a_i') la base duale de a_i. Comme A est une algèbre, on définit des constantes de structure ainsi : pour tout 3-uplet $(i,j,\ell) \in [1,m]^3$; il existe $\gamma_{ij}^{\ell} \in \mathbb{Q}$ tels que $a_i\, a_j = \sum_{\ell} \gamma_{ij}^{\ell}\, a_\ell$; on a $\gamma_{ji}^{\ell} = (-1)^{\alpha_i\alpha_j}\, \gamma_{ij}^{\ell}$, et $\gamma_{ij}^{\ell} = 0$ si $\ell \leq j$ ou $\ell \leq i$.

On pose $(SZ)^k = \bigoplus_{j-i=k} [\mathrm{Hom}(A^i,\mathbb{Q}) \otimes Z^j]$ et $SZ = \bigoplus_k (SZ)^k$.

On pose, pour $i \geq 0$, $(S_iZ)^k = \mathbb{Q}\, a_i' \otimes Z^{k+\alpha_i}$ et $S_iZ = \bigoplus_k (S_iZ)^k$, on a $S_oZ = Z$.

Comme $Z^p = 0$ si $p \leq n$ et que $\alpha_i \leq n$, alors $(S_iZ)^k \neq 0$ implique que $k \geq 1$.

On a des isomorphismes linéaires de degré $-\alpha_i$:

$$S_i : Z^{k+\alpha_i} \to (S_iZ)^k$$

définis par $S_i(z) = a_i' \otimes z$, pour tout $i \in [0,\dots,m]$, $(S_o = \mathrm{Id})$.

On peut écrire : $(SZ)^k = \bigoplus_{i=0}^{m} (S_iZ)^k$ et $SZ = Z \oplus S_+Z$, où $S_+Z = \bigoplus_{i=1}^{m} S_iZ$, $SZ = \Lambda Z \otimes \Lambda(S_1Z) \dots \otimes \Lambda(S_mZ)$.

On définit un homomorphisme de A-algèbres :

$$\varepsilon : A \otimes \Lambda Z \to A \otimes \Lambda SZ$$

par :

$$\begin{cases} \varepsilon(a \otimes 1) = a \otimes 1 & \text{pour tout } a \in A \\ \varepsilon(1 \otimes \zeta) = \sum_{i=0}^{m} a_i \otimes (a_i' \otimes \zeta) = \zeta + \sum_{i=1}^{m} a_i \otimes S_i(\zeta), & \text{si } \zeta \in \Lambda Z. \end{cases}$$

Remarque : Le fait que ε soit un homomorphisme équivaut à dire que, pour tout $\ell \in [1,\dots,m]$, l'application S_ℓ se prolonge en une application $\Lambda Z \to \Lambda Z \otimes \Lambda(S_1 Z) \otimes \dots \otimes \Lambda(S_m Z)$ telle que, pour tout $\zeta \in \Lambda Z$, pour tout $\zeta' \in \Lambda Z$, on a :

$$(1) \qquad S_\ell(\zeta\zeta') = S_\ell(\zeta)\zeta' + (-1)^{|\zeta|\cdot\alpha_\ell} \zeta\, S_\ell(\zeta') + \sum_{1\leq i,j\leq \ell-1} (-1)^{\alpha_j(|\zeta|-\alpha_i)} \gamma_{ij}^{\ell}\, S_i(\zeta)\, S_j(\zeta')$$

Soit $(A,d_A) \to (A\otimes \Lambda Z,d) \to (\Lambda Z,d_o)$ le modèle minimal du fibré, on peut écrire, pour tout $z \in Z$:

$$dz = d_o z + \sum_{i=1}^{m} a_i \otimes \theta_i(z) \quad \text{où} \quad \theta_i(z) \in \Lambda Z.$$

On définit ainsi des dérivations $\theta_i : \Lambda Z \to \Lambda Z$ de degré $-(\alpha_i - 1)$.

D'après [5], il existe une unique différentielle D sur ΛSZ telle que $\varepsilon : (A\otimes \Lambda Z,d) \to (A,d_A) \otimes (\Lambda SZ,D)$ soit un morphisme d'A.D.G.C. Par définition de ε, on a :

$$\varepsilon(dz) = \varepsilon(d_o z) + \sum_{i=1}^{m} a_i \otimes \theta_i(z) + \sum_{i=1}^{m} (a_i \otimes 1) \sum_{j=1}^{m} a_j \otimes S_j(\theta_i(z))$$

$$\varepsilon(dz) = d_o z + \sum_{i=1}^{m} a_i \otimes [S_i(d_o z) + \theta_i(z)] + \sum_{i,j,\ell} \gamma_{ij}^{\ell}\, a_\ell \otimes S_j(\theta_i(z)).$$

D'autre part, $D\varepsilon(z) = Dz + D(\sum_{i=1}^{m} a_i \otimes S_i(z))$

$$D\varepsilon(z) = Dz + \sum_{i=1}^{m} d_A a_i \otimes S_i(z) + \sum_{i=1}^{m} (-1)^{\alpha_i} a_i \otimes DS_i(z)$$

$$D\varepsilon(z) = Dz + \sum_{i=1}^{m} \sum_{j>i} \beta_{ij} a_j \otimes S_i(z) + \sum_{i=1}^{m} (-1)^{\alpha_i} a_i \otimes DS_i(z).$$

L'unique différentielle D telle que ε soit un morphisme d'A.D.G.C. est donnée par la formule :

$$(2)\quad \begin{cases} Dz = d_o z & \text{si } z \in Z \\ (-1)^{\alpha_\ell} DS_\ell(z) = S_\ell(d_o z) + \theta_\ell(z) + \sum\limits_{1\le i,j\le \ell-1} \gamma_{ij}^{\ell} S_j(\theta_i(z)) - \\ \qquad - \sum\limits_{2\le i\le \ell-1} \beta_{i\ell} S_i(z) & \text{si } z \in Z \text{ et } \ell \in [1,\ldots,m]. \end{cases}$$

Proposition 2.4. (Haefliger, [5]) : $(\Lambda SZ,D) = \Sigma_+^*(A \otimes Z),D)$ <u>est un modèle de l'espace</u> Γ <u>des sections du fibré</u> $E \to X$ <u>vérifiant les hypothèses</u> 1) et 2).

Remarque 2.5. On a un fibré algébrique

$$(\Lambda Z,d_o) \to (\Lambda SZ,D) = (\Lambda Z \otimes \Lambda S_+Z,D) \xrightarrow{q} (\Lambda S_+Z,\bar{D})$$

où q est la projection de ΛSZ sur $\Lambda SZ/\Lambda^+Z$, et $\bar{D}$ est définie par $\bar{D}q = qD$.

Soit le fibré $F \hookrightarrow E \xrightarrow{\Pi} X$ donné ; soit $x_o \in X$ fixé, on définit une application $\psi : \Gamma \to F = p^{-1}(x_o)$ par $\psi(s) = s(x_o)$ pour toute section s, alors le fibré algébrique décrit ci-dessus est un modèle au sens de Halperin, [6], de l'application ψ.

Remarque 2.6. Soient X et F des espaces connexes nilpotents tels que X a le type d'homotopie d'un complexe simplicial de dimension $n \geq 1$ et $H^+(X,\mathbb{Q}) \neq 0$, F est n-connexe, et soit F^X l'espace des applications continues de X dans F muni de la topologie compacte ouverte. Fixons $x_o \in X$ et $y_o \in F$ et notons $(F,y_o)^{(X,x_o)}$ l'espace des applications s de X dans F telles que $s(x_o) = y_o$. Alors on a un fibré :

$$(F,y_o)^{(X,x_o)} \hookrightarrow F^X \xrightarrow{\psi} F$$

où $\psi(s) = s(x_o)$ pour $s \in F^X$.

Ce fibré satisfait aux hypothèses du théorème 20.3 de [6] ou du théorème 6.4 de [2], et a pour modèle, le fibré algébrique :

$$(\Lambda Z, d_o) \hookrightarrow (\Lambda SZ, D) = (\Lambda Z \otimes \Lambda S_+ Z, D) \xrightarrow{q} (\Lambda S_+ Z, \bar{D})$$

En particulier, $(\Lambda S_+ Z, \bar{D})$ est un modèle de $(F, y_o)^{(X, x_o)}$. Ici, les formules (2) se simplifient puisque $\theta_i = 0$, pour tout $i \in [1, \dots, m]$. La différentielle D est donnée par :

$$(2') \quad \begin{cases} Dz = d_o z \\ (-1)^{\alpha_\ell} DS_\ell(z) = S_\ell(d_o z) - \sum_{2 \le i \le \ell - 1} \beta_{i\ell}\, S_i(z). \end{cases}$$

Exemple 2.7. $X = S^n$, $n \geq 1$ et F est n-connexe, alors l'algèbre de de Rham $\Omega(X)$ possède le modèle réel $(A, d_A) = (\Lambda(u)/u^2, d_A = 0)$ où $|u| = n$. L'espace F^{S^n} des applications continues de S^n dans F a pour modèle :

$$(\Lambda Z \otimes \Lambda(S_n Z), D) \qquad \text{où} \qquad (S_n Z)^k = Z^{k+n} \qquad \text{si } k \geq 1,$$

$(-1)^n D\, S_n(z) = S_n(d_o z)$ où $S_n : \Lambda Z \to \Lambda Z \otimes \Lambda(S_n Z)$ est la dérivation de degré $-n$ prolongeant l'application identique de Z dans $(S_n Z)$. Le fibré $\Omega^n F \to F^{S^n} \to F$ a pour modèle :

$$(\Lambda Z, d_o) \to (\Lambda Z \otimes \Lambda(S_n Z), D) \to (\Lambda(S_n Z), 0).$$

3. *Démonstration des théorèmes 3.3. et 3.4.*

Soit $F \hookrightarrow E \xrightarrow{\Pi} X$ un fibré nilpotent dont tous les espaces sont connexes par arcs, nilpotents, ont le type d'homotopie d'un C.W. complexe de type fini.

On suppose :

1) X a le type d'homotopie d'un complexe simplicial de dimension n et $H^+(X, \mathbb{Q}) \neq 0$.

2') F a le type d'homotopie rationnelle d'un bouquet de sphères $\bigvee_{i=1}^{r} S^{k_i+1}$ où $r \geq 2$, $\inf(k_i) \geq n$.

Soit $H^*(F,\mathbb{Q}) = H = \mathbb{Q} \oplus H^+$, la cohomologie de F et soit $\hat{H}$ l'homologie rationnelle de F.

D'après [10], [15] chapitre II, la modèle de Quillen de F est l'algèbre de Lie libre $L(s^{-1}\hat{H}_+)$ munie de la différentielle nulle (où $(s^{-1}\hat{H}_+)_n = \hat{H}_{n+1}$). Si C^* désigne le foncteur "cochaines" de la catégorie des algèbres de Lie différentielles graduées dans A.D.G.C., alors $C^*(L(s^{-1}\hat{H}_+),0) = (\Lambda Z,d_o)$ est minimale et munie d'une graduation inférieure correspondant à la suite centrale descendante de l'algèbre de Lie libre. De plus, par adjonction, on définit un morphisme surjectif $\rho : (\Lambda Z,d_o) \to H$ et on démontre, dans [15], qu'on obtient ainsi le modèle minimal bigradué de H au sens de Halperin-Stasheff, [8]. On a

$$Z = \bigoplus_{\substack{p>n \\ k\geq 0}} Z_k^p \quad , \qquad d_o(Z_k^p) \subset (\Lambda^2 Z)_{k-1}^{p+1}$$

et d_o est donnée par :

$$\langle z, s[y,y'] \rangle = \pm \langle d_o z, sy \wedge sy' \rangle$$

si $z \in Z$; $y,y' \in L(s^{-1}\hat{H}_+)$; (si y est de degré p, alors sy est l'élément égal à y mais sy est de degré $p+1$). De plus, Z_o correspond à l'espace des générateurs $s^{-1}\hat{H}_+$ sur lequel est construite l'algèbre de Lie libre $L(s^{-1}\hat{H}_+)$, et $\rho(\bigoplus_{k>0} Z_k) = 0$.

Il s'en suit que l'application $p_k \circ d_o : Z_{k+1} \to Z_o \cdot Z_k$ est injective, où p_k est la projection de $(\Lambda^2 Z)_k$ sur

$$(\Lambda^2 Z)_k \Big/ \bigoplus_{1\leq i\leq [\frac{k}{2}]} Z_i\, Z_{k-i} \quad .$$

Soit (A,d_A) un modèle de X satisfaisant aux conclusions du lemme 2.3., et $(A,d_A) \hookrightarrow (A \otimes \Lambda Z,d) \to (\Lambda Z,d_o)$ le modèle minimal du fibré de base (A,d_A). Quitte à faire un isomorphisme, on peut supposer que $(\Lambda Z,d_o)$ est le modèle minimal bigradué de H au sens de [8]. Soit $(\Lambda SZ,D)$ le modèle de l'espace des sections du fibré donné par la proposition 2.4. D'après les résultats de Shibata, [11] page 404, on peut décrire $(\Lambda SZ,D)$ ainsi : il existe une structure de A-algèbre de Lie différentielle graduée sur $A \otimes L(s^{-1}\hat{H}_+)$ telle que le complexe de cochaînes $C^*_A(A \otimes L(s^{-1}\hat{H}_+))$ est un modèle de E et le complexe de cochaînes $C^*_{\mathbb{Q}}(A \otimes L(s^{-1}\hat{H}_+))$ est exactement $(\Lambda SZ,D)$.

Lemme 3.1. [11] : Soit $(A,d_A) \to (A \otimes \Lambda Z,d) \to (\Lambda Z,d_o)$ le modèle minimal décrit ci-dessus. Alors, pour tout $z \in Z^p$, pour tout $p > 0$, il existe $\alpha_z \in (A^+ \otimes \Lambda Z)^p$ tel que si on pose $\varphi(z) = z-\alpha_z$, et $d' = \varphi^{-1} \circ d \circ \varphi$ alors $d'z = d_oz + \sum_j b_j \otimes z_j$, $b_j \in A^+$, $z_j \in Z$.

Compte-tenu du lemme 3.1, on supposera donc, dans la suite que, pour tout $z \in Z$, $dz = d_oz + \sum_{i=1}^{m} a_i \otimes \theta_i(z)$ où $\theta_i(z) \in Z$.

On définit sur $H \otimes \Lambda S_+Z$ une différentielle D' par $D'_{|H} = 0$ et $D'S_i(z) = (\rho \otimes Id)\, DS_i(z)$ si $z \in Z$ et $i \in \{1,\ldots,m\}$. Alors $\rho \otimes Id : (\Lambda SZ,D) \to (H \otimes \Lambda S_+Z,D')$ induit un isomorphisme en cohomologie. On a donc

$$H^*(\Gamma,\mathbb{Q}) = H^*(H \otimes \Lambda S_+Z,D').$$

Le but de cette étude est de minorer les dimensions des groupes de cohomologie de $(H \otimes \Lambda S_+Z,D')$.

Lemme 3.2. Il existe $N \in \mathbb{N}$ tel que pour tout $p \geq N$, tout $z \in Z^p$, on a : $D'S_1(z) \in H^+ \otimes S_1Z$, et $D'(H^+ \otimes S_1Z^p) = 0$.

Démonstration : D'après (2) page 12, on a :

$$(-1)^{\alpha_1} \cdot DS_1(z) = S_1(d_o z) + \theta_1(z) ,$$

d'où :

$$(-1)^{\alpha_1} \cdot D'S_1(z) = (\rho \otimes Id)\left[S_1(d_o z) + \theta_1(z)\right] .$$

D'après (1), page 11, S_1 est une dérivation de degré $-\alpha_1$ de ΛZ dans $\Lambda Z \otimes \Lambda(S_1 Z)$. Si $z \in Z^p_k$, $d_o z = \Sigma z_i u_i + \Sigma v_i w_i$, où $z_i \in Z_o$, $u_i \in Z_{k-1}$, v_i et $w_i \in Z_+$; on montre facilement que

$$(\rho \otimes Id)(S_1(d_o z)) \in \rho(Z_o) \otimes S_1 Z = H^+ \otimes S_1 Z$$

puisque S_1 est une dérivation et que $\rho(Z_+) = 0$. Soit

$$N = \alpha_1 + \sup\{d > 0 \mid H^d \neq 0\},$$

alors si $z \in Z^p$, $p \geq N$, on a

$$|\theta_1(z)| = 1+p-\alpha_1 \geq 1 + \sup\{d \mid H^d \neq 0\} ,$$

donc $\rho(\theta_1(z)) = 0$.

On va démontrer la croissance exponentielle des nombres de Betti de Γ dans des cas particuliers.

Théorème 3.3. Soit X un espace nilpotent ayant le type d'homotopie rationnelle d'un complexe simplicial de dimension $n \geq 1$ et tel que $H^+(X,\mathbb{Q}) \neq 0$. Soit F un espace ayant le type d'homotopie rationnelle d'un bouquet de sphères $\bigvee_{i=1}^{r} S^{k_i+1}$ où $r \geq 2$ et $\inf_i(k_i) \geq n$.

Alors si F^X est l'espace des applications continues de X dans F muni de la topologie compacte ouverte, il existe N et une constante réelle $C > 1$ tels que, pour tout $p \geq N$, on a :

$$\sum_{i=0}^{p} \dim H^i(F^X,\mathbb{Q}) \geq C^p .$$

Démonstration : D'après la remarque 2.6., un modèle de F^X sera $(H \otimes \Lambda S_+Z, D')$ où

$$(-1)^{\alpha_\ell} D'S_\ell(z) = (\rho \otimes Id)\, S_\ell(d_o z) - \sum_{2\leq i\leq \ell-1} \beta_{i\ell}\, S_i(z).$$

D'après le lemme 3.2., on a : $D'(H^+ \otimes S_1Z)^p = 0$ pour p suffisamment grand. Il existe donc N tel que, pour tout $p \geq N$ $(H^+ \otimes S_1Z)^p \big/ (\mathrm{Im}\, D' \cap H^+ \otimes S_1Z)^p \hookrightarrow H^p(\Gamma, \mathbb{Q})$, ce qui implique que

$$\dim H^p(\Gamma) \geq \dim (H^+ \otimes S_1Z)^p - \dim(\mathrm{Im}\, D' \cap H^+ \otimes S_1Z)^p.$$

Il s'agit de majorer la dimension de $(\mathrm{Im}\, D' \cap H^+ \otimes S_1Z)^p$.

On a:
$$\begin{array}{ccc} H \otimes \Lambda S_+Z & \xrightarrow{D'} & \mathrm{Im}\, D' \\ \uparrow & & \uparrow \\ S_1Z & \longrightarrow & \mathrm{Im}\, D' \cap H^+ \otimes S_1Z \end{array}$$

On va montrer que D'_1 est surjective, à partir d'un certain degré.

Soit $\phi \in H \otimes \Lambda S_+Z$, on peut décomposer ϕ de la manière suivante : $\phi = \phi'_1 + \sum_{2\leq i\leq m} \phi'_i + \phi''$ où $\phi'_i \in S_iZ$ si $i \in \{1,\dots,m\}$ et $\phi'' \in (H^+ \otimes \Lambda SZ) \otimes \Lambda^{\geq 2} SZ$.

On voit que

$$D'\phi'' \in \left(\sum_{2\leq i\leq m-1} H^+ \otimes S_iZ\right) \oplus \left(H^+ \otimes \Lambda^{\geq 2} S_+Z\right) \oplus \Lambda^{\geq 2} S_+Z$$

Ecrivons

$$(H \otimes \Lambda SZ)^+ = H^+ \otimes S_1Z \oplus (H^+ + \sum_{i\geq 2} (H^+ \otimes S_iZ) + H^+ \otimes \Lambda^{\geq 2} SZ + \Lambda^+ SZ)$$

$$(H \otimes \Lambda SZ)^+ = (H^+ \otimes S_1Z) \oplus C$$

il est clair que $D'\phi'' + \sum_{2\leq i\leq m} D'\phi'_i \in C$.

Soit maintenant $\alpha \in \operatorname{Im} D' \cap (H^+ \otimes S_1 Z)$, il existe donc ϕ tel que $\alpha = D'\phi = D'\phi'_1 + (D'\phi'' + \sum_{2\leq i\leq m} D'\phi'_i)$, ce qui entraine que $\alpha = D'\phi'_1$ et $D'\phi'' + \sum_{i\geq 2} D'\phi'_i = 0$. On a donc, pour tout $p > 0$:

$$\dim(\operatorname{Im} D' \cap H^+ \otimes S_1 Z)^p \leq \dim(S_1 Z)^{p-1}.$$

On en déduit qu'il existe $N \in \mathbb{N}$ tel que si $p \geq N$, on a :

$$\dim H^p(\Gamma,\mathbb{Q}) \geq \dim(H^+ \otimes S_1 Z)^p - \dim(S_1 Z)^{p-1}.$$

Comme $H^+ = H^+(\bigvee_{i=1}^{r} S^{k_i+1})$, on a :

$$\dim(H^+ \otimes S_1 Z)^p = \sum_{i=1}^{r} \dim(S_1 Z)^{p-k_i-1} = \sum_{i=1}^{r} \dim Z^{p+\alpha_1-k_i-1}$$

d'où, si $p \geq N$, $\dim H^p(\Gamma,\mathbb{Q}) \geq \sum_{i=1}^{r} \dim \Pi_{p+\alpha_1-1-k_i}(F) \otimes \mathbb{Q} - \dim \Pi_{p+\alpha_1-1}(F) \otimes \mathbb{Q}$.

La démonstration du théorème 3.3, à l'aide de cette minoration, est identique à celle du théorème 4.1. de [18].

Remarque. Le théorème 3.3. se généralise, de manière évidente, à l'espace des sections d'un fibré $F \hookrightarrow E \xrightarrow{\Pi} X$ où X et F vérifient les hypothèses du théorème 3.3., et le fibré possède un modèle minimal

$$(A,d_A) \hookrightarrow (A \otimes \Lambda Z, d) \to (\Lambda Z, d_o) \quad \text{où} \quad d = d_A \otimes 1 + 1 \otimes d_o .$$

Théorème 3.4. Soit un fibré nilpotent $F \hookrightarrow E \xrightarrow{\Pi} X$ ayant les propriétés suivantes : il existe $n \geq 1$ tel que

(1) X a le type d'homotopie rationnelle d'un bouquet $S^d \vee Y$ où $1 \leq d \leq n$ et Y est un complexe simplicial nilpotent de dimension $\leq n$.

(2) F a le type d'homotopie rationnelle d'un bouquet de sphères $\bigvee_{i=1}^{r} S^{k_i+1}$ où $r \geq 2$, $\inf(k_i) \geq n$. Alors, si Γ est l'espace des sections continues du fibré, il existe $N \in \mathbb{N}$ et une constante réelle $C > 1$ tels que si $p \geq N$, on a : $\sum_{i=0}^{p} \dim H^i(\Gamma,\mathbb{Q}) \geq C^p$.

Démonstration : D'après le lemme 2.3. on peut supposer que Y a un modèle (B,d_B) de dimension finie tel que $B^p = 0$ si $p > n$. D'autre part, l'algèbre commutative graduée $(\Lambda(u)/u^2, d_S = 0)$, où $|u| = d$, est un modèle de S^d. Il est classique que $(\Lambda(u)/u^2) \vee (B,d_B)$, notée (A,d_A), est un modèle de $S^d \vee Y$, on a :

$$A = \mathbb{Q} \oplus (\mathbb{Q}u/u^2) \oplus B^+, \qquad u.B^+ = 0, \qquad d_A u = 0, \qquad d_A(b) = d_B(b).$$

Si on utilise les techniques de Haefliger résumées par la proposition 2.4., on voit qu'un modèle de Γ est

$$(\Lambda Z \otimes \Lambda(\bigoplus_{i=1}^{m} S_i Z) \otimes \Lambda S_u Z, D) \quad \text{où} \quad (S_i Z)_{1 \leq i \leq m}$$

est défini à partir d'une base homogène $(b_1,\ldots,b_m)$ de B^+, et S_u est l'isomorphisme linéaire de degré $-d$ de Z sur $\mathbb{Q}u' \otimes Z$ si u' est la forme linéaire sur A définie par $u'(1) = 0$, $u'(b_i) = 0$, $u'(u) = 1$.

On a $S_u(\zeta\zeta') = S_u(\zeta).\zeta' + (-1)^{d|\zeta|}\,\zeta.S_u(\zeta')$ si $\zeta,\zeta' \in \Lambda Z$; et $(-1)^d DS_u(z) = S_u(d_o z) + \theta_u(z)$.

Une démonstration analogue à celle du lemme 3.2 montre qu'il existe $N \in \mathbb{N}$ tel que pour tout $p \geq N$, on a $D'(H^+ \otimes S_u Z)^p = 0$. On vérifie, comme pour le théorème 3.3., que pour p assez grand, l'application D' de $(S_u Z)^p$ dans $[\mathrm{Im}\, D' \cap (H^+ \otimes S_u Z)]^{p+1}$ est surjective, et donc que:

$$\text{si} \quad p \geq N, \quad \dim H^p(\Gamma,\mathbb{Q}) \geq \sum_{i=1}^{r} \dim \Pi_{p+d-1-k_i}(F) \otimes \mathbb{Q} - \dim \Pi_{p+d-1}(F) \otimes \mathbb{Q}$$

et on conclut comme dans [18], théorème 4.1.

Théorème 3.5. <u>Soit un fibré nilpotent</u> $F \hookrightarrow E_o \xrightarrow{\Pi_o} S^q$ <u>où</u> $F = \bigvee_{i=1}^{r} S^{k_i+1}$, $r \geq 2$, $q \leq \inf(k_i)$. <u>Soit</u> X <u>un complexe simplicial nilpotent de dimension</u> n <u>où</u> $q \leq n \leq \inf(k_i)$. <u>On se donne une application continue</u> $f : X \to S^q$ <u>telle que la</u> $q^{\text{ième}}$ <u>application induite en homotopie ration-</u>

nelle $(f_{\#} \otimes 1_{\mathbb{Q}})_q : \Pi_q(X) \otimes \mathbb{Q} \to \Pi_q(S^q) \otimes \mathbb{Q}$ soit non nulle. Alors, pour le fibré Π_o et le fibré image réciproque $\Pi = f_*(\Pi_o) : F \to E \to X$, la cohomologie de l'espace des sections est à croissance exponentielle.

Démonstration : Soit $A^*(\cdot)$ le foncteur de Sullivan défini de la catégorie des complexes simpliciaux dans celle des A.D.G.C.. On a un morphisme $A^*(f) : A^*(S^q) \to A^*(X)$. Soit $(\Lambda(a)/a^2, d = 0) \xrightarrow{m} A^*(S^q)$ un modèle de dimension finie de S^q (a est un générateur de degré q). Un modèle minimal du fibré Π_o est de la forme :

$$(\Lambda(a)/a^2) \to ((\Lambda(a)/a^2) \otimes \Lambda Z, d) \to (\Lambda Z, d_o)$$

On a : $dz = d_o z + a \otimes \theta_a(z)$ où $\theta_a(z) \in \Lambda Z$.

L'application $(\Lambda(a)/a^2) \xrightarrow{A^*(f)om} A^*(X)$ a un modèle

$$(\Lambda(a)/a^2) \xhookrightarrow{i} (\Lambda(a)/a^2 \otimes \Lambda U, \delta) \xrightarrow{\psi} A^*(X)$$

où ψ est un quasi-isomorphisme et $\psi \circ i = A^*(f) \circ m$. L'hypothèse que $(f_{\#} \otimes 1_{\mathbb{Q}})_q$ est surjective, entraine que pour tout $u \in U$, on a : $\delta u = \delta_o u + a \otimes \alpha(u)$ où $\delta_o u \in \Lambda U$ et $\alpha(u) \in \Lambda^+ U$. On définit alors un morphisme d'A.D.G.C. $r : (\Lambda(a)/a^2 \otimes \Lambda U, \delta) \to \Lambda(a)/a^2$ par $r(a) = a$, $r(u) = 0$ pour tout $u \in U$. On a $r \circ i = \mathrm{Id}$. Comme dans le lemme 2.3 , soit $\mathcal{I}$ l'idéal de $\Lambda(a)/a^2 \otimes \Lambda U$ engendré par les éléments de degré $> n$ et par un supplémentaire de $\mathrm{Ker}\, \delta$ en degré n . Le passage au quotient

$$\rho : (\Lambda(a)/a^2 \otimes \Lambda U, \delta) \to [(\Lambda(a)/a^2 \otimes \Lambda U)/\mathcal{I}, \bar{\delta}] = (A, d_A)$$

est un isomorphisme en cohomologie. On a :

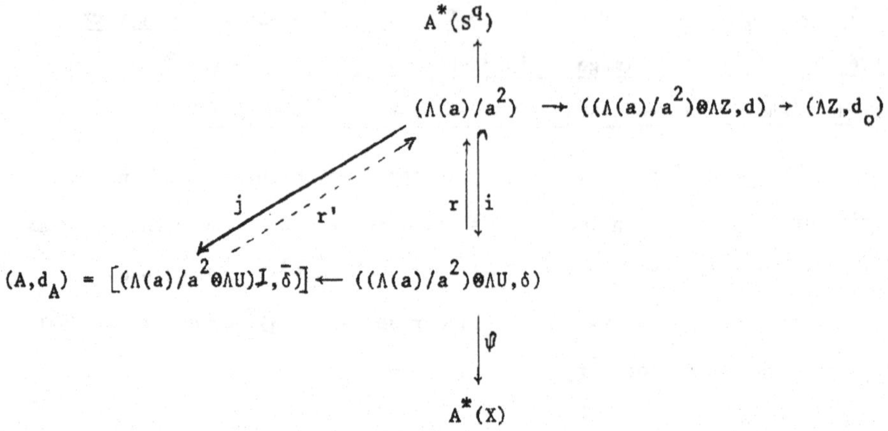

On appelle j l'inclusion déduite de i ; comme $r(\mathcal{I}) = 0$, on définit un morphisme d'A.D.G.C. $r' : (A,d_A) \to \Lambda(a)/a^2$ tel que $r' \circ j = \mathrm{Id}$.

Un modèle minimal de base (A,d_A) du fibré image réciproque $\Pi = f_*(\Pi_o)$ est :

$$(A,d_A) \hookrightarrow (A \otimes \Lambda Z, D) \to (\Lambda Z, d_o)$$

où

$$Dz = d_o z + j(a) \otimes \theta_a(z) = d_o z + a \otimes \theta_a(z).$$

Dans ces conditions, les morphismes j et r' s'étendent en des morphismes d'A.D.G.C. $j \otimes \mathrm{Id}$ et $r' \otimes \mathrm{Id}$ rendant commutatifs les diagrammes suivants :

$$\begin{array}{ccccc}
(\Lambda(a)/a^2, d=0) & \hookrightarrow & (\Lambda(a)/a^2 \otimes \Lambda Z, d) & \longrightarrow & (\Lambda Z, d_o) \\
\downarrow{\scriptstyle j} & & \downarrow{\scriptstyle j\otimes \mathrm{Id}} & & \| \\
(A,d_A) & \longrightarrow & (A \otimes \Lambda Z, D) & \longrightarrow & (\Lambda Z, d_o) \\
\downarrow{\scriptstyle r'} & & \downarrow{\scriptstyle r'\otimes \mathrm{Id}} & & \| \\
(\Lambda(a)/a^2, d=0) & \longrightarrow & (\Lambda(a)/a^2 \otimes \Lambda Z, d) & \longrightarrow & (\Lambda Z, d_o)
\end{array}$$

Il est clair que la construction Σ^* de Haefliger décrite dans le théorème 2.1 est fonctorielle, on a donc des morphismes d'A.D.G.C. $R = \Sigma^*(r' \otimes Id)$ et $I = \Sigma^*(j \otimes Id)$ tels que

$$I \circ R = Id : \quad (\Sigma^*(A \otimes Z), D) \underset{R}{\overset{I}{\rightleftarrows}} \Sigma^*(\Lambda(a)/a^2 \otimes Z, d).$$

En particulier, l'application induite par I en cohomologie est surjective ; on a donc, pour tout $n \in \mathbb{N}$:

$$\dim H^n(\Gamma, \mathbb{Q}) \geq \dim H^n(\Gamma_o, \mathbb{Q}) \qquad \text{où } \Gamma \text{ (resp. } \Gamma_o)$$

est l'espace des sections du fibré Π (resp. Π_o).

Le théorème 3.5 se déduit donc du théorème 3.4 dans lequel on prend Y égal à un point.

On est amené à énoncer la conjecture suivante

Conjecture : Soit un fibré nilpotent $F \hookrightarrow E \xrightarrow{\Pi} X$ où X a le type d'homotopie d'un complexe simplicial de dimension $n \geq 1$ et $H^+(X,\mathbb{Q}) \neq 0$, F a le type d'homotopie d'un bouquet de sphères $\bigvee_{i=1}^{r} S^{k_i+1}$ où $r \geq 2$ et $\inf(k_i) \geq n$, alors la suite des nombres de Betti de l'espace des sections du fibré est à croissance exponentielle.

Ce résultat aurait des applications intéressantes dans l'étude de la cohomologie de Gelfand-Fuchs d'une variété :

Soit M une variété C^∞ paracompacte de dimension $n \geq 1$, soit L_M l'algèbre de Lie des champs de vecteurs continus sur M, on s'intéresse à la cohomologie de l'A.D.G.C. $C^*(L_M)$ des formes multilinéaires continues sur M, appelée cohomologie de Gelfand-Fuchs de M. On considère le U_n-fibré : $U_n \to EU_n^{(2n)} \longrightarrow BU_n^{(2n)}$ restriction du fibré universel, au-dessus du 2n-squelette de la base BU_n. Soit $\hat{\gamma}_n : EU_n^{(2n)} \to EU_n^{(2n} \times_{U_n} EU_n \to BU_n$ le fibré associé au-dessus de BU_n et de fibre $EU^{(2n)}$. Le complexifié du fibré tangent de M est classifié par une application $f : M \to BU_n$.

L'image réciproque par f du fibré $\hat{\gamma}_n$ est un fibré : $EU^{(2n)} \to E \to M$.

On a le résultat suivant démontré par Haefliger :

Théorème 3.6. <u>Conjecture de Bott</u> [3] : $C^*(L_M)$ <u>est un modèle de l'espace des sections continues du fibré</u> : $EU^{(2n)} \to E \to M$.

On montre que $EU^{(2n)}$ est 2n-connexe, et a le type d'homotopie rationnelle d'un bouquet de sphères. Si $n = 1$, $EU^{(2)}$ a le type d'homotopie rationnelle de S^3 ; si $n \geq 2$, $EU^{(2n)}$ a le type d'homotopie d'un bouquet d'un nombre fini de sphères en nombre ≥ 2. Un modèle de l'espace $EU^{(2n)}$ est l'A.D.G.C., non libre : $(E(h_1,\ldots,h_n) \otimes S[c_1,\ldots,c_n]/I,d)$ où $|h_i| = 2i-1$, $|c_i| = 2i$, I est l'idéal de $S[c_1,\ldots,c_n]$ engendré par les éléments de degré $> 2n$, on a $dc_i = 0$, $dh_i = c_i$, (voir par exemple, [12]). Une base de $H^*(EU^{(2n)},\mathbb{Q})$ a été décrite par Vey [3].

Dans [3] ou [11], un modèle du fibré $EU^{(2n)} \to E \to M$ est donné à partir du modèle de $\hat{\gamma}_n$ et du modèle de f noté

$$f^* : H^*(BU_n,\mathbb{R}) = R[\bar{c}_1,\ldots,\bar{c}_n] \to \Omega^*(M)$$

où $|\bar{c}_i| = 2i$, et $f^*(\bar{c}_{2i-1}) = 0$, $f^*(c_{2i}) = \tilde{p}_i \in \Omega^{4i}(M)$ est une forme fermée représentant la classe de Pontryagin $p_i \in H^{4i}(M,\mathbb{R})$.

Il est clair, que si toutes les classes de Pontryagin sont nulles, le fibré $EU^{(2n)} \to E \to M$ possède un modèle minimal du type

$$(\Omega^*(M),d_M) \to (\Omega^*(M) \otimes \Lambda Z,d) \to (\Lambda Z,d_o)$$

où $d = d_M \otimes 1 + 1 \otimes d_o$.

On déduit, de la remarque suivant le théorème 3.3 le résultat suivant :

Théorème 3.7. Soit M une variété C^∞ connexe, compacte, nilpotente de dimension ≥ 2 et telle que $H^+(M,\mathbb{R}) \neq 0$. On suppose que toutes les classes de Pontryagin de M sont nulles, alors il existe un entier N et une constante $A > 1$ tels que, si $p \geq N$, on a

$$\sum_{i=0}^{p} \dim H^i(C^*(L_M)) > A^p.$$

Le théorème 3.7. s'applique en particulier aux sphères (de dimension ≥ 2), aux groupes de Lie compacts connexes nilpotents, et aux produits finis de telles variétés.

BIBLIOGRAPHIE

[1] GELFAND I.M. and D. FUCHS : *The cohomology of the Lie algebra on a smooth manifold*. Funct. Anal. 3 (1969) 194-210.

[2] GRIVEL P.P. : *Formes différentielles et suites spectrales*. Annales Inst. Fourier. 24 (1979) 17-37.

[3] HAEFLIGER A. : *Sur la cohomologie de l'algèbre de Lie des champs de vecteurs*. Ann. Scient. ENS, 4ème série, 9, (1976) 503-532.

[4] HAEFLIGER A. : *Sur la cohomologie de Gelfand-Fuchs*, Lectures Notes in Mat, n°484, 121-152.

[5] HAEFLIGER A. : *Rational homotopy of the space of sections of a nilpotent bundle*. Trans. Am. Math. Soc. 273 (1982) 609-620.

[6] HALPERIN S. : *Lectures on minimal models*. Mémoires de la Soc. Math. France 9/10, 1983.

[7] HALPERIN S. : *Rational fibrations, minimal models, and fibrings of homogeneous spaces*. Trans. Am. Math. Soc. 244, (1978), 199-223.

[8] HALPERIN S, STASHEFF J. : *Obstructions to homotopy equivalence*. Advances in Math. 32 (1979) 233-279.

[9] LEHMANN D. : *Théorie homotopique des formes différentielles*. Astérique 45 (1977).

[10] QUILLEN D. : *Rational homotopy theory*. Ann. of Math. 90 (1969) 205-295.

[11] SHIBATA K. : *On Haefliger's model for the Gelfand-Fuchs cohomology*. Japan J. Math. 7 (1981) 379-415.

[12] SHIBATA K. : *Sullivan-Quillen mixed type model for fibrations and the Haefliger model for the Gelfand-Fuchs cohomology*. Astérisque, 113-114, 1984, 292-297.

[13] SILVEIRA da F. : *Homotopie rationnelle d'espaces fibrés*. Thèse. Université de Genève (1979).

[14] SULLIVAN D. : *Infinitesimal computations in topology*. Publ. I.H.E.S. 47 (1977) 269-331.

[15] TANRÉ D. : *Modèles de Chen, Quillen, Sullivan*. Lecture Notes in Mathematics, 1025, 1983, Springer-Verlag, Berlin, Heidelberg, New York, Tokyo.

[16] THOM R. : *L'homologie des espaces fonctionnels*. Colloque Topo. Alg. Louvain (1956) 29-39.

[17] VIGUÉ-POIRRIER M. : *Réalisation de morphismes donnés en cohomologie et suite spectrale d'Eilenberg-Moore*. Trans. Am. Math. Soc. 265 (1981) 441-484.

[18] VIGUÉ-POIRRIER M. : *Homotopie rationnelle et croissance du nombre de géodésiques fermées*. Ann. Scient. Ecole Normale Sup., 4e série, 17, 1984, 413-431.

Micheline VIGUÉ-POIRRIER
37, Parc d'Ardenay
F. 91120 Palaiseau

Vol. 1034: J. Musielak, Orlicz Spaces and Modular Spaces. V, 222 pages. 1983.

Vol. 1035: The Mathematics and Physics of Disordered Media. Proceedings, 1983. Edited by B.D. Hughes and B.W. Ninham. VII, 432 pages. 1983.

Vol. 1036: Combinatorial Mathematics X. Proceedings, 1982. Edited by L.R.A. Casse. XI, 419 pages. 1983.

Vol. 1037: Non-linear Partial Differential Operators and Quantization Procedures. Proceedings, 1981. Edited by S.I. Andersson and H.-D. Doebner. VII, 334 pages. 1983.

Vol. 1038: F. Borceux, G. Van den Bossche, Algebra in a Localic Topos with Applications to Ring Theory. IX, 240 pages. 1983.

Vol. 1039: Analytic Functions, Błażejewko 1982. Proceedings. Edited by J. Ławrynowicz. X, 494 pages. 1983

Vol. 1040: A. Good, Local Analysis of Selberg's Trace Formula. III, 128 pages. 1983.

Vol. 1041: Lie Group Representations II. Proceedings 1982-1983. Edited by R. Herb, S. Kudla, R. Lipsman and J. Rosenberg. IX, 340 pages. 1984.

Vol. 1042: A. Gut, K.D. Schmidt, Amarts and Set Function Processes. III, 258 pages. 1983.

Vol. 1043: Linear and Complex Analysis Problem Book. Edited by V.P. Havin, S.V. Hruščëv and N.K. Nikol'skii. XVIII, 721 pages. 1984.

Vol. 1044: E. Gekeler, Discretization Methods for Stable Initial Value Problems. VIII, 201 pages. 1984.

Vol. 1045: Differential Geometry. Proceedings, 1982. Edited by A.M. Naveira. VIII, 194 pages. 1984.

Vol. 1046: Algebraic K-Theory, Number Theory, Geometry and Analysis. Proceedings, 1982. Edited by A. Bak. IX, 464 pages. 1984.

Vol. 1047: Fluid Dynamics. Seminar, 1982. Edited by H. Beirão da Veiga. VII, 193 pages. 1984.

Vol. 1048: Kinetic Theories and the Boltzmann Equation. Seminar, 1981. Edited by C. Cercignani. VII, 248 pages. 1984.

Vol. 1049: B. Iochum, Cônes autopolaires et algèbres de Jordan. VI, 247 pages. 1984.

Vol. 1050: A. Prestel, P. Roquette, Formally p-adic Fields. V, 167 pages. 1984.

Vol. 1051: Algebraic Topology, Aarhus 1982. Proceedings. Edited by I. Madsen and B. Oliver. X, 665 pages. 1984.

Vol. 1052: Number Theory, New York 1982. Seminar. Edited by D.V. Chudnovsky, G.V. Chudnovsky, H. Cohn and M.B. Nathanson. V, 309 pages. 1984.

Vol. 1053: P. Hilton, Nilpotente Gruppen und nilpotente Räume. V, 221 pages. 1984.

Vol. 1054: V. Thomée, Galerkin Finite Element Methods for Parabolic Problems. VII, 237 pages. 1984.

Vol. 1055: Quantum Probability and Applications to the Quantum Theory of Irreversible Processes. Proceedings, 1982. Edited by L. Accardi, A. Frigerio and V. Gorini. VI, 411 pages. 1984.

Vol. 1056: Algebraic Geometry. Bucharest 1982. Proceedings, 1982. Edited by L. Bădescu and D. Popescu. VII, 380 pages. 1984.

Vol. 1057: Bifurcation Theory and Applications. Seminar, 1983. Edited by L. Salvadori. VII, 233 pages. 1984.

Vol. 1058: B. Aulbach, Continuous and Discrete Dynamics near Manifolds of Equilibria. IX, 142 pages. 1984.

Vol. 1059: Séminaire de Probabilités XVIII, 1982/83. Proceedings. Edité par J. Azéma et M. Yor. IV, 518 pages. 1984.

Vol. 1060: Topology. Proceedings, 1982. Edited by L.D. Faddeev and A.A. Mal'cev. VI, 389 pages. 1984.

Vol. 1061: Séminaire de Théorie du Potentiel. Paris, No. 7. Proceedings. Directeurs: M. Brelot, G. Choquet et J. Deny. Rédacteurs: F. Hirsch et G. Mokobodzki. IV, 281 pages. 1984.

Vol. 1062: J. Jost, Harmonic Maps Between Surfaces. X, 133 pages. 1984.

Vol. 1063: Orienting Polymers. Proceedings, 1983. Edited by J.L. Ericksen. VII, 166 pages. 1984.

Vol. 1064: Probability Measures on Groups VII. Proceedings, 1983. Edited by H. Heyer. X, 588 pages. 1984.

Vol. 1065: A. Cuyt, Padé Approximants for Operators: Theory and Applications. IX, 138 pages. 1984.

Vol. 1066: Numerical Analysis. Proceedings, 1983. Edited by D.F. Griffiths. XI, 275 pages. 1984.

Vol. 1067: Yasuo Okuyama, Absolute Summability of Fourier Series and Orthogonal Series. VI, 118 pages. 1984.

Vol. 1068: Number Theory, Noordwijkerhout 1983. Proceedings. Edited by H. Jager. V, 296 pages. 1984.

Vol. 1069: M. Kreck, Bordism of Diffeomorphisms and Related Topics. III, 144 pages. 1984.

Vol. 1070: Interpolation Spaces and Allied Topics in Analysis. Proceedings, 1983. Edited by M. Cwikel and J. Peetre. III, 239 pages. 1984.

Vol. 1071: Padé Approximation and its Applications, Bad Honnef 1983. Prodeedings. Edited by H. Werner and H.J. Bünger. VI, 264 pages. 1984.

Vol. 1072: F. Rothe, Global Solutions of Reaction-Diffusion Systems. V, 216 pages. 1984.

Vol. 1073: Graph Theory, Singapore 1983. Proceedings. Edited by K.M. Koh and H.P. Yap. XIII, 335 pages. 1984.

Vol. 1074: E.W. Stredulinsky, Weighted Inequalities and Degenerate Elliptic Partial Differential Equations. III, 143 pages. 1984.

Vol. 1075: H. Majima, Asymptotic Analysis for Integrable Connections with Irregular Singular Points. IX, 159 pages. 1984.

Vol. 1076: Infinite-Dimensional Systems. Proceedings, 1983. Edited by F. Kappel and W. Schappacher. VII, 278 pages. 1984.

Vol. 1077: Lie Group Representations III. Proceedings, 1982-1983. Edited by R. Herb, R. Johnson, R. Lipsman, J. Rosenberg. XI, 454 pages. 1984.

Vol. 1078: A.J.E.M. Janssen, P. van der Steen, Integration Theory. V, 224 pages. 1984.

Vol. 1079: W. Ruppert. Compact Semitopological Semigroups: An Intrinsic Theory. V, 260 pages. 1984

Vol. 1080: Probability Theory on Vector Spaces III. Proceedings, 1983. Edited by D. Szynal and A. Weron. V, 373 pages. 1984.

Vol. 1081: D. Benson, Modular Representation Theory: New Trends and Methods. XI, 231 pages. 1984.

Vol. 1082: C.-G. Schmidt, Arithmetik Abelscher Varietäten mit komplexer Multiplikation. X, 96 Seiten. 1984.

Vol. 1083: D. Bump, Automorphic Forms on GL (3,IR). XI, 184 pages. 1984.

Vol. 1084: D. Kletzing, Structure and Representations of Q-Groups. VI, 290 pages. 1984.

Vol. 1085: G.K. Immink, Asymptotics of Analytic Difference Equations. V, 134 pages. 1984.

Vol. 1086: Sensitivity of Functionals with Applications to Engineering Sciences. Proceedings, 1983. Edited by V. Komkov. V, 130 pages. 1984

Vol. 1087: W. Narkiewicz, Uniform Distribution of Sequences of Integers in Residue Classes. VIII, 125 pages. 1984.

Vol. 1088: A.V. Kakosyan, L.B. Klebanov, J.A. Melamed, Characterization of Distributions by the Method of Intensively Monotone Operators. X, 175 pages. 1984.

Vol. 1089: Measure Theory, Oberwolfach 1983. Proceedings. Edited by D. Kölzow and D. Maharam-Stone. XIII, 327 pages. 1984.

Vol. 1090: Differential Geometry of Submanifolds. Proceedings, 1984. Edited by K. Kenmotsu. VI, 132 pages. 1984.

Vol. 1091: Multifunctions and Integrands. Proceedings, 1983. Edited by G. Salinetti. V, 234 pages. 1984.

Vol. 1092: Complete Intersections. Seminar, 1983. Edited by S. Greco and R. Strano. VII, 299 pages. 1984.

Vol. 1093: A. Prestel, Lectures on Formally Real Fields. XI, 125 pages. 1984.

Vol. 1094: Analyse Complexe. Proceedings, 1983. Edité par E. Amar, R. Gay et Nguyen Thanh Van. IX, 184 pages. 1984.

Vol. 1095: Stochastic Analysis and Applications. Proceedings, 1983. Edited by A. Truman and D. Williams. V, 199 pages. 1984.

Vol. 1096: Théorie du Potentiel. Proceedings, 1983. Edité par G. Mokobodzki et D. Pinchon. IX, 601 pages. 1984.

Vol. 1097: R.M. Dudley, H. Kunita, F. Ledrappier, École d'Éte de Probabilités de Saint-Flour XII – 1982. Edité par P.L. Hennequin. X, 396 pages. 1984.

Vol. 1098: Groups – Korea 1983. Proceedings. Edited by A.C. Kim and B.H. Neumann. VII, 183 pages. 1984.

Vol. 1099: C.M. Ringel, Tame Algebras and Integral Quadratic Forms. XIII, 376 pages. 1984.

Vol. 1100: V. Ivrii, Precise Spectral Asymptotics for Elliptic Operators Acting in Fiberings over Manifolds with Boundary. V, 237 pages. 1984.

Vol. 1101: V. Cossart, J. Giraud, U. Orbanz, Resolution of Surface Singularities. Seminar. VII, 132 pages. 1984.

Vol. 1102: A. Verona, Stratified Mappings – Structure and Triangulability. IX, 160 pages. 1984.

Vol. 1103: Models and Sets. Proceedings, Logic Colloquium, 1983, Part I. Edited by G.H. Müller and M.M. Richter. VIII, 484 pages. 1984.

Vol. 1104: Computation and Proof Theory. Proceedings, Logic Colloquium, 1983, Part II. Edited by M.M. Richter, E. Börger, W. Oberschelp, B. Schinzel and W. Thomas. VIII, 475 pages. 1984.

Vol. 1105: Rational Approximation and Interpolation. Proceedings, 1983. Edited by P.R. Graves-Morris, E.B. Saff and R.S. Varga. XII, 528 pages. 1984.

Vol. 1106: C.T. Chong, Techniques of Admissible Recursion Theory. IX, 214 pages. 1984.

Vol. 1107: Nonlinear Analysis and Optimization. Proceedings, 1982. Edited by C. Vinti. V, 224 pages. 1984.

Vol. 1108: Global Analysis – Studies and Applications I. Edited by Yu.G. Borisovich and Yu.E. Gliklikh. V, 301 pages. 1984.

Vol. 1109: Stochastic Aspects of Classical and Quantum Systems. Proceedings, 1983. Edited by S. Albeverio, P. Combe and M. Sirugue-Collin. IX, 227 pages. 1985.

Vol. 1110: R. Jajte, Strong Limit Theorems in Non-Commutative Probability. VI, 152 pages. 1985.

Vol. 1111: Arbeitstagung Bonn 1984. Proceedings. Edited by F. Hirzebruch, J. Schwermer and S. Suter. V, 481 pages. 1985.

Vol. 1112: Products of Conjugacy Classes in Groups. Edited by Z. Arad and M. Herzog. V, 244 pages. 1985.

Vol. 1113: P. Antosik, C. Swartz, Matrix Methods in Analysis. IV, 114 pages. 1985.

Vol. 1114: Zahlentheoretische Analysis. Seminar. Herausgegeben von E. Hlawka. V, 157 Seiten. 1985.

Vol. 1115: J. Moulin Ollagnier, Ergodic Theory and Statistical Mechanics. VI, 147 pages. 1985.

Vol. 1116: S. Stolz, Hochzusammenhängende Mannigfaltigkeiten und ihre Ränder. XXIII, 134 Seiten. 1985.

Vol. 1117: D.J. Aldous, J.A. Ibragimov, J. Jacod, Ecole d'Été de Probabilités de Saint-Flour XIII – 1983. Édité par P.L. Hennequin. IX, 409 pages. 1985.

Vol. 1118: Grossissements de filtrations: exemples et applications. Seminaire, 1982/83. Edité par Th. Jeulin et M. Yor. V, 315 pages. 1985.

Vol. 1119: Recent Mathematical Methods in Dynamic Programming. Proceedings, 1984. Edited by I. Capuzzo Dolcetta, W.H. Fleming and T. Zolezzi. VI, 202 pages. 1985.

Vol. 1120: K. Jarosz, Perturbations of Banach Algebras. V, 118 pages. 1985.

Vol. 1121: Singularities and Constructive Methods for Their Treatment. Proceedings, 1983. Edited by P. Grisvard, W. Wendland and J.R. Whiteman. IX, 346 pages. 1985.

Vol. 1122: Number Theory. Proceedings, 1984. Edited by K. Alladi. VII, 217 pages. 1985.

Vol. 1123: Séminaire de Probabilités XIX 1983/84. Proceedings. Edité par J. Azéma et M. Yor. IV, 504 pages. 1985.

Vol. 1124: Algebraic Geometry, Sitges (Barcelona) 1983. Proceedings. Edited by E. Casas-Alvero, G.E. Welters and S. Xambó-Descamps. XI, 416 pages. 1985.

Vol. 1125: Dynamical Systems and Bifurcations. Proceedings, 1984. Edited by B.L.J. Braaksma, H.W. Broer and F. Takens. V, 129 pages. 1985.

Vol. 1126: Algebraic and Geometric Topology. Proceedings, 1983. Edited by A. Ranicki, N. Levitt and F. Quinn. V, 523 pages. 1985.

Vol. 1127: Numerical Methods in Fluid Dynamics. Seminar. Edited by F. Brezzi, VII, 333 pages. 1985.

Vol. 1128: J. Elschner, Singular Ordinary Differential Operators and Pseudodifferential Equations. 200 pages. 1985.

Vol. 1129: Numerical Analysis, Lancaster 1984. Proceedings. Edited by P.R. Turner. XIV, 179 pages. 1985.

Vol. 1130: Methods in Mathematical Logic. Proceedings, 1983. Edited by C.A. Di Prisco. VII, 407 pages. 1985.

Vol. 1131: K. Sundaresan, S. Swaminathan, Geometry and Nonlinear Analysis in Banach Spaces. III, 116 pages. 1985.

Vol. 1132: Operator Algebras and their Connections with Topology and Ergodic Theory. Proceedings, 1983. Edited by H. Araki, C.C. Moore, Ş. Strătilă and C. Voiculescu. VI, 594 pages. 1985.

Vol. 1133: K.C. Kiwiel, Methods of Descent for Nondifferentiable Optimization. VI, 362 pages. 1985.

Vol. 1134: G.P. Galdi, S. Rionero, Weighted Energy Methods in Fluid Dynamics and Elasticity. VII, 126 pages. 1985.

Vol. 1135: Number Theory, New York 1983–84. Seminar. Edited by D.V. Chudnovsky, G.V. Chudnovsky, H. Cohn and M.B. Nathanson. V, 283 pages. 1985.

Vol. 1136: Quantum Probability and Applications II. Proceedings, 1984. Edited by L. Accardi and W. von Waldenfels. VI, 534 pages. 1985.

Vol. 1137: Xiao G., Surfaces fibrées en courbes de genre deux. IX, 103 pages. 1985.

Vol. 1138: A. Ocneanu, Actions of Discrete Amenable Groups on von Neumann Algebras. V, 115 pages. 1985.

Vol. 1139: Differential Geometric Methods in Mathematical Physics. Proceedings, 1983. Edited by H.D. Doebner and J.D. Hennig. VI, 337 pages. 1985.

Vol. 1140: S. Donkin, Rational Representations of Algebraic Groups. VII, 254 pages. 1985.

Vol. 1141: Recursion Theory Week. Proceedings, 1984. Edited by H.-D. Ebbinghaus, G.H. Müller and G.E. Sacks. IX, 418 pages. 1985.

Vol. 1142: Orders and their Applications. Proceedings, 1984. Edited by I. Reiner and K.W. Roggenkamp. X, 306 pages. 1985.

Vol. 1143: A. Krieg, Modular Forms on Half-Spaces of Quaternions. XIII, 203 pages. 1985.

Vol. 1144: Knot Theory and Manifolds. Proceedings, 1983. Edited by D. Rolfsen. V, 163 pages. 1985.